Community Ecology

Second edition

Gary G. Mittelbach
Michigan State University, USA

Brian J. McGill
University of Maine, USA

OXFORD
UNIVERSITY PRESS

OXFORD
UNIVERSITY PRESS

Great Clarendon Street, Oxford, OX2 6DP,
United Kingdom

Oxford University Press is a department of the University of Oxford.
It furthers the University's objective of excellence in research, scholarship,
and education by publishing worldwide. Oxford is a registered trade mark of
Oxford University Press in the UK and in certain other countries

© Gary G. Mittelbach & Brian J. McGill 2019

The moral rights of the authors have been asserted

First Edition published in 2012
Second Edition published in 2019
Impression: 1

Published in the United States of America by Oxford University Press
198 Madison Avenue, New York, NY 10016, United States of America

British Library Cataloguing in Publication Data
Data available

Library of Congress Control Number: 2018968492

ISBN 978–0–19–883585–1 (hbk.)
ISBN 978–0–19–883586–8 (pbk.)

DOI: 10.1093/oso/9780198835851.001.0001

Printed in Great Britain by
Bell & Bain Ltd., Glasgow

Links to third party websites are provided by Oxford in good faith and
for information only. Oxford disclaims any responsibility for the materials
contained in any third party website referenced in this work.

Preface to 2nd Edition

The first edition of this book was published in 2012 and since that time there have been major advancements in the field of community ecology. Many of these advances are centered on the ever-expanding knowledge of how local and regional processes interact to shape ecological communities, as well as the growing recognition that ecological and evolutionary processes are inexorably linked in structuring communities on both short and long time scales. Our goal in producing this second edition was to highlight these and other exciting new developments in the study of community ecology (e.g., trait-based analyses, modern coexistence theory). In addition, we have worked hard to incorporate the many suggestions from instructors and students who used the first edition of *Community Ecology* in their courses and who have generously provided us with their feedback. Finally, up-to-date coverage of the literature was central to the success of the first edition. In this edition, we have done our best to update all the topics covered in the first edition. Even the most foundational topics in ecology (e.g., population regulation) continue to see advancements in both ideas, data and synthesis.

I (Mittelbach) imagine that writing the second edition of a book is a little like childbirth—you have to let the pain of the first experience wear off before thinking about doing it again. The biggest factor for me in minimizing the pain and making the experience profoundly rewarding was having Brian McGill join in writing this second edition. And for me (McGill) being asked to be involved in a book that Gary had already done so well on his own was a bit intimidating, but also a very exciting opportunity We bring different approaches and expertise to the study of community ecology, but share a common love for this area of ecology that straddles the boundary between populations and ecosystems, and that focuses squarely on questions related to the generation, maintenance, distribution, and preservation of Earth's biodiversity. Michael Rosenzweig (Brian's PhD advisor and Gary's friend) once wrote that "I am not ashamed to be a puzzle solver". Neither are we! Together, we hope that this second edition both highlights the "puzzles" that abound in the study of ecological communities and the ways that ecologists have marshalled theory, experiments, and observations to solve many of ecology's puzzles, while (of course) generating new ones.

The field of scientific publishing continues to change at a rapid pace and a major change affecting the second edition of *Community Ecology* was the acquisition of Sinauer Associates (first edition publisher) by Oxford University Press. The wisdom and good-natured advice of Andy Sinauer and Ian Sherman (Oxford Press), both with years of publishing experience under their belts, made the transition as smooth as possible and resulted in a book that we can all be proud of. We hope that you, the readers, find it interesting and useful, and we welcome your feedback. Another change in publishing has been the increasing use of online supplements to provide a more dynamic interaction than a book can provide. We are excited to incorporate a website that has dynamic versions of several of the mathematical models and associated figures described in this book. Each model allows the viewer to change parameters and observe how the model outcomes change. To access a model use the URLs found in the figure legend, or visit the website http://communityecologybook.org

Preface to 1st Edition

The community is…the hierarchical level where the basic characteristics of life—its diversity, complexity and historical nature—are perhaps most daunting and challenging.

Michel Loreau, 2010: 50

For communities…the task of choosing which topics to emphasize and which to elide [omit]…is necessarily quirky.

Robert M. May et al., 2007: 111

Well those drifter's days are past me now
I've got so much more to think about
Deadlines and commitments
What to leave in, what to leave out

Bob Seger, Against the Wind

"What to leave in, what to leave out"? Much of writing well about any subject comes down to this simple question. (I sometimes think much of life comes down to this simple question.) But, as Robert May notes, the task of choosing which topics to emphasize in community ecology is "necessarily quirky." Our discipline is broad and there is no clear roadmap. The organizational path I have chosen for this book is the one that works best for me when teaching this subject to graduate students. I begin with an exploration of patterns of biodiversity—that is, how does the diversity of life vary across space and time? Documenting and understanding spatial and temporal patterns of biodiversity are key components of community ecology, and recent advances in remote sensing, GIS mapping, and spatial analysis allow ecologists to examine these patterns as never before. Next, with species disappearing from the Earth today at a rate unprecedented since the extinction of the dinosaurs, what are the consequences of species loss to the functioning of communities and ecosystems? This question drives the very active research area of biodiversity and ecosystem functioning discussed in Chapter 2.

With the patterns of biodiversity at local and regional scales firmly in mind and with an appreci-ation for the potential consequences of species loss, I next shift gears to focus on what I call the "nitty-gritty" of community ecology: population regula-tion and species interactions, including predation, competition, mutualism, and facilitation. The goal here is to understand in some detail the mechanics of species interactions by focusing on consumers and resources in modules of a few interacting spe-cies. From these simple building blocks we can assemble more complex ecological networks, such as food webs and mutualistic networks, which involves exploring the importance of indirect effects, trophic cascades, top-down and bottom-up regulation, alternative stable states, diversity–sta-bility relationships, and much more. In the section on spatial ecology, I focus on the processes that link populations and communities across space (metap-opulations and metacommunities) and on the con-sequences of these local and regional links for species diversity.

The interplay between local and regional pro-cesses is a prominent theme throughout the book. Likewise, the interplay between ecology and evolu-tion—what is termed "eco-evolutionary dynamics" or, more broadly, evolutionary community ecology—is an important new area of research. I explore evolutionary community ecology, along with the impacts of variable environments on species interactions, in the book's final section on changing environments and changing species. Applied aspects of community ecology (e.g., resource management and harvesting, invasive species, diseases and parasites, and community restoration) are treated throughout the book as natural extensions of basic theoretical and empirical work. The emphasis, however, is on the basic science. Theoretical concepts are developed using simple equations, with an emphasis on the graphical presentation of ideas.

This is a book for graduate students, advanced undergraduates, and researchers seeking a broad coverage of ecological concepts at the community level. As a textbook for advanced courses in ecology, it is not meant to replace reading and discussing the primary literature. Rather, this book is designed to give students a common background in the principles of community ecology at a conceptually advanced level. At Michigan State University, our graduate community ecology course draws students from many departments (zoology, plant biology, fisheries and wildlife, micro-biology, computer science, entomology, and more) and students come into the course with vastly different exposures to ecology. I hope this book helps students from varied academic backgrounds fill in the gaps in their ecological understanding, approach a new topic more easily, and find an entry point into the primary literature. I'd be doubly pleased if it can do the same for practicing ecologists.

When teaching community ecology, I try to show students how seemingly differently ideas in ecology have developed over time and are linked together. This is important and hopefully useful to students who are just beginning to sink their teeth deeply into the study of ecology. At least, I believe it is useful. An early reviewer of this book wrote, "It is obvious that Mittelbach has a deep understanding and respect for the literature." I take this as a great compliment. We do, after all, stand on the shoulders of giants, and it's important to acknowledge where ideas come from. Moreover, an appreciation for the historical development of ideas and for how concepts are linked together helps deter us from recycling old ideas under new guises. However, another early reviewer suggested that students today aren't all that interested in the history of ideas and that a modern textbook on community ecology should focus on what's new, particularly on how community ecology can inform and guide conservation biology and the preservation of biodiversity. I appreciate this advice as well. I have worked to include cutting edge ideas and to provide case studies from the most recent literature, along with some of the classics. Hopefully, the balance between old and new contained herein is one that works. I recognize, however, that more could be done to illuminate the links between community ecology and conservation biology. Perhaps someone else will take up this call.

Finally, I want to say a few words about the use of mathematical models and theory in this book. Robert May (2010: viii) wrote that "mathematics is ultimately no more, although no less, than a way of thinking clearly." May also pointed out that one of the most celebrated theories in all of biology, Darwin's theory of evolution by natural selection, is a verbal theory. In most cases, however, the ability to express an idea mathematically makes crystal clear the assumptions and processes that underlie an explanation. I have used mathematical models here as a way to "think more clearly" about ecological processes and the theories put forth to explain them. The mathematical models in this book are simple, heuristic tools that, combined with graphical analyses, can help guide our thinking. Readers with limited mathematical skills should not be anxious when they see equations. I am a mathematical lightweight myself, and if I can follow the models presented here, so can you. On the other hand, readers with a strong background in mathematics and modeling will quickly recognize that I have stuck to the very basics and that much more sophisticated mathematical treatments of these topics abound. I have tried to point the way to these treatments in the references cited.

This book is a labor of love that has stretched out for over five years. I always knew that the ecological literature was vast, but I never truly appreciated its scale until I started this project. OMG! It's humbling to spend weeks reviewing the literature on a topic, only to stumble across a key paper later (and purely by accident). I know that I have missed much. I apologize in advance to those scientists whose excellent work I passed by (or simply missed in my ignorance) in favor of studies that were better known to me. Please don't be shy in telling me what I missed.

The first ecology textbook I purchased as an undergraduate in the early 1970s (at the hardcover price of $7.00!) was Larry Slobodkin's Growth and Regulation of Animal Populations. In his preface to this marvelous little book, Slobodkin wrote, "Every reader will find some material in this book that appears trivially obvious to him. I doubt, however, that all of it will appear obvious to any one person or that any two readers will be in agreement as to which parts are obvious. Bear with me when I repeat, in a naïve-sounding way, things you already know" (Slobodkin 1961: page v). Ditto.

Gary G. Mittelbach
March, 2012

Acknowledgments to 2nd Edition

I once read an interview with a professional bull rider. The interviewer asked, "How do you get ready to ride the bull?" To which the rider replied, "You're never really ready, it just becomes your turn." This simple answer applies to many things—you put the wheels in motion and then you see what happens. Producing the second edition of *Community Ecology* was like that for me. The first step was to ask Andy Sinauer if he would be interested in publishing a second edition and he said "YES" (thankfully). I owe Andy a great debt for his encouragement and support with the first edition of *Community Ecology* and with this edition as well, even though Sinauer Publishing became part of Oxford University Press just as we were getting rolling and Andy has now retired. At Oxford Press, Ian Sherman and Bethany Kershaw were instrumental in guiding us through the production process and in smoothing the transition from Sinauer to Oxford. We gratefully acknowledge their always-cheerful help and support.

For the second edition, I wanted to enlist a co-author as I felt there were many aspects of community ecology where I lacked expertise. I also hoped to find someone interested in carrying the book forward after I retired. Brian McGill was my first choice, hands down. Fortunately, he agreed, jumping in with both feet, even though he was busy writing his own book on macroecology and doing a million other things. Writing this book with Brian was my great pleasure and the product is much richer and broader in scope because of his efforts. Thank you, Brian, for your friendship, your smarts, and for your deep commitment to the field of ecology.

I thank all those who used the first edition of *Community Ecology* in their classes and who sent me their comments and feedback, especially Saara DeWalt, Kyle Harms, Bob Holt, Craig Osenberg, Todd Palmer, Rob Pringle, and others who I know I am forgetting. To every graduate student who has thanked me for writing a book that helped them study for their comprehensive exams, thank you—your kind words often made my day. As always, I am grateful to my colleagues at KBS/MSU for reading chapters and sharing ideas, especially Jen Lau and the Lau Lab, Chris Klausmeir and Elena Litchman, and their post-docs and students, and to my own graduate students, especially Pat Hanly. Doug Schemske has been a wonderful friend and colleague at MSU, and his unseen hand is on many parts of this second edition.

Finally, I thank my family and especially my wife Kay, for everything.

Gary G. Mittelbach

I am extremely grateful to Gary for inviting me to be involved in the 2nd edition of *Community Ecology*. I have been teaching a graduate community ecology class using the 1st edition, and it has been one of my all-time favorite academic books. Not only because community ecology is one of my favorite topics, but because it was so well executed. Even though there were a lot of good reasons to say no (there always are for writing a book), the opportunity to contribute in some small ways to a book that was starting already from such a strong, elevated position was way too good an opportunity to pass up. So was the opportunity to work again with Gary who has been a wonderful mix of mentor, colleague, and friend all the way back to my postdoc days. It has been more fun than I can describe to bounce ideas about the state of community ecology around with Gary.

I would second Gary's appreciation for the partnership with Ian Sherman and Bethany Kershaw at Oxford University Press. This is the second of what I hope will soon be three books for me with Ian. He has a wonderful way of humanizing the increasingly corporate world of publishing. I am also very grateful to two locations where I did a lot of my writing. The Monteverde Institute provided a wonderful place to write while on sabbatical (my "office" was a rocking chair on a porch in front of the library looking out across a tropical rain forest). The Orono Public Library has always been welcoming, even when I have ducked over there many hundreds of afternoons to find a quiet place for focused work. The University of Maine Library has also been ace at tracking down obscure references for me. Libraries and librarians are just plain awesome, and they don't get the appreciation they deserve!

I could not begin to list all of the colleagues that have taught me so much about community ecology through our conversations and co-authored papers, but thank you all. Finally, a great debt of gratitude to my wife, Sarah, and two sons, Eli and Jasper, for just smiling and saying "that's nice" when I told them I had started working on two books simultaneously. Living with an author is not always easy!

Brian J. McGill

We are very grateful to Mark McPeek who generously shared with us computer code from his book *Evolutionary Community Ecology*, allowing us to add a website with dynamic models and figures to this second edition of *Community Ecology*.

Finally, Gary and Brian would both like to dedicate this book to Michael Rosenzweig. Mike was Brian's PhD adviser, and Gary's friend and colleague. Mike recently retired and over his five decades of work, he has had a broad and deep impact on community ecology. The Rosenzweig–MacArthur predator–prey equation covered in Chapter 5 was co-invented by him. He has also done important work on optimal choice (especially habitat choice), biodiversity measurement, the species–area relationship and the latitudinal diversity gradient (Chapter 2), the interface of ecology and macroevolution (Chapter 15), and founded two key journals in the field of evolutionary ecology. He is also a wonderful teacher. It is hard to have a conversation with Mike without learning something new about science and how scientists think. Mike is also a wonderful human being and a generous scientist. Community ecology will be less rich without him, but we both wish Mike a happy retirement.

Gary G. Mittelbach & Brian J. McGill

Acknowledgments to 1st Edition

I have had the privilege of teaching a graduate course in population and community ecology at Michigan State University for 25 years. At some point it occurred to me that I should take what I have learned from teaching this course and put it in a book. No doubt the thought of leaving something behind drives many an author to write a book, for as Peter Atkins notes, textbooks capture a mode of thinking. I have focused on that part of ecology that I find most exciting—community ecology. It was only after I was well into the writing that I realized how woefully ignorant I was about my chosen field. I still am ignorant, but less so now. Writing this book has made me a better student of ecology and a better teacher. I will count myself lucky if it helps others in the same way.

I am fortunate to have had five excellent co-instructors in our "Pop and Com" course at MSU over the years: Don Hall, Katherine Gross, Doug Schemske, Elena Litchman, and Kyle Edwards. I thank you all for helping make teaching a fun and rewarding experience. Thanks also to our students (500+ and counting). You listened and challenged, and I hope you will recognize your many contributions in these pages.

Special thanks to the many people who read and commented on early drafts of chapters: Peter Abrams, Andrea Bowling, Stephen Burton, Peter Chesson, Ryan Chisholm, Kristy Deiner, Jim Estes, Emily Grman, Jim Grover, Sally Hacker, Patrick Hanly, Allen Hurlbert, Sonia Kéfi, Jen Lau, Mathew Leibold, Jonathan Levine, Nancy McIntyre, Brian McGill, Mark McPeek, Carlos Melián, Sabrina Russo, Dov Sax, Oz Schmitz, Jon Shurin, Chris Steiner, Steve Stephenson, Katie Suding, Casey ter-Horst, Mark Vellend, Tim Wootton, and Justin Wright. I owe a particularly large debt to Peter Abrams, who piloted an early draft of this book in his graduate course at the University of Toronto and who provided many insightful comments in his usual, no-holds-barred style.

Interactions with Doug Schemske, Kaustuv Roy, Howard Cornell, Jay Sobel, David Currie, Brad Hawkins, and Mark McPeek have been instrumental in helping me think about broad-scale patterns of biodiversity. Likewise, conversions with Kevin Gross, Armand Kirus, Chris Klausmeier, Kevin Lafferty, Jonathan Levine, Ed McCauley, Craig Osenberg, Josh Tewksbury, Earl Werner, and the "2010–2011 cohort" of postdocs at the National Center for Ecological Analysis and Synthesis (NCEAS) had a significant impact on the book. Thanks to my former graduate students for so many things and to my current graduate students for putting up with an advisor far too preoccupied with writing a book. Thanks also to Colin Kremer and Mark Mittelbach for their mathematical help. I gratefully acknowledge colleagues and staff at the Kellogg Biological Station for many years of support and friendship. I don't dare start naming names now, because there are too many people to thank. You know who you are and you know why you make KBS such a special place and that's enough. How was I ever so lucky to land here and somehow make it stick for a career?

This book has had a long gestation. When I first approached Andy Sinauer with a book proposal, my one request was that he not put time constraints on me, because I knew this would take awhile (and, secretly, I questioned whether I could pull it off at all). Andy graciously agreed, and he and the staff at Sinauer have been extraordinarily encouraging and helpful in every step of the processes. In particular, I thank my terrific editors, Carol Wigg and Norma Sims Roche, as well as art and production director Chris Small. I appreciate that Michigan State University granted me sabbatical leaves in 2001–2002 and 2010–2011, the first of which helped inspire this book; the second allowed me to finish it (almost).

Large parts of both sabbaticals were spent at the National Center for Ecological Analysis and Synthesis (NCEAS), a center funded by the National Science Foundation, the University of California at Santa Barbara, and the State of California. NCEAS provided the ideal environment for thinking and writing. This book would never have happened without its support. I am particularly grateful to Jim Reichman, Ed McCauley, Stephanie Hampton, and the wonderful NCEAS staff for their friendship and support. In spring 2011, Kay and I spent a short but magical time at EAWAG research institute on the shores of Lake Lucerne, Switzerland, where I worked on the final chapters of this book. I thank Ole Seehausen, Carlos Melián, and the scientists and staff at EAWAG Kastenienbaum for their hospitality and for making our brief stay productive and memorable.

I have enjoyed writing this book. My fond hope is that you enjoy reading it and will find it useful.

Contents

CHAPTER 1

Community ecology's roots

Every genuine worker in science is an explorer, who is continually meeting fresh things and fresh situations, to which he has to adapt his material and mental equipment. This is conspicuously true of our subject, and is one of the greatest attractions of ecology to the student who is at once eager, imaginative, and determined. To the lover of prescribed routine methods with the certainty of "safe" results, the study of ecology is not to be recommended. **Arthur Tansley, 1923: 97**

If we knew what we were doing, it wouldn't be called research. **Albert Einstein**

The diversity of life on our planet is remarkable. Indeed, among the biggest questions in all of biology are: How did such a variety of life arise? How is it maintained? What would happen if it were lost? Community ecology is that branch of science focused squarely on understanding Earth's biodiversity, including the generation, maintenance, and distribution of the diversity of life in space and time. It is a fascinating subject, but not an easy one. Species interact with their environment and with one another. As we will see in the pages that follow, these interactions underlie the processes that determine biodiversity. Yet, unlike the interacting particles studied by physicists, species also change through time—they evolve. This continual change makes the study of interacting species perhaps even more challenging than the study of interacting particles.

In his 1959 address to the American Society of Naturalists, G. Evelyn Hutchinson posed a simple question: "Why are there so many kinds [species] of animals?" Hutchinson's question remains as fresh and relevant today as it was half a century ago. This book will explore what ecologists understand about the processes that drive the distribution of animal and plant diversity across different spatial and temporal scales. In order to appreciate the current state of community ecology it is important to know something about its history, particularly the development of ideas. This first chapter provides a brief summary of that history. Those of you familiar with the field may skip ahead, while those of you interested in learning more should consult the books and papers by Hutchinson (1978), Colwell (1985), Kingsland (1985), May and Seger (1986), McIntosh (1980, 1985, 1987), and Ricklefs (1987, 2004). Many of these "histories" were written by ecologists actively involved in the field's development, for community ecology is a relatively young science.

What is a community?

A **community** is a group of species that occur together in space and time (Begon et al. 2006). This definition is an operational one. Any limits on space and time are arbitrary, as are any limits on the number of species in a community. For example, the study of "bird communities" or "fish communities" might be referred to, in order to delimit the assemblage of interest, recognizing that it is impossible to study all the species that occur together in the same

Community Ecology. Second Edition. Gary G. Mittelbach & Brian J. McGill, Oxford University Press (2019).

place at the same time. Although most ecologists would be happy with this definition, the concept of what a community is and how is it organized has changed widely through time (Elton 1927; Fauth et al. 1996).

The first community ecologists were botanists who noted what appeared to be repeated associations between plant species in response to spatial and temporal variation in the environment. Frederic Clements, the pioneer of North American plant ecology, viewed these plant associations as a coherent unit—a kind of superorganism—and he presumed that plant communities followed a pattern of succession to some stable climax community (Clements 1916; see discussion of ecological succession in Chapter 15). Limnology, the study of lakes, also adopted a superorganism view of communities. In 1887, the limnologist Stephen Forbes published a famous paper entitled "The lake as a microcosm," in which he stated that all organisms in a lake tend to function in harmony to create a system in balance. Thus, these early ecologists tended to view communities as unique entities, and they became preoccupied with classifying plant and lake communities into specific "types."

This superorganism concept of communities did not sit well with everyone. It was soon questioned by a number of plant ecologists, most notably Henry Gleason (1926) and Arthur Tansley (1939). Gleason asserted that species have distinct ecological characteristics, and that what appear to be tightly knit associations of species on a local scale are, in fact, the responses of individual species to environmental gradients. Gleason's individualistic concept of communities was ignored at first, but later asserted itself, and led to a rejection of Clements's hypotheses in favor of a continuum or gradient theory of plant distributions (Whittaker 1956).

The debate between Clements and Gleason over the nature of communities may seem like a historical footnote today, but at its core is a question that is very much alive: To what extent are local communities—the collections of species occurring together at a site—real entities? Ricklefs (2004) suggested that "ecologists should abandon circumscribed concepts of local communities. Except in cases of highly discrete resources or environments with sharp ecological boundaries, local communities do not exist. What ecologists have called communities in the past should be thought of as point estimates of overlapping regional species distributions." This focus on the interplay between local and regional processes in determining species associations is a theme that we will return to often in this book.

In contrast to plant ecology, the study of animal communities grew out of laboratory and field studies of populations. Animal population biologists, resource managers, and human demographers were concerned with the factors that regulate the abundance of individuals over time (birth, death, migration). Charles Elton, one of the pioneers of animal community ecology, worked for a time as a consultant for the Hudson's Bay Company, and his thinking was strongly influenced by the fluctuations he observed in the abundance of Arctic animals. Elton was opposed to the "balance of nature" concept espoused by Forbes and others, and in a book entitled *Animal Ecology* (1927), he discussed such important ideas as food webs, community diversity and community invasibility, and the niche. In another book, *The Ecology of Animals*, Elton (1950, p. 22) proposed that communities have limited membership, stating that in any prescribed area, "only a fraction of the forms that could theoretically do so actually form a community at any one time." Elton went on to note that, for animals as well as humans, it appears that "many are called, but few are chosen."

Elton's idea of limited membership was a significant insight, and it meshed well with the concurrent development of mathematical theories of population growth and species interactions. In the 1920s, mathematical ecologists Alfred Lotka (1925) and Vito Volterra (1926) independently developed the now famous equations that bear their names, which describe competition and predation between two or more species. These mathematical models showed that two species competing for a single resource cannot coexist. Gause (1934) experimentally tested this theory with protozoan populations growing in small bottles on a single resource. He found that species grown separately achieved stable densities, but that when pairs of species were grown together in a simple environment, one species always won out and the other species became extinct (Figure 1.1). Other "bottle experiments" with fruit flies, flour beetles, and annual plants produced similar results. The apparent generality of these results led to the formulation of what became known as Gause's **competitive**

(A)

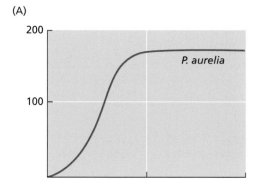

(B)

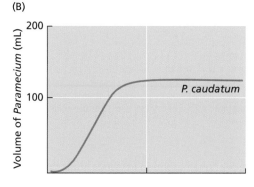

(C)

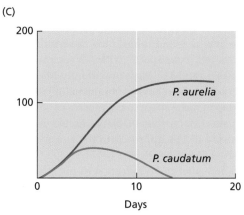

Figure 1.1 Results of Gause's competition experiments with two *Paramecium* spp. (*P. aurelia* and *P. caudatum*) grown separately and together in small containers in the laboratory. (A, B) Each species reached a stable population size (carrying capacity) when grown in isolation. (C) When grown together, however, one species always outcompeted and eliminated the other. After Gause (1934).

exclusion principle, which can be stated as "two species cannot coexist on one limiting resource." The competitive exclusion principle had a profound effect on animal ecology at the time and, in a modified

form, became a cornerstone of the developing field of community ecology.

In the 1940s and 1950s, there was vigorous debate over the competitive exclusion principle and whether populations were regulated by density-dependent or density-independent factors. Important figures in this debate were Elton, Lack, and Nicholson on the side of competition and density-dependent regulation, and Andrewartha and Birch on the side of density independence. In 1957, a number of ecologists and human demographers met at the Cold Spring Harbor Institute in Long Island, New York, to debate the issues of population regulation, with little consensus. However, this symposium did lead to one remarkable result. At the end of the published conference proceedings is a paper by G. E. Hutchinson (1957), modestly entitled "Concluding Remarks." In this paper, Hutchinson formalized the concept of the niche and ushered in what might well be considered the modern age of community ecology.

The ecological niche

The concept of the **niche** has a long history in ecology (see Chase and Leibold 2003a for an excellent summary). Grinnell (1917) defined the niche of an organism as the habitat or environment it is capable of occupying. Elton (1927) independently defined the niche as the role a species plays in the community. Gause (1934) made the connection between the degree to which the niches of two species overlap and the intensity of competition between them. Each of these concepts of the niche was incorporated into Hutchinson's thinking when he formalized the niche concept and connected it to the problem of species diversity and coexistence (Hutchinson 1957, 1959). In his "Concluding Remarks," Hutchinson showed how we might quantify an organism's niche, including both biotic and abiotic dimensions of the environment, as axes of an *n*-dimensional hypervolume (Figure 1.2).

Hutchinson (1957) went on to distinguish between an organism's **fundamental** (or *pre-interactive*) **niche** and its **realized** (or *post-interactive*) **niche**. The fundamental niche encompasses those parts of the environment that a species could occupy in the absence of interactions with other species, whereas the realized niche encompasses those parts of the environment that a species actually occupies in the

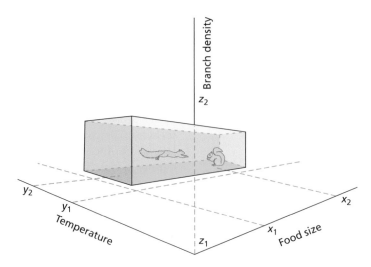

Figure 1.2 Hutchinson's visualization of the niche as an *n*-dimensional hypervolume. In this hypothetical example, the fundamental niche of a squirrel species is shown along three environmental dimensions. One axis (here labeled *y*) might define the range of temperatures tolerated by the species, another dimension (*x*) might describe the range of seed sizes (e.g., acorns) eaten, and a third axis (*z*) the range of tree branch densities (diameter, volume) that support this squirrel species. Subscripts 1 and 2 represent the lower and upper limits for each niche dimension. After Hutchinson (1978).

presence of interacting species (e.g., competitors and predators). In Hutchinson's view, a species' realized niche was smaller than its fundamental niche, due to negative interspecific interactions. However, positive interactions between species (mutualisms, commensalisms) can result in a species occupying portions of the environment that were previously unsuitable; in other words, it is possible for the realized niche to be *larger* than the fundamental niche (Bruno et al. 2003). The fact that positive interactions were not explicitly considered in Hutchinson's niche concept shows how completely the ideas of competition and predation permeated ecological thinking at the time.

Hutchinson's definition of the niche provided the framework on which ecologists would build a theory of community organization, based on interspecific competition. First, however, they needed to make Hutchinson's concept more workable. An "*n*-dimensional niche" is fine in the abstract, but empirically, it is impossible to measure an undefined number of niche dimensions. It took one of Hutchinson's students, Robert MacArthur, to make the concept operational. MacArthur's approach (1969) was to focus on only a few critical niche axes—those for which competition occurs. If, for example,

interspecific competition for seeds limits the number of seed-eating birds in a community, then we should focus our study on some measure of seed availability to define a species' niche (e.g., seed size). Thus, it became possible to examine the distribution of species in "niche space" within a community (see Figures 8.3 and 8.4). More importantly, MacArthur's approach showed how we might quantitatively link overlap in niche space to the process of competitive exclusion (MacArthur 1972). We discuss the modern extension of this approach to the study of species coexistence in Chapter 8.

Shortly after the publication of his "Concluding Remarks" in 1957, Hutchinson (1959) provided another key insight in his published presidential address to the American Society of Naturalists, entitled "Homage to Santa Rosalia, or why are there so many kinds of animals?" Here, Hutchinson pondered a question that goes a step beyond Gause's competitive exclusion principle: If the competitive exclusion principle is true and interspecific competition limits the coexistence of species within the same niche, then how dissimilar must species be in their niches in order to coexist? Hutchinson suggested that the answer might be found in the seemingly regular patterns of difference in body size among members of an **ecological guild**—co-occurring species that use the same types of resources. He noted that species that were similar in most ways, except the sizes of prey eaten, tended to differ by a constant size ratio: a factor of about 1.3 in length and 2.0 in body mass. Such regular differences in size among coexisting species were found in many ecological guilds (one example is shown in Figure 1.3) and became known as **Hutchinsonian ratios**.

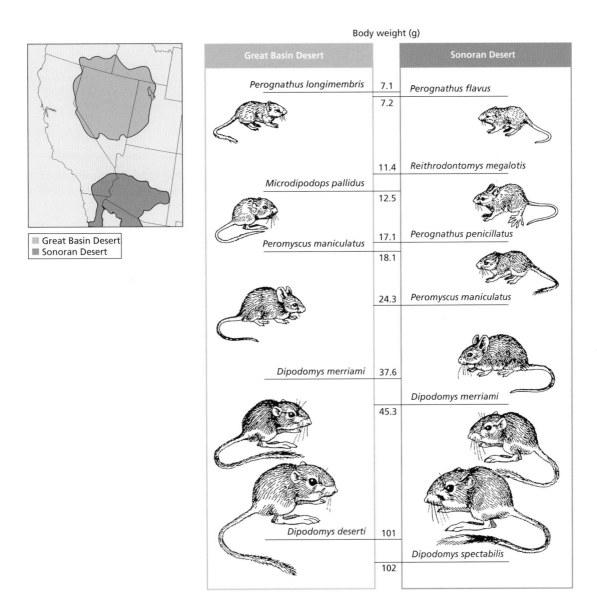

Figure 1.3 Hutchinsonian ratios among desert rodents found in the Great Basin and Sonoran Deserts of the western USA. Differences in body size reflect differences in diet and habitat use (niche differences) between these species. The pattern of body-size spacing observed in these two desert rodent ecological guilds is more regular than would be expected by chance. From Brown (1975).

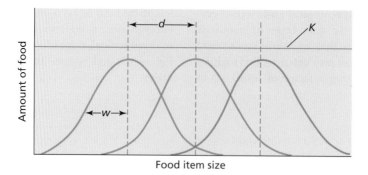

Figure 1.4 The concept of limiting similarity illustrated for three species utilizing a continuum of food resources. *K* represents a resource continuum (for example, the amount of food as a function of food size). Each species' niche is represented by the mean and the standard deviation (*w*) of its resource utilization curve, and *d* is the distance between the mean resource uses of the closest pair of species. MacArthur and Levins (1967), and May and MacArthur (1972) showed that the minimal niche separation required for the coexistence of competing species (under very specific conditions) is $d/w \approx 1$. After May and MacArthur (1972).

MacArthur and Levins (1967) built on these ideas and introduced the concept of **limiting similarity**, which specified the minimal niche difference between two competing species that would allow them to coexist. In MacArthur and Levins's theory, species are arrayed linearly along a resource (niche) axis, and each species' resource use is represented by a normal (bell-shaped) utilization curve (Figure 1.4). The overlap between adjacent utilization curves can be used (under specific assumptions) as a measure of the competition coefficients (α's) in the Lotka–Volterra model of interspecific competition (described in Chapter 7). Using this model of competition, MacArthur and Levins (1967) were able to specify the minimum niche difference required for two species to coexist. Later, May and MacArthur (1972) and May (1973b) used a different approach, based on species in fluctuating environments, to arrive at a very similar outcome: the limiting similarity between two competing species is reached when $d/w \approx 1$, where d is the separation in mean resource use between species and w is the standard deviation in resource use (see Figure 1.4).

Whither competition theory?

In less than 50 years, animal community ecology progressed from the simple recognition that species too similar in their niches cannot coexist to the development of a theoretical framework, poised to predict the number and types of species found in natural communities based on a functional limit to the similarity of competing species. This was an enormous leap forward, and community ecology seemed well on its way to becoming a more quantitative and predictive science. The heady optimism of the times is reflected in Robert May's (1977a, p. 195) comment that "the question of the limits to similarity among coexisting competitors is ultimately as deep as the origin of species itself: although undoubtedly modified by prey–predator and mutualistic relations, such limits to similarity are probably the major factor determining how many species there are." In the end, however, the theory of limiting similarity failed to achieve its promise. What happened?

First, there were strong challenges to the idea that interspecific competition is the only, or even the primary, factor structuring communities. Much of the evidence for the importance of interspecific competition in communities was based on descriptive patterns, such as regularly spaced patterns of body size among coexisting species (see Figure 1.3) or "checkerboard" distributions of species on islands (Diamond 1973, 1975). When examined more closely, however, many of these patterns turned out to be indistinguishable from those predicted by models that did *not* include interspecific competition as an organizing force—that is, by null models (Strong et al. 1979; Simberloff and Conner 1981; Gotelli and

Graves 1996). Secondly, predictions of limiting similarity between species turned out to be model-dependent (Abrams 1975, 1983b). That is, even though most mathematical models of interspecific competition predict some limit to how similar species may be in their resource use and still coexist, that limit varies widely depending on the assumptions and structure of the model. We now know there are no universal (or "hard") limits to similarity as originally envisioned by MacArthur and Levins (1967). However, recent work has shown that, while there are no formal limits to similarity, the more tightly packed a community is in terms of niche space, the more fragile is species coexistence (Meszéna et al. 2006; Barabás et al. 2012, 2013). The extreme fragility of tightly packed communities suggests a reinterpretation of the limiting similarity principle, rather than its complete abandonment.

These and other challenges caused ecologists to look beyond interspecific competition, and to consider the plurality of factors that might determine species diversity. In contrast to the unbridled optimism that characterized community ecology in the 1960s and early 1970s, the next decade was a period of soul-searching, as ecologists struggled to find a conceptual framework to replace what had seemingly been lost (McIntosh 1987). In the end, however, the idea that communities are organized around strong interspecific interactions was not so much wrong as it was overly simplistic.

New directions

The "failure" of simple competition-based models to explain community diversity led to important new directions in community ecology, and many of these directions continue to influence how we study ecology. For example, the "null model debate" of the 1970s led directly to an increased emphasis on using field experiments to test ecological hypotheses. Many of these field experiments focused on studying interspecific competition. However, pioneering experiments by Paine (1966), Dayton (1971), and Lubchenco (1978), all working in the marine intertidal zone, also showed that the presence or absence of predators could have dramatic effects on species diversity. These experiments set the stage for a wealth of future work on food webs, trophic

cascades, and top-down effects. We will consider these topics in detail in subsequent chapters, as well as more recent approaches to characterizing food webs and other types of ecological networks (see Chapters 10 and 11).

The experimental studies cited above demonstrated how competition and predation may interact to affect species diversity and composition (for example, diversity is increased when predators feed preferentially on a competitive dominant), again setting the stage for subsequent empirical and theoretical work on keystone predation and competition–predation trade-offs. Over time, the accumulation of results from multiple field experiments fostered the application of **meta-analysis** in ecology, in which the outcomes of many experiments are combined and synthesized to arrive at general conclusions (Gurevitch et al. 1992; Borenstein et al. 2009; Schwarzer et al. 2015). Today, ecologists rely heavily on meta-analysis, and there will be many times in this book when the authors will look to the results of meta-analyses to evaluate the importance of a process in ecology.

The "failure" of a single process (interspecific competition) to account for many of the patterns in species diversity observed in the 1960s and 1970s led ecologists to take a more pluralistic approach to their science (Schoener 1986; McIntosh 1987). A pluralistic ecology recognizes that multiple factors may interact to determine the distribution and abundance of species. The difficultly with pluralism, however, is that it can quickly lead us into a morass. In his thought-provoking article, Vellend (2010, p. 183) suggested that "despite the overwhelmingly large number of mechanisms thought to underpin patterns in ecological communities, all such mechanisms involve only four distinct kinds of processes— selection, drift, speciation, and dispersal." In Vellend's framework, "selection" encompasses the processes that determine the relative success of species within a local community (e.g., competition, predation, disease), whereas "drift" refers to changes in species' relative abundances due to chance or other random effects, and "dispersal" is the movement of individuals and species, into and out of local communities. "Speciation" operates over spatial scales larger than the local community, and it is the process that ultimately generates diversity in regional species

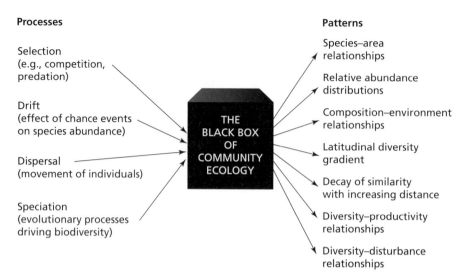

Figure 1.5 A conceptual view of the functioning of community ecology, in which four basic processes (selection, drift, speciation, and dispersal) combine to determine the biodiversity patterns listed on the right. The "black box of community ecology" refers to the fact that there are many ways in which the four processes listed at the left may combine to produce the patterns listed at the right. After Vellend (2010).

pools. Vellend (2010, 2016) suggests that conceptual synthesis in community ecology can be achieved by focusing on these four major drivers of species diversity patterns at different spatial and temporal scales (Figure 1.5). We agree with Vellend's suggestion. Moreover, we believe that community ecologists are further challenged to illuminate the interior of the "black box" in Figure 1.5, and to better understand how the four basic processes of community ecology interact to determine patterns of biodiversity.

The recognition that local communities bear the footprint of historical and regional processes (Ricklefs and Schluter 1993) is an important insight that grew out of the narrow, local community focus of the 1960s and 1970s. Interestingly, MacArthur (1972) anticipated this paradigm shift, but he died too young to be a part of it (see discussion in McIntosh 1987). Simply put, few communities exist in isolation. Instead, the diversity of species within a community is a product of their biotic and abiotic interactions (i.e., species sorting or "selection," together with drift), the dispersal of species between communities, and the composition of the regional species pool (a function of biogeography and evolutionary history). Therefore, it is necessary to consider the

processes that regulate diversity on a local scale, as well as the processes that link populations and communities into metapopulations and metacommunities, and the processes that ultimately generate diversity at regional levels. This is a tall order. Chapter 2 will use broad-scale diversity gradients, particularly the latitudinal diversity gradient, as a vehicle to begin to think about the processes that generate regional diversity. At geographic scales of regions or continents, biodiversity is a function of evolutionary processes that may play out over millions of years. In addition, chance events in Earth's history can influence a region's size, geomorphology, climate, and the amount of time available for speciation. These historical factors conspire to make the study of the processes that determine regional biodiversity challenging. Of course, there is also little opportunity to do experiments at such vast scales of time and space. However, as we will see, recent advances in molecular biology, phylogenetics, paleontology, and biogeography have greatly facilitated the study of broad-scale diversity patterns, and these new tools are providing the key to understanding the factors that generate biodiversity at regional scales.

PART I

The Big Picture: Patterns, Causes, and Consequences of Biodiversity

CHAPTER 2

Patterns of biological diversity

Biodiversity, the variety of life, is distributed heterogeneously across the Earth.

Kevin Gaston, 2000, p. 1

To do science is to search for repeated patterns, not simply to accumulate facts.

Robert MacArthur, 1972, p. 1

Why do different regions of the Earth vary so dramatically in the number and types of species they contain? This question has challenged ecologists and evolutionary biologists for a very long time. The early naturalists were struck by the remarkable diversity of life they found in the humid tropics (e.g., von Humboldt 1808; Darwin 1859; Wallace 1878). As an example, there are approximately 600 tree species in all of North America (Currie and Paquin 1987), whereas a tropical rainforest may contain 600 tree species in just a few hectares (Pitman et al. 2002), and the New World tropics, as a whole, support an estimated 22,000 tree species (Fine and Ree 2006; Slik et al. 2015). For the past 200 years ecologists have sought to explain why the species richness of most taxa increases dramatically from the poles to the equator (Figure 2.1). In his review of almost 600 published studies, Hillebrand (2004) documents that this **latitudinal diversity gradient** exists for nearly all taxonomic groups. The strength of the latitudinal diversity gradient does not differ between plants and animals, nor between marine and terrestrial environments, nor between warm- and cold-blooded organisms (Figure 2.2).

The striking, repeatable increase in species richness from the poles to the tropics is, without doubt,

the prime example of what Robert MacArthur (1972) referred to when he said that ecologists should "look for repeated patterns" in nature. Robert Ricklefs called it "the major, unexplained pattern in natural history…one that mocks our ignorance" (quoted in Lewin 1989, p. 527). This chapter will examine the factors thought to drive variation in species diversity across broad spatial scales, using the latitudinal diversity gradient as a focus. First, however, it considers some of the challenges associated with assessing biological diversity at different spatial scales, focusing attention on one key aspect of biological diversity—species richness. It then examines three of the best-studied patterns in species richness; the relationship between species richness and area, the distribution of species abundances, and the relationship between productivity and species richness. The chapter concludes with a detailed exploration of the most dramatic of Earth's biodiversity patterns— the latitudinal diversity gradient. These patterns of biodiversity constitute much of what community ecology seeks to explain about nature (see Figure 1.5); their study provides a foundation from which to explore the mechanisms of species interactions, and to understand the processes that drive variation in species numbers and their distribution.

Community Ecology. Second Edition. Gary G. Mittelbach & Brian J. McGill, Oxford University Press (2019).
© Gary G. Mittelbach & Brian J. McGill (2019). DOI: 10.1093/oso/9780198835851.001.0001

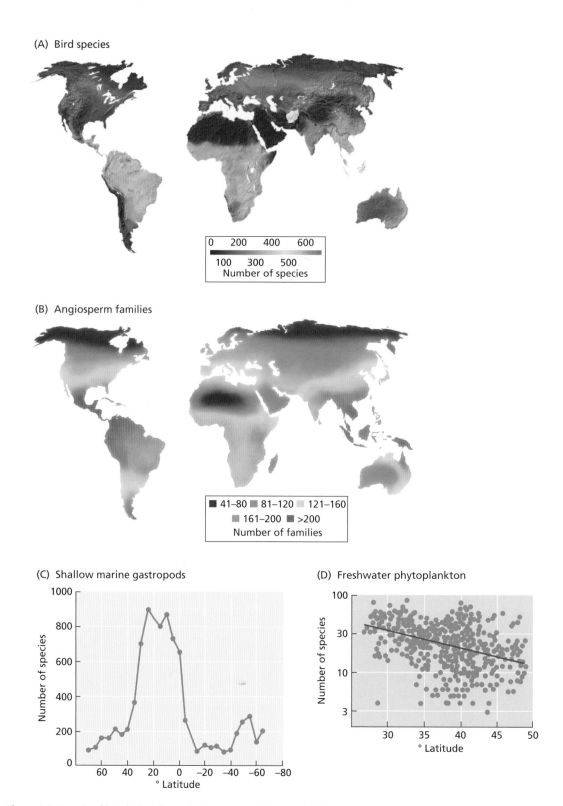

Figure 2.1 Examples of latitudinal gradients in biodiversity. Most biologists are familiar with polar–tropical gradients in large, conspicuous organism, such as birds, mammals, and trees. However, recent work documents that latitudinal diversity gradients exist for a wide variety of taxa, including microorganisms, and that these gradients occur in terrestrial, marine, and freshwater habitats. (A) Bird species richness. (B) Angiosperm family richness. (C) Marine gastropod species richness. (D) Freshwater phytoplankton species richness. (A) taken from Hawkins et al. (2007); (B) from Francis and Currie 2003; (C) after Rex et al. 2005; (D) after Stomp et al. 2011.

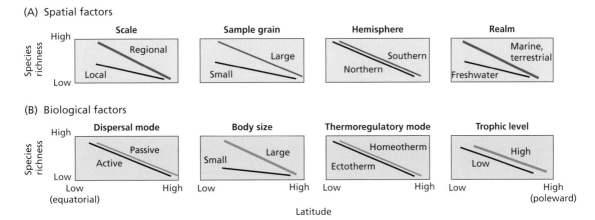

Figure 2.2 Data compiled from nearly 600 studies of species richness show the trend of a latitudinal gradient—i.e., increasing species richness closer to the equator—for spatial (A) and organismal (B) factors. The graphs illustrate in stylized fashion the observed change in species richness from low to high latitude for eight variables, as well as the relative impact (shown by the difference in the two slopes) and correlational strength (shown by the relative line thickness) of each variable with the latitudinal gradient. After Hillebrand (2004).

Assessing biological diversity

The simplest concept of diversity is species richness, the number of species found in a local community. Even this seemingly simple concept can be surprisingly complicated to measure and use correctly. For starters, just as the boundaries of a community are arbitrary (Chapter 1), the boundaries for measuring species richness are arbitrary. They often end up being simply the boundaries of the quadrat, transect, net sweep, or other sampling unit used. This is problematic because species richness is strongly correlated with sample size, making our choice of sampling scale consequential. To make matters worse, richness scales non-linearly with sample size. This means that there is no simple means to compare an observation of 10 tree species on a 1-ha plot with 20 tree species on a 5-ha plot. One might be tempted to convert these numbers to "species density" and say that the first plot has 10 species/ha, while the second plot has $20/5 = 4$ species/ha and is thus lower in richness. This type of standardization often works for abundance or biomass, which scale roughly linearly with area, but such an approach is completely erroneous for species richness, because of the non-linear relationship between richness and sample size—it should never be used (Gotelli and Colwell 2001).

Another issue is that species richness is only one aspect of biodiversity. Other important aspects of biodiversity include genetic diversity (how much genetic variation is found within a community), functional diversity (how the species in a community fill out the n-dimensional space of traits), and phylogenetic diversity (how much distinct evolutionary history is present in a community). These three additional aspects of diversity are discussed in detail in Magurran and McGill (2011). While the other measures of diversity are often positively correlated with taxonomic diversity (i.e., species richness), they are not always. For example, small-scale functional diversity peaks at the equator, just as taxonomic diversity does, but larger-scale regional functional diversity peaks at temperate latitudes, unlike taxonomic diversity (Lamanna et al. 2014).

The core tool for rigorously measuring, comparing, and understanding species richness is known variously as the species–individual curve, the species accumulation curve, a rarefaction curve, or a collector's curve (Figure 2.3(A)). Imagine a biologist walking in a straight line through a community, counting the number of species she/he encounters. The dashed curve in Figure 2.3(A) is what would be expected to be seen. Initially, every new individual encountered is likely to be a different species, but at some point species will be encountered that have

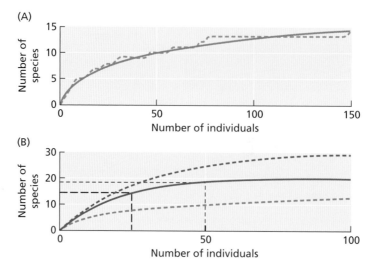

Figure 2.3 (A) An example of a species–individual curve, which plots the number of species encountered versus the number of individuals encountered. The green dashed curve represents a single instance and is jagged. The blue, smooth curve represents an average across many randomizations of order of encounter of individuals and so is smoother. (B) The solid red curve represents a species–individual curve for 1000 individuals with 20 species, with each species having an equal abundance of 50. The other curves show how variation in the structure of how individuals are distributed between species and across space, affects the species–individual curve and thus alpha diversity. The dashed blue line assumes a more typical highly uneven distribution of abundances across the species with many rare and few common species (the line increases more slowly because most of the individuals encountered are from the common species and only rarely does one encounter a new rare species). The dashed red line assumes that the regional (gamma) diversity is doubled (40 species). The dashed horizontal and vertical black lines show how the number of species observed increases in a non-linear fashion with the number of individuals observed (which can change either due to changing the observation area or the density of individuals).

already been seen and the line becomes horizontal for a while, then a new species is encountered and the line goes up by one species again. If many such curves are averaged across many random orderings of individuals or analytical formulas are used (Hurlbert 1971) a smooth curve is obtained (the solid line in Figure 2.3(A)). All communities show a curve of this general shape (i.e., an increasing decelerating function). It looks roughly like the square root function, but in fact, there is a great deal of variation in the details of the shape. In situations where the abundance of individuals is not recorded, but instead simply the presence or absence of species across many sites is recorded, a related technique, known as sample-based rarefaction, can be used, where standardization by the number of samples on the x-axis occurs, instead of the number of individuals (Gotelli and Colwell 2001). Species–individual curves not only emphasize the scale dependence of species richness, but in fact, provide the solution. It is now possible to compare diverse systems by standardizing for sample size (number of individuals) and comparing the height of the species–individual curves (Figure 2.3(B)).

Alpha, beta, and gamma diversity

Given the importance of scale when talking about richness, it is important to have a shared vocabulary when discussing diversity. Robert Whittaker (1960, 1972) provided the foundation for measuring diversity at different spatial scales when he proposed that species diversity could be expressed in three ways:

1. **Alpha (α) diversity**, the number of species found at a local scale (i.e., within a habitat or local site).
2. **Beta (β) diversity**, a measure of the difference in species composition, or species turnover, between two or more habitats or local sites within a region.
3. **Gamma (γ) diversity**, a measure of species richness in a region.

Region X
5 species

Region Y
6 species

Site 1
5 species

Site 2
3 species

Site 3
5 species

Site 4
3 species

Figure 2.4 An illustration of alpha (α), beta (β), and gamma (γ) diversities, as measured at four sites and in two regions (X and Y). Alpha diversity quantifies diversity at the local scale (i.e., within a given site or habitat). In this example, alpha diversity is greatest at sites 1 and 3 (five species each) and lower at sites 2 and 4 (three species each). Beta diversity measures the change in diversity between sites—that is, the amount of species turnover within a given region. In this example, region Y has higher beta diversity than region X (i.e., the same five species are found at both sites in Region X, whereas there is no overlap between the three species present at each of the two sites comprising Region Y). Gamma diversity measures the total diversity within a region, assessed across all sites. In this example, gamma diversity is higher in region Y (six species) than in region X (five species). After Perlman and Anderson (1997).

These cross-scale comparisons can be informative about how diversity is structured. Whitaker defined beta diversity by $\beta = \gamma/\alpha$ or $\alpha * \beta = \gamma$. The simple example in Figure 2.4 can be illustrative. In Region X (green sites), alpha diversity is 5 (average of the richness of 5 across the two sites), gamma diversity for Region X is 5 (there are 5 total species found across all sites, so beta diversity is 1 (γ/α) In contrast, Region Y (yellow sites) has an alpha diversity of 3 (average of 3 species across both sites), a gamma diversity of 6, and so a beta diversity of 2. We can see that Region Y has a higher gamma diversity due to a higher beta diversity, despite lower alpha diversity.

You might ask, "What do we mean by a region?" That is a very good question. Ecologists use the concept in a number of different ways, depending on the question of interest (Table 2.1). A **region** is generally considered to be large in spatial scale, extending over many square kilometers, and containing a large number of habitats and communities. In addition, ecologists often use "region" to refer to an area from which species, over time, have a good probability of colonizing a local community of interest. The species likely to colonize a local community are collectively referred to as the **regional species pool**. Finally, ecologists studying global patterns of biodiversity often think about regions as parts of the globe that have distinct evolutionary histories—that is, geographic areas that are isolated by major physical boundaries. Wallace (1876) noted

Table 2.1 Some spatial designations in community ecology[a]

Term	Definition
Biogeographical region/realm	Extremely large spatial extent, roughly corresponding to continental and subcontinental areas. Spatial limits coincide with major geological and climatic barriers that are the result of plate tectonics; these barriers (e.g., oceans, mountains, deserts) limit the dispersal of organisms over evolutionary time. Functionally defined as portions of the globe that have a shared evolutionary history and that, consequently, contain endemic and related taxa. There is a rapid turnover in species at the boundaries of biogeographic regions
Province	Similar to a biogeographical region/realm
Biome or ecoregion	Very large spatial extent. Terrestrial biomes are functionally defined by climate and the vegetation and animals associated with specific climatic conditions, wherever on the globe those climatic conditions occur. Examples: desert, tropical rain forest, temperate grassland, tundra. Aquatic biomes, like terrestrial biomes, are characterized by assemblages of organisms adapted to similar conditions; they are defined not by climate, but by a variety of factors including salinity, water temperature and movement, light penetration, and substrate composition. Examples: River, lake, salt marsh, intertidal, open ocean
Region	Large spatial extent, covering many square kilometers and containing a large number of habitats and communities. Sometimes used synonymously with biogeographic region or province, but more commonly used to define a somewhat smaller spatial scale. Functionally defined as an area in which the processes of speciation and extinction operate to affect biodiversity and from which species, over time, have a good probability of colonizing a local community. For example, we may refer to the "regional species pool" as the collection of species likely to colonize a local community of interest
Local	Small spatial extent, < 1 m to a few square kilometers. Spatial extent is determined, in part, by the size of organisms within a community; e.g., from microorganisms living within a tree hole or a pitcher plant leaf, up to trees in a tropical forest. Functionally, a local area or local community is defined by the likelihood of species interactions. Within a local area or local community, species have a high probability of interacting with each other and influencing each other's dynamics

[a]The terms listed here are widely used by community ecologists to describe spatially dependent relationships and processes. Although the definitions given reflect common usage, strict definitions of these terms do not exist. It is important to note that the terms are defined both by their spatial extents (approximate area/size) and by their functional roles in community ecology (see Cornell and Lawton 1992).

that Earth's biota show pronounced dissimilarities at major geographic boundaries, such as continental margins, major mountain ranges, or climatic breaks. These biologically and climatically distinct regions are termed **biogeographic regions** or **biogeographic realms** (Kreft and Jetz 2010; Holt et al. 2013; Ficetola et al. 2017). In the study of global patterns of diversity, biogeographic regions are often used to delineate regional diversity.

It is also useful to think of alpha and gamma diversities as being defined by the processes that are dominant at each spatial scale. "Thus, the γ-scale is that at which evolutionary process (specifically speciation and global extinction) determine biodiversity and variation, is driven by broad-scale gradients such as climate…area…and biogeographical, historical contingencies.…In contrast, α-diversity is the scale at which processes traditionally studied in community ecology such as dispersal limitation, microclimate, and species interactions predominate" (McGill 2011, p. 482). Although we might wish to have more precise definitions for the three spatial components of diversity (alpha, beta, and gamma), they are inherently relative concepts. The main

challenge with assessing alpha and gamma diversity often boils down to agreeing on what size areas to measure. Assessing beta diversity, however, is more complex and reaching a consensus has been challenging. On the one hand, beta diversity is conceptually simple—it is how quickly the composition of species changes across sites within a region. As such, it is a measure of turnover or a rate of change. Beyond that, the consensus quickly falls apart. Whittaker's $\beta = \gamma/\alpha$ formula (1972) provides one way to measure beta diversity. Others have proposed an additive model ($\beta = \gamma - \alpha$), which is easier to work with statistically (Crist et al. 2003).

An entirely different way to measure beta diversity is to quantify the change in species composition between pairs of local communities. Similarity indices, like Sorenson and Jaccard calculate the percentage overlap in species composition between communities. More complex similarity indices like Morisita–Horn and Bray–Curtis consider the relative abundance of different species (Koleff et al. 2003). Calculating one minus similarity gives a measure of dissimilarity for a pair of communities and the average of dissimilarity across all possible pairs of communities is a measure of beta diversity. So is a calculation of the rate at which similarity between pairs of communities declines with the distance between communities (Nekola and White 1999). At this point, dozens of methods for measuring beta diversity have been developed (Koleff et al. 2003; Anderson et al. 2011). More recently, there have been efforts to calculate a measure of beta diversity that controls for gamma diversity using randomization null models (Kraft et al. 2011). Those authors found that once gamma diversity was controlled for, beta diversity no longer varied along latitudinal gradients. This inspired a number of critiques and alternative proposals for controlling beta diversity for either alpha or gamma diversity, or both (Qian et al. 2013; references in Ulrich 2017). In the end, however, it is believed that these efforts are misguided (Ulrich 2017). Whittaker's original formulation (1960) shows beta diversity is intimately interrelated with alpha and gamma diversity. The notion that they can be decoupled is wrong. Ecologically, alpha, beta, and gamma are all just limited views onto a larger complex underlying reality of how individuals of different species are arrayed across space. The fact that each view captures only a piece of the total picture should not fool us into thinking they are independent.

Let us for a minute return to the vision of a biologist walking through a community, counting the total number of species that she/he encounters, and consider how different aspects of the distribution of species might affect the assessment of alpha and beta diversity. For example, if we move from assuming every species is equally abundant (solid red line in Figure 2.3.B) to a more realistic scenario, where most species are rare, and a few are very common (i.e., more uneven), the species individual curve becomes lower (blue dashed line in Figure 2.3.B), indicating a lower alpha diversity. This is because the common species quickly show up in our sample, and new individuals encountered are very likely to be the same already encountered common species. It can take a long time to encounter a rare species that has only a few individuals in the region. Thus, the net effect of increasing the inequality of abundance of species is to lower alpha diversity. However, it increases beta diversity (because the few rare species found in a local community are different from the rare species found in another local community). Using a smaller plot size equates to sampling fewer

individuals, which equates to shifting left along the species–individual curve, resulting in lower alpha diversity (shown by the dashed horizontal and vertical lines in Figure 2.3.B; and again, increased beta diversity). Equally, lowering the density of individuals across the region, while holding plot area constant has the same effect of lowering the number of individuals in a plot, and decreasing alpha and increasing beta diversity. A more subtle effect is to no longer scatter the individuals at random across space, but to distribute them so that individuals of the same species are more likely to be found in proximity to each other than by chance (Condit et al. 2000). This effect is very similar to the effect of increasing unevenness. From the lens of a single plot, some species become more common because they are in the focal plot and clumped together, increasing their apparent abundance, and others become rarer because they are clumped together in some other plot and thus rarer from the point of view of the target plot. Thus, in contrast to a random distribution, clumping individuals of a species together decreases alpha diversity and increases beta diversity (dashed blue line Figure 2.3B). Finally, if we increase the regional species richness (dashed red line in figure 2.3.B), then both the alpha and the beta diversity will go up. Therefore, there are five key factors (Table 2.2) that change the alpha and beta diversity:

- the unevenness of the distribution of abundances;
- the clumped vs spatially random distribution of individuals within a species;
- the total regional (gamma) richness;
- the area of what is called a local plot;
- the density of individuals across the region.

Table 2.2 How different aspects of the distribution and abundance of species impact alpha and beta diversity

	Alpha diversity	Beta diversity
More uneven distribution of abundance	↓	↑
Spatially clumped individuals within a species	↓	↑
Higher regional (gamma) diversity	↑	↑
Smaller local plot area	↓	↑
Higher density of individuals	↑	↓

This analysis shows us several things. First, we can see that species richness is a summative variable, which accumulates the balance of many different factors affecting a community. It can also be seen how α, β, and γ diversity are all just pieces, albeit highly interconnected, of the complex underlying reality. This verbal model is also a step towards a predictive model of local species richness or alpha diversity. For example, if we wonder what effect an invasive species might have on species richness, we can break it out into analyzing how the invasive species would affect the five factors of distribution listed above. In several well-studied systems, invasive species primarily act by decreasing the density of native individuals and increasing the dominance of the one common, invasive species. Both of these factors tend to decrease alpha richness, but have relatively little effect on regional richness (Powell et al. 2013), thereby increasing beta diversity as well.

Patterns of biological diversity

With an understanding of how biological diversity is measured at different spatial scales, this chapter now turns to examining some of the better-known biodiversity patterns in nature, and their potential causes.

Area and species richness

Large areas contain more species than small areas. Organisms on islands provide the most prevalent examples of this **species–area relationship (SAR)**, but it exists for mainland and marine areas as well. Like the latitudinal diversity gradient, the SAR is pervasive, its discovery dates to the mid-nineteenth century, and it was one of the first diversity patterns quantified in ecology (Arrhenius 1921; Gleason 1922; Rosenzweig 1995).

The relationship between species richness and area can be described by a power function of the form

$$S = cA^z \qquad \text{[Eqn 2.1]}$$

where S is the number of species, A is the area, and c and z are fitted constants. This relationship yields a straight line on a log–log plot (Figure 2.5). Note that the shape of the SAR (Figure 2.5 inset) is similar to the

individual species accumulation curve (Figure 2.3A), which is not a coincidence since the number of individuals is usually proportionate to area. Other functions (e.g., logarithmic) also have been used to fit the SAR, and there is some evidence that the power function may not be its best descriptor (Kalmar and Currie 2007). However, most studies describe SARs with a power function, and the fitted constants c and z provide a useful way to compare relationships among studies (see Connor and McCoy 1979; Drakare et al. 2006).

Why do larger areas contain more species? The answer has two parts. First, larger areas typically encompass a greater variety of habitats. This increase in habitat diversity with area contributes to the SAR because different species have different habitat affinities (Figure 2.6). However, even in areas of relatively uniform habitat, larger areas can be expected to harbor more species because of demographic processes. Thus, the second part of the answer is that larger areas can support larger populations, which have a lower probability of extinction. MacArthur and Wilson (1967) developed a simple model of island biogeography to illustrate this effect (Figure 2.7). Simberloff (1976) tested this model experimentally by reducing the sizes of small mangrove islands off the coast of Florida with chain saws. Islands reduced in size decreased in species richness over time, whereas control islands showed no effect (Simberloff 1976). These observations and MacArthur and Wilson's (1967) theory have important implications for conservation and the design of nature preserves, as will be discussed in more detail in Chapter 13.

Although the basic processes driving the species–area relationship are understood, there are many aspects of SARs that continue to challenge our understanding. For example, the slope of the SAR (the fitted constant z in Equation 2.1) is of fundamental interest because it describes the rate at which new species are encountered as sampling area increases (and is thus capturing one aspect of beta diversity). Rosenzweig (1995), following earlier work by Preston (1960), noted that z varies with spatial scale, resulting in a triphasic SAR (Figure 2.8, see also Shmida and Wilson 1985; Hubbell 2001; McGill 2011a). The SAR is steepest (approximating unity) at the interprovincial scale (i.e., when diversities are compared

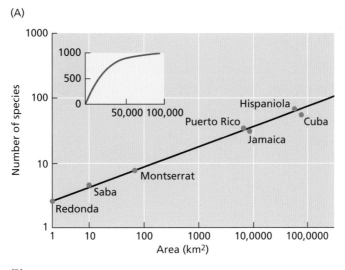

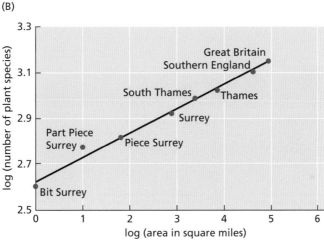

Figure 2.5 Two species–area curves. Note that the axes are logarithmic and that the relationships are well fitted by a straight line based on Equation 2.1 ($S = cA^z$, where c and z are fitted values; the arithmetic relationship is depicted in the inset in panel (A)). (A) MacArthur and Wilson plotted the number of reptile and amphibian species present against the size of several islands in the Greater and Lesser Antilles. (B) One of the first species–area curves was constructed by H. C. Watson in 1859 for plant species in Great Britain. He constructed the curve based on different-sized regions within Great Britain and the island as a whole. (A) After MacArthur and Wilson (1967), data from Darlington (1957); (B) after Rosenzweig (1995).

among biological provinces) and is shallower within provinces (at regional scales). Provinces correspond to biogeographical regions with separate evolutionary histories (see Table 2.1). Species richness should be expected to change dramatically when moving between areas with separate evolutionary histories. Moreover, Allen and White (2003) and McGill (2011a) showed that the slope of the SAR will approach a value of 1 when the spatial scale of study is larger

than most species' ranges. McGill (2011a) provides a very general explanation for the triphasic SAR shown in Figure 2.8, based on sampling distributions and species ranges.

The majority of species–area studies have focused at the regional scale, and at this scale z-values cluster around 0.25 and generally range from 0.15 to 0.40 (Rosenzweig 1995). Early attempts to explain this apparent consistency in z-values focused on the

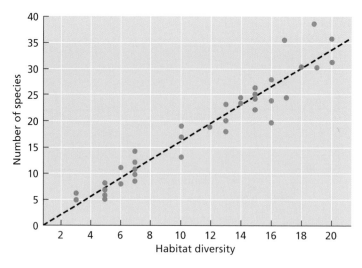

Figure 2.6 Relationship between habitat diversity and the number of terrestrial isopod species inhabiting the central Aegean Islands. Each data point represents an island; the line is the best-fit linear regression. After Hortal et al. (2009).

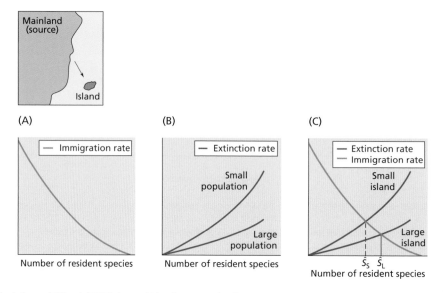

Figure 2.7 MacArthur and Wilson's (1967) theory of island biogeography, illustrating the effect of island size on equilibrium species richness. In MacArthur and Wilson's model, species richness on an island is a function of the rate of immigration of new species from the mainland species pool and the loss of island species due to extinction. (A) The immigration rate declines as the number of resident species on the island increases, going to zero when the island contains all the species in the mainland source pool. (B) The extinction rate increases with the number of resident species on the island because (1) there are more species to go extinct, and (2) the number of individuals per species decreases as the total number of resident species increases, and small populations are more likely to go extinct than large populations. (C) The extinction rate on large islands is lower than on small islands because large islands have more resources and should, therefore, support more individuals of all species. Thus, the equilibrium number of species $\hat{S}$, found at the intersection of the immigration and extinction curves, is greater on large islands than on small islands.

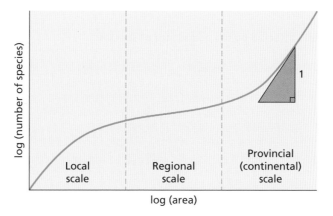

Figure 2.8 The triphasic species–area relationship (Rosenzweig 1995; Hubbell 2001). The SAR curve is steepest at the provincial scale (i.e., when diversities are compared between biogeographical regions) and is shallower at regional scales (i.e., within provinces). Allen and White (2003) and McGill (2011a) showed that the slope of the SAR will approach a value of 1 when the spatial scale of study is larger than most species' ranges.

properties of the lognormal species-abundance curve (Preston 1948; May 1975). However, recent work shows that z-values may vary considerably between study systems and that this variation is strongly linked to climate. For example, a comparison of bird species richness on islands and continents world-wide shows that z-values increase significantly with temperature and precipitation (Kalmar and Currie 2006, 2007). Likewise, in a meta-analysis of almost 800 SARs obtained from the literature, Drakare et al. (2006) found that the slope of the SAR increases when moving from the poles to the equator. Interestingly, neither Drakare et al. (2006) nor Kalmar and Currie (2007) found any difference in the form of the SAR between island and mainland areas, once the effect of distance-based isolation was taken into account. Drakare et al. (2006, p. 221) conclude from their meta-analysis that "SAR show extensive, systematic variation—they are far from being an invariant baseline…Instead, there are important differences between parameters of the SAR based on the sampling scheme (nested versus independent), across latitudes and sizes of organisms, and between different habitats."

The fact that species richness accumulates faster for a given increase in an area when sampled in the tropics than when sampled in temperate zone (or in warm, wet areas compared with cool, dry areas) provides an interesting link to the latitudinal diversity gradient and other climate–diversity relationships. Moreover, it is consistent with what we know

about changes in species–abundance distributions across latitude. To see this, we must first delve into another prominent diversity pattern in ecology.

The distribution of species abundance

Most communities contain a few species that are common and many species that are rare. If the number of species in each abundance class is plotted, a **species–abundance distribution** (Figure 2.9) is obtained. Statistical models to describe species–abundance distributions (SADs) were developed in the 1930s and 1940s, first by Motomura (1932), then by the renowned statistician Sir Ronald Fisher (working in collaboration with two lepidopterists; see Fisher et al. 1943), and also by Frank Preston, a ceramic engineer and amateur ecologist, who published a classic series of papers on the topic (Preston 1948, 1960, 1962; see Box 2.1).

Based on large samples of Lepidoptera they collected in England, Fisher and his colleagues (1943) observed that the numbers of individuals per species followed a modified negative binomial frequency distribution. Based on this distribution, they proposed that the number of species (F_n) having n individuals each can be expressed as

$$F_n = \alpha \frac{x^n}{n} \text{ for } n > 0 \qquad \text{[Eqn 2.2]}$$

where X is an arbitrary scaling parameter and α is a fitted, empirical constant. Equation 2.2, referred to as

---

Box 2.1 How is diversity affected by the abundance of species

So far this chapter has used only the number of species as a measure of diversity, although it is acknowledged that genetic, functional, and phylogenetic diversity are also important. However, the realization that common species in a community may be 10–1000 times more abundant than the rare species raises an interesting question. Should a rare species with one individual really count the same towards diversity as an abundant species with 1000 individuals? The answer, of course, depends on the question or goals. However, there is a substantial body of literature that works under the assumption that the answer is no. This literature gives the word diversity a much more specific meaning where diversity represents measures that count rare species less towards diversity and, thus, specifically contrast diversity with richness (which weights all species equally). In this context, diversity is an increasing function of both richness and evenness. Evenness refers to the notion of how equal the abundances of different species in a community are. In a perfectly even community, every species would have the same abundance (i.e. N/S, where N is the total number of individuals in the community and S is the number of species). It is a curious, but absolutely universal fact of ecological systems that communities are highly uneven (as demonstrated by species abundance distributions), but still some communities are more even (really less uneven) than others.

There is a long tradition of studying evenness using two measures borrowed from physics. Let p_i be the relative abundance of a species i in a community (i.e. $p_i = N_i/N$, where N_i is the abundance of the one species and $N = \Sigma_i N_i$ is the total abundance in the community). Then, Shannon diversity $H = \Sigma_i p_i \ln(p_i)$, and the Simpson index is $\lambda = \Sigma_i p_i^2$. As defined λ goes in the wrong direction to create an index of diversity so Simpson diversity $D = 1/\lambda = 1/\Sigma_i p_i^2$ is commonly used. Both H and D increase as species richness increases, but also as evenness increases. Since the maximum possible value (when all species are equally abundant) for H is $\ln(S)$ and for D is S, we can divide these metrics by their maximum to give metrics of evenness ($E_H = H/\ln(S)$ and $E_D = D/S$). Both evenness indices range from 0 to 1 (1 = maximally even, 0 = maximally uneven). A closely related metric to Simpson diversity is Hurlbert's PIE $= 1 - \Sigma_i p_i^2$. Here, PIE stands for probability of an interspecific encounter (the odds that if one randomly picks two individuals, they will be different species). PIE has been shown to directly link to the species-individual curve as being the slope of the curve at $N = 1$ individual (Olszewski 2004). Another metric of evenness is Berger Parker dominance (N_1/N) or McNaughton dominance ($(N_1 + N_2)/N$, where N_1 is the abundance of the most common species in the community and, similarly, N_2 for the second most common. A high dominance means most of the individuals are concentrated in one or two species, which is the opposite of evenness. The main problem with all of these metrics is that there is little idea of what a diversity of $H = 2.78$ or PIE $= 0.4$ means or how to compare them.

MacArthur (1965), Hill (1973) and Patil and Taillie (1982) provided a nice conceptual unification of these concepts, which has recently been popularized and advocated for by several researchers (Jost 2006; Chao and Jost 2015). MacArthur introduced the concept of equivalent species. He noted that richness has clear units (number of species), but the problem with Simpson and Shannon diversity is that they have no clear units. For Shannon diversity, MacArthur showed that if we look at $\exp(H)$ instead of H, then $\exp(H)$ now has units of species again. In particular, $\exp(H)$ tells us how many species of equal abundance would be needed to be as diverse as the community being examined. Thus, every real-world highly uneven community is converted back into units of the number species in a perfectly even community, which would be needed to have the same degree of diversity. A nice side benefit of this transformation is that if we double the number of species in a community of perfect evenness, then $\exp(H)$ also doubles as one would expect. MacArthur showed for Simpson diversity that if we use $D = 1/\Sigma_i p_i^2$, instead of the PIE form, then D also is in units of equivalent species. So, S, $\exp(H)$ and D all are in units of equivalent species. Hill then showed that all of these diverse metrics were special cases of a single formula:

$$^qD = \left[\sum_{i=1}^{S} p_i^q \right]^{1/(1-q)}$$

Then $^0D = S$, $^1D = \exp(H)$ (in the limit as $q \to 1$), $^2D =$ Simpson's D, and $^\infty D =$ Berger Parker dominance. If the rarest species is a singleton (only one individual in the community, as happens often), then at $q = -\infty$ we have $^{-\infty}D = N$ the total number of individuals in a community. Thus, all the metrics for diversity are just special cases of Hill's formula. In general, for values of $q > 1$, as q gets bigger, the dominant species have increasingly more effect on the value of ^qD, and for $q < 1$, rare species have increasingly more effect on the value of ^qD. This realization gives a special prominence to Shannon diversity (or at least the exponential transformation of it) at $q = 1$, where all abundances of species are equally weighted. To calculate evenness instead of diversity in the Hill context we would look at a ratio of two Hill diversities, typically, $^1D/^0D = \exp(H)/S = E_H'$ (where E_H' is an alternative to Shannon Evenness E_H) or $^2D/^0D = 1/D/S = E_D$. Thus, ^qD is a very general form of diversity that unifies most of the independently developed and seemingly unrelated measures for diversity, as long as we switch to using exp(H) instead of H and focus on $1/D$ instead of PIE. All the answers are in units of equivalent species. We also now have formally that $^1D = SE'_H$ so that diversity is exactly richness × evenness, giving a mathematical basis to what had previously just been a verbal intuition.

An interesting question is whether the tools developed above can capture and inform about differences between communities, for example, along an environmental gradient of productivity or elevation. Kempton (1979) explicitly asked this question and found that $q = 0-0.5$ (i.e., richness or something close to richness) provided the best ratio of low variance across years within a site and high variance across sites (known to differ). However, Kempton also found that other metrics that focused on neither the most common nor

the rarest species, but species of intermediate abundance had even better discrimination than any Hill number (a result also found by Gray (1979)). This matches the authors' own experience that evenness tends to try to do too much and ends up explaining little. More work in this area is needed. In particular, it seems likely that metrics that break apart changes in rare, common, and intermediate species separately are promising. However, in the meantime, it seems that simultaneously using multiple metrics, such as richness (S), total number of individuals in a community (N), and a metric of evenness (E_H or E_D), together captures much of the variation in differences between communities in simple fashion. The use of these metrics assumes a constant sampling effort across the communities. If sampling effort varies, see McGill (2011) for recommendations for different sets of metrics.

Fisher's log series distribution, describes a distribution in which singleton species (i.e., those with only one individual in the sample) always make up the most abundant class (Figure 2.9(A)). Because α depends on the shape of the SAD, but not on the size of the sample, it is a useful descriptor of diversity and is termed Fisher's α (Magurran 2004). The total number of species S in a sample depends on the number of individuals N as

$$S \cong \alpha \ln\left(1 + \frac{N}{\alpha}\right) \qquad \text{[Eqn 2.3]}$$

This is one particular form of the species–individuals curve assuming the abundance of species are distributed according to Equation 2.2. As already noted, the relative proportions of rare and common species determines the exact shape of the species–individual curve and, thus, has an important role in determining alpha diversity. Thus, the species–individuals curve links the notions of alpha diversity, species–area relationships, and species–abundance distributions.

Preston (1948) challenged the adequacy of Fisher's log series distribution, noting that many empirical SADs display an internal mode, rather than a peak in the singleton class. He further noted that these SADs were non-normal, but could be normalized when log-transformed (i.e., they are lognormal; see Figure 2.9(B)). Preston illustrated these lognormal curves using an x-axis on which each successive abundance class represents a doubling in species

abundance, which is equivalent to taking the logarithm of species abundance to base 2. Preston further predicted that SADs that failed to display an internal mode were the result of incomplete samples and that with increased sampling these SADs would reveal an internal peak. Preston's prediction was confirmed when more sampling years were added to the Rothamsted moth data (Williams 1964) used in the original development of Fisher et al.'s log series SAD (see Hubbell 2001).

Fisher's and Preston's models are statistical descriptors of species abundance patterns. They were developed without any real consideration of underlying biological mechanisms. What followed in the 1960s was a decade of development of ecological models of SADs based on niche theory, the most famous of which is MacArthur's "broken stick" model and its relatives (Sugihara 1980). (Excellent accounts of the development of these models can be found in Tokeshi 1999 and Magurran 2004.) Interest in these "niche-based" models and in SADs in general waned in the late 1980s and 1990s, as ecologists focused their attention more on experimental studies and less on descriptive patterns. This situation changed dramatically, however, with the publication of Stephen Hubbell's book *The Unified Neutral Theory of Biodiversity and Biogeography* (2001). Hubbell proposed a radical new model that could account for both Fisher's log series and Preston's lognormal SADs, and that fit the empirical data remarkably well. Hubbell's neutral theory revived interest in

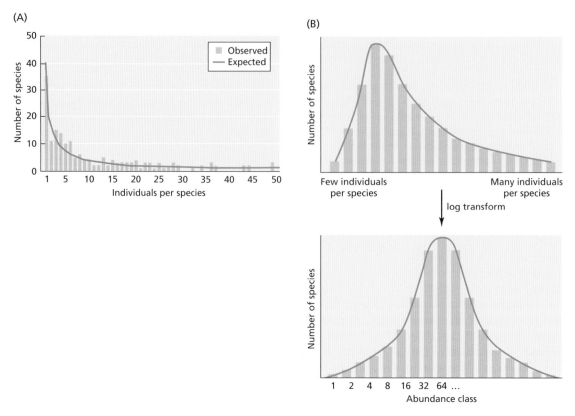

Figure 2.9 (A) An example of Fisher's logseries distribution fit to data on species abundances in moths collected at light traps over a 4-year period at Rothamsted, England. (B) A hypothetical example of Preston's lognormal distribution, showing how the distribution of species abundances can be normalized when log-transformed, in this case by using an x-axis, where each successive abundance class represents a doubling in species abundance (log$_2$ scale). (A) After Hubbell (2001).

species–abundance distributions, which are currently enjoying a lively renaissance (Shipley et al. 2006; McGill et al. 2007; McGill 2011b). Hubbell's model and what it predicts about SADs will be discussed in Chapter 14. However, a key understanding that came out of this activity is that many different processes can create very similar patterns. This makes it impossible to distinguish which processes are occurring purely from examining the shape of the species–abundance distribution (McGill et al. 2007).

Another important development in recent years has been to realize that observed species–abundance distributions are imperfect reflections of the community and are best treated as samples. One can take a theoretical species–abundance distribution (like the lognormal) as indicating the underlying probability of a species having that relative abundance, but this can then be used as the frequency parameter in a Poisson distribution to indicate the actual observed number of individuals in an empirical sample (Pielou 1975; Green and Plotkin 2007). Among other issues, this resolves the problem that occurs when a species–abundance distribution predicts a species will have a fractional number of individuals (e.g., 3.7), which cannot be observed in the real world. In particular, the Poisson–lognormal distribution is increasingly used to represent what a sample from a lognormal would look like.

Improved data have also shown us much about SADs. A meta-analysis of 558 species–abundance distributions already in the literature found that completely sampled distributions tend to be well fitted by the lognormal distribution, while incompletely sampled distributions are often better fit by the logseries (Ulrich et al. 2010). In a 21-year dataset of intensively sampled estuarine fish in the UK, fish were divided into two categories (Magurran and Henderson 2003). Resident species appeared in at least 10 of the 21 years, while transient species appeared less often. The resident species alone showed a lognormal distribution, while the transient species showed a logseries distribution. Sampling larger areas can also change the shape of a SAD from logseries to lognormal. In an enormous dataset on coral reefs, both corals and fish show a pattern where local communities show a logseries-like distribution, which shifts to looking increasingly lognormal at the whole reef and regional scales (Connolly et al. 2005). However, it turns out that a model of sampling from a lognormal (Poisson–lognormal) actually fits the data better than the logseries at all scales. Thus, a mathematically appropriate model of sampling neatly resolves the debate between Preston and Fisher. Similarly, Hubbell's very different model of sampling (see Chapter 13, for discussion of neutral theory) has a parameter, m, for migration rate into the local community that allows the shape of the distribution from lognormal-like (low migration) to logseries-like (high migration). Returning to the coral reef dataset, the distribution of biomass for fish or percentage cover for corals was lognormal at all scales, suggesting that resource use is lognormal and it is only small samples of individuals that distort this.

Thus, a very solid picture of the range of variability of SADs in nature has been developed. Although the veil-line has been replaced with more sophisticated notions of sampling distributions; basically, Preston was right. Most SADs fall somewhere on a spectrum between something close to lognormal to something close to logseries. Communities with many transient species tend to be more logseries-like with the mode at the smallest abundance. Whether the transient species are transient because of small spatial or temporal sample sizes or because of genuine transient ecological dynamics doesn't matter and is actually a matter of perspective. Communities without sampling artefacts or genuine transients look lognormal. With this view, it is not surprising in hindsight that Fisher's study of

migrating insects flying to light traps was logseries while Preston's study of nesting birds looked lognormal.

Productivity and species richness

Species richness, at broad spatial scales, tends to increase with an increase in primary productivity (Currie 1991; Hawkins et al. 2003). Areas that have high productivity (e.g., the humid tropics) tend to have high species richness. Although the productivity–diversity relationship at large spatial scales is generally positive, it may be decelerating (i.e., the rate of increase in species richness declines with increasing productivity). At smaller spatial scales, the relationship between productivity and species richness is more varied; both positive and negative relationships occur, as do hump-shaped and U-shaped relationships (Figure 2.10; Mittelbach et al. 2001; Gillman and Wright 2006; Partel et al. 2007; Pu et al. 2017). Ecologists once thought that hump-shaped productivity–diversity relationships were nearly ubiquitous at local to regional scales (e.g., Tilman and Pacala 1993; Huston and DeAngelis 1994; Rosenzweig 1995). This is now known to not be true. However, hump-shaped relationships occur often enough to demand an explanation, and this explanation is far from obvious. While the ascending limb of the hump may be the result of increased productivity allowing more species to exist at minimum viable population sizes, the descending limb is problematic. As Rosenzweig (1995, p. 353) notes, "to me the decrease phase presents the real puzzle: why, past a certain point, does enhanced productivity tend to reduce the number of species?"

Literally dozens of hypotheses have been proposed to explain the causes of productivity–diversity relationships, particularly the hump-shaped curve (reviewed in Abrams 1995; Rosenzweig 1995), yet there is no consensus explanation. Some of the more prominent hypotheses include:

1. Grime's "hump-back model" (Grime 1973a), which proposes that species richness is limited by abiotic stress in unproductive environments and by competitive exclusion in very productive environments.

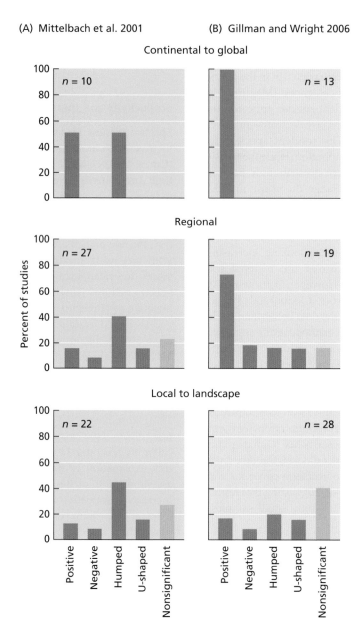

Figure 2.10 Types of productivity–diversity relationships in plant communities at different spatial scales. Statistically significant productivity–diversity relationships were classified into four types: positive, negative, hump-shaped, and U-shaped. (A) Mittelbach et al.'s (2001) initial analysis of the literature showed that hump-shaped relationships were most common at smaller spatial scales, whereas positive and hump-shaped relationships predominated at the largest spatial scale. (B) Gillman and Wright's (2006) subsequent literature review concluded that positive, hump-shaped, and U-shaped relationships were equally represented at small spatial scales, and that positive relationships predominated at large spatial scales.

2. A resource-based hypothesis proposing that there is a shift from nutrient limitation to light limitation at high productivity, resulting in a decline in plant species richness in very productive environments (Tilman and Pacala 1993).
3. The hypothesis that environments of intermediate productivity are more common across the landscape than are high- and low-productivity environments (especially in the temperate zone); therefore, more species have evolved to utilize intermediate productivities, creating a larger species pool for those environments (Denslow 1980; Partel and Zobel 2007; Partel et al. 2007).
4. Communities are more likely to occur in alternative states at intermediate productivities; therefore, summing species richness across these alternative-state communities leads to more species at intermediate productivity (Chase and Leibold 2003b).
5. Two hypotheses involving predation:
    (a) high productivity increases predator density, magnifying the impact of apparent competition, leading to a loss of species richness in very productive environments (Abrams 2001);
    (b) more productive environments lead to greater coevolutionary diversification of predators and prey, resulting in humped and positive productivity-diversity relationships (Pu et al. 2017).

Each of these hypotheses has some empirical and theoretical support, but none is a clear winner.

There continues to be a great deal of controversy over the shape and strength of productivity–diversity relationships in nature (e.g., Mittelbach 2010; Whittaker 2010; Adler et al. 2011; Grace et al. 2016; Fraser et al. 2015). Moreover, the amount of variance in species richness explained by productivity is often quite low (even though the relationship may be statistically significant). This has led some researchers to suggest the abandonment of the study of bivariate productivity–diversity relationships in favor of "… fresh, mechanistic approaches to understand the multivariate links between productivity and richness" (Tredennick et al. 2016). Interestingly, as will be seen in the Chapter 3, species richness and productivity may also feedback on each other (i.e., experimental plots sown with more plant species

tend to produce more plant biomass). Thus, the relationship between productivity and species richness is multifaceted.

The latitudinal diversity gradient

The tropics teem with a diversity of life richer than elsewhere on the planet and the dramatic increase in species richness when moving from the poles to the equator, the latitudinal diversity gradient (LDG), is the Earth's predominant biogeographic pattern. Dozens of hypotheses exist for the latitudinal diversity gradient (Pianka 1966; Mittelbach et al. 2007; Fine 2015) and these hypotheses can be grouped into four general categories that encompass the major drivers of regional diversity:

1. "Null-model" explanations based on geometric constraints on species ranges distributed across the globe.
2. Ecological hypotheses that focus on an area's carrying capacity for species.
3. Historical explanations based on geological history and the time available for diversification.
4. Evolutionary hypotheses that focus on rates of diversification (speciation minus extinction).

These four classes of explanations will be considered next, recognizing that regional diversity may be determined by a combination of factors working in concert; the critical question is the relative importance of the different processes.

A null model: geometric constraints and the "mid-domain effect"

A **null model** in ecology attempts to specify how a relationship or pattern in nature should look in the absence of a particular process or mechanism (for example, how might two species be distributed across a habitat in the absence of interspecific competition). Gotelli and Graves (1996, p. 3) give a more formal definition:

[A null model can be seen as] a pattern-generating model that is based on randomization of ecological data or random sampling from a known or imagined distribution. Certain elements of the data are held constant, and others are allowed to vary stochastically to create new assemblage patterns. The randomization is designed to produce a pattern that would be expected in the absence of a particular mechanism.

The power of the null model approach is that we can specify the probability that an observed pattern in nature differs from the predicted "null pattern," much as we can specify how an observed experimental result may differ significantly from a random expectation in a statistical model. Null models have a rich history in community ecology. They were instrumental, for example, in focusing debate about the importance of interspecific competition during the late 1970s and early 1980s (see Chapter 1; see also Gotelli and Graves 1996).

Colwell and Hurtt (1994) proposed that the latitudinal gradient in species diversity may simply reflect the outcome of placing species ranges on a bounded domain (the globe). They reasoned as follows—imagine a collection of different-sized pencils within a pencil box (Figure 2.11). Each pencil's length represents the range of a given species, and the sides of the box represent constraints placed on the distribution of these ranges (for example, by the harsh environment of Earth's poles). When the box is shaken, we find that the pencils overlap most in the center of the box. By analogy, species ranges are expected to overlap more at the equator than at the poles and, therefore, species richness should show a gradient with latitude. This explanation for the latitudinal diversity gradient has been termed the **mid-domain effect**, and Colwell and colleagues suggest that it offers an appropriate null model, against which other explanations for the latitudinal

gradient should be measured (e.g., Colwell and Hurtt 1994; Colwell et al. 2004, 2005).

The mid-domain effect has generated considerable debate among ecologists. It has strong supporters as well as critics. Its critics have focused on three main issues:

1. Whether the model is truly "neutral" to the processes that it attempts to rule out (e.g., because observed species ranges reflect the influence of climate, history, and species interactions).
2. Technical issues of model development and testing with empirical data (McCain 2007). Recent studies suggest that the mid-domain effect provides a reasonable explanation for diversity gradients within certain bounded regions, such as mountains (Watkins et al. 2006; Brehm et al. 2007; but see McCain 2007) and rivers (e.g., Dunn et al. 2006).
3. Whether the mid-domain works equally as well in the east–west direction as it does in the north–south direction. The mid-domain effect predicts highest richness in the longitudinal center of continents as well, but this is rarely found (Bokma and Monkkonen 2000; Hawkins and Diniz-Filho 2002).

However, by itself, the mid-domain effect appears unlikely to account for diversity gradients on a global scale. For example, using the global distribution of bird species, for which ecologists have remarkably complete data, Storch et al. (2006) predicted how bird species diversity should vary with latitude based on random dynamics of species ranges (using

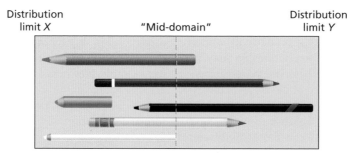

Figure 2.11 The mid-domain effect illustrated by a simple thought experiment with a pencil box. Species' ranges are represented by the pencil lengths, and limits to species distributions (by analogy, Earth's poles) are represented by the ends of the pencil box. Shaking the box so as to randomly distribute species ranges between their upper and lower limits results in ranges overlapping more in the middle of the box than at the ends; species richness is highest in the center (by analogy, near the equator).

a generalized spreading dye model; see Jetz and Rahbek 2001). Storch et al. found that the mid-domain effect, when applied globally, explained less than 2% of the variation in avian richness. When applied within biogeographic regions, the mid-domain model did much better. However, when used in this way, the model *a priori* incorporates regional (i.e., tropical/temperate) differences in species richness and, thus, provides a weak test. Most interesting, however, was the observation that mid-domain type models that allowed the locations of species ranges to be positively influenced by available energy (i.e., species ranges spread preferentially into more productive areas) provided a good fit to the global bird data set. Thus, while the mid-domain effect per se appears inadequate to explain the global latitudinal diversity gradient (but see recent theory by Gross and Snyder-Beattie 2016), linking climate to species range dynamics, suggests a promising avenue for future research (Gotelli et al. 2009).

Ecological hypotheses: climate and species richness

A layperson's response to the question, "Why are there more species in the tropics than at the poles," might be, "It's the climate, stupid." Climatic variables (e.g., mean and variance in temperature and precipitation) do explain much of the global variation in taxonomic richness (see O'Brien 1998; Kreft and Jetz 2007). Hawkins et al. (2003) found that the best single climatic variable affecting a given region explained, on average, about 64% of the variance in species richness across a broad range of taxa, and measures of energy (solar radiation, temperature), water (precipitation), or water–energy balance consistently explained variation in species richness better than other climatic and non-climatic variables. Water variables usually represented the strongest predictors of species richness in the tropics, subtropics, and warm temperate zones, whereas energy variables (for animals) and water–energy balance variables (for plants) were the strongest predictors in high latitudes (Hawkins et al. 2003; Whittaker et al. 2007; Eiserhardt et al. 2011).

Contemporary climate and species diversity are strongly correlated, especially at broad spatial scales (Field et al. 2009). However, what is the mechanism

behind this correlation? Of the many ecological hypotheses that have been offered to explain why there are more species in the humid tropics than elsewhere, perhaps the most prominent is the **more-individuals hypothesis (aka the species-energy hypothesis**; Currie et al. 2004**).** The more-individuals hypothesis (MIH) postulates that species richness varies with climate because of two factors. First, the number of individuals an area can support should increase with an increase in primary productivity/available energy (greatest where it is warm and wet). Secondly, species richness should be higher in areas that can support more individuals because a greater number of individuals can become divided among more species with viable population sizes. Therefore, there should be more species in climate zones with higher productivity (Hutchinson 1959; Brown 1981; Wright 1983).

Currie et al. (2004), Brown (2014) and Storch et al. (2018) examined the predictions of the species–energy hypothesis in detail. They found that the number of individuals within a taxon (e.g., trees, birds) tended to increase with primary productivity and that the number of species tended to increase with the number of individuals. Thus, the two basic components of the species–energy hypothesis were supported. However, the combination of the two relationships was insufficient to account for the observed increase in species diversity with decreasing latitude. In other words, for well-studied taxa, species accumulate much more quickly, as one moves from the temperate zone to the equator, than would be predicted based on a moderate increase in the number of individuals (see also Evans et al. 2005, 2008). Currie et al. (2004) derived their predictions from the famous species–abundance distributions determined by Fisher et al. (1943) and Preston (1948, 1962), which were discussed earlier in this chapter. More recently, Šímová et al. (2011) used an expanded global set of 370 "Gentry-style" (0.1-ha) forest plots to examine more closely how tree species richness varies as a function of the number of individual trees per plot. They concluded that tree species richness does not simply follow the total number of individuals, and that variation in tree diversity worldwide is not simply the result of more productive areas supporting more individuals and therefore more species.

Why are there more species in the tropics than would be expected, based on a fixed relationship between productivity, number of individuals, and number of species? At one level, the answer appears simple. There is growing evidence that the relationship between the number of individuals sampled and the number of species accumulated—**the species–individuals curve**—varies predictably with climate, and that warmer regions (with higher productivity) support more species per number of individuals sampled (Figure 2.12). Note that the sampling sites for Sanders's (1969) surveys of marine benthic diversity span a broad range of latitude (Figure 2.12(A)) and that the species–individuals curves become steadily steeper as one moves from

the poles to the tropics. Thus, for a given number of individuals sampled, many more species will be found at low latitudes than at high latitudes. Recall that the slope of the species–area curve (z-value) also increases with a decrease in latitude (Drakare et al. 2006; Kalmar and Currie 2006, 2007). Here, again, it can be seen that more species are encountered for the same area sampled in the tropics than in the temperate zone. The steepness of the species–individuals curve and of the species–area curve should vary similarly with latitude if the number of individuals per unit area ρ (density of individuals) remains approximately constant, such that

$$I = A \times \rho \qquad \text{[Eqn 2.4]}$$

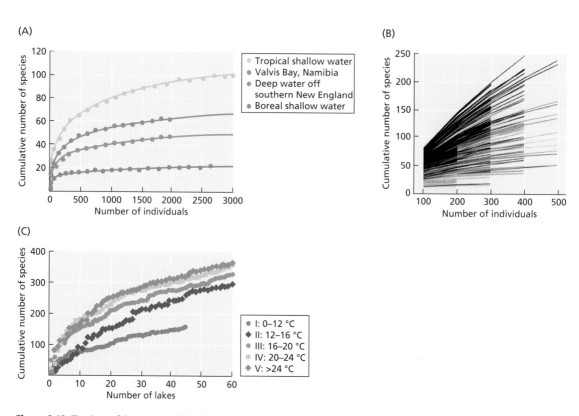

Figure 2.12 The shape of the species–individual curve changes over broad climatic gradients. (A) The number of benthic invertebrate species found in samples collected from four marine sites ranging from boreal to tropical waters shows that many more species are encountered for the same sampling effort in warm tropical waters than in cold boreal waters. (B) Rarefaction curves for tree species richness in forest plots worldwide (constructed for sampling efforts of 100, 200, 300, 400, and 500 individual trees) show that more species are encountered for the same sampling effort at high-productivity sites (darker shading) than at low productivity sites (lighter shading); productivity is measured as AET. (C) The number of freshwater phytoplankton species found in North American lakes in samples from five different temperature bands shows that more species are encountered for the same sampling effort (number of lakes sampled) in regions of warm temperatures than in regions of cold temperatures. (A) after Hubbell (2001), redrawn from May (1975), original data from Sanders (1969); (B) after Šímová et al. (2011); (C) after Stomp et al. (2011).

where A is area and I is the number of individuals. Surprisingly, almost nothing is known about how Equation 2.4 varies geographically (e.g., with latitude) for most taxonomic groups. Currie et al. (2004) show that for trees and birds, the number of individuals per sampling site (area) increases with a decrease in latitude. However, the change is relatively small, and Enquist and Niklas (2001) found no change in tree biomass with latitude.

If species can persist at smaller population sizes in warmer, more equable climates, then we would expect similar sampling efforts to yield more species in tropical than in temperate climates. However, based on the relationships in Currie et al. (2004), a ten-fold increase in richness would require a 6600-fold decrease in minimum viable population size (m). Thus, it seems unlikely that variation in m could account for the orders-of-magnitude changes in species richness observed across latitude. Reed et al. (2003), in fact, observed no latitudinal variation in m for vertebrates, although more work on latitudinal variation in minimum viable population sizes is needed. All of the above lead Storch et al. (2018, p. 15) to conclude that "The MIH can be therefore mostly rejected in its strict formulation ...". However, climate and energy availability have profoundly shaped the development of the LDG over evolutionary time and some of the mechanisms of this will be explored next.

Historical hypotheses: the time-integrated area hypothesis and the concept of tropical niche conservatism

Wallace (1878) first proposed that the tropics are more diverse because they have had more time to accumulate species than temperate regions. Wallace reasoned that the relatively harsh climate found at high latitudes, including periodic "ice ages" and glaciations, repeatedly caused species extinctions, effectively setting back the diversity clock. On the other hand, more stable and benign tropical environments have steadily accumulated species. While early statements of the "time hypothesis" focused on the impact of the Pleistocene glaciations (e.g., Fischer 1960), it is now known that the latitudinal diversity gradient has roots far deeper in time than the Pleistocene, both on land and in the oceans

(Crame 2001; Powell 2009; Crame et al. 2018). Therefore, more recent explorations of the "time" hypothesis have looked at the extent of tropical and temperate climates throughout the Cenozoic (65 million years ago to present), a time period following the last major global extinction and during which the Earth's current LDG formed and strengthened. This **"time-integrated area hypothesis"** focuses on the combined effects of time and area on the development of the LDG (Fine 2015).

The Earth was much warmer early in the Cenozoic than it is today. Tropical floras existed as far north as London and southern Canada, warm (>18°C) surface waters extended to the Arctic, and the equator-to-pole temperature gradient was much reduced (Sluijs et al. 2006; Graham 2011). This thermal maximum was followed by a cooling trend beginning around 45 million years ago and continuing to the present. The greater spatial extent of the tropics during the early Cenozoic may have contributed to higher rates of diversification due to:

1. The positive effect of area on speciation rates (Kisel and Barraclough 2010; Schluter and Pennell 2017).
2. The decline in the rate of extinction with area (Terborgh 1973; Rosenzweig 1995; Chown and Gaston 2000).

A number of lines of evidence provide indirect support for the **time-integrated area hypothesis**. For example, Fine and Ree (2006) show that variation in current tree species richness among boreal, temperate, and tropical biomes is significantly correlated with the area that each biome has occupied integrated over time since the Eocene. Jetz and Fine (2012) further showed that current-day vertebrate species richness in the world's terrestrial bioregions is best explained by a model that integrates bioregion area over geological time (productivity and temperature also contribute).

For the time-integrated area hypothesis to account for the LDG, the dispersal of lineages between tropical and temperate regions needs to be limited (i.e., lineages tend to remain and diversify within their region of origin). This appears to be true for many organisms based on evidence from species distributions and phylogenetic relationships. For example, there is a nearly complete turnover in flora and fauna between tropical and extratropical terrestrial bioregions (Hawkins 2010; Jetz and Fine 2012; Fine 2015). Fossils also reveal a tropical origin for many lineages of terrestrial plants and animals with limited transitions between regions (Wiens and Donoghue 2004; Wiens et al. 2010; Kerkhoff et al. 2014). This has led ecologists to hypothesize that the transition from tropical to temperate environments presents an important physiological hurdle. For terrestrial plants, this physiological hurdle is strongly associated with evolving frost tolerance (e.g., Farrell et al. 1992; Latham and Ricklefs 1993; Kleidon and Mooney 2000; Kerkhoff et al. 2014).

The tendency of lineages to originate in the tropics and to remain tropical due to the retention of climatic niches over evolutionary time has been termed **tropical niche conservatism** (Wiens and Donoghue 2004; Wiens et al. 2009) and is common in terrestrial organisms (but see Jansson et al. 2013; Boucher-Lalonde et al. 2015 for counter arguments). Tropical niche conservatism is less evident in the ocean (Roy et al. 1996; Jablonski et al. 2006), where, at least for marine mollusks, the pattern is one of tropical origin, then dispersal into temperate and polar regions (the "out of the tropics" hypothesis; see Jablonski et al. 2006, 2017; Kiessling et al. 2010).

Current latitudinal gradients in taxonomic diversity bear the clear imprint of history. However, few of the species that comprise these current diversity gra-

dients date back 40–60 million years, to when tropical and subtropical environments dominated the Earth (Antonelli and Sanmartin 2011). Thus, whereas the relative age and extent of tropical environments may account for much of the latitudinal gradient in higher taxonomic diversity (order, family, genus), what is the link between historical climates and current species diversity? Has species richness increased steadily through time, or has it followed an irregular course of advance and retreat? The palynological (fossilized pollen) record suggests that tropical tree species richness in the Amazon lowlands was high during the Eocene Optimum, decreased with late Eocene and early Miocene cooling, but rebounded again in the late Miocene or Pliocene under different climatic conditions (Jaramillo et al. 2006; Mittelbach et al. 2007). Crame (2001, p. 182) also notes that for many taxa "the Cenozoic diversification event continued well past the prolonged late Paleocene–middle Eocene interval of global warming." Finally, fossil data show that the latitudinal diversity gradient for marine mollusks has strengthened and weakened numerous times far back in Earth's history (Figure 2.13; Powell 2007). Thus, while tropical habitats were once very widespread, the evidence suggests that time-integrated area by itself may not provide a complete explanation for the current latitudinal gradient in species richness (Antonelli and Sanmartin 2011; Pigot et al. 2016).

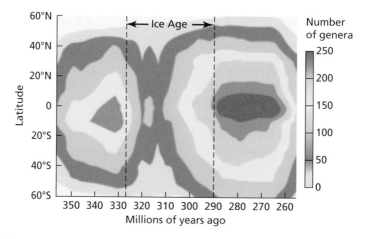

Figure 2.13 Latitudinal diversity gradient for marine brachiopod genera through late Paleozoic time, shown as a contour plot. Note that the diversity gradient weakened during the late Paleozoic ice age and then re-formed. After Powell (2007).

Evolutionary hypotheses: do rates of diversification differ across latitude?

Do diversification rates (speciation minus extinction) differ between temperate and topical regions, and if so, could such variation in evolutionary rates produce the latitudinal diversity gradient? This question has puzzled evolutionary biologists for a very long time (e.g., Dobzhansky 1950; Fischer 1960), and it was G. L. Stebbins (1974) who colorfully asked whether the tropics are a "cradle" for the generation of new taxa (i.e., higher rates of speciation) or a "museum" for the preservation of existing diversity (i.e., lower rates of extinction). Currently, there is much debate about whether diversification rates are higher in the tropics than in the temperate zone and the evidence is conflicting (e.g., Jetz and Fine 2012; Rolland et al. 2014; Marin and Hedges 2016; Pulido-Santacruz and Weir 2016). Paleontologists and evolutionary biologists use two different methods to measure evolutionary rates—studying fossils and phylogenies. Both methods let us look back in time, although in each case the view is imperfect.

The fossil record shows that diversification rates tend to be higher in the tropics than in temperate and polar regions, and the evidence suggests that this difference is due both to higher rates of origination and lower rates of extinction in the tropics (reviewed in Mittelbach et al. 2007; Valentine and Jablonski 2010). However, the fossil evidence is limited to a relatively few taxonomic groups, with much of the data coming from marine invertebrates. For example, Buzas et al. (2002) showed that planktonic foraminifera had higher gains in species richness over the last 10 million years in tropical than in temperate seas (Figure 2.14). Jablonski (1993) and Jablonski et al. (2006) showed that orders of marine invertebrates and genera of marine bivalves, respectively, preferentially originate in the tropics, as do species of fossil foraminifera (Allen and Gillooly 2006). Krug et al. (2009) applied a different approach to compare the diversification rates of tropical and temperate/Arctic bivalves. Using a large data set of some 854 living bivalve genera and subgenera collected from around the globe, whose ages they determined from the fossil record, they found that the rate of increase in the number of bivalve genera

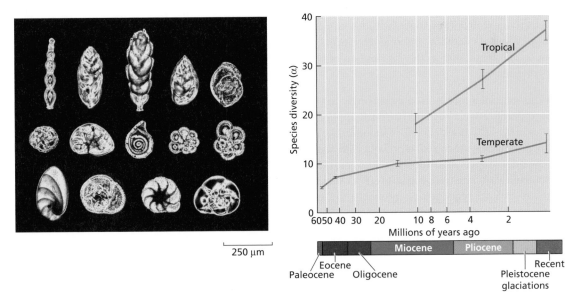

Figure 2.14 Unicellular foraminiferans secrete calcium carbonate shells that fossilize well. Over 10 million years, increase in foraminiferan diversity was 150% greater on the tropical Central American isthmus than on the temperate Atlantic coastal plain. After Buzas et al. (2002); photograph © Science Photo Library.

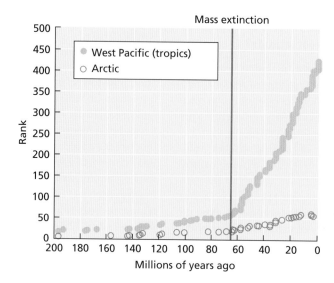

Figure 2.15 Age frequency distributions of genera of marine bivalves. The curves plot the ages of genera, ranked from youngest to oldest, in the tropical West Pacific and Arctic bioregions. The vertical line marks the mass extinction at the Cretaceous–Paleogene boundary (65 million years ago). The slopes of the two provinces differ slightly before the extinction event, but diverge strongly thereafter, as many more new species evolved in the tropics than in the Arctic. After Crame (2009), based on data from Krug et al. (2009).

was much higher in tropical than in temperate or polar waters (Figure 2.15). Recently, Shiono et al. (2018) showed that, for woody angiosperm genera, the LDG rapidly steepened post-Pliocene due to:

- selective extinction at high latitudes of older tropical genera with low freezing tolerance;
- an equatorward shift in the distribution of temperate genera.

Phylogenetic inference is used increasingly to estimate diversification rates and has been applied to studies of the LDG (Ricklefs 2007; Morlon 2014). The findings, however, are mixed, especially with regard to how rates of speciation and extinction contribute to diversification rates across latitude. For example, analyses of bird phylogenies (probably the best studied of all taxonomic groups) have found either:

- higher speciation rates at low latitudes (Ricklefs 2006b; Pulido-Santacruz and Weir 2016, for New World birds);
- higher speciation rates at high latitudes (Weir and Schluter 2007; Pulido-Santacruz and Weir 2016, for Old World birds);
- little difference in speciation rates across latitude (Jetz et al. 2012; Rabosky et al. 2015).

Schluter (2016) and Schluter and Pennell (2017) suggest that this disagreement may be the result of estimating speciation rates via different phylogenetic methods. Estimates of speciation rates, based on data at or near the tips of the phylogenetic tree (which measure "recent" speciation events) tend to be equal or higher in the temperate zone than in the tropics. In contrast, speciation rates estimated using data that integrate the recent and distant past for large clades, tend to show higher speciation rates in the tropics. This suggests that speciation rates in the temperate zone relative to those in the tropics have changed from the past to the present (Schluter and Pennell 2017). However, there are many challenges to estimating geographical variation in speciation and (even more so) extinction rates from phylogenetic data (Morlon 2014; Rabosky and Goldberg 2015; Meyer and Wiens 2017). Evolutionary biologists continue to develop new methods to address these challenges (e.g., Rabosky and Huang 2016), but it may be some time before it can confidently be concluded whether speciation rates vary predictably with latitude based on phylogenetic data.

Recently, Hanly et al. (2017) used a different (non-phylogenetic) approach to address whether latitude affects the probability of speciation and the extent of diversification in freshwater fish by examining the global distribution of endemic species (fish species found in only one lake or only one river drainage in the world). Hanly et al. (2017) reasoned that the presence of an endemic species is evidence of at least one speciation event at the site and that the number of endemic species is an estimate of the extent of diversification (speciation minus extinction) at the site.

For both freshwater lakes and river basins, the probability of speciation and the extent of diversification increased with decreasing latitude. For lake fish, the significant effect of latitude on speciation was independent of the positive effects of lake age and area, and was of similar magnitude (Figure 2.16), complimenting a large number of studies of endemic species on islands that also show positive effects of island age and area on speciation (reviewed in Warren et al. 2015).

Comparisons of extinction rates between temperate and tropical regions are few, but they tend to support lower rates of extinction at low latitudes than at high, both in the recent and distant past (e.g., Pulido-Santacruz and Weir 2016). Lower extinction rates in the tropics would fit Stebbins's analogy of the tropics as a "museum" for diversity and would match our expectation that extinction rates should be lower in mild and stable tropical environments compared with the harsh and less stable climates of temperate and polar regions. But, why might speciation rates differ between regions—that is, why might the tropics also be a "cradle" for diversity? Fedorov (1966) proposed that the forces of genetic drift, acting on small populations in the tropics, might contribute to rapid speciation. However, this mechanism requires that tropical communities be diverse already, with many species existing in small

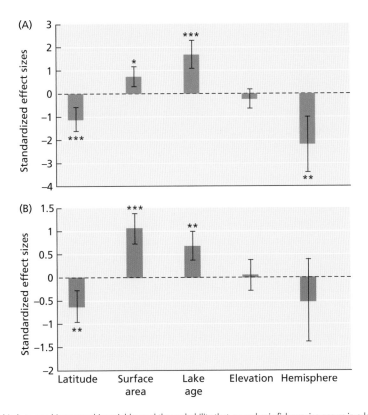

Figure 2.16 Relationship between biogeographic variables and the probability that an endemic fish species occurs in a lake (A) and the number of endemic fish species in a lake (B). Boxes represent the direction and magnitude of the standardized effect size of each variable with associated normal-based 95% confidence intervals. Asterisks denote significant predictor variables: *P < 0.1; **P < 0.001; ***P < 0.0001. After Hanley et al. (2017).

populations, and therefore cannot explain the *origins* of diversity (Schemske 2002). Other authors have suggested that opportunities for geographic isolation may be greater in tropical regions (Janzen 1967; Rosenzweig 1995), whereas Rohde (1992) and Allen et al. (2002) have suggested that higher mutation rates and shorter generation times due to the kinetic effects of temperature could contribute to higher speciation rates in the tropics. Schemske (Schemske 2002; Schemske et al. 2009), building on earlier ideas of Dobzhansky (1950), suggested that stronger biotic interactions (e.g., predator–prey, host–parasite, symbioses), and a greater potential for coevolution in the tropics could lead to higher rates of speciation. Evidence points to more rapid molecular evolution in some tropical taxa than in temperate taxa (e.g., Wright et al. 2003, 2006; Davies et al. 2004; Allen et al. 2006; Gillman et al. 2009, 2010), but a general link between nucleotide substitution rates and speciation rates remains to be demonstrated. Likewise, although many biotic interactions appear to be stronger and have a greater effect on species dynamics in the tropics than in higher latitudes (Schemske et al. 2009), do such interactions really facilitate speciation? More research on these questions is clearly needed.

The latitudinal diversity gradient is one of the most striking features of our natural world, and the pervasiveness of this pattern across taxa, among environments, and through time suggests that it may have a general explanation. Ecologists and evolutionary biologist are closing in on the answer. In all likelihood, multiple processes have contributed to the pattern, and sorting out their relative contributions is a major challenge. Ultimately, large-scale biodiversity patterns are the result of geographic patterns of speciation, extinction, and migration (Swenson 2011). In the end, our continuing efforts to understand large-scale gradients in biodiversity, such as the latitudinal diversity gradient, will provide valuable insights into the processes that govern the development of regional biotas and species pools.

Conclusion

As has been seen in this chapter, the diversity of life varies across space and time, and there are patterns to this variation. Recent advances in technology—from satellite imagery and remote sensing, to DNA sequencing and phylogenetic analysis, to GIS mapping and spatial analysis—have allowed ecologists to better document these patterns at all spatial scales and to search for their interrelationships (e.g., McGill 2011a). As ecologists, we are interested not only in "searching for repeated patterns" (MacArthur 1972), but also in understanding the processes that underlie these patterns. This chapter has highlighted some of the mechanisms thought to underlie patterns in biodiversity, focusing on those hypotheses that have received the most theoretical and empirical support. Our ability to explain spatial and temporal patterns in biodiversity is the litmus test of what we discover from detailed studies of ecological mechanisms.

Chapters 4–9 will focus on interactions between species (competition, predation, mutualism, and facilitation), and how they affect species coexistence and diversity in local communities. For a time, we will put on "blinders," if you will, to the impact of regional processes on the structure of local communities, as we focus on species interactions in small modules. There will be a return to regional processes and their effects on local communities later in the book. Before getting to the "nitty-gritty" of species interactions, however, it is necessary to explore one more important aspect of biodiversity—"How does biodiversity affect the functioning of ecosystems, and what are some of the consequences of lost biodiversity to the services that ecosystems provide?"

Summary

1. Species diversity is measured at different spatial scales and can be expressed in three important ways. Alpha (α) diversity is the number of species found at a local scale. Beta (β) diversity is the difference in species composition, or species turnover, between two or more habitats or local sites within a region. Gamma (γ) diversity is a measure of regional species richness. Species-individual curves help to control for the strong effects of sample size on diversity and allow comparisons of richness.

2. Larger areas contain more species, as they typically contain a greater variety of habitats and can support larger populations. The slope of a species–area curve varies across spatial scales and across latitude.

3. Most species are rare and a few are abundant, a relationship that can be expressed as a species–abundance distribution. Most empirical species–abundance distributions vary between roughly lognormal and roughly logseries, with the latter being a result of transient species occurring due to sampling effects and/or ecological processes.

4. At broad spatial scales, species richness generally increases with productivity. At local to regional scales, productivity–diversity relationships are more varied, and tend to be either positive or hump-shaped. Several hypotheses have been proposed to explain the descending limb of humped-shaped productivity–diversity relationships (where productivity is high, but diversity is low).

5. Species diversity differs between biogeographic regions. For similar-sized areas, there are many more species in the tropics than in the temperate zone. Hypotheses proposed to explain this latitudinal diversity gradient can be grouped into four general categories:

- Null-model explanations based on geometric constraints on species ranges. Null models generate patterns based on randomized data; such patterns (e.g., the mid-domain effect) would be expected in the absence of a particular mechanism and are rejected (and mechanisms hypothesized) if observed data do not match the null model.

- Ecological hypotheses, such as the species–energy hypothesis, which postulates that because the tropics tend to be highly productive, these regions can support more individuals per unit of area. A greater number of individuals presumably leads, over evolutionary time, to a greater diversity of species.

- Historical explanations, such as the time-integrated area hypothesis based on geological history—that is, because tropical environments are older and have been more widespread than extra-tropical environments, there has been more opportunity for diversification to occur in the tropics.

- Evolutionary hypotheses focus on rates of diversification (speciation minus extinction). Speciation might be higher ("cradle" effect) in the tropics because of the effects of uniformly warm temperatures, and/or because there are more opportunities for coevolution there. Extinction rates might be lower ("museum" effect) for a number of reasons, including the more stable and equable climate.

CHAPTER 3

Biodiversity and ecosystem functioning

Gains and losses of species from ecosystems are both a consequence and driver of global change, and understanding the net consequences of this interchange for ecosystem functioning is key.
David A. Wardle et al., 2011, p. 1277

[O]ur well-being is far more intertwined with the rest of the biota than many of us would be inclined to believe.
Thomas Lovejoy, 1986, p. 24

We have seen that species richness varies with latitude, climate, productivity, and a host of other physical and biological variables. Community ecologists have long focused on understanding the mechanisms driving these patterns of diversity. Recently, however, ecologists have become interested in how biodiversity, in turn, may affect the functioning of ecosystems. This is no casual question. Species are being lost at a rate unprecedented since the mass extinction that claimed the dinosaurs; some scientists speculate that by the end of the current century, as many as 50% of Earth's species may have disappeared due to habitat destruction, overharvesting, invasive species, eutrophication, global climate change, and other effects of human activities (Wake and Vredenburg 2008; Pereira et al. 2010; Ceballos et al. 2017, but see Loehle and Eschenbach 2012). Can we, as ecologists, predict the consequences of this lost biodiversity? We know that Earth's biota moves hundreds of millions of tons of elements and compounds between the atmosphere, hydrosphere, and lithosphere every year. Living organisms obviously play a major role in the planet's dynamics, but what role does species

diversity per se have in the functioning of ecosystems? Or, to put it simply, how many species do we need to ensure that ecosystems continue to function normally? Two groundbreaking papers published in 1994 opened the door to this question.

Tilman and Downing (1994), working in a grassland in east-central Minnesota, found that plots with high plant species diversity were better buffered against the effects of a severe drought; those plots maintained higher plant productivity during the drought than did low-diversity plots (Figure 3.1). Their results showed an apparent link between species richness, ecosystem productivity, and ecosystem stability; however, Tilman and Downing studied a set of communities in which plant species richness had been changed by adding fertilizer, where diversity decreased with increased fertility. Therefore, the presence of this additional factor (added soil fertility) may have clouded their interpretation of a causal relationship between plant species richness, and ecosystem productivity and stability (Huston 1997).

In a second study, also published in 1994, Naeem and colleagues directly manipulated species richness

Community Ecology. Second Edition. Gary G. Mittelbach & Brian J. McGill, Oxford University Press (2019).

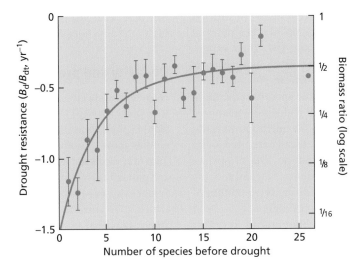

Figure 3.1 Tilman and Downing performed one of the first studies showing that plant species richness has a positive effect on ecosystem functioning. They showed that drought resistance was positively associated with plant species richness in four grasslands at the Cedar Creek Ecosystem Science Reserve, Minnesota. Drought resistance was measured as the loss in plant community biomass during a drought year as compared with plant biomass in the previous year (highest drought resistance is thus at the top of the y axis). Biomass ratio expresses the proportional loss in biomass due to the drought. Data points are means ($\pm$ 1 SE) for plots of a given species richness. Plant species richness varied among the 207 study plots as the result of previous and ongoing nutrient manipulations. After Tilman and Downing (1994).

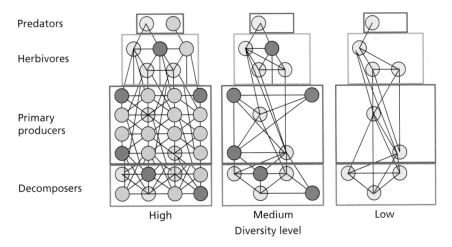

Figure 3.2 The experimental design used by Naeem and colleagues to study the effects of species losses at multiple trophic levels on ecosystem functioning. In their experiment, model communities containing four trophic levels—primary producers (annual plants), herbivores (molluscs and insects), predators (parasitoids), and decomposers (collembolids and earthworms)—were established in 14 large environmental chambers in the "Ecotron" at Silwood Park, UK. Three diversity levels were established, such that each of the lower-diversity treatments contained a subset of the species used in the higher-diversity treatments, to mimic the uniform loss of species from high-diversity ecosystems. After Naeem et al. (1994).

in a set of small, replicated ecosystems constructed in 14 large, controlled environmental chambers at the "Ecotron" at Silwood Park, UK. Their experiment included multiple trophic levels at three levels of diversity, such that the medium-diversity treatment contained a subset of the species in the high-diversity treatment, and the low-diversity treatment a subset of the species in the medium-diversity treatment (Figure 3.2). The researchers used these treatments to mimic the uniform loss of species from high-diversity ecosystems. Naeem and his colleagues found that high-diversity communities consumed more CO_2 (i.e., had higher respiration rates) than low-diversity communities and that higher plant diversity resulted in greater primary productivity (Naeem et al. 1994).

The experiments of Naeem et al., and Tilman and Downing, paved the way for a veritable flood of studies that have examined relationships between biodiversity and **ecosystem functioning**. The sheer volume of this work and the rapid pace at which it is advancing make it difficult to summarize in a single chapter. This chapter highlights some of the findings with respect to the effects of biodiversity on four important ecosystem functions:

• productivity;
• nutrient use and nutrient retention;
• community and ecosystem stability;
• invasibility.

The second half of this chapter discusses some of the many questions about the biodiversity–ecosystem function relationship that remain unresolved and highlight directions for future research.

Diversity and productivity

Does higher species diversity lead to higher ecosystem productivity? The answer, at least for low-stature plant communities, is yes. Three large-scale studies show this effect quite clearly. In 1993, David Tilman and colleagues (Tilman et al. 1997, 2001) set up 168 plots (9 m × 9 m) at the Cedar Creek Ecosystem Science Reserve in Minnesota (Figure 3.3), into which they introduced seeds from 1, 2, 4, 8, or 16 grassland–savanna perennial species. The species composition for each of the different diversity treatments was randomly generated from a pool of 18 species, and species were introduced to the plots by adding 15 g of seed per m² for each species. A small army of workers weeded the plots to maintain the diversity treatments, and Tilman and colleagues followed the effects of species diversity and plant functional group diversity on ecosystem productivity, nutrient dynamics, and stability for more than 20 years. In Europe, researchers employed a similar experimental design, repeated across seven countries from Sweden in the north to Greece in the south, to look at the effects of plant diversity on ecosystem functioning (Hector et al. 1999). In this experiment, known as BIODEPTH, the regional species pool differed between sites, allowing rare insight into the potential generality of the results across different environments. A third major experiment along

Figure 3.3 An aerial view of the study plots at the Cedar Creek Ecosystem Science Reserve, located in east-central Minnesota, USA. Photo from Tilman (2001).

these lines, the Jena Biodiversity Experiment, was established in 2002 near Jena, Germany (Weisser et al. 2017). As in the other two experiments, plant diversity was manipulated in a large number of experimental plots, in order to study above- and below-ground productivity and nutrient use efficiency. In addition, the Jena Experiment had a strong focus on examining the effects of plant diversity on multiple trophic levels and in understanding the mechanisms by which diversity affects ecosystem functioning (e.g., Marquard et al. 2009; Hector et al. 2011; Weisser et al. 2017).

Species richness had a strong, positive effect on primary production in each of these experiments. At Cedar Creek, productivity (measured as plant biomass) increased rapidly in response to the addition of a few plant species, but the rate of increase then slowed in an asymptotic fashion as the number of species increased (Figure 3.4A). The positive effect of species richness on plant biomass increased over time in the Cedar Creek experiment and in the others (Tilman et al. 2014; Weisser et al. 2017). Results from the pan-European BIODEPTH experiment showed variation among study sites in the strength of the relationship between biodiversity and productivity, but the general response was quite consistent—in seven of the eight study sites, productivity increased significantly with an increase in species richness (Figure 3.4B; note that the *x*-axis in this plot is on a log scale). The Jena Experiment also showed a positive species richness–biomass relationship, as well as a positive functional group richness–biomass relationship (Figure 3.4C). Hector et al. (2011) re-analyzed the results of these three experiments to address whether biomass production is more strongly affected by species richness or by species composition (i.e., the overall variation between the different species compositions used). Counter to some earlier predictions (e.g., Hooper et al. 2005), Hector et al. found that the number and types of species (functional groups) present in experimental grassland communities were of similar importance in determining aboveground productivity.

Although much less studied than species richness, species evenness (Chapter 2) may also have important effects on primary productivity. An early study by Wilsey and Potvin (2000) found that higher evenness led to higher productivity in an old-field plant community. However, the possible mechanisms for

evenness to affect productivity are diverse and ensuing results have been mixed (Hillebrand et al 2008).

The positive effects of species richness on biomass production observed in grasslands (reviewed in Tilman et al. 2014) can be extended to a more general conclusion—increasing species richness increases biomass production in primary producers. A meta-analysis of 368 independent experiments, which manipulated plant or algal species richness in a variety of terrestrial, freshwater, and marine ecosystems, found that the most diverse polycultures attain, on average, 1.4 times more biomass than the average monoculture (Cardinale et al. 2011). This result was consistent for both aquatic and terrestrial systems. Cardinale and colleagues went on to fit a variety of mathematical functions to the observed diversity–productivity relationships and found that 79% of the relationships were best-fit by a function that was positive and decelerating. These results suggest that some fraction of species can be lost from communities with minimal loss in productivity, but that beyond a certain level of species loss, productivity declines rapidly. The researchers then attempted to calculate the fraction of species needed to maintain maximum producer biomass by fitting a saturating function (Michaelis–Menten curve) to the data. They concluded that it would take about 90% of the maximum number of species used in experiments to maintain 50% of maximum producer biomass. However, this conclusion is based on the problematic assumption that a fitted saturating function can be extrapolated to estimate maximum biomass production as species richness goes to infinity. Cardinale et al. (2011) caution against taking their estimates too literally. Similarly, it is not clear if the steep increases in productivity as richness moves from one to six species or the relatively slow increases from about eight species and up (Figure 3.4) is more biologically relevant in the field. At this point we simply do not know what fraction of species may be lost from a community before we see a specified decrease in productivity.

Mechanisms underlying the diversity–productivity relationship

Why might productivity increase with species richness? Ecologists have focused on two likely mechanisms: niche complementarity and species

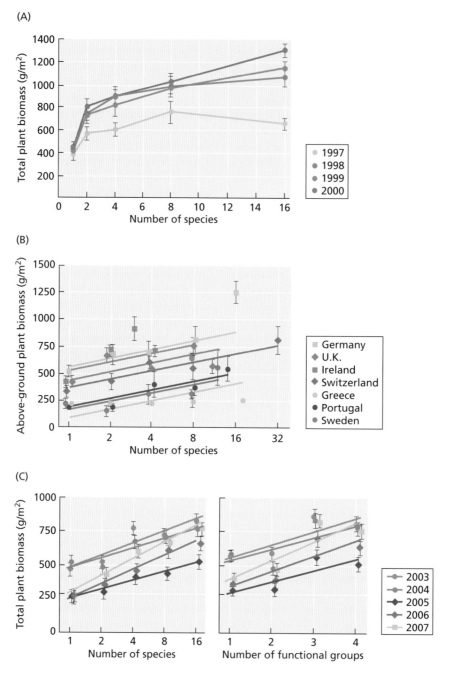

Figure 3.4 Three large-scale field experiments have shown that plant community biomass (and thus primary production) increase with increased species richness. (A) Average total plant biomass from the Cedar Creek experiment for the years 1997–2000. (B) Annual aboveground plant biomass as a function of species richness in the BIODEPTH experiment, conducted at eight sites in Europe (data from the two U.K. sites, at Sheffield and Silwood Park, have been combined). Note that the x axis displays species richness on a log scale ($\log_2$). (C) Total plant community biomass as a function of species richness (x-axis, $\log_2$ scale) and plant functional group richness in the Jena Biodiversity Experiment for the years 2003–2007. Data points in all graphs are means ± 1 SE. (A) modified from Loreau (2010b), original data from Tilman et al. (2001); (B) modified from Loreau (2010b), original data from Hector et al. (2002); (C) modified from Marquard et al. (2009).

selection (Huston 1997; Loreau and Hector 2001; Tilman et al. 2014). **Niche complementarity** may occur if species differ in the way they use limiting resources. Plant species, for example, may differ in their phenology, physiology, rooting depths, or nutrient requirements. If a community consists of species that differ in their niches, the overall efficiency of resource use by the community may increase with an increase in species richness, leading to higher overall productivity. Complementarity effects may also result from facilitation (e.g., the presence of nitrogen-fixing legumes may increase the growth rates of non-nitrogen-fixing species). **Species selection** (also called the **sampling effect**) will lead to increased productivity with increased species richness if diverse communities are more likely to contain more productive species that come to dominate the community (Aarssen 1997; Huston 1997; Wardle 1999; Loreau and Hector 2001). One way to look for evidence of niche complementarity is to test whether experimental plots containing species mixtures yield more biomass than <u>any</u> monoculture plot (termed **transgressive overyielding**; Loreau and Hector 2001). If so, then we know that the higher productivity of mixed species plots is not only due to species selection.

Niche complementarity and species selection can act in concert to increase ecosystem functioning, and the relative contributions of these two mechanisms can be partitioned statistically using the methods outlined by Loreau and Hector (2001). Such statistical partitioning is possible when we can identify the contributions of individual species to ecosystem functioning within a mixture of species. In terrestrial plant communities, it is possible to measure individual species contributions to total primary production (by measuring plant standing biomass). For many other ecosystem attributes, however, it is impossible to identify how individual species contribute to the functioning of polycultures. Therefore, most of what we know about the relative effects of niche complementarity and species selection on ecosystem functioning comes from studies of productivity in plants.

Studies of terrestrial plant communities suggest that species selection (sampling) effects may be the dominant drivers of ecosystem functioning early in an experiment, whereas niche complementarity effects become stronger as communities mature

(Cardinale et al. 2007; Reich et al. 2012). In the Cedar Creek experiment, the percentage of plots exhibiting transgressive overyielding increased with increasing species richness in later years (Tilman et al. 2001, 2014). Thus, there is evidence of niche complementarity late, but not early, in the experiment. Fargione et al. (2007) and Marquard et al. (2009) applied the methods of Loreau and Hector (2001) to partition the effects of niche complementarity and species selection in this same experiment. They found that niche complementarity effects increased over time, whereas species selection effects decreased over time (Figure 3.5). Results from the Jena experiment also show that the effects of species richness on ecosystem functioning become stronger over time (Meyer et al. 2016). Interestingly, this temporal strengthening of biodiversity effects in the Jena experiment was due to a combination of increased ecosystem functioning at high biodiversity and deteriorating ecosystem functioning at low biodiversity (see Guerrero-Ramírez et al. 2017 for a similar conclusion for grasslands in general). Deteriorating productivity in low diversity treatments (including monocultures) may be due to negative plant–soil feedbacks (discussed in more detail later), whereas complementary effects may become stronger over time as the result of selection for increased niche differentiation between species (e.g., Zuppinger-Dingley 2014), resulting in increased functional trait diversity.

A meta-analysis of 44 plant community studies (Cardinale et al. 2007) showed that niche complementarity can be as important as species selection (based on Loreau and Hector's 2001 methods for statistical partitioning of effects) and that the importance of niche complementarity increases over time (see also Cardinale et al. 2011). In experiments lasting 2–5 plant generations, species mixtures out-yielded the most productive monocultures in about 12% of cases. A more extensive meta-analysis covering primary producers in aquatic and terrestrial ecosystems found transgressive overyielding in 37% of the studies, which means that 63% of the time, diverse polycultures yielded less biomass than the single highest-yielding monoculture (Cardinale et al. 2011). On average, the most diverse polyculture in each study yielded 0.87 times the biomass of the highest-yielding monoculture. Thus, Cardinale and colleagues (2011, p. 581) concluded that "there is presently little evidence to support the hypothesis

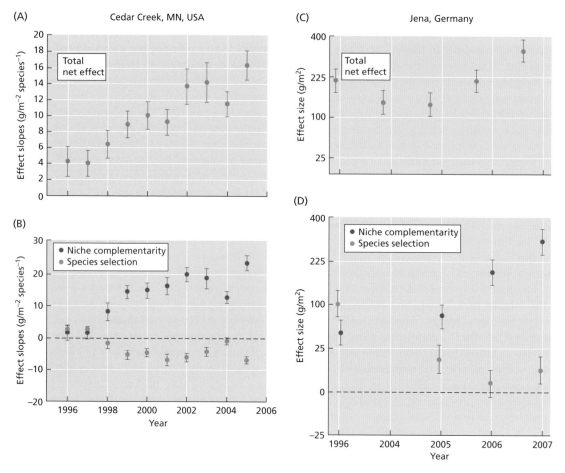

Figure 3.5 Experimental evidence that the contributions of niche complementarity and species selection may change over time. Data points in all graphs are means ± 1 SE. (A, B) Species richness/biomass relationship in the Cedar Creek experiment, 1996–2006. (A) The net effect of species richness on community biomass production (i.e., the slope of the relationship) increased over the 10-year period. (B) Contributions of niche complementarity and species selection to the overall slope of the species richness/productivity relationship. The complementarity effect increases over time, whereas the selection effect decreases. (C, D) Contributions of niche complementarity and species selection to the total net effect of biodiversity on production in the Jena Experiment, 2003–2007. The y axis is plotted on a square root scale. (A,B) modified from Fargione et al. (2007); (C,D) after Marquard et al. (2009).

that diverse polycultures out-perform their most efficient or productive (constituent) species" (but see Tilman et al. 2014).

To date, niche complementarity effects have been demonstrated almost exclusively in terrestrial plant communities, most often when focal species differ qualitatively in their functional traits (e.g., nitrogen-fixing versus non-nitrogen-fixing plants; Spehn et al. 2002; Heemsbergen et al. 2004; Fargione et al. 2007; Marquard et al. 2009). A few studies have attempted to use quantitative differences among species (measured along some niche axis) to predict the impact of niche complementarity on ecosystem functioning (e.g., Norberg 2000; Wojdak and Mittelbach 2007). However, as Ives et al. (2005, p. 112) state, "We know of no conclusive evidence from empirical studies that resource partitioning [niche complementarity] is the mechanism underlying an effect of consumer diversity on resource or consumer density." The limited evidence for such a clear and overriding effect of niche complementarity on ecosystem functioning is somewhat surprising, given its central place in the theory (e.g., Tilman et al. 1997). The next section, however, discusses two novel experiments, which better elucidate the role of niche complementarity in driving the positive effect of species richness on nutrient retention.

The accumulation of species-specific enemies (e.g., soil pathogens) at low diversity can act in concert with niche complementarity at high diversity to produce a strong biodiversity-productivity relationship. For example, experiments that varied plant species richness growing in natural soil versus soil that was sterilized to kill all microbes (via gamma radiation) or fungi (via fungicides), found that a pronounced positive relationship between plant species richness and plant productivity depended on having living microorganisms present in the soil (Maron et al. 2011; Schnitzer 2011; Hendriks et al. 2013). The authors propose that soil-borne herbivores and pathogens have strong negative density-dependent effects on plant productivity at low species richness because the plants are likely to be growing near (infected) conspecific neighbors (a Janzen–Connell effect, Chapter 4). In more diverse plant communities, however, species-specific soil-borne diseases or herbivores will have less of a negative effect on productivity because a given plant is less likely to be in close contact with an infected individual of the same species. The authors conducted experiments to document the species-specific, negative density-dependent effects of soil pathogens and conclude that "niche-based/competitive process and soil pathogen effects…may act in concert" to drive the diversity–productivity relationship (Maron et al. 2011, p. 40).

Diversity, nutrient cycling, and nutrient retention

If species exhibit resource partitioning and niche complementarity, then more diverse communities should use resources more efficiently, and the amount of unused resources in an ecosystem should decline as species richness increases. These predictions have been supported in a number of studies that involved manipulating plant diversity in terrestrial ecosystems, where nitrogen availability limits primary productivity (e.g., Tilman et al. 1996; Hooper and Vitousek 1997, 1998; Niklaus et al. 2001; Scherer-Lorenzen et al. 2003; Oelmann et al. 2007). Oelmann et al. (2007) presented a particularly complete analysis of nitrogen availability in plant communities of varying species richness. They showed that an

increase in plant species richness (which varied from 1 to 16 species) reduced available nitrogen (NO_3^-) concentrations in the soil, and decreased dissolved organic nitrogen (DON) and total dissolved nitrogen (TDN). Nitrogen contained in the aboveground vegetation also correlated positively with species diversity, indicating that total N uptake increased with increasing diversity. A similar study by Dijkstra et al. (2007), again varying species richness from 1 to 16 species, found that an increase in plant species richness reduced leaching loss of dissolved inorganic nitrogen (DIN) as a consequence of more efficient nutrient uptake by the plant community. Dijkstra and colleagues found that DON, which does not dissolve easily and may not be directly available for plant uptake (Neff et al. 2003), was lost at a higher rate in more species-rich plots.

The weight of the evidence shows a positive effect of plant species richness on nitrogen use efficiency in terrestrial plant communities. Cardinale et al. (2011) found in their review of 59 studies (56 on grasslands and 3 on freshwater algae) that species-rich polycultures reduced nutrient concentrations, on average, 48% more than the mean monoculture. However, they found little evidence that the polycultures used resources more efficiently than the most efficient monocultures. Standing nutrient concentrations in soil or water were higher, rather than lower, in polycultures than in the most efficient monocultures (Cardinale et al. 2011).

Two studies (Northfield et al. 2010; Cardinale 2011) employed novel experimental designs to test whether species-rich polycultures use resources more efficiently than the most efficient monocultures and, if so, whether this more efficient resource use was due to niche complementarity. Most studies of the effects of species richness on ecosystem functioning use a "substitutive" experimental design, where the number of species is varied among treatments, but the total number of individuals is held constant. Northfield and colleagues (2010) crossed this design with another that varied the number of individuals within each species to generate a full "response surface design." Their new design allowed them to test for complementarity in resource use by different species of predatory insects (i.e., bugs, parasitic wasps) feeding on aphids that infect crop plants. They reasoned that, if different predator species use the prey resource differently (e.g., by feeding in different

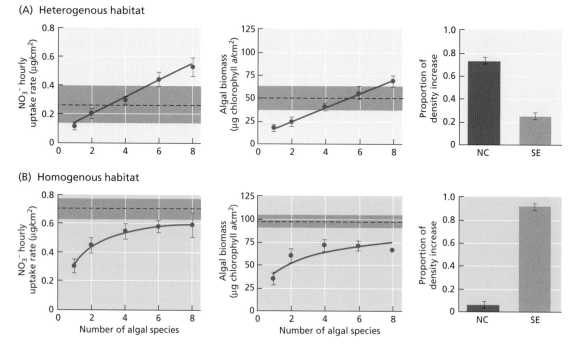

Figure 3.6 Effects of algal species richness on nitrogen uptake and algal biomass in streams differ depending on whether or not the different species can partition resources. (A) Streams with heterogenous habitats allow for resource partitioning by the 8 algal species in the study. (B) Where stream habitats are homogeneous, opportunities for resource partitioning are limited and a single species becomes dominant. Data points are means ±1 SE, with best-fit functions plotted as solid lines. Dashed horizontal lines and shaded areas show means ±1 SE for *Stigeocloneum*, the species that achieved the highest values in monoculture. The rightmost panels of each sequence show the proportion of the increases in algal density driven by niche complementarity (NC) or by selection effects (SE, the effect of the dominant species, in this case *Stigeocloneum*). After Cardinale (2011).

ways or in different microhabitats), then the total resource use of any single predator species should plateau (saturate) with increasing predator density, but the total resource use of more diverse predator communities should not plateau, or should plateau at higher levels (Northfield et al. 2010). They found that a diverse predator community exhibited better aphid control than any single predator species. Moreover, they showed that "no single consumer species, even at high density, was capable of driving resources to as low a level as could be achieved by a diverse mix of consumers—the hallmark of resource partitioning" (Northfield et al. 2010, pp. 346–7).

Cardinale used a very different experimental design to show that niche complementarity plays a significant role in the way biodiversity in stream algae improves water quality (Cardinale 2011). Nitrogen run-off into streams and rivers is a significant source of pollution in many areas. In labora-

tory streams with high habitat heterogeneity, an increase in algal species richness (up to 8 algal species) dramatically reduced the amount of dissolved nitrogen (NO_3^-) in the stream water. However, when the stream environment was made more homogeneous, Cardinale found that the loss of niche opportunities for the different algal species led to reduced algal diversity and a reduction in nitrogen uptake. In streams with diverse habitats, most of the positive effect of algal species richness on water quality (by reducing nitrogen levels) was due to niche complementarity (Figure 3.6).

Diversity and stability

The notion that diversity begets stability in ecosystems has a long and venerable history in ecology. In the 1950s, MacArthur (1955), Elton (1958), Odum (1959), and others suggested that communities containing more species should be better buffered against the effects of species extinctions, species invasions, or environmental perturbations. Theoretical development of these ideas followed shortly thereafter (e.g., Levins 1970; May 1971, 1972; MacArthur 1972), culminating at the time with May's 1973a monograph, *Stability and Complexity in Model Ecosystems*. In this book, May reached the surprising conclusion that diversity does not promote ecosystem stability; instead, the more species a community of competitors contains, the less likely it is to return to equilibrium after a perturbation. The topic of diversity and ecosystem stability has been controversial ever since (McNaughton 1977, 1993; Pimm 1984; Ives and Carpenter 2007). In recent years, experimental explorations of the effect of species diversity on ecosystem functioning have shed new light on the question of whether diversity begets stability. The answer, it turns out, depends in part on how we view stability.

Temporal stability

The consistency of a quantity (e.g., species abundance) over time—has been the primary focus of empirical ecologists studying diversity–stability relationships (McNaughton 1977; Pimm 1984). Operationally, temporal stability is calculated as the variance in species abundance (usually biomass) measured over time and scaled to the mean abundance. This measure of stability can be applied to the entire community or to its constituent species. The temporal stability of an individual species (S_i) depends on the species' mean abundance (μ_i) and the standard deviation in its abundance (σ_i) as μ_i/σ_i, whereas the temporal stability of the total community (S_T) can be expressed as

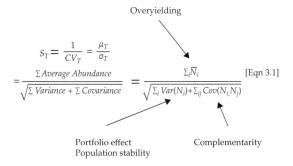

where the summations are taken over all species in the community (Lehman and Tilman 2000).

As Equation 3.1 shows, diversity can have a positive effect on temporal community stability, either by increasing total species abundance, by decreasing the summed variance in species' abundances, by decreasing the summed covariance in species' abundances, or by any combination of the above. We have already seen that increased species richness may result in an increase in total community biomass (overyielding). Therefore, all else being equal, overyielding will contribute to an increase in temporal stability by increasing the numerator of Equation 3.1. Alternatively, if diversity does not affect total community biomass, and species fluctuate randomly and independently in their abundances over time (i.e., no covariance in Equation 3.1), then an increase in species richness will reduce the summed variance in Equation 3.1 and increase temporal stability purely on statistical grounds (Doak et al. 1998). Tilman et al. (1998) referred to this phenomenon as the "portfolio effect" because the same principle applies to a fixed financial investment that is diversified across a portfolio of stocks. Finally, if species show a correlated response to environmental fluctuations, then any negative covariance in response will lead to a positive effect of diversity on community temporal stability, whereas positive covariance will have the opposite effect. Interspecific competition would be expected to result in negative covariances among species, whereas facilitation or mutualism should result in positive covariances. Thus, we can visualize the portfolio effect as operating through a reduction in summed variances among species, whereas the covariance effect results from a reduction in the summed covariances (Lehman and Tilman 2000).

Loreau (2010a; see also Loreau and de Mazancourt 2008) have taken issue with the above approach of partitioning the positive effect of species richness on temporal stability into components resulting from species' variances and covariances in abundance, arguing that this approach is too dependent on specific assumptions of how species interact. Instead, they suggest, a more general conclusion is that the main mechanism driving the stabilizing effect of species diversity on community abundance (numbers or biomass) is simply the asynchrony of species responses to environmental fluctuations and overyielding (de Mazancourt et al. 2013).

Does diversity affect temporal stability and, if so, can we interpret its effects in light of asynchronous responses of species to the environment? In the Cedar Creek experiment, temporal stability in community biomass increased dramatically with an increase in species richness (Tilman et al. 2006, 2014). On aver-

age, the treatment plots with the highest diversity were about 70% more stable in their biomass production over time than were monocultures (Figure 3.7A). On the other hand, the stability of individual plant species declined significantly with an increase in the number of species in the community (Figure 3.7B). Therefore, diversity had a strong positive effect on the temporal stability of the total plant community,

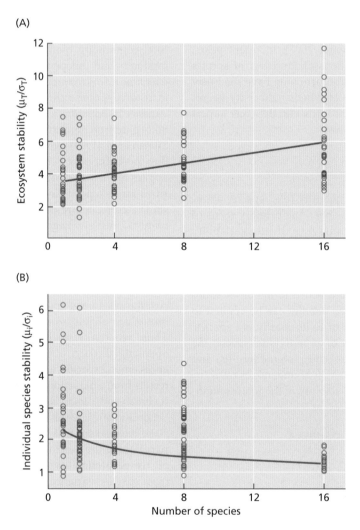

Figure 3.7 Temporal stability in ecosystem functioning (total community biomass in this case) increases with species richness. (A) Temporal stability, measured as the ratio of mean total plant biomass per plot (μ) to its temporal standard deviation (σ), as a function of species richness in the study plots at Cedar Creek from 1996 to 2005. (B) Unlike community temporal stability, the temporal stability of individual plant species was a negative function of species richness for the period 2001–2005. After Tilman et al. (2006).

but a negative effect on the stability of individual species abundances. Gross et al. (2014) came to a similar conclusion after reanalyzing the results of 27 biodiversity experiments conducted with primary producers. In grasslands, community biomass is stabilized over time because species richness increases mean biomass more strongly than its variance (as in Equation 3.1), whereas in algal communities species richness has a minimal effect on community stability because richness affects the mean and variance in biomass similarly. Compensatory interactions between species (probably due to interspecific competition) also contributed to greater stability at higher diversity (Gross et al. 2014). In North American birds, both overyielding and population stability increased community stability while the complementarity effect decreased it (Mikkelson et al 2011). In this study, species evenness played an even larger role than species richness. Other studies also have shown a positive effect of species richness on the temporal stability of total community productivity (McNaughton 1977) and other ecosystem properties (McGrady-Steed et al. 1997). Although overyielding seems to be consistently part of the relationship between diversity and stability, the contribution of both portfolio and complementarity effects have varied between studies. There are also some significant examples where no relationship between diversity and stability was found (Pfisterer and Schmid 2002; Bezemer and van der Putten 2007).

Diversity may also serve to stabilize yields of economically important species when those species are composed of populations (or stocks) that differ in their ecologies and life histories (i.e., within-species diversity). Schindler et al. (2010) provide a dramatic example of the stabilizing consequences of stock diversity to the yield of sockeye salmon (*Oncorhynchus nerka*) harvested from Bristol Bay, Alaska (a multibillion dollar natural resource). The salmon returning to Bristol Bay each year comprise multiple stocks that spawn in the nine major rivers flowing into the bay, and each river stock contains tens to hundreds of locally adapted populations distributed among tributaries and lakes. Variation in fish life histories means that Bristol Bay sockeye spend anywhere from 1 to 2 years in freshwater and 1 to 3 years in the ocean as they complete their life cycles. As a consequence of this local adaptation and life history diversity, the different sockeye populations of the Bristol Bay fishery tend to fluctuate asynchronously in abundance. Schindler et al. (2010) use 50 years of population data to show that variability in annual Bristol Bay sockeye returns is two-fold less than it would be if the system consisted of a single homogeneous population. Furthermore, a homogeneous sockeye population would probably see 10 times more frequent fisheries closures than the existing diverse salmon population. These results provide a dramatic demonstration of the importance of maintaining population diversity for stabilizing ecosystem services and the economies that depend on them.

Diversity and invasibility

Are more diverse communities more resistant to invasion by exotic species? The answer, it seems, is yes, but some important questions remain. Fifty years ago, Elton (1958) proposed that successful invaders must overcome resistance from the resident community, and that this "biotic resistance" should increase with the number of resident species. Elton (1958, p. 117) reasoned that more diverse communities are more likely to repel invaders due to increased interspecific competition: "This resistance to newcomers can be observed in established kinds of vegetation, [due to] competition for light and soil chemicals and space." Indeed, Fargione and Tilman (2005), using the Cedar Creek study plots, found that both the biomass and the number of invading plant species declined as the richness of the resident plant community increased (Figure 3.8A,B; see also Levine 2000; Seabloom et al. 2003). Fargione and Tilman then went on to demonstrate the importance of interspecific competition, showing that increased resident species richness led to higher root biomass (Figure 3.8C) and lower soil nitrate levels (Figure 3.8D). This higher nutrient use efficiency was due to both niche complementarity and a species selection effect, in that treatments with higher resident species richness tended to contain more warm-season, C_4 grasses, and C_4 grasses caused the largest reduction in soil nitrogen levels. Findings from the Cedar Creek experiment suggest that established species exert their greatest inhibitory effects on invaders that are functionally similar to them, a result supported by the Jena biodiversity experiment, where invading species tended to be from functional groups not represented in the existing community (Petermann et al. 2010).

Most experimental studies on diversity and invasibility have been conducted in low-stature plant communities. However, there are a number of studies with animals that support the general conclusion that invasibility decreases with increased diversity. For example, Stachowicz et al. (1999) experimentally manipulated the species richness of sessile, suspension-feeding marine invertebrate communities ("marine fouling communities") raised on tiles, and looked at the establishment success of an invasive species, the sea squirt (ascidian) *Botrylloides diegensis*. They found that the survival of *Botrylloides*

recruits decreased with an increase in the native community's species richness, and that this result was consistent with a reduction in the space available for colonization as community species richness increased. Open space was consistently lower in communities with more species due to natural cycles in the abundances of individual species (i.e., a kind of portfolio effect). Shurin (2000) also found that the plankton communities of freshwater ponds in Michigan were less invasible by zooplankton species the higher the number of resident species. By experimentally reducing the density of the resident species, Shurin showed that interspecific competition from the resident community was a significant factor limiting colonization by introduced species.

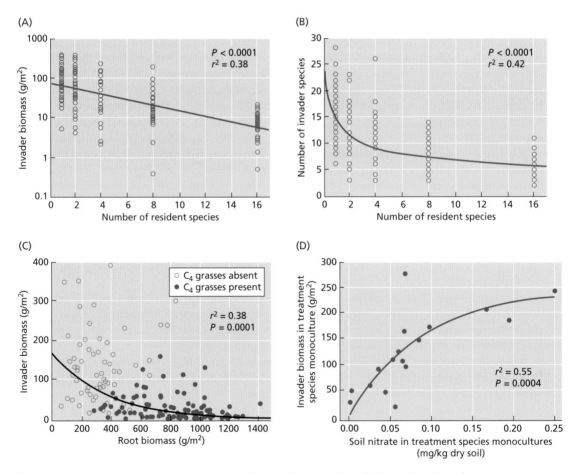

Figure 3.8 Invasion resistance increases with species richness (1–16 resident species) in study plots at the Cedar Creek Ecosystem Science Reserve. Invasion resistance is measured as (A) invader biomass per plot and (B) number of invading species per plot. The mechanisms behind this effect appear to be (C) an increase in root biomass and (D) a decrease in available nitrate in plots with greater species richness. After Fargione and Tilman (2005).

(Additional examples of reduced invasibility in diverse communities can be found in Stachowicz et al. 2002; France and Duffy 2006; Maron and Marler 2007.)

The studies discussed above support the hypothesis of Elton (1958) and others (e.g., MacArthur 1970; Case 1990) that resistance to invasion increases as resident species richness increases, and that the mechanism underlying this response is a reduction in resource availability as species richness increases. What exactly do we mean by "resistance to invasion", however? Are species-rich communities non-invasible, or do invasions of these communities simply take longer or require greater propagule pressure compared with species-poor communities? Levine and colleagues addressed this question with a meta-analysis of resistance to exotic plant invasions (Levine et al. 2004). Reviewing 13 experiments from 7 studies, they found that "although biotic resistance significantly reduces the seedling performance and establishment fraction for individual invaders, it is unlikely to completely repel them" (p. 976). They based their conclusion, in part, on the observation that the most diverse natural plant communities are also those with the highest number of invaders (Levine et al. 2004; see also Robinson et al. 1995; Wiser et al. 1998; Lonsdale 1999; Stohlgren et al. 1999; Levine 2000), a paradox we return to in the section on "Unanswered Questions".

Biodiversity and ecosystem multifunctionality

We have seen how biodiversity affects many different ecosystem functions (e.g., the production of biomass, the retention of nutrients, resistance to invasion, and the stability of biomass over time). Considered individually, these ecosystem processes often show a positive, but saturating (or decelerating) relationship with species richness. The general form of this biodiversity ecosystem functioning (BEF) relationship suggests that there is some redundancy in the contribution of species to ecosystem functioning, such that some fraction of species can be lost before we see a major change in an ecosystem process. However, what if some species have the greatest effects on one ecosystem function (e.g., biomass production), whereas different species make large contributions to a different ecosystem function (e.g., nutrient retention)? In this case, it seems that the number of species needed to preserve multiple ecosystem functions would be greater than that expected based on individual BEF relationships. This logic underlies recent studies on the effects of species richness on **ecosystem multifunctionality** (Hector and Bagchi 2007).

A number of studies from both terrestrial and aquatic ecosystems suggest that the effects of biodiversity (species richness) on ecosystem functioning

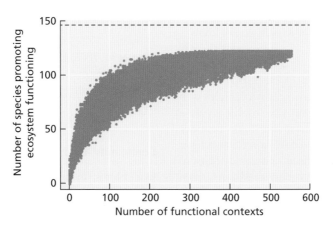

Figure 3.9 The number of species promoting ecosystem functioning increases with the number of functional contexts considered in a meta-analysis conducted by Isbell et al. (2011). The total number of species studied (147) is indicated by the dashed line which provides an upper limit on the observed response. The *x*-axis includes variation across years, sites, functions, environmental change scenarios and species pools. After Isbell et al. (2011).

become more important as more ecosystem functions are considered (e.g., Hector and Bagchi 2007; Gamfeldt et al. 2008; Zavaleta et al. 2010; Lefcheck et al. 2015). For example, Isbell et al. (2011) found, in an analysis of 17 grassland biodiversity experiments, that the number of plant species promoting ecosystem functioning increased with the number of functional contexts considered (Figure 3.9). However, Gamfeldt and Roger (2017) recently questioned the logic behind the argument that the effects of biodiversity become more important when we consider more ecosystem functions. In their analysis, Gamfeldt and Roger (2017) showed that increasing the number of ecosystem functions does not, by itself, change the nature of the relationship between biodiversity and ecosystem multifunctionality. Rather, the effect of species richness on ecosystem multifunctionality simply equals the average effect of richness on single ecosystem functions. Thus, they caution against the expectation that the effects of biodiversity are stronger when multiple ecosystem functions are considered compared with single functions.

Unanswered questions (revisited)

In the first edition of *Community Ecology* (Mittelbach 2012), this chapter ended with six "Unanswered questions" concerning how biodiversity affects the functioning of ecosystems. It is interesting to see how much research since that time has focused squarely on those six "Unanswered questions". We revisit them here, discussing what ecologists have learned since 2012 and what we still do not know.

Multiple trophic levels

Most ecological communities contain a complex web of interacting species at multiple trophic levels—detritivores, herbivores, carnivores, and primary producers. Yet most biodiversity experiments have focused on studying the effects of diversity at a single trophic level (usually primary producers, often in grasslands). How representative of other systems are the results of these experiments? We do not know the answer to this question; what we *do* know is that trophic-level interactions are important in the functioning of ecosystems, and that a change in the number and/or kind of top-level consumers can

have significant consequences that cascade down the food chain (see Chapter 11). Moreover, we know that the number of species per trophic level tends to decline from the bottom to the top of the food web, and that species are more likely to be lost from higher than from lower trophic levels (due to smaller population sizes, longer generation times, and greater risk of over-exploitation; Petchey et al. 2004; Brose et al. 2016). If biodiversity is lost preferentially from higher trophic levels, this loss may result in cascading species extinctions and large effects on ecosystem functioning (Petchey et al. 2004; Thebault et al. 2007; Estes et al. 2011; Brose et al. 2016). For example, O'Connor et al. (2016) showed that slope of the species richness-productivity (biomass) relationship differs between trophic levels (Figure 3.10), suggesting that the same percentage of species lost at different trophic levels would have different effects on productivity. Brose et al. (2016) showed that the greater probability of losing species from higher trophic levels can lead to stronger indirect effects on ecosystem functioning that would be expected if species are simply lost at random from the food web. The presence of higher trophic levels (consumer species) can also change the effect of species richness on productivity. Seabloom et al. (2017) showed that consumption of plant biomass by arthropod herbivores and fungi reduces how much plant species richness affects biomass production because consumers removed a constant proportion of total plant biomass at all levels of plant diversity. More studies are needed to address the effects of multiple trophic levels on ecosystem functioning (e.g., Downing 2005; Duffy et al. 2005; Steiner et al. 2006; Vogt et al. 2006; Long et al. 2007; Northfield et al. 2010; Barnes et al. 2018), as well as better estimates of the probability of species loss at different levels in the food web.

Community assembly or species loss?

Our interest in understanding how species diversity contributes to ecosystem functioning is driven, in part, by the unprecedented rate at which species are being lost from the biosphere. This rapid loss of diversity has led some ecologists to argue for more experiments that examine the effects of species removals from ecosystems, instead of (or in add-

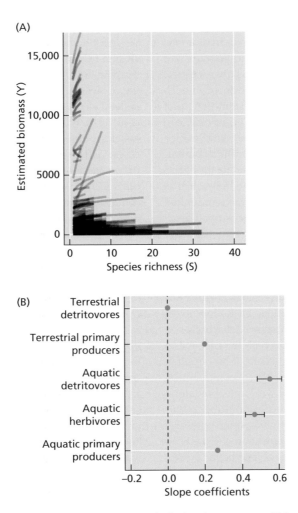

Figure 3.10 The biomass-species richness relationship varies across trophic levels and ecosystem type. (A) Standing stock biomass as a function of species richness for each entry in the meta-analysis database (*n*=558). Each entry was analyzed in a hierarchical mixed effects model using a linearized power function; relationships shown with a gray line and overlapping lines appear darker. (B) Mean (+1 SE) slope of the linearized power function of biomass and species richness for different trophic groups in terrestrial and aquatic ecosystems. After O'Conner et al. (2016).

ition to) manipulating biodiversity by constructing assemblages of varying species richness drawn randomly from a regional species pool (Diaz et al. 2003; Wootton and Downing 2003; Wardle 2016). Removing species from natural ecosystems, however, raises many ethical and practical concerns. Thus, there have been relatively few experiments using species removals to probe the link between biodiversity and ecosystem functioning. In general, such studies (e.g., Symstad and Tilman 2001; Smith and Knapp 2003; Wardle and Zackrisson 2005; Suding et al. 2006; Valone and Schutzenhofer 2007; Walker and Thompson 2010) have shown natural ecosystems to be rather resilient to decline in diversity. However, a recent study by Fanin et al. (2017) looking at plant species removals from replicated islands in northern Sweden found strong and consistent effects of biodiversity loss (plants and associated fungi) on ecosystem multifunctionality. Unlike most studies of artificially assembled plant communities, however, the effects of species loss on plant biomass decreased over time in this system (Kardol et al. 2018).

We know from decades of previous work that the loss of particular species can result in dramatic ecosystem changes (see discussion in Wootton and Downing 2003; Estes et al. 2011). Thus, removal studies raise the perplexing question of "which species we should remove," given that most natural communities are diverse and that it is impossible to remove all species combinations. If we knew which species were most likely to be lost from a community, we could, more realistically, simulate the effects of species extinctions in nature. Some traits, such as habitat specialization, small population size, limited geographic range, large body size, high trophic level, and low fecundity, make species more vulnerable to extinction (Pimm et al. 1988; Tracy and George 1992; McKinney 1997; Purvis et al. 2000; Cardillo et al. 2005a; Payne et al. 2016). Some of these traits may be associated with particular ecological functions within a community or positions in the food web. For example, Solan et al. (2004) showed that body size in marine invertebrates affects both the probability of species extinction and the extent of bioturbation of marine sediments by invertebrate communities. Species at the top of the food web (apex predators) are especially vulnerable to extinction due to harvesting and other human activities, often with dramatic impacts on ecosystem functioning (Estes et al. 2011). Future research may be able to better predict the consequences of species loss if we know how particular traits affect both a species' probability of extinction and the role that species plays within a community (Gross and Cardinale 2005; Srinivasan et al. 2007).

Global extinction and local ecosystem functioning: is there a mismatch of spatial scales?

The accelerated loss of species globally due to human activities is often cited as the reason we need to better understand how biodiversity affects ecosystem functioning, but there is something of a mismatch of spatial scales in this argument. While the global loss of species is a tragedy of unknown consequences, what matters to ecosystem functioning is not how many or what percentage of species are lost globally. Rather, what matters to ecosystem functioning is how biodiversity is changing at local or regional scales; these are the scales at which species interact and

where we expect biodiversity to most strongly affect the functioning of ecosystems. The surprising conclusion from a number of recent studies looking at temporal trends is that species richness **is not declining** on average in contemporary communities (e.g., Vellend et al. 2013, 2017a; Dornelas et al. 2014; Elahi et al. 2015, Hillebrand et al. 2017) because species losses are often countered by species gains. In many cases, native species are being lost, but different native species or non-native species are taking their place. Thus, while there is clear evidence that the composition of local communities is changing over time due to species turnover (temporal beta diversity), alpha diversity (the total number of species) often does not to show a significant temporal trend (Dornelas et al. 2014, Magurran et al. 2018; but see Iknayan and Beissinger 2018 for a dramatic example of bird species loss in desert communities due to climate change over the past century).

These conclusions need to be tempered by the fact that wholesale habitat destruction due to urbanization or modern agriculture does lead to significant species loss. Murphy and Romanuk (2014) estimate a 25% loss of local species richness due to major land-use change (transformation of natural habitat to either urban or agricultural habitat). In a comprehensive analysis of species richness at >10,000 sites worldwide, Newbold et al. (2015) estimated the average loss of species richness due to land-use change is 8–14% compared with undisturbed vegetation, with worst-affected habitats showing even greater species loss. The debate over the extent of human-caused species loss at local and regional scales continues (e.g., Gonzalez et al. 2016; Velland et al. 2017b) and criticisms have been levied against the studies that first reported little evidence for declining local species richness. However, as the editors of the journal *Biological Conservation* recently noted (Primack et al. 2018, pp. A1–A3), "these … criticisms do not discount the value of the analyses that Vellend et al. (2013), Dornelas et al. (2014), and others have done. Their analyses are reasonable and convincing, given the available data, and the weaknesses, which the authors acknowledge, are difficult or impossible to avoid." It is also worth emphasizing that whatever the general trend in local species richness over time, previous experimental work to quantify the effects of richness on ecosystem functioning remains of value. Ecological knowledge derived from carefully constructed experiments is foundational to our understanding of the role of species in ecosystems. However, a clear challenge to the study of biodiversity and ecosystem functioning in the future is to address other important aspects of biodiversity besides species richness, such as species composition, species turnover, species relative abundances, and species traits. These aspects of biodiversity appear to be changing the most at local and regional scales. Another important challenge is to develop theory or empirical tests of how the plot-scale studies on ecosystem function vs richness scale up to landscape scales.

The invasion paradox

In nature, areas of high native plant diversity also contain the greatest number of invading plant species (Herben et al. 2004; Fridley et al. 2007; Figure 3.11). This positive correlation between native and exotic species richness runs opposite to that found in most experimental manipulations of biodiversity, in which the most species-rich communities show the lowest rate of invasion by exotics (e.g., Fargione and Tilman 2005). Fridley et al. (2007) refer to this mismatch as the "invasion paradox." How do we reconcile the observational and experimental patterns?

The most likely possibility is that, in nature, the same environmental factors that support high native plant diversity (e.g., habitat heterogeneity, resource availability) also support high exotic plant diversity, leading to positive covariation in native and exotic species richness. Small-scale experimental studies, on the other hand, seek to control this environmental variation and focus only on the effects of native diversity on invasion success. Experiments in river shore plant communities in California suggest that exactly this type of scale mismatch is occurring (Levine 2000). Despite experimentally showing negative effects on invaders in local plots at the landscape scale invasion success was positively correlated with richness in the same system. While this explanation is plausible, the reason for the "invasion paradox" is far from settled, and there is considerable debate over the observed patterns (Rejmanek 2003; Taylor

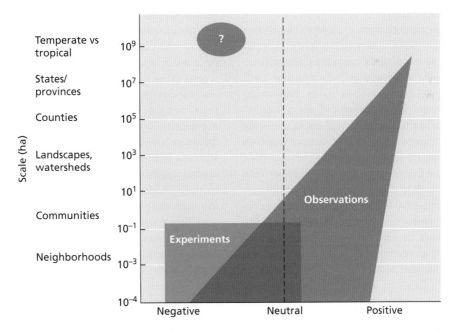

Figure 3.11 Conceptual diagram of the "invasion paradox." Most broad-scale observations in nature find positive correlations between native and exotic species richness (the tropics may be an exception to this pattern, indicated by the circle at upper left). On the other hand, local-scale studies (most of which are experimental) tend to find negative correlations between native species richness and the number of species of exotic invaders. After Fridley et al. (2007).

and Irwin 2004; Harrison 2008; Stohlgren et al. 2008). Nevertheless, as Levine et al. (2004, p. 982) note, if environmental factors "override any negative effect of diversity [on invasibility], that diversity effect cannot be strong, let alone absolute." As with the richness-productivity debate, scale should not be ignored going forward.

How important are diversity effects in nature?

Studies of synthetic communities constructed in the field and in the laboratory show that increasing species diversity often leads to greater and more stable biomass production, more efficient resource use, and reduced invasibility. Species removal experiments are more equivocal in their findings and tend to show smaller impacts of diversity change (e.g., Smith and Knapp 2003; Wardle and Zackrisson 2005; Suding et al. 2006; Wardle 2016). Nevertheless, the over-whelming conclusion from more than 20 years of experimental work following the pioneering papers of Naeem et al. (1994) and Tilman and Downing (1994) is that biodiversity has a significant impact on the functioning of ecosystems (Cardinale et al. 2011; Tilman et al. 2014).

Yet despite evidence from experimental studies, some ecologists have questioned whether species diversity per se has a significant effect on the func-tioning of natural ecosystems (e.g., Grime 1997; Huston 1997; Lepš 2004; Wardle 2016). They argue that experimental manipulations of species diver-sity tend to have their strongest effects at relatively low species richness, often at much lower numbers of species than are found in natural communities. They also argue that there is a big difference between demonstrating that a process operates in a (con-trolled) experiment and showing that the same pro-cess is a potent force in nature. It is possible that the effects of species diversity on ecosystem function-ing are simply overwhelmed by other interactions with the abiotic and biotic environment in nature. How do we address this question? One approach is to compare the strength of species richness effects on ecosystem functioning with other well-known drivers of ecosystem functioning. For example, Tilman et al. (2012) examined data from their long-term biodiversity experiments at Cedar Creek and concluded that biodiversity impacts ecosystem

productivity (plant biomass) as strongly as resources, disturbance, or herbivory (Figure 3.12).

Another approach is to compare the patterns observed in experimental manipulations of biodiver-sity with those observed across natural diversity gra-dients and ask if there is congruence. For example, we have already seen that there is a distinct mis-match between observational patterns and experi-mental results with regard to community invasibility (Figure 3.11). Experimentally constructed communi-ties that are more diverse are more resistant to inva-sion, whereas natural areas with higher native species richness also tend to have more invasive species (Fridley et al. 2007). With regard to community stability and resource use efficiency, natural and experimental communities appear to show similar patterns of increased nutrient use efficiency and increased ecosystem stability with increased species richness (e.g., White et al. 2006; Ptacnik et al. 2008; Hautier et al. 2015). Finally, productivity often increases with species richness in nature, as well as in experiments. For example, Liang et al. (2016) collected data from 777,126 forest biodiversity plots from 44 countries to construct a global relationship for tree species richness and productivity. They found that productivity was positively correlated with tree species richness (Figure 3.13), such that a 10% decrease in tree species richness (from 100% to 90%) would cause a 2–3% decline in productivity. If only one randomly selected tree species occurred in a forest plot (e.g., a monoculture), the expected decline in productivity was between 26% and 66%. In a recent literature review, Duffy et al. (2017) searched for all studies examining the effects of biodiversity on eco-system productivity in nature and concluded that biomass production increases with species richness in a wide range of wild taxa and ecosystems. More-over, they found that positive relationships between species richness and biomass production in natural systems were more likely to be significant when studies statistically controlled for the effects of the external environment. This would seem to counter criticisms that the impact of species diversity on eco-system functioning in nature is unlikely to be strong because of other overriding environmental factors. However, it is not clear how Duffy et al's analysis considers the most obvious confounding effect of the abiotic environment, e.g., the potential two-way

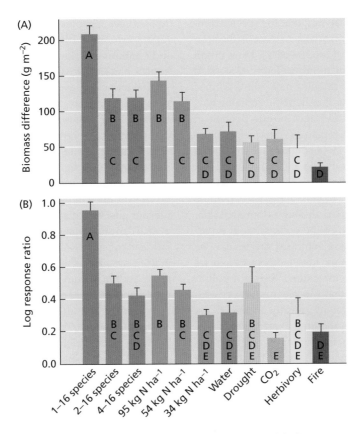

Figure 3.12 Relative influences of biotic and abiotic factors on plant productivity in the Cedar Creek biodiversity experiments. (A) Biomass differences (e.g., between different levels of species richness or abiotic factors). (B) Relative change in biomass as measured by the log response ratio. Bars with the same letter within each panel are not significantly different (P < 0.01 level; corrected for multiple comparisons). After Tilman et al. (2012).

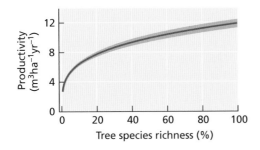

Figure 3.13 Primary productivity increases with tree species richness as determined from > 700,000 permanent forest plots distributed worldwide (mean plot size = 0.04 ha). Red line fit to the data with pink bands representing 95% confidence interval. After Liang et al. (2016).

causality between biodiversity and productivity (discussed in the next "unanswered question").

An important difference between studying the effects of species richness on ecosystem functioning in experiments and in nature is that experimentally constructed communities try to isolate the effects of species richness per se by selecting species at random for a common species pool and by initializing the experimental community with equal abundances of each species (an even species abundance distribution). However, as has already been seen in Chapter 2, species are not equally abundant in nature. Instead, natural communities usually consist of a few abundant (dominant) species and many rare ones. The impact of these dominant species is often more important in determining the functioning of natural ecosystems than is the total number of species (e.g., Vázquez et al. 2005; Fauset et al. 2015; Kleijn et al. 2015; Winfree et al. 2015, 2018; Lohbeck et al. 2016).

We also know from Chapter 2 that the species composition of a community varies across space (the beta-diversity component of biodiversity). How might beta-diversity affect the number of species needed to ensure ecosystem functioning regionally? Winfree et al. (2018) explored this question by examining how bee species diversity determines the effectiveness of pollination in three crop plants (watermelon, blueberry, and cranberry) in 48 commercial crop fields scattered across Pennsylvania and New Jersey. For a given threshold of pollination success (e.g., 50% or 75% of the mean total pollination per site), Winfree et al. (2018) showed that, although dominant species contribute to most of the pollination at a particular site, because these dominant species may be in low abundance at other sites, many more species are needed regionally to ensure successful pollination. They concluded that species turnover between communities (beta-diversity) greatly increased the total number of species required to achieve a given-level of ecosystem functioning for all communities in a region compared with what we might conclude from experimental communities at any one site. Of course, at very broad spatial scales (where there is nearly complete turnover in species composition) this conclusion is true by definition. To see this, consider a hypothetical example where we measure the annual biomass produced by tree species in a series of forest plots from Canada to Brazil. A forest plot in Canada shares no tree species in common with a forest plot in Brazil. Thus, to achieve a given-level of biomass production in all the forest plots along the full temperate to tropical gradient requires many more species than would be needed to achieve the same level of biomass production at any one site. Clearly, evaluating the consequences of spatial turnover in species (beta-diversity) to ecosystem functioning locally and regionally is an important question, but it requires a careful consideration of the spatial scales involved.

The two sides of diversity and productivity

Ecologists have struggled to reconcile the positive effect of species richness on productivity (biomass) found in many biodiversity experiments with observations from natural communities showing that richness and productivity often are not positively related (see Chapter 2). For example, Grace et al. (2007) examined diversity–biomass relationships in 12 natural grassland ecosystems using structural equation models to determine significant pathways of interaction between four variables—species richness, plant biomass, abiotic environment, and disturbance. They found that abiotic environment and disturbance had significant effects on species richness and biomass in a number of ecosystems, and that biomass had significant effects on species richness in 7 of the 12 ecosystems studied. However, in no case did they find that species richness had a significant effect on plant biomass. Grace et al. (2007, p. 680) concluded that "these results suggest that the influence of small-scale diversity on productivity in mature systems is a weak force, both in absolute terms and relative to the effects of other controls on productivity." More recently, Grace et al. (2016) analyzed data from >1000 grassland plots spanning 39 sites across five continents to examine the links between species richness and productivity (plant biomass), again using structural equation modeling. They detected clear signals of multiple underlying mechanisms linking richness and productivity. Most importantly they found:

1. Biomass production increased with increasing species richness (a strong effect when analyzed across the 39 global study sites, but a weak effect when analyzed at the 1-m^2 plot scale).
2. Strong and independent influences of macroclimate and soils on species richness and productivity.
3. Clear evidence that the accumulation of plant biomass negatively affects species richness (consistent with resource competition theory, Chapter 7).

Thus, Grace et al. (2016, p. 390) conclude that in nature "…productivity and richness are jointly controlled by a complex network of processes".

As shown in Chapter 2, there is a long history of looking at patterns of productivity and diversity in nature. The two most commonly observed patterns are a hump-shaped relationship, in which species richness peaks at intermediate productivities, and a monotonic (but often decelerating) increase in species richness with productivity (e.g., Liang et al. 2017; see reviews in Mittelbach et al. 2001; Gillman and

Wright 2006; see Figure 2.12). Efforts to reconcile these patterns theoretically is ongoing. For example, Loreau (2000, 2010b) and Schmid (2002) hypothesized that resource supply and species traits determine the maximum number of competing species that can coexist at a site, whereas the number of species at a site determines biomass production and the efficiency of resource use. This combination of drivers can generate different forms of the productivity–diversity relationship across sites. Yet, for all those forms, controlled experiments would show that species richness has a positive effect on productivity within a site.

More recently, Gross and Cardinale (2007) developed a mathematical model in which species compete for resources locally, and in which their regional abundance is determined by their dynamics across all local communities (i.e., in a metacommunity; Chapter 13). In this model, biodiversity is a hump-shaped function of resource supply in the local community, and both total biomass production and the efficiency of resource use in the metacommunity increase as diversity increases. Thus, Gross and Cardinale's model is able to account for some of the joint patterns seen in nature and in experimental manipulations of diversity because biomass and species richness are both responding to spatial variation in resource supply. However, it remains to be seen whether these simple models of competition for resources can account for the broad-scale variation in biodiversity seen across large spatial scales in the natural world.

Conclusion

BEF experiments provide perhaps the best example in all of ecology of multiple research groups conducting replicated field experiments to address the same, fundamental question - how does biodiversity (species richness) affect ecosystem functioning. At least four of these BEF experiments are very large in scale, long-term, and were conducted at different sites across the globe (NA and Europe, but none in the tropics; Clark et al. 2017). Taken as a whole the findings from these experiments (and hundreds of other smaller-scale experiments) confirm the positive impact of species richness

on productivity, ecosystem stability, and nutrient retention. Thus, we can confidently conclude that biodiversity matters to the healthy functioning of ecosystems. Yet, important questions remain, particularly with regard to how biodiversity relates to ecosystem functioning in nature (Wardle 2016; Winfree in press).

This chapter began with a simple question: "How many species are needed to ensure the successful functioning of ecosystems?" Clearly, we do not yet know the answer to this question, despite the great strides that ecologists have made in a very short time. This much we do know:

- species play a fundamental role in the functioning of ecosystems;
- the species composition of communities worldwide is changing at an unprecedented rate.

Taken together, these two facts make the question of how biodiversity affects ecosystem functioning one of vital importance. In addition, the study of biodiversity and ecosystem functioning continues to play a vital role in bring together the often disparate fields of community and ecosystem ecology (Loreau 2010a).

Summary

1. Higher species diversity is generally associated with higher primary production. It is still unknown what fraction of species could be lost from a community before we would see a major decrease in primary production.
2. Niche complementarity and species selection are two mechanisms that may tie species richness to production:
   - Niche complementarity may occur if species differ in the way they use limiting resources, in which case the overall efficiency of resource use by the community may increase with an increase in species richness, leading to higher production. Complementarity effects may also result from facilitation.
   - Species selection (also called the sampling effect) may lead to increased productivity if species-rich communities are more likely to contain more productive species that come to dominate the community.

3. A number of studies have shown that species-rich communities use resources more efficiently than less diverse communities, provided the species exhibit resource partitioning.

4. Temporal stability of communities (total biomass or total species abundance) tends to be greater in more diverse communities due to asynchrony in species responses to environmental fluctuations.

5. Species-rich experimental communities are more resistant to invasion by exotic species than are less diverse communities, but this "biotic resistance" does not allow them to repel invaders indefinitely or at larger spatial scales.

6. More research into the effects of species diversity on ecosystem functioning is needed in several areas, including:
   - effects at multiple trophic levels;
   - how particular traits affect both a species' probability of extinction and its functional role within a community;
   - how biodiversity is changing at local and regional scales;
   - the factors underlying the "invasion paradox";
   - the strength of species richness effects relative to those of other ecological interactions;
   - how diversity and productivity are functionally related across different spatial scales.

PART II

The Nitty-Gritty: Species Interactions in Simple Modules

CHAPTER 4

Population growth and density dependence

It is becoming increasingly understood by population ecologists that the control of populations, i.e., ultimate upper and lower limits set to increase, is brought about by density-dependent factors.

Charles Elton, 1949, p. 19

[Elton's] statement was somewhat premature.   **Peter Turchin, 1995, p. 19**

The reproductive capacity of biological organisms is amazing. A single bacterium dividing every 20 minutes under optimal conditions would cover the Earth knee-deep in less than 48 hours. The world population of our own species (*Homo sapiens*) has grown from fewer than 10,000 breeding couples in Africa some 70,000 years ago (Ambrose 1998), to about 300 million people in the first century CE, to 1 billion sometime around 1880, to about 7.6 billion in 2018. The current growth rate adds roughly 80 million people per year to the planet, leading to a projected world population size of 9.8 billion by 2050, according to the United Nations. Thomas Malthus (1789) recognized that this tremendous capacity for population increase could not be maintained in an environment with limited resources. His writings had a profound influence on Charles Darwin, who appreciated the far-reaching consequences of the fact that many more individuals are born than can possibly survive. Since the time of Malthus and Darwin, the study of population growth and population regulation has occupied a central place in ecology. As William Murdoch (1994, p. 272) put it, "Population growth underlies most...ecological problems of interest, such as the dynamics of

diseases, competition, and the structure and dynamics of communities.... [Population] regulation is essential to long-term species persistence and...is still the central dynamical question in ecology."

The first part of this chapter reviews the basic mathematics of population growth under conditions where resources are unlimited (the exponential growth model) and where resources are limited (the logistic growth model). For most readers, this material will be familiar. However, it is useful to see how these simple models of population growth are derived from first principles, as they form the foundation for more complex models of species interactions; this will be covered in the following chapters on predation, competition, and mutualism. In deriving the exponential and logistic growth equations, we have borrowed rather shamelessly from the excellent presentation in Gotelli (2008).

The second part of this chapter examines the role of density dependence in population regulation. Ecologists have debated the importance of density dependence in natural populations for almost 100 years, and it is important to know something of this debate. Ecologists also have long debated the shape of the density dependence curve. In addition,

Community Ecology. Second Edition. Gary G. Mittelbach & Brian J. McGill, Oxford University Press (2019).
© Gary G. Mittelbach & Brian J. McGill (2019). DOI: 10.1093/oso/9780198835851.001.0001

the study of density dependence in single-species populations leads naturally to the question of **community-level regulation**—the idea that species richness or the total abundance of individuals in a community may be regulated just like abundance in a single-species population. This concept is the basis for most theories of community stability and ecosystem resilience (Holling 1973; Gonzalez and Loreau 2009). Finally, links have been made recently between the subject of this chapter, density dependence, and the maintenance of rare vs common species and species richness within communities (Chapter 2). After we have examined these topics, related to density dependence of a single species and a single community, we will be ready to delve more deeply into the study of species interactions, including a mechanistic look at some of the processes underlying population regulation, such as competition for limited resources (Chapters 7 and 8) and density-dependent mortality due to predators (Chapters 5 and 6).

Exponential population growth

Consider a population growing in an unlimited environment, such as a bacterium introduced into a culture flask or a pregnant rat arriving on an oceanic island free of predators. If we let N_t represent the number of individuals in a population at time t, then the size of the population at the end of the next time interval N_{t+1} will depend on the number of individuals born B, the number that die D, the number that immigrate into the population I, and the number that emigrate from the population E during that time interval (Figure 4.1).

Thus, the population size at time $t + 1$ can be written as

$$N_{t+1} = N_t + B + I - D - E$$

and the change in population size from time t to time $t + 1$ as

$$N_{t+1} - N_t = B + I - D - E$$

For simplicity, a closed population has been assumed (i.e., no immigration or emigration), in which case

$$N_{t+1} - N_t = B - D.$$

It is also assumed that population growth is continuous—i.e., that time steps are infinitely small, and that births and deaths happen continuously. This assumption allows us to use the mathematics of differential equations (i.e., calculus). If you have forgotten most of your calculus or never understood it all that well in the first place, don't worry. All you really need to recognize here is that dN/dt is the rate of change in the number of individuals N with regard to time (Figure 4.2).

The rate of population growth dN/dt can now be written as

$$dN/dt = B - D$$

It is useful at this point to express birth and death rates on a per capita (i.e., per individual) basis. Therefore, we will define a constant b as the per capita birth rate and a constant d as the per capita death rate, such that

$$dN/dt = bN - dN$$

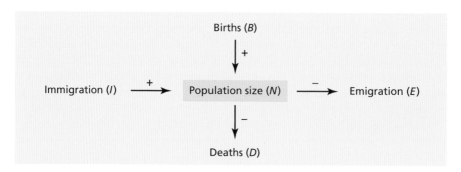

Figure 4.1 Factors affecting the size of a local population. After Case (2000).

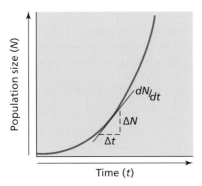

Figure 4.2 Population size (*N*) plotted as a function of time (*t*) for a population growing exponentially. The rate of population change (Δ*N*) during a time interval (Δ*t*) is equal to Δ*N*/Δ*t*. If we apply the methods of calculus and assume an infinitely small time step, then the rate of population growth can be written as *dN/dt*, which is equal to the slope of the line tangent to the curve.

or

$$dN/dt = (b - d)N$$

Letting $r = (b - d)$,

$$dN/dt = rN \qquad \text{[Eqn 4.1]}$$

Equation 4.1 thus describes how a population will grow in an unlimited environment (see Figure 4.2), where birth and death rates are unaffected by population density. The rate of population growth is governed by the constant *r*, which is the **instantaneous rate of increase**. When resources are unlimited, *r* is at its maximum value for a species, referred to as a species' **intrinsic rate of increase**. However, as we will see later, birth and death rates may vary with population density.

To predict the size of the population at any given time *t*, we need the integral form of Equation 4.1, which is

$$N_t = N_0 e^{rt} \qquad \text{[Eqn 4.2]}$$

where N_0 is the initial population size at time zero and *e* is the base of the natural logarithms (≈2.718).

Equations 4.1 and 4.2 describe a population undergoing **exponential growth**, where numbers increase without bound (see Figure 4.2). Or if *r* is negative the population will decrease exponentially fast towards an asymptote at zero. Populations are expected to grow exponentially when resources are unlimited, such as may occur when a species colonizes a new environment. Perhaps the most famous example of exponential growth began with the introduction to Australia of about 20 European rabbits (*Oryctolagus cuniculus*) by English settlers in 1859. In their new environment, the rabbits multiplied like, well, like rabbits. Within less than a decade, there were literally millions of rabbits in Australia; in 1907, the government constructed a 1800-km "rabbit-proof" fence in an attempt to limit their spread, but to little avail. There are many other examples of introduced or recolonizing species showing exponential population growth (e.g., Van Bael and Pruett-Jones 1996; Madenjian et al. 1998); however, we do not expect such growth to be sustained in a limited environment. For example, the current estimated mean value of the net growth rate for Australia's rabbit population is zero (Hone 1999).

One piece of good news is that exponential growth means that populations can recover quickly from the negative impacts of humans if the impacts are removed. A clear example of this is the northern elephant seal (*Mirounga angustirostris*). These seals were hunted to near extinction (less than 100 individuals living on just a single island off the coast of Mexico) by 1900. They also are among the slowest growing populations of mammals alive (one estimate is *r*=0.08) due to their large size (over a metric ton) and long life span (a decade or two). Despite this low reproductive potential, once hunting was banned, their population grew exponentially to almost 50,000 individuals by 1977 covering much of the Pacific coast of California and Mexico. Only then did elephant seals began to encounter resource limitations that slowed the rate of population growth to less than exponential (but still reaching a current population of close to 200,000 (Stewart et al. 1994).

Logistic population growth

As a population increases in size, its growth rate will, at some point, begin to decrease due to the depletion of resources or due to the direct negative effects of a high population density, such as the accumulation of toxic waste, or aggression between individuals. This negative feedback between population size and per capita growth rate is the essence of what ecologists call **density dependence** (Abrams 2009a). A good working definition is found in Turchin

(2003, p. 155): "density dependence is some (non-constant) functional relationship between the per capita rate of population change and population density, possibly involving lags." Note the emphasis here on the *per capita* rate of population change.

In a closed population, the negative feedback between population density and population growth must occur through changes in per capita birth rate, changes in per capita death rate, or both. The simplest way to model these changes is to assume that per capita birth rates or death rates change linearly with population size. Following Gotelli (2008, p. 26), we can modify the equation for exponential population growth (Equation 4.2) to

$$dN/dt = (b' - d')N \qquad \text{[Eqn 4.3]}$$

where b' and d' now represent the per capita density-dependent birth rate and per capita density-dependent death rate, respectively. We can express the linear change in the birth rate b' with population size N as

$$b' = b - aN$$

and the linear change in the death rate d' with population size N as

$$d' = d + cN$$

where a and c are constants (Figure 4.3).

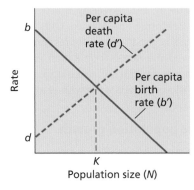

Figure 4.3 Density-dependent per capita birth (b') and death (d') rates in the logistic model of population growth. The per capita birth rate declines linearly with population size ($b' = b - aN$), and the per capita death rate increases linearly with population size ($d' = d + cN$). Where the two curves intersect (i.e., when birth rate equals death rate), population size is unchanging and the population is at its carrying capacity (K).After Gotelli (2008).

The equation for population growth in an unlimited environment (Equation 4.1) can now be modified to include the effects of density dependence on per capita birth and death rates by changing

$$dN/dt = (b - d)N$$

to

$$dN/dt = (b' - d')N$$

or, by substitution,

$$dN/dt = [(b - aN) - (d + cN)]N \qquad \text{[Eqn 4.4]}$$

(Note that, in the above formulation, birth rates could become negative at large values of N, a nonsensical result.)

Gotelli (2008, p. 27) showed how the terms in Equation 4.4 may be rearranged and simplified to

$$dN/dt = rN\left[1 - \frac{(a+c)}{(b-d)}N\right] \qquad \text{[Eqn 4.5]}$$

where $r = (b - d)$. Because a, b, c, and d are all constants, we can define a new constant K as

$$K = \frac{(b-d)}{(a+c)}$$

Thus, Equation 4.5—the equation for density-dependent population growth—can now be written as

$$dN/dt = rN\left(1 - \frac{N}{K}\right) \qquad \text{[Eqn 4.6]}$$

Alternatively, it is sometimes written as

$$dN/dt = rN\left(\frac{K-N}{K}\right) \qquad \text{[Eqn 4.7]}$$

Equation 4.6 (or 4.7) is the well-known **logistic equation** for population growth in a finite environment. First introduced by Pierre François Verhulst (1838), the logistic equation has played an important role in both the study of population dynamics and the development of models of species interactions (Lotka 1925; Volterra 1926). Because of its importance to ecology, we will explore the properties of the logistic growth equation in more detail here.

The logistic equation predicts a sigmoidal pattern of population growth through time (Figure 4.4A), with population size reaching a stable maximum at **K**, the **carrying capacity**. We can determine the number of individuals N in the population at

(A)

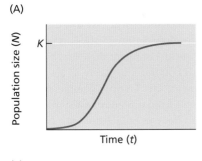

(B)

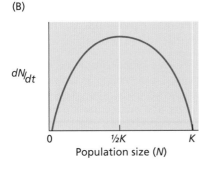

(C)

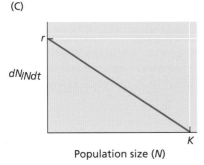

Figure 4.4 Properties of a population exhibiting logistic growth. (A) Change in population size *N* as a function of time. (B) The population growth rate, *dN/dt*, peaks at an intermediate population size (equal to half the carrying capacity, or ½*K*). (C) The per capita population growth rate, *dN/Ndt*, decreases linearly with an increase in population size, declining from a maximum value equal to the species' intrinsic growth rate, *r*, and reaching zero at the species' carrying capacity *K*.

any time t from the integral form of Equation 4.6, which is

$$N_t = N_0 \frac{K}{(K - N_0)e^{-rt} + N_0} \qquad \text{[Eqn 4.8]}$$

It is useful to visualize how the population growth rate dN/dt and the per capita population growth rate dN/Ndt vary as functions of population size. As Figure 4.4B shows, the population growth rate dN/dt is a peaked function of N, reaching its maximum

when $N = \frac{1}{2}K$. This maximum growth rate corresponds to the inflection point in the curve shown in Figure 4.4A. The per capita growth rate dN/Ndt behaves differently, however: it declines linearly with N, starting at a maximum value equal to r and declining to zero when $N = K$ (Figure 4.4C).

The above analysis shows how the logistic growth model may be derived from the exponential growth model by assuming linear density dependence in per capita birth and death rates, and a maximum population size of K. In fact, we get the same result if only the birth rate or the death rate is linearly density-dependent.

The logistic equation is a very simple model for population growth and, as such, it makes four important assumptions that are rarely met in nature (Turchin 2003; Gotelli 2008):

1. It is assumed that the per capita growth rate is a linear function of N (see Figure 4.4C). [Sibly and Hone (2002) provide empirical examples, showing both linear and non-linear density dependence.]
2. It is assumed that population growth rate responds instantly to changes in N (i.e., there are no time lags).
3. It is assumed that the external environment has no influence on the rate of population growth.
4. It is assumed that all individuals in the population are equal (i.e., there are no effects of individual age or size).

Despite these unrealistic assumptions, the logistic equation continues to be taught and used in ecology because, as Turchin (2003, p. 49) notes, "it provides a simple and powerful metaphor for a regulated population, and a reasonable starting point for modeling single-population dynamics, since it can be modified to address all four criticisms listed above."

This short review of population growth has (hopefully) served to refresh your memory of the basic tenets underlying exponential and logistic growth. We will build on this foundation and on the insights gained from the properties of the logistic equation (see Figure 4.4) in the remainder of the book. There is a wealth of knowledge on the dynamics of single-species populations that will not be discussed in this book, including the effects of time lags, and age or size structure on population dynamics, as well as the complex dynamics that may be exhibited by populations in models and in nature (e.g., limit cycles and chaos). Interested readers should consult Case (2000), Turchin (2003), Vandermeer and Goldberg (2013), and Gotelli (2008) for an introduction to these more complex dynamics.

The logistic equation can be readily modified to incorporate the effects of additional species. We will make use of a modified logistic model (e.g., the Lotka–Volterra equations for predator–prey dynamics and for interspecific competition) when we dig into the "nitty-gritty" of species interactions in the next few chapters. However, before moving on to the study of interspecific interactions, let's take a closer look at the evidence for density dependence in nature and at some of its consequences.

The debate over density dependence

Although the existence of density-dependent population regulation is widely accepted today, it was once the topic of one of the most acrimonious and long-running debates in ecology. As noted in Chapter 1, during the early twentieth century, there were two opposing schools of thought on the matter. On the one side, Nicholson, Lack, and Elton believed strongly that for populations and species to persist without growing to unimaginably large numbers, they must be subject to density-dependent regulation. On the other side, Andrewartha and Birch saw no evidence for density dependence in natural populations, and they questioned whether density dependence was even necessary to explain the persistence of species. This debate over the factors regulating populations continued for decades. In the 1980s, Dempster (1983) and Strong (1986) argued that populations spend most of their time fluctuating in a density-independent manner, and that it is only when their numbers become very large or very small that density-dependent factors come into play, effectively setting floors and ceilings to population size.

At the same time, some of the debate over density dependence and population regulation shifted subtly to include the question of whether populations exhibit equilibrium or non-equilibrium dynamics. The observation that species abundances in many ecosystems seemed to fluctuate caused ecologists to question whether population models that predict a stable equilibrium density, such as the logistic growth model developed previously, had any relevance to the real world. For many, there was broad acceptance that most communities were in a non-equilibrium state. One of our favorite commentaries from this time comes from a review of a symposium entitled "The Shift from an Equilibrium to a Non-equilibrium Paradigm in Ecology," held during the 1990 annual meeting of the Ecological Society of America. In this review, Murdoch (1991, p. 50) gives a flavor for the dispute over equilibrium and non-equilibrium dynamics that existed at the time.

Certainly the early proponents of this idea [nonequilibrium dynamics] were viewed as radicals. The ultimate punishment awaiting iconoclasts, however, is canonization, and the title and contents of this symposium, together with talks in several others, suggest that a similar fate is now befalling those few hardy souls who long ago threw in their lot with Andrewartha. For a significant segment of the ESA, it seems, the Reformation is accomplished and, to confuse some history, there is a new pope in Avignon.

Murdoch goes on to express his doubts, however, about the wisdom of embracing this paradigm shift from equilibrium to non-equilibrium ecology:

Perhaps it is the certitude with which the equilibrium point of view was discarded that impels me to its defense, even though my Society of Ecological Anarchists membership card bears a low number. Is it really true that there is no evidence for stable equilibria in Nature? That an equilibrium view can never be fruitful? Surely not. Convergence in density following experimental perturbation has been demonstrated in at least three systems (perhaps the only ones tested). Is it easy to explain the amazing constancy of the Thames heron population over 50 years without some reference to equilibrium? Regular cycles are surely not examples of stable point equilibria, but an equilibrium-centered view certainly leads to the closely related idea of a stable-limit cycle, which offers at least a plausible explanation for many apparently oscillating populations. It's messy, but I'm afraid there will need to be more than one pope.

Murdoch's comments, besides displaying a keen wit, bring up an important issue that we want to address before going further into our examination of density dependence. The question relates to the usefulness of simple models for understanding the real-world dynamics of populations and species. We believe that simple models, despite their "unrealistic assumptions," are very useful for developing our insight into the factors that govern population growth and determine the outcome of species interactions. As mentioned earlier, we will build on the logistic equation to develop models for predator–prey interactions and for interspecific competition. In addition, we will look for equilibrium solutions to these models when we examine the factors that promote (or destabilize) species coexistence. This is not to say that we believe these models to be literal descriptions of nature or that we expect the world to be at equilibrium. All populations are affected by exogenous factors, and their densities fluctuate to various degrees. However, as Turchin (1995, pp. 25–6) notes (and it is worth quoting him at length):

[T]he notion of equilibrium in ecological theory [can be] replaced with a more general concept of *attractor* (e.g., Shaffer and Kot 1985). In addition to point equilibria, there are periodic [limit cycles], quasiperiodic, and chaotic attractors (e.g., Turchin and Taylor 1992). The notion of attractor can be generalized even further, so that it is applicable to mixed deterministic/stochastic systems. . . . In the presence of noise, a deterministic attractor becomes a stationary probability distribution of population density (May 1973b), which is the defining characteristic of a regulated dynamical system. Put in intuitive terms, the equilibrium is not a point, but a cloud of points (Wolda 1989; Dennis and Taper 1994). Thus, if we define equilibrium broadly as a stationary probability distribution, then *being regulated* and *having an equilibrium* are one and the same thing. The whole issue of equilibrium versus nonequilibrium dynamics becomes a semantic argument (see also Berryman 1987).

Taking a probability distribution view of equilibrium is not only visually descriptive, but it actually opens up a mathematical toolbox for asking precise questions such as how much variance is there in a regulated population and how is this variance partitioned across different sources of variability such as demographic vs environmental stochasticity (e.g. Engen et al. 2002).

Evidence for density dependence in nature

Although the early ecologists debated whether density dependence was necessary for population regulation, it is now widely accepted that regulation cannot occur in the absence of density dependence (Murdoch and Walde 1989; Hanski 1990a; Godfray and Hassell 1992; Turchin 1995). Thus, the way to detect population regulation is to test for density dependence. Furthermore, density dependence leading to population regulation must take a specific form (Turchin 1995, 2003). First, it must operate in the right direction (there must be a tendency for the population to return to its previous size after a disturbance). Secondly, this return tendency must be strong enough and quick enough to overcome the effects of disturbance. Most tests for density dependence have examined a time series of estimated population densities or manipulated density experimentally. Detecting density dependence in nature, however, is surprisingly difficult.

The most common approach to looking for density dependence is to examine how the growth rate of a population (r) varies with population density in a long time series of data. Turchin (1995, p. 28) proposed a simple population model for use in such a test:

$$r_t \equiv \ln N_t - \ln N_{t-1} = f(N_{t-1}) + \varepsilon_t \quad \text{[Eqn 4.9]}$$

where f is typically a linear function of either N_{t-1} (e.g., Dennis and Taper 1994) or log N_{t-1} (e.g., Pollard et al. 1987) and ε_t is a term representing random density-independent factors. The above model may be criticized on many grounds, including the fact that the time interval of the censuses used, often does not correspond to the time scale at which density dependence works (it is often continuous). A further challenge is that it can be hard to determine the shape of f, if the population only fluctuates around a small range of values of N_t near the equilibrium (Clark et al. 2010). Still, as Turchin (1995) notes, the above model is usually adequate to address the simple question of whether there is evidence for regulation in the population. Pollard and colleagues proposed looking for regulation by testing the correlation between r_t and log N_{t-1}, whereas Dennis and Taper's test regresses r_t on the untransformed population density N_{t-1}. Sibly and Hone (2002) provide a good review of the different approaches ecologists have taken to studying density dependence (i.e., testing for the existence of density dependence vs testing for the mechanisms causing density dependence).

Using the methods outlined above, Woiwod and Hanski (1992) analyzed almost 6000 time series of aphid (94 species) and moth (263 species) populations recorded at various sites in Britain. They found density dependence in 69% of aphids and 29% of moth time series. However, if short time series (<20 years) are removed from the analysis, the percentage of studies showing density dependence jumps to 84% and 57%, respectively. Holyoak (1993) and Wolda and Dennis (1993) also found that the percentage of studies showing significant density dependence increased with the length of the time series. Furthermore, Turchin and Taylor (1992) found density dependence in 18 of 19 forest insect populations and in 19 of 22 vertebrate populations, using tests of delayed, as well as direct density dependence, and they observed cyclic and chaotic, as well as stable dynamics (Murdoch 1994). Zimmermann et al. (2018) detected density-dependent recruitment in 70% of fish populations from the North Atlantic Ocean. Finally, Brook and Bradshaw (2006) conducted a meta-analysis of long-term abundance data from 1198 species spanning a broad range of taxa (vertebrates, invertebrates, and plants). Using a variety of analytical methods, including multi-model inference, as well as the traditional methods outlined above, they found "convincing evidence for pervasive density dependence in the time series data" and "little discernible difference across major taxonomic groups" (Brook and Bradshaw 2006, p. 1448). Moreover, they found a convincing trend toward a zero net growth rate (implying equilibrium) as the number of generations in the time series increased (Figure 4.5).

All of these tests are complicated by the fact that measurement error also tends to show a return tendency (overestimates due to measurement error tend to show a lower population in the next time period, simply because it is unlikely for measurement error to repeat in the same direction, and vice versa). The evidence for density dependence weakens as the amount of measurement error increases. Knape & de Valpine (2012) found only 45% of the same database studied by Sibly et al. (2005) showed evidence for density dependence when measurement error was accounted for. Nonetheless, the overall weight of the evidence shows that, when examined with sufficiently long time series, density dependence

is very often found for many different types of species (Turchin 1995). Still mostly unanswered is the relative importance of density dependence vs environmental stochasticity in real world population dynamics, which is exactly the original debate between Nicholson and Andrewartha.

Given the statistical complexity involved in detecting density dependence in observational time series, probably a stronger test is to experimentally alter the densities of natural populations and look for density-dependent changes in birth rate, death rate, and population size (Nicholson 1957; Murdoch 1970). Harrison and Cappuccino (1995) reviewed 60 experimental studies that tested for density dependence in this way in a wide range of animal taxa. They found that of the 84 effects examined in the 60 studies, almost 80% showed negative density dependence (positive density dependence was detected twice). Thus, they concluded that experimental manipulations of population density corroborate the results from time-series analysis. In other words, density-dependent population regulation is common and widespread among animal taxa, although Harrison and Cappuccino caution that their conclusions are based on a relatively small number of experimental studies.

It is worth noting that most models of density dependence (including the logistic model) do not explicitly incorporate resources, even though competition for resources is assumed to be the major factor driving negative density dependence in most systems. In the logistic model, density dependence acts in a linear fashion—that is, per capita population growth declines linearly with population density (see Figure 4.4C) due to (unspecified) linear effects of density on per capita birth rates and/or per capita death rates (see Figure 4.3). However, other popular models of single species dynamics, such as the Ricker (1954) and Beverton and Holt (1957) models assume a convex-down functional form. How might we expect the functional form of density dependence to look in the real world? Using a modification of the logistic model known as the theta-logistic fitted to a large database of population time series, Sibly and colleagues (2005) found that there was a strong preponderance of convex-down functional forms (consistent with the Ricker model). However, this result has been questioned

(A)

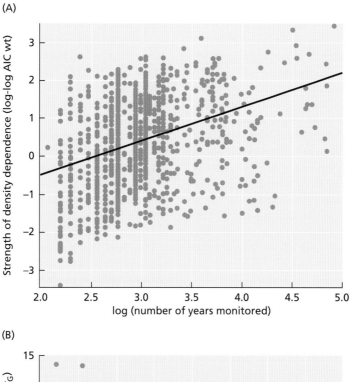

(B)

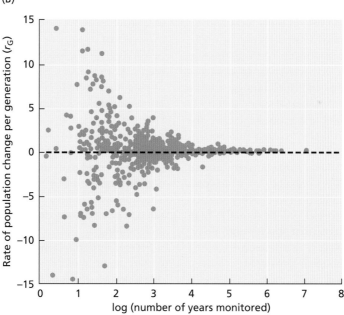

Figure 4.5 The results of a meta-analysis examining the evidence for density dependence in population time-series data for 1198 species in nature. (A) The strength of the evidence for density-dependent regulation, measured as log–log transformation of summed model fits Akaike's Information Criterion, or AIC weights, of three density-dependent models, increases as the length of the population time series increases. (B) The average rate of per-generation population change as a function of the number of generations monitored, showing that the observed net rate of population change tends toward zero (dashed line) as the number of generations monitored increases. After Brook and Bradshaw (2006).

on methodological grounds (Clark et al. 2010). Using a different method, North American bird populations also fit a Ricker (convex-down) form of density dependence better than a linear logistic model (McGill 2013). Theoretical considerations suggest that the form of density dependence in most consumer–resource interactions will be highly non-linear and even sigmoidal (with both convex and concave phases; Abrams 2009a,b). However, these models assume density dependence is driven by bottom-up forces, which is not always true. It may seem surprising that we do not have a definitive answer to such a basic question, but empirical study of the functional form of density dependence requires even longer time series than tests for the existence of density dependence and also encounters the same thicket of statistical issues.

Bottom-up or top-down density dependence? and other questions

A growing area of research is what causes density dependence (i.e. the decline in birth rates and/or the increase in death rates when densities are high). A common assumption motivated by the derivation of the logistic equation is that the change in rates is due to intraspecific competition for limiting resources ("bottom-up forces"). An alternative possibility is that the natural enemies of a species increase disproportionately when there are high numbers of the victim species ("top-down forces"). One very specific version of top-down is the **Janzen-Connell hypothesis** (Janzen 1970; Connell 1971), which proposes that specialist herbivores and pathogens that attack one particular species will be found in high abundance near adults of that species and, therefore, suppress offspring of the same species landing near to its parent. Although originally framed as an explanation for species diversity, the Janzen–Connell hypothesis is fundamentally a statement of negative density dependence via natural enemies.

To search for the mechanisms of density dependence, Ghedini et al. (2017) measured the energy intake and basal metabolic energy expenditure of colonial bryozoans at varying densities. Both energy intake and basal metabolic energy expenditure declined at higher densities, but the energy intake declined faster, meaning that under higher densities the bryozoans

had less and less spare energy to devote to growth, reproduction, and maintenance. This provides a nice experimental confirmation of a mechanistic role of bottom-up resource limits in density dependence. More broadly, Harrison and Cappuccino (1995) found in their survey that negative density dependence was more commonly the result of competition (bottom-up) than the result of control by predators and other natural enemies. In contrast, recent studies in plants (especially tropical trees) are building an increasingly strong case that density dependence in these systems is a result of natural enemies (top-down as hypothesized by Janzen and Connell) and specifically by soil pathogens. Observational studies (e.g., Comita et al. 2010), as well as field experiments that manipulated conspecific densities or distance from parents (meta-analysis in Comita et al. 2014) all show support for Janzen–Connell. Some of the strongest evidence comes from recent greenhouse experiments (Klironomos 2003; Mangan et al. 2010), where seedlings were planted in soil cores taken from under a conspecific adult, a heterospecific adult, or from sterilized soil. These studies all showed that conspecific soil greatly reduced growth or survival in comparison to the other two treatments. Klironomos separated the soil-based effects of potentially beneficial mycorrhiza and detrimental pathogens and found the pathogen effect was stronger. Mangan and colleagues (2010) also used field experiments to directly contrast above-ground herbivory vs below-ground herbivores and pathogens, and found that, although both created density dependence, the below-ground effects were stronger. Thus, below ground pathogens are strongly implicated in plant density dependence, at least for seeds and seedlings.

Several important questions remain open in the empirical study of density dependence. We would like to know which demographic parameters (e.g., reproduction, survival, the b and d parameters in the derivations above) respond most strongly to changes in population size, as well as how the strength of density dependence varies with organism age, body mass, and ecological context (Bonenfant et al. 2009). Whether the results mentioned previously in support of Janzen–Connell extend beyond seedlings and saplings to adult trees is a consequential example of this question that is not currently known. Other important foci in the study of population regulation are the relative impacts of endogenous (e.g. density dependence) vs exogenous (e.g. variable environment) forces on population change, and whether regulation is most often the result of local processes or the outcome of metapopulation dynamics (Murdoch 1994; Turchin 1999). For many, the answers to these questions are far more meaningful than wondering whether density dependence does or does not occur in a population (Krebs 1995).

Positive density dependence and Allee effects

Up to this point, we have focused our discussion of density dependence on those factors that cause the per capita population growth rate to decline with density, because negative density dependence is crucial for population regulation. However, a number of factors may result in *positive* density dependence (i.e., a positive correlation between density and per capita population growth), particularly at very low population densities (Case 2000). These factors are commonly referred to as **Allee effects**, named after the zoologist W. C. Allee (Allee et al. 1949). For example, at low population densities, the inability to find a mate may limit the ability of a species to colonize new environments (Gascoigne et al. 2009). A nice example of an Allee effect comes from the work of Sarnelle and Knapp (2004) and Kramer et al. (2008), who studied the recolonization success of zooplankton in high alpine lakes in the Sierra Nevada of California. Two large-bodied zooplankton species, the copepod *Hesperodiaptomus shoshone* and the daphnid cladoceran *Daphnia melanica*, were native to these lakes prior to their being stocked with trout in the early to mid-1900s. Large-bodied zooplankton, however, are a preferred prey for trout, and following the introductions, *Hesperodiaptomus* and *Daphnia* disappeared. When trout were removed from some study lakes in the late 1990s, *Daphnia* rapidly recolonized. *Hesperodiaptomus*, however, failed to reappear in many of the study lakes where they were once abundant. Sarnelle and Knapp (2004) hypothesized that the copepods were unable to recover from low densities because of a strong Allee effect. Copepods and daphnids both produce resting eggs that may remain dormant in lake sediments for many years, and these resting eggs are

the most likely source of colonists in these isolated alpine lakes. *Daphnia* can re-establish a population from only a few colonists (potentially a single, hatched resting egg) because they can reproduce asexually (by parthenogenesis), as well as sexually. Copepods, however, are obligately sexual and must find a mate in order to reproduce. Kramer et al. (2008) hypothesized that the inability of *Hesperodiaptomus* to find mates and reproduce at very low densities was the reason they were unable to recolonize lakes following trout removal. Experimental introductions of copepods at varying densities supported their hypothesis and provided one of the few experimental tests of Allee effects in nature.

Other factors that may lead to positive effects of population density on per capita growth rates include group protection from predators, amelioration of a harsh environment by neighbors (see examples in Chapter 9), and increased foraging efficiency in groups (Dugatkin 2009). Kramer et al. (2009) provide a recent review of the evidence for Allee effects, in which they conclude that density often has positive effects on various components of individual fitness, particularly at low densities. However, the total impact of density on per capita population growth must include the combined effects of both positive and negative density-dependent factors. This recognition has led ecologists to distinguish "component Allee effects," in which density has a positive impact on some component of fitness, from "demographic Allee effects," in which the total impact of density on per capita population growth rate is positive (Stephens et al. 1999). Kramer et al. (2009, p. 341) concluded that "although we find conclusive evidence for Allee effects due to a variety of mechanisms in natural populations of 59 animal species, we also find that existing data addressing the strength and commonness of Allee effects across species and population is limited." In a recent analysis of demographic Allee effects, Gregory et al. (2010) found only limited evidence for such Allee effects in 1198 natural populations. Using a best-fit model approach, they concluded that positive density feedbacks might influence the growth rate of about 1 in 10 natural populations. In their review of mate-finding Allee effects, Gascoigne et al. (2009) also found only a few examples of demographic Allee effects. However, Perälä and

Kuparinen (2017) and others caution that detecting demographic Allee effects with time series data from natural populations is notoriously difficult due to the challenges raised by sampling low-abundance populations. Still, most evidence suggests that the demographic consequences of density on population growth are largely negative and that we are on safe ground in focusing on negative density dependence when constructing simple models of population growth (such as the logistic equation). We will build on the logistic model as we begin our study of species interactions in resource-limited environments in Chapters 5, 7, and 9.

Community-level regulation of abundance and richness

The concept of density dependence in a population may be extended to the community level. Either the total abundance of individuals (of all species) in a community or the number of species (species richness) could show regulation in the same way that the abundance of a single species population does (i.e., a probabilistic cloud centered around an attractor). As a general phenomenon this has been called "community-level regulation" (Gotelli et al. 2017) or "zero-sum dynamics" (Ernest et al. 2008). There are at least three broad categories of mechanisms that could cause community-regulation. The first are neutral or non-neutral processes that posit decreasing birth and increasing death rates with total population size across a community (or decreasing colonization and increasing extinction rates with richness). Such dynamics cause populations to have a noisy equilibrium where density causes the birth and death rates to be equal such as is found in MacArthur and Wilson's island biogeography theory of species richness (1967, Chapter 2). Another purely probabilistic mechanism, at least for community regulation of abundance, is portfolio theory (Schindler et al. 2015, McGill et al. 2015). The portfolio effect notes that the variance of a sum of random variables (e.g., the total population of a community summed across all species) will be less than the variance of each individual variable due to statistical averaging. The third mechanism posits that the species studied come from a single trophic guild and share a common limiting resource, and use roughly equal amounts of resource per individual. Then the total population size is set by the amount of resource (i.e., there is a community level carrying capacity). In this scenario, if one species increases in density, one or more other species must decrease. This has been called **density compensation** (MacArthur et al. 1972) or **compensatory dynamics** (Gonzalez & Loreau 2009) or zero-sum dynamics (Hubbell 2001). Gonzalez and Loreau (2009) provide an excellent overview of some of the causes and consequences of compensatory dynamics in communities. Just as in the debate over density dependence within single-species populations, the evidence for community-level regulation has been hotly contested.

Both experimental manipulations and time-series analyses have been used to test for community-level regulation. An excellent example of community-level regulation of richness in a natural community comes from the work of James Brown and his colleagues, who conducted a long-term experiment on rodents, plants, and ants in the Chihuahuan Desert near the town of Portal, Arizona. In 1977, Brown and Kodric-Brown established 24 fenced plots (50 m × 50 m), to which various rodent exclusion treatments were applied (including control plots to which all rodents had access). Gates in the plot fences excluded or permitted access by different rodent species. Some of these gates excluded the largest rodent species (members of the genus *Dipodomys*, the kangaroo rats). In their study of compensatory dynamics, Brown and colleagues focused primarily on comparing the natural dynamics of the rodent communities in the control plots with those in the kangaroo rat removal plots. This system is an especially good one for studying compensatory dynamics, because the rodent species are all granivores and are, therefore, supported by the same resource base.

Importantly, this long-term experiment has been conducted against a background of environmental change—specifically in precipitation levels—which has resulted in a vegetation shift from grasses to woody shrubs, and a concomitant shift in rodent species composition. Ernest et al. (2008) found evidence for community-level regulation at the study site. Over time, the rodent community shifted from dominance by grassland species to dominance by shrubland species (Figure 4.6A). However, species

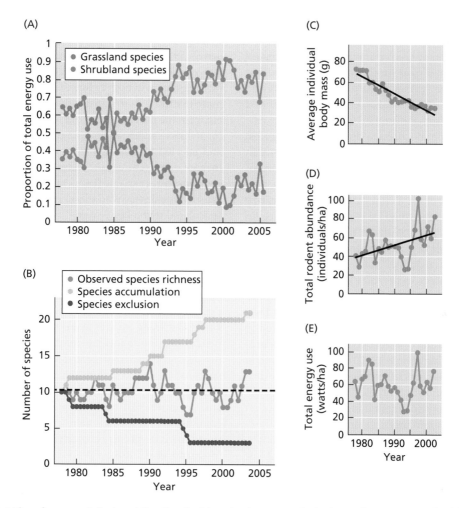

Figure 4.6 Evidence for community-level regulation of species richness in a long-term study of a desert rodent community in the Chihuahuan Desert near Portal, Arizona. (A) Temporal trends in the proportion of total energy flux through the rodent community demonstrated a shift in dominance from grassland species to shrubland species. (B) Trends in rodent species richness demonstrated that the long-term average species richness (dashed line) was essentially constant despite considerable species turnover. (C–E) Trends in mean individual rodent biomass (C), total rodent abundance (D), and summed rodent metabolic rate, an index of community-level energy use (E). Regression lines are plotted for significant relationships. After Ernest et al. (2008).

richness at the site remained quite constant despite the considerable species turnover (Figure 4.6B). Moreover, the shift in species composition of the rodent community resulted in a decrease in average rodent body size and an increase in total rodent abundance (Figure 4.6C, D). In combination, these changes in body size and abundance resulted in no change in the total energy flux (resource consumption) by the rodent community as a whole (Figure 4.6E). Thus, Ernest et al. (2008) concluded that the consistency in total energy use through time by the rodent

community showed evidence of density compensation (i.e., the resource base could support a constant total energy use, regardless of the species composition). Tanner et al. (2009) provide another example of community-level density dependence based on 27 years of data on coral community dynamics from Australia's Great Barrier Reef. Hawkins et al. (2000) also found evidence of density compensation in stream invertebrate communities in comparisons between disturbed and reference sites for more than 200 taxonomic groups and 200 study sites.

Studies looking specifically for compensatory dynamics often test whether time series of abundance of different species are negatively correlated with each other over time. This would mean that each time one species went up in abundance the other went down. Houlahan et al. (2007) examined 41 different plant and animal data sets to assess patterns of covariation among population abundances. They predicted that compensatory dynamics within communities should result in a preponderance of negative covariation in species densities. Instead, they found that 31 of the 41 data sets had more positive covariances than would be expected by chance; in other words, populations in the community were growing larger or smaller together. Based on these results, they concluded "that variability in community abundance appears to be driven more by processes that cause positive covariation among species (e.g., similar responses to the environment) than processes that cause negative covariation among species (e.g., direct effects of competition for scarce resources)" (Houlahan et al. 2007, p. 3276). Mutshinda et al. (2010) also concluded that environmental variation plays a much larger role than intra- or interspecific interactions in determining temporal variation in species abundances in an analysis of time-series data from a range of taxa, including moths, fishes, macrocrustaceans, birds, and rodents. Most of the observed environmentally driven covariation between species was positive, and Mutshinda and colleagues (2010, p. 2923) concluded that "compensatory dynamics are in fact largely outweighed by environmental forces, and the latter tend to synchronize the population dynamics." However, Ranta et al. (2008) cautioned that the lack of negative covariation in species densities over time does not imply an absence of competition. In particular, there are mathematical limits to how many species can negatively covary with each other (if A and B negatively covary and B and C negatively covary then A and C are likely to positively covary).

More recently, Gotelli and colleagues (2017) took a totally different (and mechanism-agnostic) approach to testing for community-level regulation in both total community abundance and richness. They used time series models to test whether after perturbations time series showed a random walk from the new level (the null hypothesis) or a propen-sity for returning to the previous level (suggestive of community-level regulation). They found that across 59 assemblages of diverse taxa more than 90% of the communities trended towards community regulation in both richness and abundance, and more than 50% were statistically significantly different from the null hypothesis of a random walk both for richness and total abundance. They also tested and rejected the idea that compensatory dynamics (negative covariances) and shared responses to a key environmental variable (temperature, and a possible cause for positive covariances) were involved. They speculated that a key reason the test for compensatory dynamics in abundance did not find them despite evidence for community-level regulation is that there was a great deal of turnover (loss and gain of species), which violates the assumptions of tests for compensatory dynamics. This parallels the example described above with desert rodents, where declining species were not primarily compensated for by an increase in other existing species but instead were compensated for by an increase of species new to the system. This result is also supported by two studies showing that there are strong patterns of rapid turnover in community composition even when there are no strong trends up or down in species richness (Dornelas et al. 2014; Magurran et al. 2018).

What can we conclude about the strength of compensatory dynamics in communities?

1. Strong community-level regulation has been found in a number of field experiments and observational studies.
2. Studies of species covariance used to detect specific process of compensation have shown that environmental effects tend to trump impacts of intra- or interspecific density dependence, or both.
3. Recent works suggests that it is important to consider both turnover in species composition (community level regulation of richness) and compensatory dynamics (community-level regulation in abundance) as the two processes interact with each other.

This work is important because compensatory dynamics figure into many aspects of community and ecosystem theory, including the relationship between biodiversity and ecosystem functioning

(Doak et al. 1998; see Chapter 3), species abundances in neutral models (Hubbell 2001; see Chapter 14), species coexistence in variable environments (Chesson 2000a; see Chapter 15), and predator–prey interactions in multi-prey communities (Abrams et al. 2003; see Chapter 6).

Density dependence, rarity, and species richness

An exciting recent line of research has linked density dependence within a single species to its relative abundance in a community (whether a species is common or rare) and ultimately with species richness, two of the key questions of Chapter 2. Simply stated, are common species or rare species more likely to experience density dependence? One could imagine that common species are more abundant and, thus, more likely to experience the effects of density dependence. Or one could imagine that rare species are rare because they experience stronger density dependence. Henderson and Magurran (2014) employed an unusually well sampled and long-term (33-year) study of 81 fish species from a single estuary in the UK to addresses this question. They found that common species showed little variation in abundance and almost always showed statistically significant density dependence. Rare species, however, were transitory (colonizing and going extinct in the system, sometimes multiple times). None of these transient species showed density dependence (which is not surprising since their abundances were often zero). This suggests that common species experience more density dependence.

A different approach has resulted in a seemingly contradictory answer. Rather than testing if density dependence exists or not (a binary measurement) in rare or common species, the alternative approach tests to see if the strength of density dependence (a continuous measurement) is correlated with species abundance. Two definitions of "strength of density dependence" have been used. Observational studies have looked at time series data and fit equation (Equation 4.9) to a linear functional form (Figure 4.7A). The steeper the downwards slope of this line, the stronger the density dependence. As shown in Figure 4.7A the slope of the line is related

to the parameters r and K in the logistic growth equation (Equation 4.6). Intuitively, it seems that larger values of K (and, hence, larger equilibrium population sizes) should lead to shallower slopes and less density dependence (although one can devise scenarios where r and K covary in certain patterns, where this is not true). In experimental studies, the strength of density dependence has been measured by contrasting the amount of suppression to seedling growth or survival when plants are grown in soil taken from under a conspecific adult vs the amount of growth or survival when grown in soil from other sources (Figure 4.7B). More suppression in conspecific soil is equated with stronger density dependence. A negative correlation between the strength of density dependence and species abundance has been found in:

- observational studies in plants (figure 4.7C from Comita et al. 2010; figure 4.7D from LaManna et al. 2017);
- a meta-analysis of a diverse collection of organisms (Yenni et al. 2017);
- experimental studies (Klironomos 2002; Mangan et al. 2010).

These results suggest stronger density dependence in rarer species—the opposite of Henderson and Magurran (2014). The resolution would appear to be that most of these studies have either implicitly (by studying regularly present species) or explicitly removed the transient species, and only studied common and non-transient rare species (e.g. Yenni et al. 2017 removed over 80% of the species as transient or otherwise unsuitable). Thus, it appears that, when comparing common to transient rare species, the common species experience more density dependence, but when comparing common species to permanent rare species the common species experience less density dependence.

The idea that common species experience different amounts of density dependence than rare species has important implications. If rare species experience stronger density dependence, then this provides an explanation for how so many rare species can avoid stochastic extinction (Yenni et al. 2012; Chisholm & Muller-Landau 2011). While it is true that strong density dependence prevents a species from becoming very abundant, it also means the population

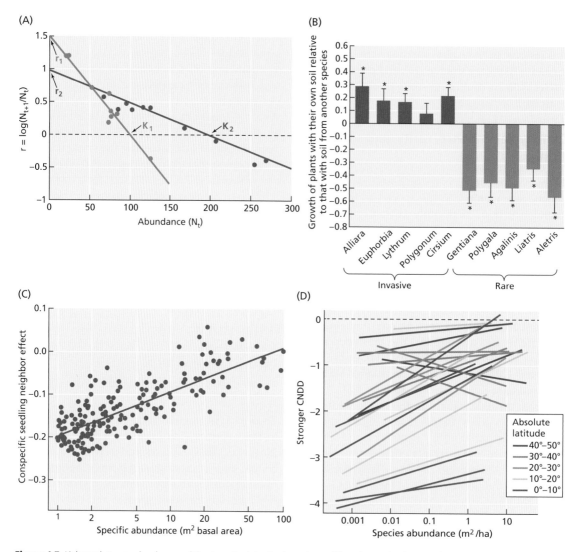

Figure 4.7 Linkages between abundance and the strength of density dependence. (A) an observational approach to assessing strength of density dependence. Plot annual population growth rates vs. population size and fit a line (or Ricker curve). The steepness of the line is the strength of density dependence. Note that a linkage between equilibrium abundance, K, and the slope of the line is likely to be negatively correlated unless there are some very particular forms of variation in r. (B) an experimental approach. Show the relative increase or decline in growth (or survival) for a target seedling planted in soil taken from under a conspecific adult vs. growth of a target seedling taken form a different soil source (Klironomos 2002). (C) plot the strength of density dependence vs abundance (Comita et al. 2010). (D) Each line represents one community. Each line is a plot of strength of density dependence vs relative abundance within that community. Different lines are for different forests. The color of the line shows the latitude of that forest (LaManna et al. 2017).

growth rate increases very quickly when the population declines even a small amount. Indeed, the notion of the strength of density dependence relates directly to the notion of stabilization mechanisms in coexistence theory (Chapter 8). Invasive species also may be successful invaders because they experience weaker negative (or even positive) density dependence (Klironomos 2002, Figure 4.7B). The fact that density dependence is stronger for rare species has been invoked as a general mechanisms promoting species richness, as any mechanism that allows rare species to persist is likely to increase species richness (since most species are rare). The average (across species within a community) strength of density dependence is stronger in forests closer to the equator (Figure 4.7D) and also in forests with more species richness (LaManna et al. 2017; but see criticisms by Chishom and Fung 2018; Hülsmann and Hartig 2018). Moreover, in the tropics rare species experience much stronger density dependence than common species, but this difference between rare and common disappears at higher latitudes with common species experiencing as much or more density dependence than rare species (Figure 4.7D). Overall, common temperate species experience the least density dependence. A parallel result was found in a single forest in Missouri along resource availability gradients (LaManna et al. 2016). Density dependence was stronger in resource-rich environments. This link between density dependence and species richness is exciting, but caution is needed. The precise mechanisms explaining the links are not clear and most of the studies (except Klironomos 2002) have been done with juvenile stages of trees. As Figure 4.7D shows the nature of these relationships is complex. In the Missouri forest, there was a complex 4-way interaction between ontogenetic stage, common vs rare, resource level, and strength of density dependence.

Conclusion

Understanding the factors that regulate population growth is fundamental to understanding all aspects of ecology. This chapter examines how simple mathematical models of population growth (the exponential and logistic equations) may be derived from basic principles, and we discussed the significance of density dependence to population regulation and the persistence of species. We will build on this foundation as we move on to study species interactions and the mechanisms that underlie predator–prey dynamics, interspecific competition, and mutualism/facilitation. The debates over whether density dependence is necessary for population regulation and whether density dependence is common in nature have been largely settled—affirmative on both accounts. However, important questions remain regarding the factors that contribute most to density

dependence (i.e., top-down vs bottom-up forces), how and when these factors operate, and what form density dependence may take under different environmental conditions and with different types of species interactions (e.g., Abrams 2009a,b, 2010). The question of whether or not there is regulation at the community level has been less studied. Initial results suggest it exists, but the more special case of compensatory dynamics is rare. This may be due to species turnover happening simultaneously with changes in species abundance. There are a growing number of studies finding strong linkages between the strength of density dependence, and the abundance of species and the richness of communities.

Summary

1. In exponential growth, a population's size increases without bound at an instantaneous rate of increase, r. Exponential growth can occur when resources are unlimited. As a population increases, however, its per capita growth rate will begin to decrease due to the depletion of resources or other negative effects of high population density. This negative feedback is referred to as density dependence.

2. The logistic growth equation models the linear effects of density dependence on population size. It predicts a sigmoidal pattern of population growth, with population size reaching a stable maximum at K, the carrying capacity.

3. The most common approach to detecting density dependence in natural populations is to examine how r varies with population density in a long time series of data. Experimental studies of density-dependence alter a population's density and then look at changes in birth rate (b) and/or death rate (d).

4. A number of factors, such as the inability to find a mate at low population densities, may result in positive density dependence. These factors, whose results are commonly referred to as Allee effects, are particularly important at very low densities.

5. "Component Allee effects," in which higher density has a positive impact on some component of fitness, are relatively common in nature,

but the overall demographic consequences of high density for population growth are largely negative.

6. Density dependence at the community level is referred to as community-level regulation. Community-level regulation of both species richness and total abundance seems to be fairly common, but studies looking for compensatory dynamics (negative covariances of species) have usually found the opposite—positive covariances presumably driven by species responding similarly to environmental fluctuations.

7. Examining the strength of density dependence has provided linkages to the relative abundance of species and species richness of a community. When all species are included, common species show density dependence and rare species are too transitory to show density dependence. When only permanently resident rare species are studied, we often (but not always) find a stronger density dependence in rare species. This result has important linkages to most areas of community ecology, although more work is needed to determine its generality.

CHAPTER 5

The fundamentals of predator–prey interactions

… prey–predator systems, be they plants and herbivores, animal prey and four- or six-legged predators, insects and parasitoids… all such systems share certain general properties…

Robert May and Charlotte Watts, 2009, p. 431

Models are tools to understand the real world and…they can sharpen our intuition about ecological mechanisms. Even the simplest model, which may match no living system, can be useful.

William Murdoch et al., 2003, p. 3

In any community, species exist within a network of other potentially interacting species. Each species in this network consumes resources and is itself consumed by other species. The **consumer–resource link** is the fundamental building block from which increasingly complex food chains and food webs may be constructed (Figure 5.1; Murdoch et al. 2003). In this and the following four chapters, we will focus on the consumer–resource link to develop a mechanistic understanding of species interactions, including predation, interspecific competition, and mutualism/facilitation. Here, we start with what is arguably the simplest consumer–resource interaction—one predator species feeding on one species of prey (Figure 5.2A). As the quote from May and Watts (2009) above suggests, we can define "predators" and "prey" broadly to include herbivores and plants, parasites and hosts, as well as more "traditional" notions of predators and prey. From this simple beginning, we will move on (in Chapter 6) to consider predators feeding on multiple prey species and how prey adaptations and behavioral responses may affect the predator–prey interaction (Figure 5.2B).

Next, we will examine consumer–resource interactions from the perspective of competition (Chapters 7 and 8) and mutualism and facilitation (Chapter 9), after which we will put the pieces together by exploring interactions between multiple consumers and multiple resources using the constructs of ecological networks and food webs (Chapters 10 and 11). Each increase in complexity allows us to consider additional properties of multispecies consumer–resource interactions (Holt 1996).

In this chapter, we encounter two important concepts that should be part of every ecologist's toolkit for the study of species interactions. The first is the Lotka–Volterra model (here formulated for predator and prey); the second is isocline analysis. The Lotka–Volterra model for two interacting species is the simplest possible consumer–resource model, and ecologists have had a love–hate relationship with it for a long time. Its critics would toss it in the dustbin of ecology as overly simplistic, outdated, and just plain wrong, but its supporters see it as a useful heuristic tool for gaining insight into the processes that govern the outcome of species interactions. Isocline

Community Ecology. Second Edition. Gary G. Mittelbach & Brian J. McGill, Oxford University Press (2019).
© Gary G. Mittelbach & Brian J. McGill (2019). DOI: 10.1093/oso/9780198835851.001.0001

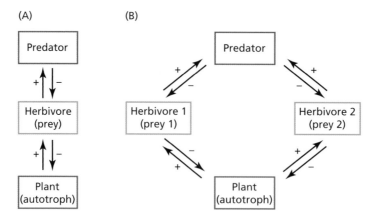

Figure 5.1 Simple consumer–resource interaction networks: a food chain (A) and a food web (B). Consumer–resource links (represented by arrows) are the building blocks of these networks. Plus signs next to the arrows indicate positive effects on consumer species from eating a resource species; minus signs indicate negative effects on the resource species being consumed.

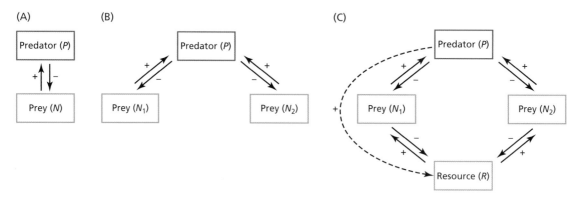

Figure 5.2 Predator–prey networks of increasing complexity. In this chapter, we consider the simplest predator–prey interaction: one predator species feeding on one prey species (A). In Chapter 6, we will add multiple prey species (B) and allow those prey species to compete for resources (C). The dashed arrow in (C) indicates an indirect positive interaction between trophic levels (discussed in Chapters 6, 10, and 11).

analysis is introduced here as a means to visually interpret the outcomes of the Lotka–Volterra model. The usefulness of isocline analysis, however, extends well beyond the analysis of Lotka–Volterra type models, and we will make use of the technique again in considering consumer–resource models of inter-specific competition and in the study of food webs.

This chapter is different from most others in this book in that it focuses almost exclusively on developing and analyzing simple mathematical models, with very few examples from nature. We have de-emphasized empirical examples here because we want to move quickly beyond the study of a single predator species feeding on a single species of prey (which is a relatively rare interaction in nature). However, the study of such a simple predator–prey system is a useful place to start our exploration of the nitty-gritty of species interactions. In the next chapter, we will examine more complex and realistic interactions between predators and their prey, including many empirical examples that tie theoretical concepts of predation to the natural world.

Predator functional responses

A fundamental property of any predator–prey interaction is the predator's **feeding rate** (the number of prey consumed per predator per unit time). This rate is expected to increase as prey density increases; however, the relationship may take different forms. For example, at low prey densities, a predator's feeding rate should be directly proportional to prey density (Turchin 2003), such that

$$\text{feeding rate} = aN \qquad \text{[Eqn 5.1]}$$

where N is the population density of the prey and a is a constant of proportionality, known as the predator's per capita attack rate. However, at high prey densities, a predator's feeding rate should reach some maximum set by the time required to capture and consume an individual prey item—referred to as that prey's **handling time**. Holling (1959) termed the relationship between prey density and predator feeding rate the predator's **functional response**, and he classified predator functional responses into three basic types (Figure 5.3).

The **type I functional response** presumes that a predator's feeding rate (measured in units of number of prey consumed per individual predator per unit time) increases linearly with prey density (Equation 5.1; Figure 5.3A). Although mathematically simple, the type I functional response is unrealistic for most predators, as it assumes that a predator can consume a constant fraction of the prey population per unit time regardless of prey density. Consequently, the per capita death rate of the prey does not vary with its own density (Figure 5.3D). Some filter-feeding organisms, such as zooplankton may fit this relationship over a range of prey densities (Jeschke et al. 2004), but the consumption rate of most predators will approach a maximum set by the prey's handling time. A maximum feeding rate could be incorporated into the type I functional response by allowing feeding rate to increase linearly with prey density up to a point where it abruptly becomes

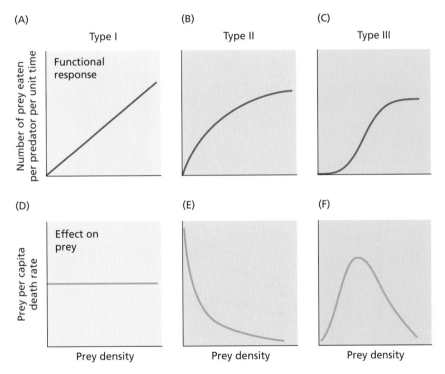

Figure 5.3 The predator's functional response and its impact on the prey's per capita death rate. The top row (A–C) shows the three types of functional responses described by Holling (1959). In each case, the number of prey eaten per predator per unit time increases with increasing prey density. The increase in predation rate with prey density is linear for the type I response, saturating for the type II response, and sigmoidal for the type III response. The bottom row (D–F) shows how the per capita death rate of the prey population changes with prey density under each type of predator functional response.

constant (Holling 1959). However, a more realistic scheme is to model the predator's functional response as a curve that smoothly approaches an asymptote.

Holling's **type II functional response** (Figure 5.3B) describes such a relationship, where

$$\text{feeding rate} = \frac{aN}{1 + ahN} \qquad \text{[Eqn 5.2]}$$

and where h is the handling time (see Gotelli 2008 pp. 136–137 for a simple derivation of Equation 5.2). The type II functional response has been shown to fit the feeding rates of a wide variety of predators (Figure 5.4; Hassell 1978; Jeschke et al. 2004), and it is to be expected whenever a single predator is foraging without interference for a single type of prey (Turchin 2003). Note that, at very low prey densities, the ahN term in the denominator is small and the feeding rate approximates a type I functional response. At very high prey densities, the feeding rate approaches a maximum rate equal to the inverse of the prey's handling time ($1/h$). Note, too, that under a type II functional response, an individual prey's

probability of death declines as prey density increases (Figure 5.3E). It is not important that the type II functional response fit Equation 5.2 exactly; rather, the defining property of the type II functional response is that it increases at a decreasing rate (see Abrams 1990 for examples).

Holling characterized the **type III functional response** as a sigmoidal relationship (Figure 5.3C), in which a predator's feeding rate accelerates over an initial increase in prey density but then decelerates at higher prey densities. A useful formulation for the type III functional response is

$$\text{feeding rate} = \frac{cN^2}{d^2 + N^2} \qquad \text{[Eqn 5.3]}$$

where $c = h^{-1}$, the maximum feeding rate, and $d = (ah)^{-1}$, the half-saturation constant (the prey density at which the feeding rate is half the maximum; see Turchin 2003 for alternative formulations and further discussion of the functional response). With a type III functional response, the prey's per capita death rate peaks at an intermediate prey density (Figure 5.3F).

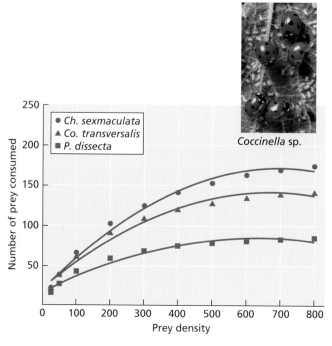

Figure 5.4 The number of aphids consumed in 24 hours by three species of adult ladybird beetles (*Cheilomenes sexmaculata, Coccinella transversalis,* and *Propylea dissecta*) feeding at different aphid densities in the laboratory. All three beetle species exemplify the type II functional response. After Pervez and Omkar 2005; photograph © Dr J. Burgess/Photo Researchers, Inc.

Holling (1959) suggested that the type III functional response might best apply to vertebrate predators (which might develop a search image for abundant prey), whereas invertebrates might best fit type II. However, it is now thought that the tendency of predators to fit either the type II or type III functional response is more a function of foraging mode. Predators feeding on a single prey type (specialists) commonly exhibit a type II functional response, whereas predators that switch between feeding on different prey types (generalists) tend to exhibit a type III functional response. Predators may also have a type III functional response if they adjust their foraging effort adaptively because it is costly to look for prey when there are very few around (see Abrams 1982, for theory, and Sarnelle and Wilson 2008, for a likely example). Because the current focus is on the interaction between a predator and a single prey species, discussion of prey switching and the type III functional response will be deferred to Chapter 6, where the interactions between selective predators and multiple prey will be considered.

In all the formulations of the functional response mentioned above, a predator's feeding rate is determined only by the density of its prey, and feeding rates are unaffected by the number of other predators present (except as mediated through changes in prey density). In other words, these formulations are **prey-dependent**. However, in some situations, predators may interfere with one another's ability to capture prey, or they may facilitate one another's success by group foraging. In such cases, the formulation of the functional response should include the density of predators, as well as the density of prey. A simple way to handle this requirement is to divide the number of prey available by the number of predators (P) and substitute the quantity N/P for N in the above formulations of the functional response (Equations 5.1–5.3). There is a sizable literature on this form of **ratio-dependent predation** (e.g., Akçakaya et al. 1995; Arditi and Ginzburg 2012), and there has been considerable debate as well about which form of the functional response (prey-dependent or ratio-dependent) provides the best starting point to construct simple predator–prey models (Arditi and Ginzburg 2012; Abrams 2015). Most predator–prey models incorporate prey-dependent processes,

perhaps because prey density must affect the functional response, whereas predator dependence may or may not be important. Here, the traditional prey-dependent formulations (Equations 5.1–5.3) will be used to examine the factors affecting the abundance and stability of predator–prey populations.

The Lotka–Volterra model

The simplest model describing the population dynamics of a single predator species feeding on a single prey species was developed independently in the 1920s by Alfred Lotka (1925) and Vito Volterra (1926). In the **Lotka–Volterra model**, the prey population (N) is assumed to grow exponentially in the absence of predators:

$$dN/dt = rN \qquad \text{[Eqn 5.4]}$$

where N is the number of individuals in the prey population and r is the prey's per capita population growth rate. If the predator has a linear (type I) functional response (i.e., the number of prey killed is directly proportional to prey density), and if predation is the only source of mortality for the prey population, then the net rate of change in prey numbers is

$$dN/dt = rN - aNP \qquad \text{[Eqn 5.5]}$$

Note that the units for a, the predator's per capita attack rate, are the number of prey eaten per prey per predator per unit time, and that a corresponds to the slope of the predator's linear (type I) functional response (see Figure 5.3A).

In the absence of prey, the predator population is assumed to decline exponentially as

$$dP/dt = -qP \qquad \text{[Eqn 5.6]}$$

where q is the predator's per capita mortality rate (which is density-independent); the units for q are the number of deaths per predator per unit time. Growth of the predator population depends on the number of prey consumed per unit time, which is equal to aNP, and on the predator's efficiency at turning prey eaten into new predators, which is assumed to be a simple constant (f). Thus, the birth rate of the predator population is equal to $faNP$, and the dynamics of the predator population is described by

$$dP/dt = faNP - qP \qquad \text{[Eqn 5.7]}$$

Together, Equations 5.5 and 5.7 make up the Lotka–Volterra predator–prey model.

Isocline analysis

The properties of the Lotka–Volterra model can be investigated by determining the zero growth isocline for each population. **Zero growth isoclines** define conditions in which a population is neither growing nor declining. For a prey population whose dynamics are governed by Equation 5.5,

$$dN/dt = 0 \text{ when } rN = aNP$$

Thus, the prey population will be at equilibrium (neither growing nor declining) when the density of predators P equals the quantity r/a. Likewise, for the predator population,

$$dP/dt = 0 \text{ when } faNP = qP$$

and the predator population will be at equilibrium when the density of prey N equals q/fa.

It is useful to plot these zero growth isoclines on a graph where the x-axis is prey density (N) and the y-axis is predator density (P). Because q, f, a, and r are constants in the Lotka–Volterra model, the zero growth isoclines for the predator and the prey are straight lines. Looking first at the prey isocline (Figure 5.5A), it can be seen that at predator densities less than r/a, the prey population increases, and at predator densities greater than r/a, the prey population decreases. Thus, a specific number of predators (equal to r/a) will maintain the prey population at equilibrium (zero growth). Looking next at the predator isocline (Figure 5.5B), it can be seen that at prey densities less than q/fa, the predator population decreases, and at prey densities greater than q/fa, the predator population increases. Thus, a specific number of prey (equal to q/fa) will maintain the predator population at equilibrium. Figure 5.5 illustrates an important, but potentially confusing point—it is the number of predators present that determines whether the prey population is at equilibrium, $dN/dt = 0$. Likewise, it is the number of prey present that determines whether or not the predator population is at equilibrium, $dP/dt = 0$.

Next, let's plot the predator and the prey isoclines on the same graph (Figure 5.5C). Note that the two isoclines intersect at a single point; only at this

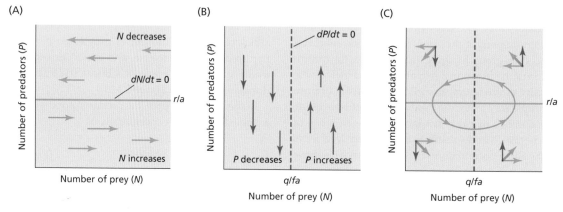

Figure 5.5 Zero growth isoclines for the Lotka–Volterra predator–prey model. (A) The prey's zero growth isocline is a horizontal line. The prey population will increase at predator densities below the line and decrease at predator densities above the line. At a predator density of r/a, the prey population will not change ($dN/dt = 0$). (B) The predator's zero growth isocline is a vertical line. The predator population will increase at prey densities to the right of the line and decrease at prey densities to the left of the line. At a prey density q/fa, the predator population will not change ($dP/dt = 0$). (C) Predator and prey zero growth isoclines plotted on the same graph. The intersection of the two isoclines defines four regions in predator–prey state space, and the vectors (narrow-lined arrows) show the response of the predator and prey populations in each region. The heavy diagonal arrows show the combined trajectories of the predator and prey populations in each region (i.e., the resultant vectors). Over time, the predator and prey populations will cycle continuously, as illustrated by the ellipse. Visit http://communityecologybook.org/predprey.html to see a dynamic version of this model.

equilibrium point will prey and predator populations remain constant. Away from the intersection of the predator and prey isoclines, the numbers of predator and prey will change over time. How they change is illustrated by the arrows (vectors of population change) in each of the four quadrants in Figure 5.5C. For example, in the upper right-hand quadrant, both prey and predators are abundant. The predator population increases in this region because prey numbers are above those needed to support zero growth in the predator population (r/a). In this same region, however, the prey population is declining because predator abundance is above that required to hold the prey population in check (q/fa). As a result, the joint trajectory of the two populations (illustrated by the heavy arrow) will move the system (over time) into the upper left-hand quadrant. In the upper left-hand quadrant, both predator and prey numbers will decline. A little reflection on the projected dynamics of predator and prey in each quadrant of Figure 5.5C should convince you that the population sizes will track counter clockwise in state space (i.e., all possible abundances of predators and prey) along an approximate ellipse.

How does this tracking in state space in Figure 5.5C correspond to changes in the population growth curves for predator and prey over time? The answer can be seen in Figure 5.6A. Both predator and prey populations cycle smoothly in abundance over time, with the amplitude and period (time required for a full oscillation) remaining unchanged. The predator population always lags behind the prey population by about one-quarter of a cycle, whereas the amplitude of the cycles is determined by the initial population densities (densities that start farther from the equilibrium point result in cycles of larger amplitude).

The Lotka–Volterra model is neutrally stable. That is, its predator and prey populations will cycle indefinitely, with the same amplitude and period; however, if the system is perturbed in any way (e.g., something comes along to change predator or prey numbers), that perturbation will cause the system to cycle with a similar period but a new amplitude determined by the size of the perturbation (Figure 5.6B,C). A model displaying such initial condition-dependent oscillations cannot be a good representation of the real world, where perturbations are common. Moreover, as Murdoch et al. (2003) note,

(A)

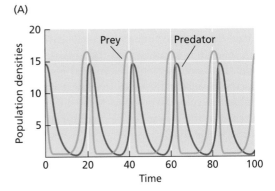

(B)

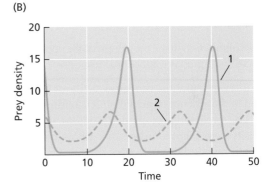

(C)

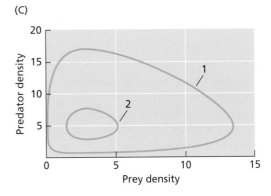

Figure 5.6 (A) In the Lotka–Volterra model, predator–prey population dynamics show continuous cycles over time. The predator population lags one-fourth of a cycle behind the prey population. (B) An external perturbation to the system (here, a change in the initial densities of predator and prey, but no change in model parameter values) results in a change in the amplitude of the cycles (only the prey cycles are shown here for clarity). The prey densities from Figure 5.6A are shown (1) along with the densities resulting from a change in initial densities (2). (C) Prey densities plotted in state space (i.e., the density of prey at time *t* plotted against the density of predators at time *t*) as predicted by the Lotka–Volterra model, with different starting densities (1, 2) but the same parameter values. After Murdoch et al. (2003).

the Lotka–Volterra model is "structurally unstable," and almost any change in the model's structure changes the predicted dynamics. Such sensitivity is often leveled as a criticism at the Lotka–Volterra model. However, sensitivity can also be a useful property because it allows us to explore how adding different ecological processes to the simple model will change the dynamics of predator and prey (Murdoch et al. 2003).

Adding more realism to the Lotka–Volterra model

Notably absent from the simple Lotka–Volterra model are:

- self-limitation in the prey or predator population;
- a realistic predator functional response.

No prey population can grow without bounds, even in the absence of predation. Therefore, a more realistic predator–prey model needs to include a density-dependent term specifying how the prey's growth rate declines with prey density. One possibility [originally proposed by Volterra (1931)] is to model prey growth in the absence of predation using the logistic equation (see Chapter 4). In this case, the prey's equation becomes

$$dN/dt = rN\left(1 - \frac{N}{K}\right) - aNP \qquad [\text{Eqn 5.8}]$$

In Equation 5.8, there are now two factors regulating the growth of the prey population—predator density and prey density. The effect of adding prey self-limitation to the model is to rotate the prey zero growth isocline in a clockwise direction (Figure 5.7A) so that it has a negative slope everywhere. Intuitively, this makes sense; as prey density increases, it takes fewer predators to keep the prey population in check (zero predators are required when the prey population is at K).

Adding self-limitation to the prey population dramatically changes the stability of the predator–prey interaction. The intersection of the predator and prey isoclines in Figure 5.7A defines an equilibrium point (where $dN/dt = 0$ and $dP/dt = 0$), just as it did in Figure 5.5C. However, if predator or prey densities are perturbed, the populations will recover and return to the equilibrium point (population oscillations will damp out over time; Figure 5.7B).

(A)

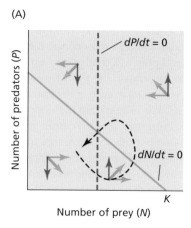

(B)

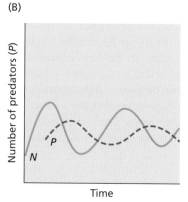

Figure 5.7 Predator–prey isoclines (A) and predator–prey dynamics (B) for a model of the Lotka–Volterra type when prey self-limitation (logistic growth) is included. Visit http://communityecologybook.org/predprey.html to see an online dynamic version of this model.

Thus, self-limitation in the prey population acts to stabilize the predator–prey interaction.

The stability of the predator–prey equilibrium can be determined by formal stability analysis (see Murdoch et al. 2003, for a good introduction to local stability analysis). However, when the predator's per capita growth rate depends on prey density (and not on its own density), a simple rule of thumb can be applied to judge the stability of the equilibrium point marked by the intersection of predator and prey isoclines. If the predator isocline intersects the prey isocline on its negative slope (i.e., Figure 5.7A), the predator–prey equilibrium is stable. Intersections on a positively sloped prey isocline are unstable. This rule of thumb glosses over some important complexities, but it will suffice for our heuristic treatment of the factors that affect predator and prey dynamics in simple models.

The Rosenzweig–MacArthur model

The analyses presented above have assumed a linear (type I) predator functional response, where each predator consumes a constant fraction of the prey population per unit time. A type I functional response might be a good approximation at low prey densities, but it cannot be true at high prey densities. When prey are abundant, the predation rate will reach a maximum determined by handling time. Clearly, in order to build a more realistic predator–prey model, a more realistic functional response is required. Rosenzweig and MacArthur (1963) were the first to incorporate the more realistic type II

functional response into a predator–prey model that also includes self-limitation in the prey, where

$$dN/dt = rN\left(1 - \frac{N}{K}\right) - \frac{aNP}{1+ahN} \quad \text{[Eqn 5.9]}$$

$$dP/dt = \frac{faNP}{1+ahN} - qP \quad \text{[Eqn 5.10]}$$

This model may well be the simplest representation of an actual predator–prey interaction and it is often used in theoretical ecology (Turchin 2003).

In Rosenzweig and MacArthur's model, the prey zero growth isocline has a hump (Figure 5.8). It is important to understand why this is so. Recall that with a type II functional response, the prey's per capita death rate declines with each increase in prey density (see Figure 5.3). Therefore, as prey density

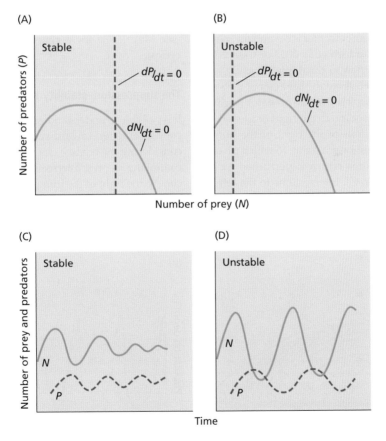

Figure 5.8 Predator and prey isoclines for the Rosenzweig–MacArthur model. Note that under this model the prey zero growth isocline is humped rather than the straight line of the Lotka–Volterra model seen in Figure 5.5A (see text for explanation). Left-hand panels illustrate stable predator–prey interactions; right-hand panels illustrate unstable interactions (limit cycles). Visit http://communityecologybook.org/predprey.html to see an online dynamic version of this model.

goes up, the number of predators required to keep the prey population in check must also increase. This relationship generates the initial ascending limb of the humped prey isocline. However, self-limitation in the prey population will eventually take over, and at the prey's carrying capacity (K), no predators are required to control the prey population—hence, the descending limb of the humped prey isocline. Thus, adding a bit more biological realism to the Lotka–Volterra model (i.e., a maximum predator feeding rate and self-limitation in the prey population) changes the shape of the prey isocline and (as will be seen later) allows us to better represent the observed dynamics of real predator–prey populations.

The Rosenzweig–MacArthur model exhibits a wider and more realistic range of dynamics than does the Lotka–Volterra model, including conditions in which predator and prey coexist at a stable equilibrium point, or in a stable limit cycle, or one or both populations go extinct. A formal analysis of the conditions leading to these outcomes is beyond the scope of this book (see Murdoch et al. 2003 or Case 2000 for a good introduction to this topic). Luckily, we can understand much about the behavior of the system simply by noting whether the predator isocline intersects the prey isocline to the right or the left of the hump. Intersections to the right of the hump result in a stable equilibrium point, to which predator and prey densities return if perturbed away from it (Figure 5.8A, C). Intersections to the left of the hump do not have a stable equilibrium point. Rather, predator and prey populations end up in a

limit cycle, in which predator and prey densities cycle endlessly (Figure 5.8B,D). It is important to note that limit cycles are a common feature of predator–prey models of this type (May 1973a) and that they are fundamentally different from the predator–prey cycles generated by neutral stability in the Lotka–Volterra model (Murdoch et al. 2003). A predator–prey system in a limit cycle responds to perturbation by returning (over time) to its original cycle (Figure 5.9), whereas a predator–prey system displaying neutral stability responds to perturbation with a change in the amplitude of the cycle (see Figure 5.6). If, however, it is assumed that some minimum population density for the persistence of predator or prey (due to finite population size), then systems displaying limit cycles may not recover from perturbation, which may lead to the extinction of the predator, or to the extinction of both predator and prey. Such outcomes are possible when predator and prey isoclines intersect well to the left of the hump (see May 1973a for a general discussion of limit cycles).

The suppression–stability trade-off

It is possible to understand what drives the predator–prey interaction toward stability or instability by considering the relative efficiency of the predator at controlling its prey. Intersections of the predator isocline to the right of the hump (Figure 5.8A) describe a situation in which the predator is relatively inefficient at consuming prey or turning prey into new

(A)

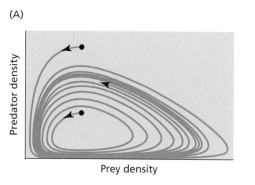

(B)

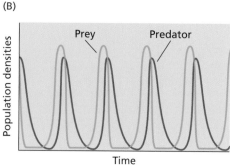

Figure 5.9 An example of a stable limit cycle for predator and prey. (A) In a stable limit cycle, the densities of predator and prey cycle perpetually in a counterclockwise direction (marked by the heavy closed loop), tracing out the densities over time shown in (B). If we perturb the populations, pushing them to a point either above or below the limit cycle (black dots in A), they return to the cycle. After Murdoch et al. (2003).

predators. In this case, the prey population exists at a high density relative to its carrying capacity and is predominantly self-limited. Intersections to the left of the hump (Figure 5.8) describe an efficient predator that holds its prey population at a low density relative to its carrying capacity. Thus, stability is more likely the closer the prey population is to its own carrying capacity and when the predator is relatively inefficient (and is, therefore, less likely to overexploit the prey population and drive it to low densities).

Rosenzweig (1971) recognized a surprising consequence of the fact that the predator–prey interaction becomes less stable the more responsible the predator is for controlling its prey. Presumably, we might expect that an increase in the prey's food supply (and, therefore, an increase in its carrying capacity, K) would benefit the prey population and help it better withstand the impact of predation. However, as Figure 5.10 shows, increasing the prey's carrying capacity may have exactly the opposite result. Increasing carrying capacity (K) may cause the predator isocline to intersect the prey isocline to the left of the hump, destabilizing the equilibrium and potentially causing one or both species to go extinct. Rosenzweig (1971) termed this result the **paradox of enrichment**, and he suggested that it could explain the loss of species in ecosystems experiencing increased nutrient input (e.g., eutrophication in aquatic ecosystems). Although the term "paradox of enrichment" is entrenched in the literature, Murdoch et al. (2003) suggest that this phenomenon might better be termed the **suppression–stability trade-off**, to emphasize the fact that the further a predator suppresses the prey equilibrium below the prey's carrying capacity, the more likely the system is to be unstable.

Density-dependent predators

Up to this point, we have modeled a predator population whose growth rate is dependent only on the density of its prey. In the Lotka–Volterra model (with its type I functional response),

$$dP/dt = faNP - qP$$

In the Rosenzweig–MacArthur model (with its type II functional response),

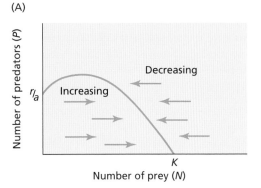

(A)

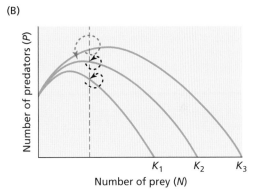

(B)

Figure 5.10 An illustration of the "paradox of enrichment" (also known as the "suppression-stability trade-off"; see Murdoch et al. (2003)) using the Rosenzweig–MacArthur predator–prey model. (A) The prey's hump-shaped zero growth isocline and regions of predator densities where the prey population is increasing or decreasing. (B) Intersection of the vertical predator isocline with a series of prey isoclines for three increasing values of prey carrying capacity (K_1, K_2, K_3). Increasing the prey's carrying capacity shifts the intersection of the predator isocline from the right to the left of the prey isocline's hump. Intersections to the right of the hump are stable, and perturbations to the system will return to a stable equilibrium point over time (black dashed arrows). Intersections to the left of the hump are unstable (red dashed arrow) and may result in the extinction of the predator or of both predator and prey.

$$dP/dt = \frac{faNP}{1+ahN} - qP$$

The predator isocline is a straight, vertical line in both models. We can see this by setting $dP/dt = 0$ and solving for N, which yields $N = q/fa$ and $N = q/fa - ahq$, respectively.

As Rosenzweig (1977, p. 373) notes, the concept of a vertical predator isocline can be confusing. "If the victim [prey] isocline has provided difficulties, the vertical predator isocline has been positively mystifying not only to students, but also to some theoretical ecologists." The key to understanding the vertical predator isocline comes from recognizing that:

slope in isoclines is generated only if the density of the species in question directly and instantaneously affects its own birth or death rates. The victim [prey] isocline slopes because more victims mean less resources per victim per instant, and less loss to the same number of predators per victim per instant. Are such things true about every predator? No. Some predators just collect food and ignore each other. At any instant, their rate of success depends only on the amount of food that exists. Of course, in the next instant, far less food will be present if there are 10^6 predators catching it, compared to 10^4. But that is taken care of by a change in *food density*. The larger population finds itself in the next instant with less food and may well decline. In order for the predator isocline to slope positively, some predators must actually encounter the same particle of resource at the same instant, so that at most but one gets fed. Or they must both need the same den or territory.

Predators that interfere with one another are expected to show direct density dependence (Arditi and Ginzburg 2012) and have a "bent" predator isocline that curves to the right (Figure 5.11). Predator interference can be defined broadly to include the energetic costs of defending a territory or chasing away other predators, as well as physically fighting with another predator over a food item. The energetic costs of interference result in a rightward bend in the predator isocline. A simple way to think about this is to recognize that, as predator density increases, each predator will expend more energy interacting with other predators. Therefore, each predator must encounter a higher density of prey to obtain enough energy to replace itself. Predator interference may also arise if the presence of more predators scares prey, so that prey spend more time in a refuge when more predators are present, thus effectively reducing prey abundance (Abrams 1982). An important outcome of including direct density dependence in the predator population is that it tends to stabilize the predator–prey interaction (i.e., the predator isocline is more likely to intersect the prey isocline to the right of the hump; see Figure 5.11).

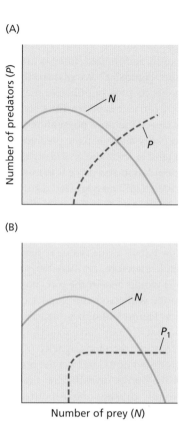

(A)

Number of predators (P)

(B)

Number of prey (N)

Figure 5.11 Self-limitation in the predator population results in a rightward bend in the predator isocline. (A) The type of bent predator isocline expected if the predator population exhibits interference competition. (B) A steeper bend in the predator isocline is expected if the predator population becomes limited by a second resource (e.g., nesting sites).

Herbivory and parasitism

As noted in the introduction to this chapter, predator–prey interactions can be viewed broadly as any relationship where one species consumes another species. **Herbivory** describes an interaction where a predator (e.g., a grazer) eats all or part of a plant. Examples of herbivory include sheep grazing grass, caterpillars eating leaves, birds foraging on berries, and aphids feeding on phloem. **Parasitism** describes an interaction where a parasite consumes all or part of its host (hosts are often both food and habitat for a parasite). Examples of parasitism include fleas on a dog (an ectoparasite), tapeworms in a mammal gut (an endoparasite), viruses and protozoans that cause disease (e.g., plasmodium causing malaria), plants that are parasites on other plants (mistletoe), and much more. A shared feature of herbivory and parasitism is that herbivores and parasites often do not kill their prey, but instead consume some fraction of the prey's biomass, reducing prey growth and fecundity. This partial consumption of prey influences how we model predatory–prey interactions for herbivores and parasites. Herbivory can be modeled using a mathematical framework very similar to that already discussed in this chapter, but with some minor modifications. Parasitism models, on the other hand, may take different forms.

Herbivory

Herbivores that feed on tiny plants (e.g., phytoplankton (algae) in freshwater and marine systems, or small annual plants on land) often consume the entire individual. In this case, the dynamics of the plant population can be represented by the number of plants present (as in the previous models in this chapter). However, when plants are larger and longer-lived (perennial), herbivores (grazers) may consume only a portion of a plant without killing it. In this case, we can model the plant population using plant biomass (V) instead of the number of individual plants, and herbivore abundance can be modeled as either the number of individuals or as herbivore biomass (H). For example, the Rozenzweig–MacArthur model for plant and herbivore populations could take the form

$$dV/dt = rV\left(1 - \frac{V}{K}\right) - \frac{aVH}{1 + ahV} \qquad \text{[Eqn 5.11]}$$

$$dH/dt = \frac{faVH}{1 + ahV} - qH \qquad \text{[Eqn 5.12]}$$

where "prey" and "predator" abundances are measured in biomass and the constants in the model are expressed as "per unit biomass," rather than "per capita". The herbivore's functional response can be modeled as either a Type I, Type II (as in Equation 5.12), or Type III; all three functional response types have been observed in nature (e.g., Lundberg 1988; Turchin and Batzli 2001; Sarnelle and Wilson 2008).

In general, grazing systems that follow the Rosenzweig–MacArthur model are predicted to show large amplitude oscillations between grazer and plant abundance, except in the most unproductive

systems (Noy-Meir 1975; Turchin 2003). Turchin and Batzli (2001) argued, however, that it may be unrealistic to model plant population growth with a logistic model in systems where herbivores consume only the above-ground portion of plant biomass. Instead, they proposed that, when plants regrow from energy stored in the roots, their growth function is better modeled as a hyperbolic curve, rather than a logistic curve (Figure 5.12). An important consequence of modeling plant population growth as regrowth, rather than logistic growth, is that is that it greatly stabilizes plant-grazer dynamics (Turchin and Batzli 2001). Intuitively, this increase in stability is due to the fact that logistic growth has an inherent time lag built in—the more vegetation is eaten, the longer it takes to grow back, whereas the regrowth function has no time lag.

Plants employ a variety of chemical and physical defenses against herbivores that work to reduce herbivore feeding, growth, survivorship, or fecundity (Feeny 1976; Coley 1983; Züst and Agrawal 2017). Plant defenses can be incorporated into standard predator–prey models (such as the Rosenzweig–MacArthur model) by modifying the herbivore's functional response, where the functional response equation (Type II) is changed to incorporate either:

- a decrease in herbivore feeding rate as a behavioral response to ingested toxins; or

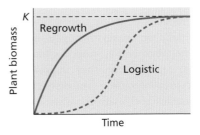

Figure 5.12 Two models for plant population growth. Logistic growth follows the equation $\frac{dV}{dt} = uV\frac{K-V}{K}$, where V is vegetation biomass (per unit area), u is the per capita rate of plant growth at V near zero, and K is vegetation carrying capacity. The regrowth model follows the equation $\frac{dV}{dt} = u\frac{K-V}{K}$, where u is the initial regrowth rate (at V near 0). The regrowth equation is often used to model nutrient dynamics in a laboratory chemostat and in theoretical treatments of species competing for abiotic resources. Figure and equations from Turchin and Batzli (2001).

- a negative effect of toxins on herbivore ingestion rate or growth (Feng et al. 2008; Feng and DeAngelis 2017).

The results of Feng et al.'s analysis demonstrate that plant defenses, and how predators respond to these defenses, can have both quantitative and qualitative effects on the outcome of the predator–prey interaction, including effects on stability and coexistence.

Plants may also enlist the help of mutualists to counteract herbivory. A classic example is the ant/acacia mutualism, where members of acacia tree family provide food and shelter to various species of ants in return for protection by the ants against insect and mammal herbivores (Janzen 1966; Palmer et al. 2008). Examples of these mutualistic defense mechanisms will be discussed in more detail in the chapter on beneficial interactions (Chapter 9). When attacked by herbivores, plants often increase their level of defense (referred to as inducible defense) above that expressed by the plant in the absence of herbivory (referred to as constituent defense). Likewise, animals show constituent and inducible defenses (e.g., longer spines, hardened shells) against their predators, as well as exhibiting antipredator behaviors. The consequences of these inducible defenses and antipredator behaviors will be explored in the next Chapter 6 and Chapter 11.

Parasitism and disease

The interactions between animal hosts and their viral, bacterial, protozoan (microparasites) infections or their helminth and arthropod parasites (macroparasites) represent another important type of predator–prey interaction. The distinction between microparasites and macroparasites is that microparasites multiply (increase in abundance) in their definitive host, whereas macroparasites do not (they need additional intermediate hosts). One can model host-pathogen dynamics using two different frameworks. One class of models, generally applied to macroparasites, is similar to the predator–prey models already discussed in that it keeps track of both resource (host) and consumer (parasite) abundances. The foundational host-parasite models were developed by Roger Anderson and Robert May in the 1970's and 1980's (Anderson and May 1978;

May and Anderson 1978). They showed that, in general, if the parasite has a more significant effect on host fecundity than on host survival, the system will have a propensity to oscillate (May and Anderson 1978). Additional factors tending to stabilized host-macroparasite interactions are parasite aggregation and host logistic growth (Rosà et al. 2006). The second type of host-pathogen models doesn't track pathogen numbers directly, but instead simply keeps track of the number of healthy vs infected hosts, and the mode and rate of transmission of the pathogen. This type of model is used in epidemiology to study diseases caused by microparasites such as viruses and bacteria. In Chapter 13, we highlight some of the parallels between epidemiology models and models of metapopulations; e.g., the importance of patch/host densities and dispersal/transmission rates in sustaining a population or a disease epidemic.

Turchin (2003) provides a nice discussion of the essential features of various predator–prey, herbivore–plant, and host–pathogen models, noting their similarities and differences. As Turchin notes, each of these consumer-resource systems is prone to oscillatory dynamics ("prey" and "predator" numbers tend to cycle through time) and stability is enhanced by the addition of density-dependent population growth in the "prey" or the "predator". One difference between microparasite models and standard predator–prey models (i.e., Rosenzweig–MacArthur) is that transmission rate in most microparasite models is often assumed to follow the equivalent of a linear Type I functional response. This will tend to make epidemic models more stable than predator-prey models containing a Type II functional response (Turchin 2003).

In the previous brief discussion, we have sought to emphasize the features shared in common when considering the dynamical properties of all predator–prey systems, be they herbivores and plants, parasites and hosts, or wolves and rabbits (i.e., ecology's more "traditional" notion of predators and prey). There are, however, unique aspects of each of these interactions that have important consequences for how ecologists develop more refined theories to describe their dynamics and predict their outcomes. Those interested in learning more about the details of modeling herbivory and parasitism will find a rich literature at their disposal (e.g. Crawley 2007). Now we need to move on from simple one predator–one prey systems to consider what happens when predators feed on multiple species of prey and when prey species may respond to predators by adaptively altering their behaviors or their phenotype.

Summary

1. The relationship between prey density and predator feeding rate is referred to as the predator's functional response. Predator functional responses can be classified into three basic types:
   - In a type I functional response, a predator's feeding rate increases linearly with prey density (unrealistic for most predators).
   - In a type II functional response, a predator's feeding rate increases with prey density, but at a steadily decreasing rate. This relationship can be modeled as a curve that approaches an asymptote.
   - A type III functional response is a sigmoidal relationship in which a predator's feeding rate accelerates over an initial increase in prey density, then decelerates at higher prey densities.
2. The Lotka–Volterra model is a simple model of the dynamics of a single predator species feeding on a single prey species. It assumes that the prey population will grow exponentially in the absence of predators; that the feeding rate of predators follows a type I functional response; and that growth of the predator population depends on the number of prey consumed per unit time.
3. The Lotka–Volterra model produces coupled oscillations in predator and prey populations. Adding self-limitation to the prey population (e.g., incorporating a carrying capacity K) acts to stabilize the system, which returns to a stable equilibrium point after being perturbed.
4. The Rosenzweig–MacArthur model of predator–prey dynamics incorporates the more realistic type II functional response, which causes the prey isocline to be humped. If the predator isocline intersects the prey isocline on the right of the hump, the intersection defines a stable equilibrium point, and the system will return to this equilibrium point following a perturbation. If the predator isocline intersects the prey isocline on the left of the hump, the system is unstable (predator and prey populations oscillate in a limit cycle); perturbations will cause the amplitude of the cycle to change, and one or both species may go extinct.
5. Predator–prey systems are more likely to be stable when the predator is relatively inefficient; this phenomenon is called the suppression–stability trade-off. Increasing the carrying capacity for prey may decrease the stability of the system, a phenomenon known as the paradox of enrichment.
6. Adding predator self-limitation (e.g., interference from other predators or limiting resources for predators) will cause the predator isocline to bend to the right and tends to stabilize the predator–prey system.
7. Herbivory can show most of the same dynamics as predator–prey systems, but in cases where the predator consumes only part of the vegetative structure of an individual plant, regrowth can happen quickly, which can help to stabilize dynamics.
8. Likewise, parasitism and diseases models often show dynamics similar to predator–prey systems. One important difference is that infections may not be fatal, leading to models more similar to metapopulation (patch colonization) models. Another important difference is that the "attack rate" is now an infection rate, which may not show handling time limitations and thus may be best modeled as Type I functional response.

CHAPTER 6

Selective predators and responsive prey

[M]ost carnivores do not confine themselves rigidly to one kind of prey; so that when their food of the moment becomes scarcer than a certain amount, the enemy no longer finds it worth while to purse this particular one and turns its attention to some other species instead.

Charles Elton, 1927. p. 122

The food of every carnivorous animal lies therefore between certain size limits, which depend partly on its own size and partly on other factors. There is an optimum size of food which is the one usually eaten and the limits actually possible are not usually realized in practice.

Charles Elton, 1927, p. 60

The previous chapter focused on the simplest possible predator–prey interaction—one predator species feeding on a single species of prey. Although this is a logical place to start, such extreme specialization is rare in nature (Williamson 1972; Schoener 1989). In reality, most predators feed on a variety of prey types. The fact that a predator may consume different prey species has important consequences, both from the perspective of the predator and from that of its prey. From the predator's point of view, there are several interesting questions. Are predators selective? Do they prefer some prey types while ignoring others? Does their preference change with prey density or prey behavior? How should we expect a predator's diet to change under different ecological conditions? From the prey's vantage point, selective predation has important consequences for mortality rates, thereby affecting prey behaviors, adaptations, and population dynamics, as well as the coexistence and diversity of prey species. This chapter explores some of these issues, starting from the predator's

point of view and asking why predators prefer some prey types over others.

Predator preference

A predator's preference for a given prey type can be defined as the difference between the proportion of that prey type in a predator's diet compared with the proportion of that prey type present in the environment. Thus, preference can be positive (a prey is selected for) or negative (a prey is selected against). It is useful to express predator preference in a way that allows comparisons between prey types, or between predators under different environmental conditions. Therefore, ecologists have developed a number of indices to measure preference (Manly 1985). The most widely used of these is the index first proposed by Manly (1974) and expanded on by Chesson (1978, 1983). The Manly–Chesson index calculates predator preference for prey type i as

Community Ecology. Second Edition. Gary G. Mittelbach & Brian J. McGill, Oxford University Press (2019).
© Gary G. Mittelbach & Brian J. McGill (2019). DOI: 10.1093/oso/9780198835851.001.0001

$$\alpha_i = \frac{d_i / N_i}{\sum_{j=1}^{k} \left(d_j / N_j \right)} \qquad \text{[Eqn 6.1]}$$

where $i = 1, 2, ..., k$, and where k is the number of prey categories, d_i is the number (or proportion) of prey of type i in the predator's diet, and N_i is the number (or proportion) of prey of type i in the environment. The index α_i ranges from 0 to 1. Prey types that are consumed in proportion to their abundance in the environment (i.e., no preference) have $\alpha_i = 1/k$; $\alpha_i > 1/k$ indicates positive preference for a prey type, and $\alpha_i < 1/k$ indicates negative preference for a prey type.

Why should predators prefer to eat some prey types and not others? We can approach this question by breaking down the act of predation into its component parts (Figure 6.1). Three main factors influence predator preference:

1. The probability that a prey item will be encountered (e.g., its "visibility" in the broadest sense).
2. The probability that an encountered prey item will be attacked (the predator's choice to go after a prey item or not).
3. The probability that an attacked prey item will be successfully captured and eaten.

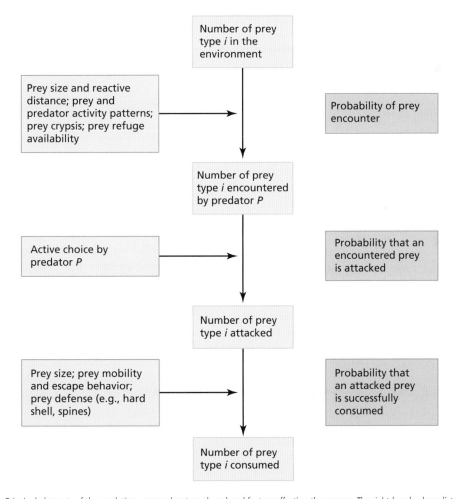

Figure 6.1 Principal elements of the predation process (center column) and factors affecting the process. The right-hand column lists the three main factors that determine the probability that a prey of type i will be consumed. Listed to the left are attributes of predators and prey that influence these probabilities. After Mittelbach (2002).

If a prey item is cryptic and hard to see, then we would expect it to be relatively rare in the predator's diet and to be selected against simply because it is "encountered" at a low rate. In general, factors that increase a prey's encounter rate or increase a predator's capture success should have a positive effect on predator preference.

What about active choice by the predator (see Figure 6.1)? Why do predators choose to go after some prey types and ignore others? Since the mid-1960s, ecologists have sought a general answer to this question by recognizing that natural selection should favor individuals that harvest food efficiently (Krebs 1978). Therefore, just as a functional morphologist might ask how natural selection has modified the jaw structure of an organism to maximize its crushing force, we can ask what set of foraging decisions should result in the most efficient energy capture. This approach to understanding diet choice has become known as **optimal foraging theory (OFT)**.

Optimal foraging theory leads to a model of predator diet choice

Consider a predator actively moving through its environment and encountering potential prey items (e.g., a bird hopping from branch to branch searching for insects). If prey are randomly distributed and encountered sequentially, and if the predator's decision to pursue and eat a prey item results in time spent handling prey that is unavailable for searching, then we can model a predator's rate of energy gain as follows.

Let E be the energy gained during a feeding period of length T, which is composed of time spent searching for prey (T_s) and time spent handling prey items (T_h). If the predator expends the same amount of energy during searching and during handling, then the predator's rate of energy gain (E/T) is

$$\text{rate of energy gain } E/T = \frac{E}{T_h + T_s}$$

Now let there be k prey types in the environment, with each prey type i having the following characteristics:

λ_i = number of prey of type i encountered per unit of search time (this means $\lambda_i T_s$ individual prey will be encountered for a given amount of search time)

p_i = probability that the predator will pursue and capture a prey of type i after it is encountered. (this

means $\lambda_i T_s P_i$ individual prey will be pursued for a given amount of search time)

E_i = energy gain from consuming one individual of prey type i

h_i = handling time for handling one individual of prey type i

The variable p_i is under the control of the predator and represents the predator's choice to attack a prey type or not. Given these properties, it follows that

$$E = \sum_i^k \lambda_i E_i T_s p_i,$$

and

$$T_h = \sum_i^k \lambda_i h_i T_s p_i,$$

then

$$E/T = \frac{\sum_i^k \lambda_i E_i T_s p_i}{T_s + \sum_i^k \lambda_i h_i T_s p_i}, \qquad \text{[Eqn 6.2]}$$

or

$$E/T = \frac{\sum_i^k \lambda_i E_i p_i}{1 + \sum_i^k \lambda_i h_i p_i}$$

Equation 6.2 may look familiar—it is of the same general form as Holling's equation for a type II predator functional response (Equation 5.2), except that Equation 6.2 includes multiple prey types, the energy gained from each prey of type i (E_i), and a parameter p_i representing the probability that a predator will pursue and capture a prey of type i.

Equation 6.2 is the standard **optimal diet model**. It was derived by several ecologists during the early 1970s (e.g., Schoener 1971; Emlen 1973; Maynard Smith 1974; Pulliam 1974; Werner and Hall 1974; Charnov 1976a) as they wrestled with the question of what rules of prey choice would yield the greatest energy gain per unit time spent foraging (i.e., the optimal diet). The model makes two general predictions and two more specific predictions about optimal diet choice. Its general predictions are that:

- foragers should prefer the most profitable prey (those that yield the most energy per unit handling time, E_i/h_i);
- an efficient forager should broaden its diet to include more low-value prey as the abundance of higher-value prey decreases (Stephens and Krebs 1986; Sih and Christensen 2001).

There is considerable evidence that predators prefer to feed on the most profitable prey when prey are abundant (Figure 6.2) and that predator diets narrow as the abundance of more profitable prey increases (Figure 6.3). Thus, the general predictions of optimal foraging theory are well supported (Sih and Christensen 2001).

The standard optimal diet model also makes two quite specific predictions about the nature of prey selection and how it should change with prey abundance. The first of these predictions is that prey types are either always eaten upon encounter ($p_i = 1$) or never eaten upon encounter ($p_i = 0$), which is known as the **zero-one rule** (Stephens and Krebs 1986). The

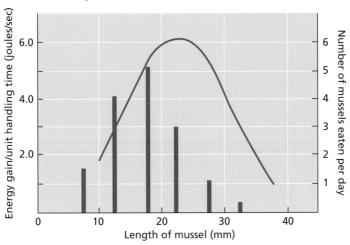

(A) Crabs feeding on mussels

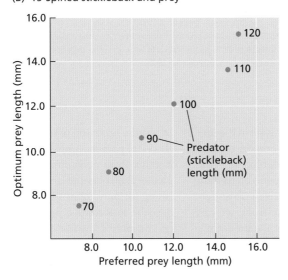

(B) 15-Spined stickleback and prey

Figure 6.2 Predators prefer to eat more profitable prey. (A) In a laboratory study of crabs feeding on mussels (Elner and Hughes 1978), prey profitability (energy gain versus handling time) peaked at intermediate mussel sizes (solid curved line), and crabs preferred to eat mussels of intermediate sizes when offered a choice (histograms). (B) The preferred size of prey consumed in the wild by a small fish, the 15-spined stickleback (*Spinachia spinachia*) plotted against the stickleback's optimal prey size, based on prey mass/handling time (see Kislaliogu and Gibson 1976). The number next to each data point denotes stickleback length; both optimal and preferred prey sizes increase with the predator's size. After Krebs (1978).

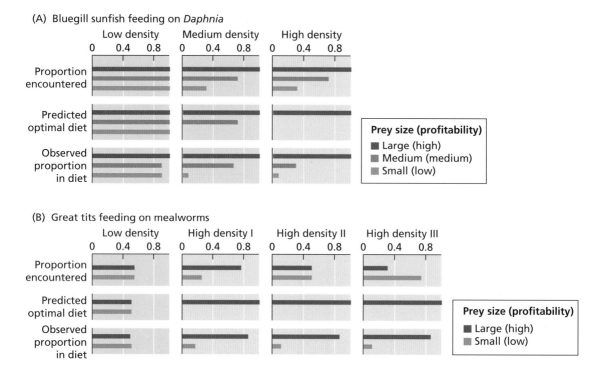

Figure 6.3 Two classic experiments tested the predictions of the standard optimal diet model. (A) Small bluegill sunfish (*Lepomis macrochirus*) in wading pools were offered a mixture of three size classes of zooplankton (*Daphnia*), and their diet choice was observed at low, medium, and high prey densities (data from Werner and Hall 1974). Ranking of prey profitability (*Daphnia* biomass/handling time) was large prey > medium prey > small prey. Histograms show the ratios of encounter rates with each *Daphnia* size class at the three prey densities, along with the predicted optimal diets and the observed bluegill diets. (B) For this experiment, Krebs et al. (1977) trained great tits (*Parus major*) to feed from a moving conveyor belt, onto which the researchers placed different-sized mealworms at different densities. The conveyor belt was then manipulated to adjust the rate at which birds encountered prey. Histograms show the proportion of large and small mealworm prey encountered; predicted to comprise the optimal diet; and observed to be eaten by the birds. After Krebs (1978).

second is that the inclusion of a prey type in the diet depends only on its profitability and on the characteristics of prey types of higher profitability (i.e., the inclusion of a prey type does not depend on its own encounter rate; Stephens and Krebs 1986). To understand where these predictions come from, we can use a simple algorithm to determine the predator's optimal diet. Following Charnov (1976a), we rank prey items from most profitable (highest E_i/h_i) to least profitable (Figure 6.4). We then add prey types to the diet in rank order (starting with the most profitable prey) until the rate of energy gain (E/T) is maximized (written as E^*/T^*). The optimal diet is the set of all prey types (i) whose value satisfies

$$E_i/h_i \geq E^*/T^* \qquad \text{[Eqn 6.3]}$$

In words, Equation 6.3 says that a predator should choose to pursue a prey item only if the rate of energy it gains from consuming that prey item (E_i/h_i) is greater than or equal to the rate of energy it gains by eating all prey items of higher rank. Note that the decision to include a prey type in the diet depends on the prey's profitability, but not on its rate of encounter, and that a prey item is either always eaten or always ignored—the two specific predictions of the optimal diet model. How well do foragers in nature match these two predictions?

Looking again at Figure 6.3, which summarizes the results of two classic tests of the optimal diet model, it can be seen that the foragers exhibited close, but not perfect, correspondence to the optimal diet model's predictions. In both studies, less profitable

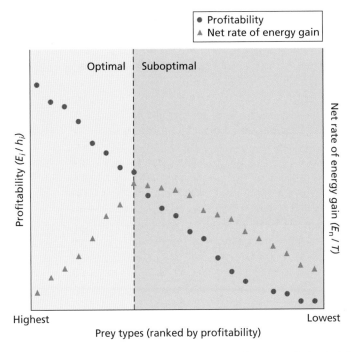

Figure 6.4 Charnov's graphic illustration of how to determine the predator's optimal diet from the standard optimal diet model (Eqn 6.2). Prey types are first ranked by their profitability (E_i/h); then the net rate of energy gain by the predator is calculated by adding prey types to the diet in rank order. The optimal diet is that set of prey types that maximizes the predator's total net rate of energy gain (E_n/T). This optimal prey set contains all prey types (red dots) whose profitabilities rank above the prey type at which E_n/T first becomes greater than E_i/h_i, denoted by the dashed line. After Charnov (1976a).

prey were dropped from the diet as the abundance of more profitable prey increased. In the case of great tits feeding on different-sized mealworms (see Figure 6.3B), increasing a bird's encounter rate with prey that were outside the optimal diet had only a minor impact on the bird's observed diet. However, in both studies, foragers included some "suboptimal" prey in their diets.

Only a few studies have attempted to test the optimal diet model in the field (e.g., Meire and Ervynck 1986; Richardson and Verbeek 1986; Cayford and Goss-Custard 1990) because of the difficulty of estimating rates of prey encounter. One of us (Mittelbach 1981) attempted such a test with bluegills in a small Michigan lake as part of their PhD research and found that bluegills preferred to eat large prey, and that their diets were similar to those predicted by the standard optimal diet model (Figure 6.5). Meire and Ervynck (1986) found a similar result for oystercatchers feeding on mussels on tidal flats in the Netherlands.

The examples given above—and nearly all other tests of the optimal diet model—have shown that

feeding efficiency and maximization of energy gain play important roles in determining predator diet choice. In their review of optimal foraging theory, Sih and Christensen (2001) concluded that 87% of studies (31 of 35) that included quantitative tests supported the predictions of the optimal diet model. Thus, optimal foraging theory has significantly increased our understanding of why some prey are selected and some ignored. However, in every study reviewed by Sih and Christensen, predators included some prey types in their diets that were not in the predicted optimal set (see also Pyke 1984; Stephens and Krebs 1986). Does this mean that optimal foraging theory is fundamentally flawed? Some would say yes, arguing that animals cannot possibility make the "optimal" choice (Gray 1986; Pierce and Ollason 1987). However, we know that some of the assumptions of the standard optimal diet model are unlikely to be true (i.e., the assumption of perfect predator knowledge of prey quality and prey density). Therefore, we would not expect the theory to make totally accurate predictions. What we need to know is how

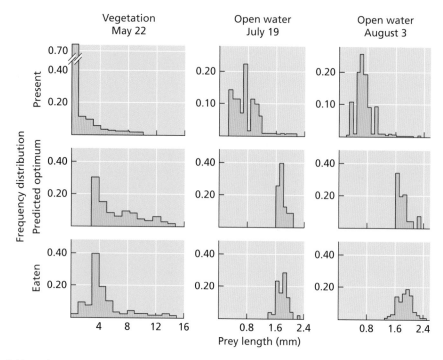

Figure 6.5 A field test of optimal diet predictions. The top row graphs the size and frequency distributions of prey found in samples from two habitats in a small Michigan lake (vegetation and open water); the predicted diets for bluegill sunfish (with an average body length of 125 mm) are in the center graphs; and the actual diets of bluegills feeding in the two habitats on the dates indicated are in the bottom row. The data support the bluegill preference for larger prey, as predicted by the optimal diet model. Note that in the vegetation habitat, the fish were able to find some larger-sized prey items than were sampled by the researcher. After Mittelbach (1981).

"imperfect knowledge" affects prey selection and what more realistic models of prey selection imply for the population dynamics of predator and prey. In particular, it seems sensible that if an organism has imperfect knowledge about handling times and energy of prey, occasional non-optimal sampling would occur to increase information, even if it slightly decreases efficiencies below the maximal.

We share Rosenzweig's (1995, p. xvii) puzzlement over the relative neglect of optimal foraging studies today:

Here is a fundamental research program (Mitchell and Valone 1990) originally suggested by no less than Charles Elton (1927). It is linking up behavior, natural selection and ecology. Yet, its critics, taking no time to understand its mathematical substance, and confusing 'optimal' with 'perfect,' consider it dead because…Well, I really don't know why they think it is dead. Students of optimality know that 'constraints' prevent perfect outcomes in the real world. But, 'constraints' are not foreign to optimality theory. They form one of its central features. Optimalists

know that, work with it, and have produced some… exciting ecology…by sticking to their rusty old guns.

There seems to be something of a recent uptick in publications on optimal foraging, suggesting that this discipline may have found new legs (e.g., Stephens et al. 2007; Calcagno et al. 2014; Garay et al. 2015). A revival of this important field linking behavior and ecology would be a welcome event.

The early developers of optimal foraging theory (MacArthur and Pianka 1966, along with Emlen 1966) envisioned that it could provide a tool to predict consumer diets, thereby providing a mechanism to better understand consumer–resource interactions and community dynamics. Unfortunately, this potential application of OFT to community ecology got lost in the rush to develop and test optimal foraging models, although a few studies (e.g., Gleeson and Wilson 1986; Fryxell and Lundberg 1994; Ma et al. 2003) showed how optimal foraging models could be used to explore predator and prey dynamics.

A more recent and exciting development is the application of OFT to predict the potential feeding links in size-based food webs (Beckerman et al. 2006; Petchey et al. 2008; Thierry et al. 2011). By modeling prey handling time as an increasing function of the ratio of prey size to predator size, and by assuming that prey energy value is a positive function of prey size, Petchey and colleagues (2008) were able to correctly predict up to 65% of the predator–prey links in four real-world food webs, based on the criteria that predators prefer to eat the most energetically rewarding prey. This application of OFT to predict who eats whom in nature comes close to achieving the goals envisioned by MacArthur and Pianka (1966) when they first introduced OFT.

The application of optimal foraging theory to food webs will be considered in more detail in Chapter 11. In the meantime, we should keep in mind two important insights gained from OFT, as we explore the interactions between predators and prey:

1. All else being equal, predators should prefer to eat the most profitable prey (those that provide the highest energy gain per handling time).
2. As the overall density of prey in the environment decreases, predator diets should broaden to include more prey types. Conversely, as the overall density of prey in the environment increases, predator diets should narrow (predators should become more specialized).

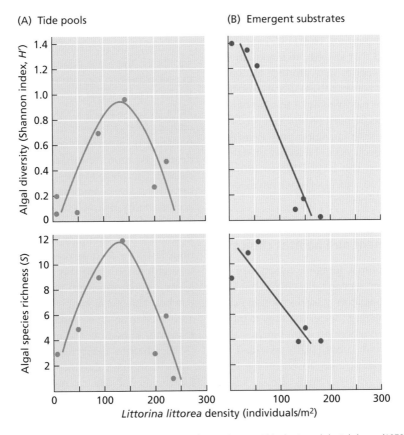

Figure 6.6 Selective predation affects the coexistence and diversity of competing prey. This classic study by Lubchenco (1978) illustrates how selective predation by the snail *Littorina littorea* may affect the diversity and species richness of algae growing along the rocky New England coast. Each point represents a single study site. (A) In tide pools, where the snails grazed preferentially on the competitively dominant algal species, both algal diversity (top) and algal species richness (bottom) peaked at intermediate snail densities. (B) On emergent substrates, where the snails grazed preferentially on the inferior algal competitors, increasing snail density had a negative effect on species richness and diversity. After Lubchenco (1978).

Consequences of selective predation for species coexistence

The fact that predators prefer to eat some prey and not others has important consequences for the structure of prey communities. For example, in a classic study of intertidal communities along the New England coast, Lubchenco (1978) showed that selective predation may increase prey diversity when predators feed preferentially on prey species that are competitively dominant, but that it decreases diversity when predation falls more heavily on inferior competitors (Figure 6.6). Leibold et al. (2017) provided a more recent example, showing that grazing by zooplankton lead to a doubling in phytoplankton (algae) species richness in experimental mesocosms. The additional phytoplankton species in grazed mesocosms were larger and, therefore, more grazer-resistant. Another example of how selective predation may influence the coexistence of prey species is **apparent competition**. Holt (1977) and Holt et al. (1994) showed that shared predation among *non-competing* prey species may produce mutually negative interactions between those species, and that the consequences for biodiversity of this "apparent competition" depend (in part) on whether predator preferences change with changing prey densities. These and other consequences of selective predation and shared predation will be considered in more detail in Chapters 7, 10, and 14, after we have examined interspecific competition.

Predator movement and habitat choice

The thinking of optimal foraging theory can be applied not only to understand diet choice, but also to understand how predators make decisions regarding where to feed (habitat choice) and how to move about in a landscape where prey are patchily distributed. Fretwell and Lucas (1970) provided an early model for habitat choice, based on the idea that (all else being equal) animals should select habitats that maximize their fitness (here, fitness is equated with maximizing feeding rate). In the simplest version of their model, termed the **ideal free distribution** (IFD), food is supplied to a foraging habitat or patch at a fixed rate, where it is consumed immediately. If predators are "free" to select among patches, have perfect or "ideal" knowledge of the rates of resource supply, and do not differ in their competitive abilities, then we expect the number of predators in a patch to be proportional to the total food input to the patch. This is called the "input-matching rule." A number of experimental studies show support for predictions of the IFD, including the classic study by Power (1984) of armored catfish (Loricariidae) feeding on periphyton (attached algae) in a stream in Panama. In Power's study, algal productivity differed among stream pools due to differences in canopy shading and the density of armored catfish in a pool was directly proportional to the percentage of open canopy over the stream (i.e., more fish in sunny, more productive pools). Moreover, algal abundance (standing crop), fish growth rates, and estimates of feeding rates were similar in both sunny and shaded pools, matching expectations of ideal free habitat selection based on consumer-resource dynamics (Lessels 1995; Oksanen et al. 1995).

Although the input-matching rule is elegant in its simplicity, there are many reasons why we might expect nature to deviate from the assumptions of the IFD. For example, IFD theory assumes that predators pay no cost in choosing habitats, hence, the epithet "free". However, in nature costs are real. One particularly important "cost" is the risk of injury or death (i.e., more rewarding habitats may also be riskier). The question of optimal habitat choice when organisms face a trade-off between foraging gain and mortality risk will be explored later, when we consider the non-consumptive effects of predators. Resources in a habitat may also be depleted over time, lowering feeding rates and causing predators to decide whether to stay in a particular patch or leave a patch in search of other, more rewarding patches. The question of whether to stay or leave is addressed in another fundamental model of optimal foraging, known as the **marginal value theorem** (MVT; Charnov 1976b; Stephens and Krebs 1986).

The marginal value theorem predicts how long a forager should stay in a resource patch before leaving in search of another patch. Two quantities influence this decision:

- the current rate of energy gain in a patch;
- the average rate of energy gain for the environment as a whole.

The MVT assumes that the rate of energy gain in a patch decreases over time as the predator depletes the number of prey in the patch, such that the curve of cumulative energy gain in a patch is decelerating (Figure 6.7A). At some point, declining energy gain within a patch dictates that a forager would be better off leaving and going to search for another patch where resource levels are higher. The question is, when is the optimal time to leave (here, we define the optimal strategy as maximizing the forager's expected rate of energy gain). If all patches are identical, the answer is quite simple and depends on the travel time between patches. By drawing a line from the average travel time between patches (point *a*) and intersecting the energy gain curve, the slope of this line defines the predator's expected rate of energy gain (i.e., energy gained in a patch divided by the sum of the time spent feeding in a patch plus the time travelling between patches). A little reflection makes it clear that the slope of this line will be greatest when the line just touches (is tangent to) the energy gain curve (as in Figure 6.7A). Any other intersection leads to a shallower slope and lower rate of energy gain. The point where the line is tangent to the curve defines the optimal time that a predator should stay in a patch before moving to another patch—any other leaving time would result in a lower rate of energy gain.

Figure 6.7B,C shows how changes in the average travel time between patches and changes in average patch quality (resource abundance) affect a predator's optimal time to leave a patch. As travel time between patches increases, predators should stay longer in a patch before leaving (Figure 6.7B). An increase in average patch quality in the environment will affect a predator's optimal residence time differently, depending on the form of the predator's functional response (Calcagno et al. 2014). Optimal leaving times should increase with increased patch quality for predators with a Type II (decelerating) functional response, but should decrease for predators with a Type III (accelerating) functional response (Figure 6.7C). When patches within a habitat differ in their resource abundance, we can no longer find the optimal solution using the graphical method shown in Figure 6.7A. However, the mathematical solution is simple enough; "The predator should leave the patch it is presently in when the marginal

capture rate in the patch [i.e., its current rate on energy gain] drops to the average capture rate for the habitat" (Charnov 1976b). If we draw lines whose slopes correspond to the average rate of energy gain across all patches in the habitat, then the point where these lines of equal slope are tangent to the energy gain curves for patches of different quality defines the optimal time to leave a patch. As Figure 6.7D illustrates, the time to leave a patch should be shorter in less-rewarding patches.

Note the parallels between the optimal diet and optimal patch-time models. Both focus on maximizing an average rate (i.e., slope or derivative) in units of energy per time. In each case the decision (to add a prey type or move to a new patch) is taken depending on how the rate gained from the current choice compares to the overall average. Add a prey if its energy/time efficiency is greater than the current diet efficiency. Leave a patch if the energy/time it provides falls below the overall landscape-wide average. These are what economists call marginal rates and they are a recurring theme in optimization problems.

Charnov's (1976b) optimal patch use model and the MVT have been extensively tested in laboratory and field experiments and as Ydenberg et al. (2007; p. 12) note, "The patch model may in fact be the most successful empirical model in behavioral ecology; its basic predictions have been widely confirmed …". Moreover, community ecologists have made good use of the insights provided by the MVT, particularly the observation that when a forager leaves a patch it reflects patch quality (e.g., Figure 6.7D). Krebs et al. (1974), Brown (1988), and others showed that the time between a predator's last prey capture in a patch and when it leaves the patch (its "giving-up time" (GUT)), and the density of prey remaining in patch when the predator leaves (the "giving-up density" (GUD)) are surrogate measures of the marginal capture rate in the patch. Thus, the easily measured quantities of GUT and GUD can be used to estimate the value of a habitat to a predator (Brown and Kotler 2004). Community ecologists have measured forager GUT's and GUD's in the laboratory and in the field to address fundamental questions related to habitat quality, interspecific competition and species coexistence (Kotler and Brown 2007). However, by far the most extensive application of these measures to

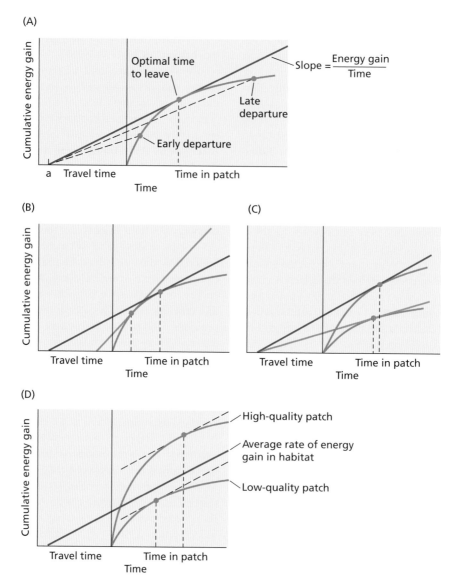

Figure 6.7 (A) A graphical model illustrating the optimal time for a predator to leave a patch based on Charnov's (1976b) MVT. Time (*x*-axis) is divided into time spent in a patch searching for (and consuming) prey and time spent traveling between patches. The predator's cumulative energy gain (*y*-axis) is expected to increase with time in a patch, but at a decreasing rate due to prey depletion. The optimal time to leave a patch can be found by drawing a line from point *a* (the average travel time between patches) that is tangent to the energy gain curve (solid-curved line). The slope of this line corresponds to the forager's maximum achievable rate of energy gain, being greater than any other line originating from point *a*, and intersecting the energy gain curve (e.g., the two dashed lines). (B) Graph showing how an increase in average travel time between patches increases the optimal time to remain in a patch. (C) Graph showing how greater patch quality within a habitat leads to an increase in the optimal time to remain in a patch (assuming a Type II functional response). (D) In this figure, there are two types of patches in a habitat; a high-quality patch and a low-quality patch (cumulative energy gain is greater in the high-quality patch than in the low-quality patch). According to the MVT, the optimal time to leave a patch is determined by the average rate of energy gain from all patches in the habitat (slope of solid line). The optimal leaving time for a patch corresponds to the point where a dashed line, whose slope is equal to the average rate of energy gain for the habitat, is tangent to the patch's energy gain curve. As panel (D) shows, predators should forage longer in high-quality patches.

community ecology has been to study the non-lethal effects of predators on their prey or, as some have more colorfully termed it, "the ecology of fear."

The non-consumptive effects of predators

Prior to 1980, most ecological theory viewed predator–prey interactions from the simple perspective of **consumptive effects** (i.e., predators kill and eat their prey). However, we now know that this view is incomplete. In recent years, ecologists have amassed a wealth of evidence showing that prey may respond to the threat of predation by changing their behaviors, morphologies, physiologies, and/or life histories. These non-lethal, or **non-consumptive effects** of predators may act in concert with the direct consumption of prey to influence prey abundance and predator–prey dynamics (Werner and Peacor 2003). Furthermore, as noted in Chapter 5, flexible antipredator traits (e.g., increased vigilance or refuge use in the presence of more predators) can make the predator's functional response predator-dependent; that is, the presence of more predators causes prey to be less "available," leading to a reduction in each predator's feeding rate. In this way, flexible antipredator traits may promote the stability of predator–prey interactions (Abrams 1982).

An excellent example of how inducible defenses may change the form of the predator's functional response is provided by Hammill et al. (2010), who studied predation on the protozoan *Paramecium aurelia* by the flatworm *Stenostomum virginianum*. In the presence of the predator, *Paramecium* reduced their average swimming speed and increased their body width (Figure 6.8A, B). These inducible defenses had a marked effect on the predator's attack rate and handling time, causing the predator's functional response to change from a type II (for undefended prey) to a type III (for defended prey). This change was brought about by low attack rates at low prey densities. Simulations using a Lotka–Volterra predator–prey model parameterized for the observed type II and type III functional responses showed that inducible defenses in the *Paramecium* population reduced the death rate at low prey densities and increased the stability of the predator–prey interaction (Figure 6.8C, D).

A variety of terms have been used to describe the consumptive and non-consumptive effects of predators on predator–prey dynamics and the interactions between predators and prey at different trophic levels. These terms include "trait-mediated indirect effects" and "density-mediated indirect effects" (Abrams 1995; Abrams et al. 1996); "trait-mediated indirect interactions" and "density-mediated indirect interactions" (TMII and DMII; Peacor and Werner 1997); and "trait-mediated interactions" (TMI; Bolker et al. 2003) and "density-mediated interactions" (DMI; Bolnick and Preisser 2005; Preisser et al. 2005). As Abrams (2007) notes, the introduction of these different terms has led to some confusion and ambiguity. Therefore, we will stick to the simpler terms "consumptive effects" and "non-consumptive effects" to describe the different impacts that predators have on their prey. The remainder of this chapter discusses a variety of non-consumptive effects, organizing them into four general categories:

- habitat use and habitat shifts;
- life history evolution;
- activity level;
- morphological changes.

We will then ask how important these non-consumptive effects are relative to the impact of predators consuming their prey. As we shall see, this is a critical question that is difficult to answer. In subsequent chapters on food webs and the indirect effects of interactions (see Chapters 10 and 11), the consequences of both consumptive and non-consumptive effects for multispecies interactions will be considered.

Habitat use and habitat shifts

The ostrich may (mythically) stick its head in the sand, but most prey show a more adaptive response when predators threaten—they seek refuge. Many studies have documented changes in habitat use by prey in the presence of predators—more than 70 studies were cited in a review by Lima (1998). These habitat shifts may be short term and have little effect on the prey's population dynamics (e.g., the mouse that hides in the grass as a hawk flies overhead). However, other types of predator-induced habitat shifts have far-reaching effects on prey dynamics and life histories. For example, small (young) individuals tend

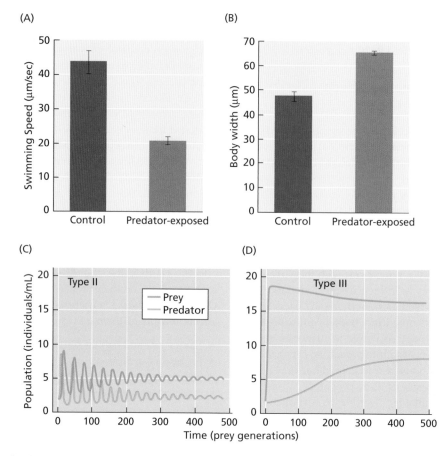

Figure 6.8 Induced antipredator defenses can stabilize predator–prey dynamics. Changes in the swimming speed (A) and body width (B) of *Paramecium* after exposure to chemical cues from a predator, the flatworm (*Stenostomum*). Means ±1 SE. Simulated prey and predator population densities from a Lotka–Volterra predator–prey model incorporating functional responses derived from the empirical data for undefended prey ((C) type II functional response) and defended prey ((D) type III functional response). Induced defenses against predators stabilized the predicted predator–prey dynamics. After Hammill et al. (2010).

to be most vulnerable to predators. As a consequence, predators may restrict vulnerable size (age) classes to protective habitats, leading to a change in the prey's habitat use and diet (if resources differ among habitats) as it grows. These **ontogenetic niche shifts** (*sensu* Werner and Gilliam 1984) may reduce competition between size (age) classes within a species (Werner et al. 1983; Mittelbach and Osenberg 1993), but may increase interspecific competition among species and size (age) classes that share a refuge (Mittelbach and Chesson 1987; Persson 1993; Diehl and Eklöv 1995).

In choosing habitats, organisms often face a trade-off between foraging gain and mortality risk

(i.e., habitats with more resources tend to be riskier). When this is the case, the "optimal" habitat choice depends on the relative costs (mortality risks) and benefits (energy gains) available in each habitat (Gilliam 1982; Gilliam and Fraser 1987; Ludwig and Rowe 1990; McNamara and Houston 1994). Moreover, if predation risk varies temporally or spatially, or changes over the life history of an organism, we would expect organisms to respond by changing their habitats. An excellent example is vertical migration by some zooplankton. In the open water of lakes and oceans, small crustaceans (zooplankton) that feed on algae (phytoplankton) show a pronounced daily migration cycle, spending the daylight hours in the dark, cold depths and migrating to warm surface waters at dusk, then returning to the depths at dawn. These diel (day–night) vertical migrations can cover tens of meters in freshwater lakes and hundreds of meters in the open ocean; such movements come at a substantial energetic cost. For years, scientists puzzled over the explanation for these migrations. We now know that, in most cases, vertical migrations are a response to temporal variation in predation risk. During the day, mortality rates from fish predation are high in the warm, well-lit surface waters, whereas mortality rates are substantially lower in the cold, dark depths. At night, predation pressure at the surface drops substantially because most fish are visual feeders. A strong vertical gradient in energy gain and developmental rate exists as well; food (phytoplankton, or algae) is often more abundant near the surface, and (more importantly) the warm surface water shortens zooplankton development times and increases birth rates, leading to higher population growth rates (Stich and Lampert 1981; Lampert 1987). Thus, by migrating vertically, zooplankton may balance the conflicting selection pressures of reducing predation risk, and increasing energy gain and reproductive rate.

Pangle et al. (2007) showed that the degree of vertical migration of zooplankton in Lake Michigan and Lake Erie was correlated with the abundance of their invertebrate predator, *Bythotrephes longimanus*. If we assume that zooplankton are migrating in response to the presence of *Bythotrephes*, then we can estimate the impact of the non-consumptive effect on population growth rate (due to a reduction in birth rate caused by migration into the cold hypolimnion) separately from the consumptive effect on population growth (due to direct mortality). In their analysis, Pangle et al. (2007) found that the non-lethal effects of the predator on zooplankton population growth were often of similar magnitude to the lethal effects.

Diel vertical migration in zooplankton is one particularly well-documented example of a very general response of prey to spatial or temporal variation in predation risk. There are literally hundreds of other examples that show how prey modify their habitat use in response to the threat of predation (reviewed by Lima 1998). Werner (1986) has extended our thinking about the non-consumptive effects of predators to include an even more dramatic type of habitat

shift—metamorphosis (described in the following section). In the examples given previously and those reviewed by Lima, the adaptive nature of habitat choice seems clear. However, it should be noted that few studies have attempted to compare an organism's observed habitat choices with predictions based on a quantitative theory of optimal habitat selection that incorporates both foraging gain and predation risk (e.g., Gilliam and Fraser 1987; Brown 1998). The challenge here is in measuring the potential benefits (e.g., energy gain) and costs (e.g., risk of mortality) of using different habitats, and expressing these costs and benefits in a common currency that can be used to estimate habitat quality. One solution to this problem is to let the organisms do it for us.

Choice experiments provide a means for getting animals to reveal the foraging costs of predation risk (Brown 1988; Brown and Kotler 2004). In such experiments, either energy reward or predation risk is continuously varied among habitats until the forager reveals a point of indifference in their habitat use. For example, an early study by Abrahams and Dill (1989) offered guppies (*Poecilia reticulata*) a choice between two patches (two sides of an aquarium) that differed in foraging gain (rate of resource supply) and predation risk (presence or absence of a predator). Not surprisingly, more guppies choose the safe half of the aquarium then the risky half, but by varying either the rate of resource supply or the risk of predation, Abrahams and Dill (1989) were able to measure how much higher a guppy's feeding rate had to be in order for it to accept a greater risk of predation. Behavioral ecologists have used the fact that animals are willing to accept higher predation risk for higher foraging gains to understand the nature of habitat selection in the field.

Recall from the discussion of Charnov's MVT that the optimal time for a forager to leave a patch is when its current rate of energy gain in the patch equals the average rate of energy gain for all patches in the habitat. Predation risk can be added to the theory of optimal patch use (e.g., Brown 1992; Houston et al. 1993), with the general result that foragers should leave patches sooner (leave at a higher current rate of energy gain) in risky habitats than in safe habitats. This prediction has been tested in the field using a simplified measure of a forager's feeding rate when it leaves a patch, which can be approximated by the amount of resources remaining in the patch at the time it leaves. This resource density, termed the "giving-up density" or GUD, should be positively correlated with the forager's rate of energy gain. Numerous studies with small mammals and birds have confirmed that patches near cover have lower GUD's than patches away from cover and that GUD's increase as the distance of a patch from a burrow or other refuge increases (reviewed in Brown and Kotler 2004; Kotler and Brown 2007). Thus, foragers perceive "open" patches and patches farther from a refuge as risker and, therefore, demand a higher foraging return for using these richer patches. Researchers have extended these findings to use measures of GUD's in a variety of habitats to characterize the "landscape of fear" perceived by a forager (Van Der Merwe and Brown 2008; Laundré et al. 2017).

Life history evolution

Life history theory predicts that organisms should undergo metamorphosis when the fitness that can be achieved in the larval habitat (e.g., by a tadpole in fresh water) drops below the fitness that could be achieved by shifting to the adult habitat (e.g., by an adult frog in terrestrial habitat; Werner 1986). In the simple case of a stable population under no time constraints, Gilliam (1982) showed that the optimal size to switch from the larval to the adult habitat occurs when μ/g in the larval habitat rises above μ/g in the adult habitat, where μ is a species' size-specific mortality rate and g is its size-specific growth rate (Figure 6.9). Intuitively, minimizing μ/g at each size permits growth at the minimal mortality cost and maximizes the probability of reaching reproductive size.

Gilliam's rule of "minimize μ/g" strictly applies only under special conditions, yet as a heuristic tool, it has proved useful in addressing a variety of questions concerning optimal behavioral decisions for organisms under predation risk, including habitat selection, activity level, and life history evolution. Ludwig and Rowe (1990), Rowe and Ludwig (1991), and Abrams and Rowe (1996) have advanced the theory to show how the optimal size at metamorphosis (or size at maturity) is affected by mortality risk in seasonal environments. They show that increased mortality risk in the juvenile stage should

favor earlier maturity, which given a fixed growth rate, also means maturing at a smaller size (Abrams and Rowe 1996). Thus, we have the general prediction that increased predation risk in the larval or juvenile stage should lead to earlier metamorphosis and metamorphosis at a smaller body size.

Bobbi Peckarsky and her colleagues tested this prediction by studying the impacts of trout predation on mayfly (*Baetis*) populations in streams in the Rocky Mountains of the western United States (Peckarsky et al. 1993, 2001, 2002). Mayflies spend most of their lives as larvae, grazing algae from the stream bed and growing through a series of instars until they metamorphose into winged adults, reproduce, and die. Adult mayflies do not feed (in fact, they lack mouthparts) and live for only 2 days. Thus, all the energy needed to develop eggs and reproduce must be gained in the larval stage. Peckarsky and colleagues noted that mayflies from streams containing trout emerged earlier, and at much smaller adult sizes, than conspecifics found in streams without trout (Peckarsky et al. 2001). Thus, their observations fit the theoretical predictions outlined previously. However, because fish prefer to eat larger prey (as we saw earlier in this chapter), an alternative hypothesis is that size-selective predation was

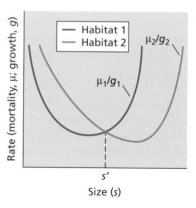

Figure 6.9 A graphic model illustrating the optimal body size at which to switch habitats as a function of individual growth rates (g) and mortality rates (μ) in each habitat. The optimal size to switch from using habitat 1 to using habitat 2 is the point at which the ratio μ/g in habitat 1 exceeds that ratio in habitat 2 (designated by s'). Predictions of this model may also be applied to the optimal size at which to metamorphose from one life stage to another (e.g., from a tadpole to a frog). After Werner and Gilliam (1984).

Table 6.1 Effect of a chemical cue ("fish cue") emitted by predatory trout on size and fecundity in mayfly (*Baetis*) populations[a]

Baetis generation	Sex	Mean size (dry mass, mg)		Reduction (%)	Mean female fecundity		Reduction (%)
		Experimental	Control		Experimental	Control	
Summer 1999	M	0.632	0.802	21	—	—	—
	F	0.927	1.161	20	357.9	547.6	35
Winter 2000	M	0.896	1.104	14	—	—	—
	F	1.259	1.455	13	638.1	838.9	24

[a]The experiment was carried out in the summer of 1999 and repeated in winter 2000. Researchers measured the size and fecundity of mayflies in two separate stream environments, neither of which contained predatory fish. In the experimental condition, "fish cue"—water from tanks containing predatory trout—was added to the stream water. The control stream received water from tanks that did not contain fish
Source: Peckarsky et al. (2002).

Table 6.2 Some estimated lethal and non-lethal effects of trout predation on mayfly larval and population growth

	Larval mortality rate	Larval final mass (mg)	Larval duration (days)	Population growth rate	Change in λ (%)
Natural mayfly population	0.0378	0.2448	42	1.993	—
Growth (non-lethal) effect removed	0.0378	0.3666	42	4.264	114.0
Mortality (lethal) effect removed	0.0300	0.2448	42	2.765	38.8
Growth and mortality effects removed	0.0300	0.3665	42	5.916	197.0

Source: After McPeek and Peckarsky (1998).

responsible for the difference in the sizes of mayflies hatching from fish-inhabited and fishless streams. To test this hypothesis, the researchers dripped water from tanks containing live trout into fishless streams and compared the size of metamorphosing mayflies in these fish-cue treatment streams with the size of mayflies from fishless streams that received water from tanks without fish (the control). The addition of the fish-cue water (containing chemical signals released by trout) resulted in a 13–21% reduction in adult mayfly size and an estimated 24–35% loss of fecundity (Table 6.1; Peckarsky et al. 2002). Moreover, demographic analysis showed that the reduction in mayfly population growth caused by the nonlethal effects of trout on mayfly size at maturation exceeded the direct effects of consumption by the predator (Table 6.2; McPeek and Peckarsky 1998). Thus, in this remarkably complete series of studies, we see a clear example of how the lethal and sublethal effects of a predator combine to affect the population growth rate of its prey.

Activity levels and vigilance

Prey also may reduce their exposure to predators by lowering their activity level (speed and extent of movement), increasing vigilance, and increasing the amount of time they spend in a refuge. A review of the literature shows that such responses are ubiquitous across taxa: almost all species studied exhibited decreased movement, increased vigilance, and/or increased refuge use in response to an increase in the risk of predation (Lima 1998; see Figure 6.10 as an example). However, this reduction in activity may come at the cost of lost feeding time, a lower rate of energy gain, and slowed growth. Thus, there is a growth rate–predation risk trade-off (Abrams 1990; Houston et al. 1993; Werner and Anholt 1993; McPeek 2004). Species differ in how they resolve this trade-off (e.g., Werner and McPeek 1994), and these interspecific differences are thought to be a major mechanism promoting species coexistence. We will explore the question of predator-mediated species coexistence in Chapters 8 and 11.

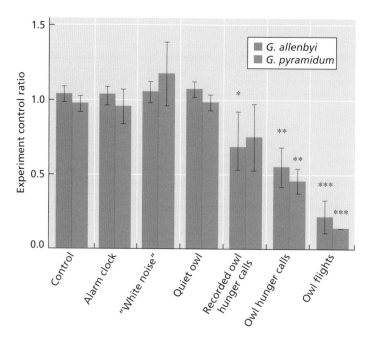

Figure 6.10 Effects of predatory owl threat cues and a variety of other treatments on the activity levels of two gerbil species, *Gerbillus allenbyi* and *G. pyramidum*, in the field. Histograms show the ratio of gerbil activity levels in control plots relative to that in treatment plots. Means ±1 SE. A ratio of 1.0 indicates no difference in gerbil activity in control and treatment plots; a ratio below 1.0 means that gerbil activity in the treatment plot was less than in the control. Note the strong inhibitory effects of owl flights and owl hunger calls on gerbil activity. After Abramsky et al. (1996). *$P < 0.05$, **$P < 0.001$, ***$P < 0.0001$.

Individuals often feed or travel in groups as a means of reducing predation risk and there is a large literature on how group size and group composition may affect an individual's risk of mortality (e.g., Lima 1995; Bednekoff 2007; Lehtonen and Jaatinen 2016). In some cases, individual may feed in mixed-species groups where certain species act as "sentinels" for the detection of predators. For example, Martinez et al. (2017) conducted a fascinating study looking at how well two different alarm-calling species (antshrikes, *Thamnomanes*) served as sentinels in mixed-species flocks (up to 70 bird species) in the Amazonian rainforest. Each mixed flock was led by a single antshrike species. By flying three different species of trained raptors towards the mixed flocks, they discovered that bluish-slate antshrikes (*T. schistogynus*), which were the sentinel species in flocks occupying forest gaps, produced more alarm calls, and provided more specific information about the size and distance of an approaching predator than did flocks in the dense forest where dusky-throated antshrikes (*T. ardesiacus*) were the sentinel species. Martinez et al. (2017) suggest that the early successional habitats occupied by bluish-slate antshrikes are more productive, but also riskier. Thus, sentinel species with different levels of vigilance may modulate the peaks and valleys in the landscape of fear.

At the intraspecific level, it is difficult to quantify the impact of reduced prey activity levels on prey population dynamics, especially in comparison to the direct consumptive effects of predators on their prey (Preisser et al. 2005; Creel and Christianson 2008). Moreover, as McPeek (2004) has shown for damselfly larvae, reduced activity in the presence of predators may not directly affect prey feeding rates, but it may affect prey growth and/or survival through physiological processes. Finally, a predator-induced reduction in feeding rate may actually increase prey population growth if this reduction in feeding rate prevents overexploitation of the prey's resource (Figure 6.11; Abrams 1992; Peacor 2002).

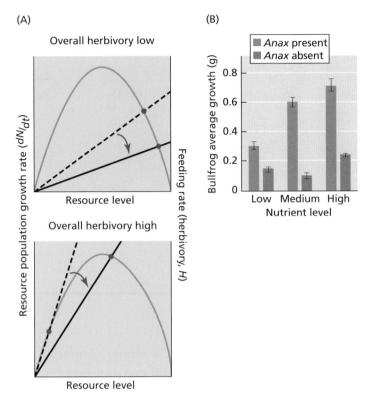

Figure 6.11 An illustration of the potential effect of a predator-induced reduction in consumer (prey) feeding rate on the equilibrium levels of the consumer's resource. (A) In this example (which assumes logistic growth for the resource), a reduction in the consumer's feeding rate (from dashed line to solid line in each panel) results in a decrease in equilibrium resource levels at a low feeding rate (top), but an increase in equilibrium resource levels at a high feeding rate (bottom). (B) Average growth rates (±SE) of consumers in an experiment in which the presence of non-lethal predators (caged dragonfly larvae, *Anax*) reduced the feeding rate of the consumers (large bullfrog tadpoles). In this experiment, the presence of *Anax* reduced tadpole feeding rates, resulting in higher standing stocks of algae (the tadpole's resource) and leading to higher growth rates of the tadpoles at medium and high nutrient levels. After Peacor (2002).

Morphology

Predators have been shown to induce a variety of morphological defenses in their prey populations, as we saw earlier for *Paramecium aurelia* (Hammill et al. 2010). These inducible defenses include changes in toxicity, color, body shape, shell hardness, and the presence of spines (Figure 6.12), all of which may reduce a prey's chances of being eaten (see reviews in Harvell 1990; Tollrian and Harvell 1999; Benard 2004). These phenotypically plastic responses can be so dramatic as to cause biologists to mistake the different body forms for different species. A case in point is the crucian carp (*Carassius carassius*). In lakes without piscivorous predators, crucian carp are found in dense populations of narrow-bodied individuals, but in lakes with piscivores (especially pike, *Esox lucius*), carp are few in number, and individuals are deep-bodied (Figure 6.13A; Brönmark and Miner 1992). The two carp morphs were originally considered separate species until transplant experiments showed them to be the same (Ekström 1838, cited in Brönmark and Miner 1992; Brönmark and Miner 1992). Ecologists then hypothesized that it was variation in resource levels and growth rates that led to the development of the two morphs. However, a series of field and laboratory studies by Brönmark and colleagues showed that the differences in body form were in fact induced by the presence of predators.

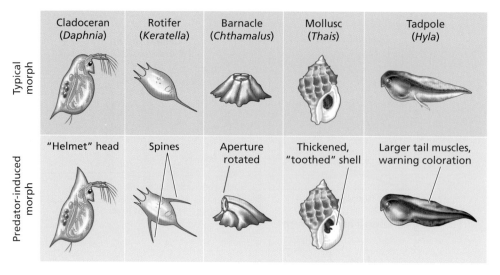

Figure 6.12 Examples of predator-induced defenses in some aquatic organisms. After Peacor (2002).

In a field experiment, Brönmark and Miner (1992) introduced young crucian carp from the same population into two ponds. Each pond was divided in half by a plastic curtain, and one half of each pond also received 15–20 pike. After 12 weeks, the carp population had diverged in body shape between the predator and no-predator treatments—in the presence of the pike, carp had significantly deeper bodies (Figure 6.13B). This shift in body form allowed most of the carp to escape being eaten by their pike predators (Figure 6.13C). Importantly, laboratory experiments showed that the observed divergence in body shape could not be accounted for by differences in resource availability and growth rates. Experiments also showed that expression of the deep-bodied form resulted in about a 30% increase in energy expended when swimming compared with the shallow-bodied, more fusiform shape (Brönmark and Miner 1992; Pettersson and Brönmark 1997). Thus, by becoming deeper-bodied, the carp reduced their vulnerability to predation, but paid a significant energetic price. In general, we expect inducible defenses to evolve when predation risk varies temporally or spatially; when prey have the ability to detect predators via reliable cues; and when the defense against predators carries a fitness cost (Brönmark and Hansson 2005).

The relative importance of consumptive and non-consumptive effects

The above discussion illustrates the wealth of non-consumptive effects that predators may have on their prey, but truth be told, we have only scratched the surface of this active and fascinating field. Despite all the excellent research in this area, however, perhaps the most significant questions remain unanswered: "How important are non-consumptive effects relative to consumptive effects in predator–prey interactions?" and "How do we incorporate nonlethal effects into our existing theory of predator–prey interactions" (Abrams 2010)? The challenge in addressing both of these questions comes from the fact that non-consumptive and consumptive effects are often measured in very different units (e.g., growth vs mortality), as well as the fact that these effects may operate on different time scales (e.g., changes in prey behaviors occur much more rapidly than changes in prey densities).

One common experimental approach to assessing the importance of non-consumptive effects relative to consumptive effects is to compare the impact of a functional predator with that of a non-lethal predator on the density or fitness (measured as fecundity or growth) of the prey population. Non-lethal predator

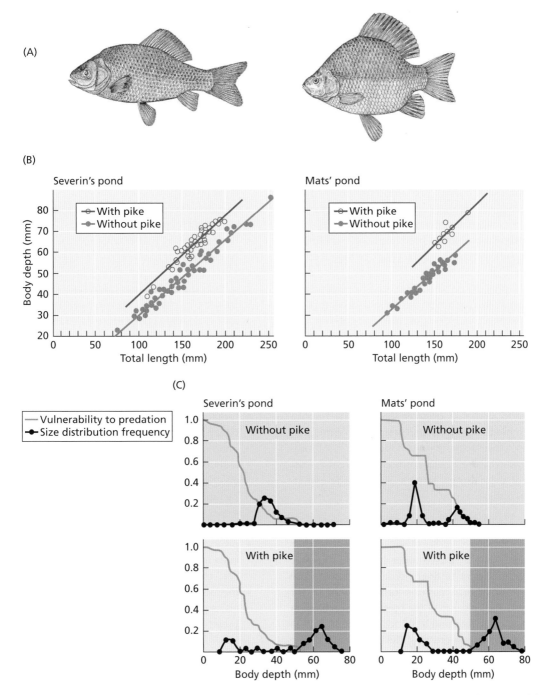

Figure 6.13 Differences in body form in the crucian carp (*Carassius carassius*) are an induced adaptation to reduce mortality by predation. (A) In the presence of predatory fish, crucian carp are very deep-bodied, whereas in the absence of predators, they are narrow-bodied. (B,C) In southern Sweden, Brönmark and Miner divided two small ponds in half with a fish-proof curtain. They introduced juvenile crucian carp and predatory northern pike (*Esox lucius*) into one half of each pond and juvenile carp only into the other half. (B) After 12 weeks, carp in the presence of pike were much deeper-bodied than carp in the absence of pike. (C) This increase in body depth reduced the carp's vulnerability to predation. Darker shaded areas denote carp that are too large to be vulnerable to predation. After Brönmark and Miner (1992).

treatments can be created using caged or disabled predators ("risk predators," such as a spider with non-functional mouthparts; Schmitz 1998) or by introducing a predator cue (e.g., water containing the chemical signal, or kairomone, of a predator). Many such experiments have been performed, and meta-analyses of the results show that the nonlethal effects of predators are often as great as or greater than the lethal effects (Bolnick and Preisser 2005; Preisser et al. 2005). These analyses give us an appreciation of the potential importance of non-consumptive effects in predator–prey interactions. However, such simple comparisons sweep a number of complications under the rug. For example, the presence of a non-lethal predator in an experiment may result in prey behaviors that reduce feeding rates and growth, resulting in weakened and even starved prey. In nature, however, such weakened prey often fall victim to functional predators. In addition, experimental exposure to non-consumptive predators can lower the subsequent survival of prey and their offspring (McCauley et al. 2011; MacLeod et al. 2018). Thus, the consumptive and non-consumptive effects of predators may interact in nature, and sophisticated experimental designs are needed to accurately estimate their relative importance in predator–prey interactions (e.g., Werner and Peacor 2003). Okuyama and Bolker (2007), Abrams (2008), and Schmitz (2010) provide excellent discussions of these issues, as well as guides to experimental design.

Looking ahead

We have seen in this chapter how the dynamic responses of prey to their predators often lead to indirect effects. In a three-link food chain, for example, the presence of the predator may result in increased vigilance and a reduction in feeding rate in the prey, thus increasing the abundance of the prey's resource. We will discuss this type of indirect interaction (a cascading effect from the predator to the prey's resource) in much more detail in Chapter 11, where we will examine top-down and bottom-up control in food chains and food webs. Before we get there, however, we need to consider consumer–resource interactions from the point of view of competition (Chapters 7 and 8), as well as

the beneficial interactions of mutualism and facilitation (Chapter 9).

Summary

1. Predator preference for a prey type can be defined as a difference between the proportion of that prey type in the predator's diet and the proportion of that prey type in the environment. Predator preference is influenced by:
   - the probability that a prey item will be encountered;
   - the probability that an encountered prey item will be attacked;
   - the probability that an attacked prey item will be captured and eaten.
2. Optimal foraging theory (OFT) is an approach to understanding predator diet choice by asking what set of foraging decisions will maximize a predator's energy gain. Optimal diet models predict a predator's diet choices.
3. The standard optimal diet model makes two general predictions and two more specific predictions about predator choice:
   The two general predictions are:
   - foragers should prefer the most profitable prey (i.e., prey that yield the most energy per unit handling time);
   - as the overall density of profitable prey decreases, an efficient forager should broaden its diet to include more of the less profitable prey.
   The two specific predictions of the standard optimal diet model are:
   - prey types are either always eaten upon encounter or never eaten upon encounter (the zero-one rule);
   - the inclusion of a prey type in the diet depends on its profitability and on the characteristics of prey types of higher profitability.
4. Tests of the optimal diet model support its predictions, but in most cases predators include some prey types in their diets that are not in the predicted optimal set.
5. The MVT predicts when to leave a patch based on how the rate of energy gain in the patch compares with the maximum obtainable overall average energy gain moving across the landscape. When the average productivity (prey availability)

of a patch goes up or the travel time between patches increases, predators will spend a longer time in one patch. One counter-intuitive prediction is that more efficient predators may leave a patch quicker than less efficient predators.

6. Prey may respond to the threat of predation by changing their behaviors, morphologies, physiologies, or life histories. These non-consumptive (or non-lethal) effects of predators act in concert with the direct consumption of prey to influence prey abundance and predator–prey dynamics.

7. In choosing habitats, organisms may face a trade-off between foraging gain and increased risk of predation. When this is the case, the "optimal" habitat choice depends on the relative costs (mortality risks) and benefits (energy gains) available in each habitat. Thus, for example, increased predation risk in the larval or juvenile stage is predicted to lead to the evolution of earlier metamorphosis and/or a smaller adult body size.

8. Many prey species reduce their activity level in the presence of predators, but these less-active prey may incur a reduction in growth (i.e., a growth rate–predation risk trade-off). Inducible morphological defenses may evolve when predation risk varies temporally or spatially; when prey have the ability to detect predators via reliable cues; and/or when the defense carries a fitness cost.

9. The relative importance of the non-consumptive and consumptive effects can be addressed by performing experiments in which the effect on prey density or fitness of a non-lethal (e.g., caged or disabled) predator or a predator cue is compared with that of a functional predator. However, such experiments do not address the complicated interactions of these effects in nature.

CHAPTER 7

The fundamentals of competitive interactions

[M]ultispecies stable equilibrial coexistence requires interspecific tradeoffs. If a species arose that was able to avoid such tradeoffs and be a superior competitor relative to all other species, it would eliminate its competitors.

David Tilman, 2007: 93

One hill cannot shelter two tigers.

Chinese proverb

Ecologists have long viewed interspecific competition as a major (perhaps *the* major) factor influencing community structure. In Chapter 1, for example, we saw how Gause's competitive exclusion principle provided a foundation for the development of later ideas about the role of interspecific competition and limiting similarity in determining the number of coexisting species. At about the same time that Gause (1934) was conducting his experiments on competitive exclusion, Alfred Lotka (1925) and Vito Volterra (1926) independently developed the first mathematical theory of interspecific competition, based on a simple extension of the logistic model of population growth. In this chapter, we will examine the working principles of interspecific competition. Much as we did in our discussion of predator–prey interactions, we will use simple mathematical theory and isocline analysis to develop this understanding. Although we will begin with the classic Lotka–Volterra (L–V) model, we will devote most of our time to studying a more recent modeling approach that explicitly considers the interactions between species and their resources. The following chapter relaxes some of the restrictive assumptions of simple competition theory and examines in more detail the

mechanisms that can promote species coexistence in competitive environments.

Defining interspecific competition

It is difficult to come up with a rigorous and succinct definition of interspecific competition. Two definitions highlight the essentials:

[Interspecific competition] is the interaction occurring between species when increased abundance of a first species causes the population growth of a second to decrease, and there is a reciprocal effect of the second on the first. (Grover 1997, p. 11.)

Interspecific competition between two species occurs when individuals of one species suffer a reduction in growth rate from a second species due to their shared use of limiting resources (exploitative competition) or active interference (interference competition). (Case 2000, p. 311)

Both definitions note that interspecific competition involves a reduction in the population growth rate of one species due to the presence of one (or more) other species. The definition by Case outlines the two general classes of mechanisms that may cause a competitive reduction in growth rate. **Exploitative**

Community Ecology. Second Edition. Gary G. Mittelbach & Brian J. McGill, Oxford University Press (2019).
© Gary G. Mittelbach & Brian J. McGill (2019). DOI: 10.1093/oso/9780198835851.001.0001

competition (also known as **resource competition**) occurs when a species consumes a shared resource that limits its and other species' population growth—thus making that resource less available to the other species. **Interference competition** (or **contest competition**) occurs when one species restricts another species' access to a limiting resource. Interference competition may involve overt aggression between individuals (e.g., territorial defense), or it may simply involve occupying a space to the exclusion of another individual. We will examine the definition of a limiting resource in more detail later in this chapter. First, however, we will consider the classic L–V model of interspecific competition.

The Lotka–Volterra competition model

Recall from the discussion of the logistic equation in Chapter 4 that we can write equations for the change in abundance of two species whose populations are growing according to a logistic model as

$$dN_1/dt = r_1 N_1 \left(\frac{K_1 - N_1}{K_1} \right)$$ [Eqn 7.1]

$$dN_2/dt = r_2 N_2 \left(\frac{K_2 - N_2}{K_2} \right)$$ [Eqn 7.2]

where (N_1, N_2) are the densities of species 1 and 2, respectively, (r_1, r_2) are their intrinsic growth rates, and (K_1, K_2) are their carrying capacities. Intraspecific competition is implicit in this formulation of population growth because the density of each species cannot exceed its carrying capacity, and the per capita rate of population growth (dN/Ndt) declines linearly with population density (see Figure 4.4C). To examine the competitive effect of species 2 on species 1 (and vice versa), Lotka (1925) and Volterra (1926) incorporated into the logistic equation a term for the density of the competing species and multiplied this density term by a constant (α_{ij}), the competition coefficient. Equations 7.3 and 7.4 represent the L–V competition model for two species:

$$dN_1/dt = r_1 N_1 \left(\frac{K_1 - N_1 - \alpha_{12} N_2}{K_1} \right)$$ [Eqn 7.3]

$$dN_2/dt = r_2 N_2 \left(\frac{K_2 - N_2 - \alpha_{21} N_1}{K_2} \right)$$ [Eqn 7.4]

The **competition coefficient** α_{ij} is the effect of an individual of species j on the per capita growth rate of species i relative to the effect of an individual of i on its own per capita growth rate (and vice versa for α_{ji}). Note that there is an implicit constant of 1 in front of the density term for the species whose growth rate is being modeled. The model can be extended to include as many species as desired by adding more terms of the form $\alpha_{ij} N_j$ in the numerator of species i's growth Equation.

Note that the L–V model does not specify the mechanism of competition between the species; it simply states that the density of one species has a negative effect on the population growth rate of the second species. Negative effects in the L–V model could be the result of direct aggression between competitors, or they could be the result of indirect interactions through consumption of a shared resource (MacArthur 1970). The L–V model is a phenomenological description of competition. Later in this chapter, we will consider a more mechanistic model of exploitative competition that explicitly incorporates resource consumption. For now, however, we can use the L–V model and simple isocline analysis to develop an understanding of the potential outcomes of two-species competition. This approach is very similar to the one used for predator–prey interactions in Chapter 5. Chapter 5 plotted the zero growth isoclines for predator and prey on a single graph, then examined the regions of population densities that yielded positive and negative growth rates to predict the outcome of the predator–prey interaction under different conditions (model assumptions). Much the same thing can be done for interspecific competition by solving Equations 7.3 and 7.4 for points where the per capita growth rates of species 1 and species 2 are zero.

The combination of densities of species 1 and species 2 that results in a zero per capita growth rate for species 1 can be found by setting $dN_1/dt = 0$ in Equation 7.3; likewise, for species 2 by setting $dN_2/dt = 0$ in Equation 7.4. Graphing this combination of densities on a plot whose axes are N_1 and N_2 yields a zero growth isocline for each species (Figure 7.1). At each point along the species' zero growth isocline, its population is at equilibrium and its density is unchanging. Taking species 1 as an example (Figure 7.1A), we know that its population

(A)

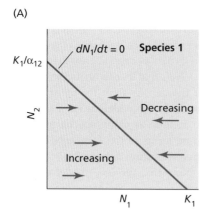

(B)

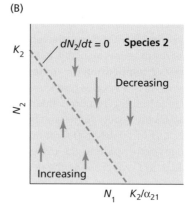

Figure 7.1 Zero growth isoclines for two-species Lotka–Volterra competition. (A) The combination of densities of species 1 (N_1; x axis) and species 2 (N_2; y axis) for which the population of species 1 shows zero growth ($dN_1/dt = 0$). For points to the left of the isocline, the population of species 1 is increasing (right-pointing arrows). Right of the isocline, the population of species 1 is decreasing (left-pointing arrows). (B) The zero growth isocline for species 2. For points below the isocline, species 2 is increasing, and for points above the isocline, species 2 is decreasing (upward and downward pointing arrows, respectively).

growth rate will be zero when its density is at its carrying capacity (K_1). Species 1's growth rate also will be zero when the density of species 2 is equivalent to the carrying capacity of species 1 (i.e., when $N_2 = K_1/\alpha_{12}$). The zero growth isocline between these endpoints is linear because per capita density dependence is linear in the model. Densities below the isocline in Figure 7.1A result in positive growth of species 1, whereas densities above the isocline result in its negative growth. You can follow the same logic to construct the zero growth isocline for species 2 (Figure 7.1B).

Plotting the zero growth isoclines for species 1 and species 2 on the same graph yields four possible outcomes (Figure 7.2). When the isocline for species 1 is above the isocline for species 2, species 1 wins in competition and excludes species 2 (Figure 7.2A). We can understand this outcome by considering the population growth trajectories of the species in each of the three regions of the graph defined by the isoclines. In the lower region, each species exhibits positive growth; in the upper region, each species exhibits negative growth. In the middle region, between the two isoclines, the population growth of species 2 is negative and the population growth of species 1 is positive—the result being that the density of species 1 reaches a stable equilibrium at K_1 and the density of species 2 goes to zero. By the same logic, when the isocline for species 2 is above

the isocline for species 1, species 1 is excluded and the density of species 2 equilibrates at K_2. In order for species to coexist, their isoclines must cross.

When the species' isoclines cross, their intersection point represents an equilibrium at which the densities of both species are positive and unchanging (Figure 7.2C,D). This equilibrium may be stable or unstable, depending on the orientation of the isoclines. When the isoclines are oriented as in Figure 7.2C, the two-species equilibrium is stable because the trajectories for both populations always point toward the equilibrium; if densities are perturbed from the equilibrium point, they will return to it. However, if the isoclines are oriented as in Figure 7.2D, the population trajectories point away from the equilibrium in two of the four quadrants. Thus, population densities in these quadrants will go to fixation at either K_1 or K_2; that is, either species 1 or species 2 will win. The outcome of competition in this case depends on the starting densities of species 1 and species 2—a result that is sometimes called **founder control** because the initially more abundant species is likely to predominate.

Another way to look at the Lotka–Volterra competition model

The isocline analysis in Figure 7.2 illustrates the four potential outcomes of two-species competition:

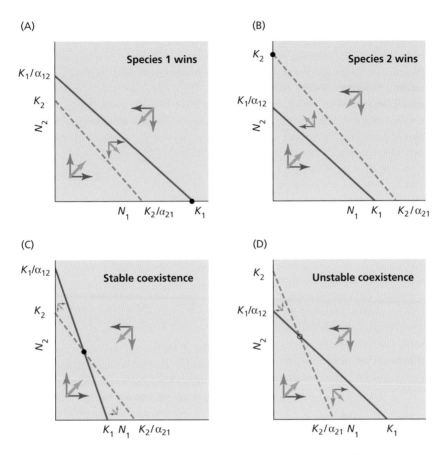

Figure 7.2 The four potential outcomes of two-species competition. Horizontal and vertical arrows show the trajectories of each species; diagonal arrows show the joint vectors of movement in the regions of state space defined by the two zero growth isoclines. (A) Species 1 outcompetes species 2. The equilibrium outcome when the species 1 isocline lies above the species 2 isocline is that the density of species 1 goes to its carrying capacity K_1 and the density of species 2 goes to zero. (B) Species 2 outcompetes species 1. When the isocline for species 2 lies above the isocline for species 1, the density of species 2 goes to its carrying capacity K_2 and that of species 1 goes to zero. (C) Stable coexistence. The isoclines cross and the population trajectories for each species point toward the equilibrium point. If densities are perturbed from the equilibrium point, they will return to it. (D) Unstable coexistence. The isoclines cross, but the population trajectories point away from the equilibrium point in two of the four quadrants. If densities are perturbed from the equilibrium point, one species or the other will win, depending on starting abundances. After Gotelli (2008). See http://communityecologybook.org/LVComp.html for an online dynamic version of this model.

species 1 wins; species 2 wins; stable coexistence; and unstable coexistence. However, many students find the L–V competition isoclines to be confusing. They are certainly less intuitive then the predator–prey isoclines we studied in Chapter 5, which is part of the reason why we considered predator–prey interactions first. Chesson (2000a) suggests that we can get a more intuitive understanding of the factors leading to species coexistence or competitive exclusion by writing the L–V model (Equations 7.3

and 7.4) in terms of absolute competition coefficients rather than relative competition coefficients. Following Chesson, we can write the two-species L–V competition Equations as

$$dN_i/dt = r_iN_i(1 - \alpha_{ii}N_i - \alpha_{ij}N_j)$$
$$\text{when } i = 1,2, j \neq i \qquad \text{[Eqn 7.5]}$$

where the quantities α_{ii} and α_{ij} are now, respectively, the **absolute intraspecific competition coefficient** and the **absolute interspecific competition coefficient** (i.e., competition coefficients that are not scaled by the species' carrying capacities). Species coexistence occurs when each species can increase from low density in the presence of its competitor; this is the basis of the commonly used **invasibility criteria** for species coexistence (Chesson 2000a. In Equation 7.5, species i can increase from low density in the presence of competitor species j if $\alpha_{jj} > \alpha_{ij}$, and species j can increase from low density in the presence of competitor species i if $\alpha_{ii} > \alpha_{ji}$. Thus, as Chesson notes, the criteria for species coexistence in this two-species competitive interaction is that $\alpha_{11} > \alpha_{21}$ and $\alpha_{22} > \alpha_{12}$ or, in words, that *intraspecific competition is greater than interspecific competition*. That is, coexistence requires that each species have a greater negative effect on its own per capita growth rate than it does on the per capita growth rate of its competitor. Coexistence criteria are more complex in multispecies communities (Barabás et al. 2016, 2018), but species coexistence will still be favored when intraspecific competition is stronger than interspecific competition for all species pairs (Adler et al. 2018).

Modification to the Lotka–Volterra competition model

The L–V model assumes that competitive effects are linear—i.e., that the per capita growth rate of each species declines linearly with increases in its own density or the densities of its competitors. However, it is unlikely that competitive effects in nature are truly linear. Curved competition isoclines may occur for a variety of reasons (Abrams 2008). Gilpin and Ayala (1973) first proposed a simple modification of the L–V model to accommodate non-linear competitive interactions, where

$$dN_i/dt = r_i N_i \left[1 - \left(\frac{N_i}{K_i} \right)^{\theta_i} - \alpha_{ij} \frac{N_j}{K_i} \right] \quad \text{[Eqn 7.6]}$$

Although Equation 7.6 (often referred to as the "θ-logistic model") produces non-linear competition isoclines when $\theta \neq 1$, the non-linear competitive effects occur only in the intraspecific portion of the interaction. A more general approach to examining the causes and consequences of non-linear competitive effects has been advanced (in a consumer–resource competition model) by Abrams et al. (2008),

who showed that non-linear competitive interactions have important consequences for predicting evolutionary responses to competition and the limiting similarities of coexisting species.

The L–V competition model may also be extended to include many competing species. The conditions for equilibrium in multispecies L–V type models are commonly written in matrix format (Levins 1968):

$$\mathbf{K} = \mathbf{A}\mathbf{N}^*$$

where **K** is a column vector of the species' carrying capacities, **N*** is a column vector of the species' equilibrium densities, and **A** is a square matrix of the interaction coefficients. For the case of three competing species, the matrices would look like this:

$$\mathbf{K} = \begin{pmatrix} K_1 \\ K_2 \\ K_3 \end{pmatrix} \mathbf{N}^* = \begin{pmatrix} N_1^* \\ N_2^* \\ N_3^* \end{pmatrix} \text{ and } \mathbf{A} = \begin{pmatrix} 1 & \alpha_{12} & \alpha_{13} \\ \alpha_{21} & 1 & \alpha_{23} \\ \alpha_{31} & \alpha_{32} & 1 \end{pmatrix}$$

The **A** matrix is commonly referred to as the **community matrix** because it specifies the intra- and interspecific per capita interaction strengths; intraspecific effects are on the diagonal and interspecific effects are on the off-diagonal (Levins 1968; Vandermeer 1970; MacArthur 1972). This matrix approach can be broadly applied to multispecies communities in which the interactions between species are governed by generalized L–V models that may include competition, predation, or positive effects (see Case 2000 for a clear presentation of this approach).

The community matrix provides an elegant way to analyze the equilibrium outcome of competition among many species (Levins 1968; Yodzis 1989) and it has been used extensively by theorists to explore the effects of interspecific competition on questions of community stability (e.g., May 1973a), community invasibility (e.g., Case 1990; Law and Morton 1996), and the equilibrium number of coexisting species (e.g., Vandermeer 1972). However, applications of the community matrix to natural systems have been rare (e.g., Seifert and Seifert 1976) for two important reasons.

First, this approach assumes that interactions between pairs of competing species are unaffected by the other species in the community (i.e., that the competition coefficients in the **A** matrix are independent of each other and thus species effects are additive). This is unlikely to be the case in nature (Neill 1974; Abrams 1983a). Non-additive (non-linear) effects are often referred to as **higher-order interactions** or **interaction modifications** (Neill 1974; Wootton 1994). The trait-mediated interactions that we examined in Chapter 6 are examples of higher-order interactions in predator–prey systems; we will discuss some of the factors causing higher-order interactions in competitive systems in later chapters on ecological networks and food webs. Secondly, in any reasonably diverse community, it is a challenge to estimate all the competition coefficients (the α's in the **A** matrix). In a community of 10 species, for example, there are 45 unique pairwise species interactions. Thus, although generalized L–V models and the community matrix continue to be useful theoretical tools, they have severe limitations when applied to natural communities. More recent models that explicitly incorporate resources into the mechanics of competition have come to the forefront, in part because they provide a stronger connection to empirical work. These **consumer–resource models** are the subject of the next section.

Consumer–resource models of competition

The L–V model shows that species coexistence and competitive effects depend critically on the relative strengths of intraspecific and interspecific competition. The L–V model does not explicitly include resources, however, and thus is of limited value in understanding exploitative competition, or in providing predictions that can be directly tested by manipulating resources. To better study exploitative competition, we need a model that incorporates how competing consumers affect their resources and how consumer population growth rates are in turn affected by resource densities.

MacArthur (1968, 1970, 1972) was the first to include resource dynamics in a model of competition for multiple resources (although Volterra had examined the case of competition for a single resource much earlier). Subsequent models were developed by May (1971), Phillips (1973), Schoener (1974, 1976), León and Tumpson (1975), and others. These pioneering efforts helped shift the focus

away from the phenomenological L–V model and toward more mechanistic models of exploitative competition.

It was the graphical portrayal of consumer–resource competition models, however, that made the results of this theory accessible to most ecologists. León and Tumpson (1975) introduced a graphical analysis of resource competition models and applied it to different types of resources, yielding criteria for species coexistence. Tilman (1980, 1982) refined and greatly expanded on this approach, and provided the first empirical tests of the model (Tilman 1976, 1977). The presentation that follows is based on a graphical analysis by Tilman (1980, 1982), which has been widely used and extended to other contexts (e.g., Grover 1997; Chase and Leibold 2003b). This simple graphical approach is less robust than the mathematical theory behind it. However, it is a useful heuristic tool, and students of community ecology should be familiar with it. For an alternative verbal explanation of the essence of exploitative competition, as well as a detailed mathematical analysis, readers should consult Abrams (1988).

What are resources?

Species compete when they share a resource whose availability in the environment limits a species' population growth. But what is a resource? Here are two definitions:

A resource [is] any substance or factor which can lead to increased [population] growth rates as its availability in the environment is increased, and which is consumed by an organism. (Tilman 1982, p. 11.)

Resources are entities which contribute positively to population growth, and are consumed in the process. (Grover 1997, p. 1)

Note the two properties that define a **resource**:

- a resource contributes positively to the growth rate of the consumer population;
- a resource is consumed (and is thus made unavailable to other individuals or species).

Resources may be further classified as biotic (living and self-reproducing) or abiotic (non-living), a distinction that is important when thinking about how resources are supplied. Examples of abiotic resources include mineral nutrients (e.g., nitrogen, phosphorus), light, and space. Biotic resources include prey (for predators), plants (for herbivores), and seeds (for granivores). Although space is often considered a resource for sessile species such as barnacles and plants, procuring space may simply be a prerequisite for obtaining other resources (e.g., light, nutrients, planktonic prey); thus, thinking about space as a resource can be complicated. Finally, some things that are *not* resources include temperature (temperature may affect population growth rates, but it is not generally altered by the organism whose growth is affected) and oxygen (oxygen is consumed by heterotrophs, but it is generally not limiting to population growth, at least not in terrestrial ecosystems).

One consumer and one resource: the concept of R^*

For most organisms, the per capita rate of population growth has an upper bound, which is approached in an asymptotic fashion as the abundances of all potentially limiting resources become high enough (Figure 7.3). This pattern of population growth can be described by a variety of mathematical functions [e.g., the Monod (1950) equation first developed for bacteria growing on organic substrates]. However, for heuristic purposes, we can skip the formal mathematics and simply graph the per capita birth rate of a consumer population as a positive, decelerating function of the availability of a resource R (Figure 7.4A). If we assume that the consumer population experiences some per capita density-independent mortality rate m (represented by the horizontal line in Figure 7.3A), then the consumer population will be at equilibrium ($dN/dt = 0$), where the lines in Figure 7.4A intersect. This intersection point specifies the resource availability (R^*) at which the consumer's per capita birth rate (b) equals its per capita mortality rate (m). Thus, $\boldsymbol{R^*}$ is the minimum equilibrium resource requirement needed to maintain a consumer population experiencing a given mortality rate (m). The R^* result is not dependent on the assumption of density-independent mortality,

but we make that assumption here for simplicity in graphic presentation.

If a consumer is introduced to an environment with mortality rate m and $R > R^*$, the consumer population will increase over time, and the abundance of the resource will decline, eventually reaching R^*, the point at which both consumer and resource are at their equilibrium values (Figure 7.4B). For abiotic resources, R^* generally defines a stable equilibrium point (Grover 1997; but see Abrams 1989 for exceptions); for biotic resources, the stability of the equilibrium depends on the specific functions used to describe the consumer and resource growth responses (Hsu et al. 1977; Armstrong and McGehee 1980). For now, we will assume that a stable equilibrium between the consumer and its resource is possible.

Two consumers competing for one resource

Now consider a second consumer species that also requires resource R. We can examine the outcome of this interaction (i.e., two species competing for one resource) by graphing the per capita birth rates of both consumer species as functions of resource availability (Figure 7.5A). For simplicity, we will assume that both species have the same per capita mortality rate (m) (we will relax this assumption in a minute). Figure 7.5B shows the population dynamics of the two species, starting with low consumer densities and abundant resources. Initially, N_1's population grows faster, because at high resource levels its net rate of population growth (birth rate minus mortality rate) is higher. However, as resources are consumed and R declines, the growth rate of N_1's population slows, reaching zero when resource levels reach R^*_1. If N_1 was the only consumer species present, the system would equilibrate when $R = R^*_1$. The situation changes, however, when N_2 is present. As we can see in Figure 7.5A, N_2's birth rate is greater than its mortality rate at $R = R^*_1$. Thus, its population will grow and continue to deplete R until $R = R^*_2$ (Figure 7.5B). When $R = R^*_2$, N_2's birth rate matches its mortality rate (Figure 7.4A), and N_2's population growth ceases. Note, however, that $R^*_2 < R^*_1$. Therefore, N_1 cannot maintain its population at a resource level equal to R^*_2, and its population declines over time.

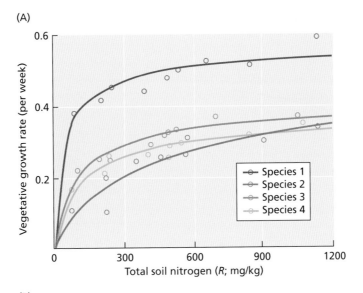

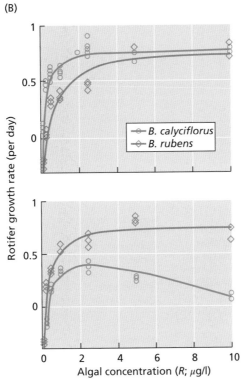

Figure 7.3 Examples of resource-dependent consumer population growth rates. Note that as resource availability rises in each case, the increase in growth rate reaches an asymptote. (A) Vegetative growth rates (measured as aboveground biomass) for four grass species as functions of soil nitrogen availability (*R*). (B) Population growth rates of two rotifer species (*Brachionus*) as functions of food (algal) concentration (*R*). *B. rubens* and *B. calyciflorus* were fed on the algae *Chlamydomonas sphaeroides* (above) and *Monoraphidium minutum* (below). (A) After Tilman and Wedin (1991a); (B) after Rothhaupt (1988).

(A)

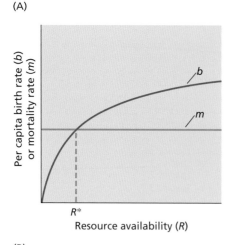

(B)

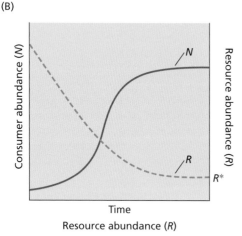

Resource abundance (R)

Figure 7.4 Graphic representation of a one consumer–one resource system. (A) The per capita birth rate (b) of the consumer is a positive, decelerating function of the availability of the resource (R), and the per capita mortality rate (m) is constant. (B) Change in consumer (N) and resource (R) abundances over time. The consumer population will increase and the abundance of the resource will decline, eventually reaching R^*.

Eventually, N_1 is eliminated by competition with N_2 (Figure 7.5B).

The analysis in Figure 7.5 suggests the following results:

- two species cannot coexist on one limiting resource;
- the species that can maintain positive population growth at the lowest resource level (lowest R^*) will win in competition.

Hsu et al. (1977, 1978) and Armstrong and McGehee (1980) showed that we need to modify these results slightly to the following:

(A)

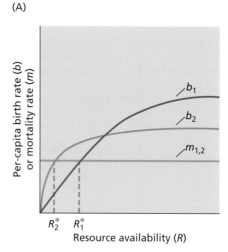

(B)

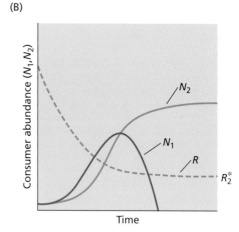

Time

Figure 7.5 Graphic representation of two consumers competing for one limiting resource. (A) The per capita birth rates (b_1, b_2) and per capita mortality rate ($m_{1,2}$) for consumer species N_1 and N_2 as a function of resource availability (R). Note that consumer species N_2 can maintain its population on a lower resource level than can consumer species N_1 ($R^*_2 < R^*_1$). (B) Change in consumer and resource abundances over time. Note that consumer species N_1 is driven extinct by competition with consumer species N_2. See http://communityecologybook.org/conres.html for a dynamic online version of this model.

- two species cannot coexist on one limiting resource *at constant population densities*;
- the species with the lowest R^* will win.

This **R^* rule** provides a powerful and intuitively satisfying way to think about the outcome of competition for a single limiting resource.

In the example above, it is assumed that each species had the same fixed mortality rate m. However, different morality rates can easily be incorporate into the graphic analysis. We can also consider how differences between consumer species' reproductive rates may affect the outcome of competition. In Figure 7.5, the species' birth rate curves cross, such that one species has a relatively greater population growth rate at low resource levels and the other species has a relatively greater population growth rate at high resource levels. These two consumer types have been termed **gleaners** and **opportunists**, respectively (Frederickson and Stephanopoulos 1981; Grover 1990). Species 2 is a gleaner because it grows relatively well when the resource is scarce. Species 1 is an opportunist—it grows relatively poorly at low resource levels but grows well when resources are abundant. Thus, the reproduction curves in Figure 7.5A illustrate a trade-off that has important consequences for species coexistence.

Coexistence on a single, fluctuating resource

In a constant environment, we would expect species 2 in Figure 7.5 to win in competition because it has the lower R^*. However, Armstrong and McGehee (1976, 1980; see also Koch 1974; Hsu et al. 1977, 1978) showed that we need to consider another possibility. If resource levels fluctuate between high and low abundance, then the two species may coexist because species 1 has a growth advantage at high resource abundance and species 2 has a growth advantage at low resource abundance. If R is an abiotic resource, then exogenous temporal variation in the physical environment (i.e., seasons) may cause fluctuations in resource abundance and allow the species to coexist (Grover 1997). If R is a biotic resource, then the interaction between consumers and resources (think predators and their prey) may generate endogenous temporal variation in resource abundance that may also allow species coexistence (Armstrong and McGehee 1980). Recall from Chapter 5 that predator and prey populations have a natural tendency to cycle, especially when predators are efficient at exploiting their prey. Under many conditions, these fluctuations in predator and prey densities may be sustained indefinitely as a stable limit cycle (May 1973a; Nisbet and Gurney 1982).

Armstrong and McGehee (1980) showed that when predator and prey densities fluctuate in a stable limit cycle, then two (or more) predator (consumer) species may coexist indefinitely on a single resource. The conditions allowing for this coexistence are:

- one of the species' resource-dependent growth functions (or equivalently, their functional responses) must be non-linear (as in Figure 7.5);
- there must be a trade-off between performance at high and low resource levels (i.e., the species with the higher growth rate at low resource densities experiences a lower growth rate at high resource densities—the gleaner–opportunist trade-off).

Chesson (2000a) refers to this coexistence mechanism as **relative non-linearity**. Abrams (2004b) further examined the effects of temporal resource fluctuations on species coexistence and showed that species coexistence is most easily promoted when the superior competitor causes fluctuations in a biotic resource, but the inferior competitor is not affected as much by those fluctuations. Levins (1979) called this "consuming the variance" because the species who benefits the most from variation in the resource must decrease this variation more than its competitor to allow coexistence. Grover (1997) reviewed some empirical examples of microorganisms that appeared to coexist via temporal fluctuations in the density of a single resource. More recently, Letten et al. (2018) showed experimentally that relative non-linearity promoted coexistence in a laboratory system of competing nectar yeasts. As it turns out, relative non-linearity is one component of a broader class of mechanisms by which environmental variation may affect coexistence, which we explore in more detail in Chapter 8.

Competition for multiple resources

We can extend the graphic analysis of exploitative competition outlined in the previous section to the case of two species competing for two limiting resources, thereby gaining insight into the nature of competition for multiple resources. The graphic treatment that follows is based on Tilman's (1982) analysis; rigorous mathematical treatments of the results can be found in Hsu et al. (1981) and Butler and Wolkowicz (1987). Tilman considered many different types of resources in his analysis, but most studies of exploitative competition have focused on either essential or substitutable resources, and we will limit our consideration to these two resource types. **Essential resources** are required for growth and one essential resource cannot substitute for another. For example, plants require about 20 different mineral resources (e.g., nitrogen, phosphorus, *and* potassium) in order to grow and reproduce. On the other hand, if one resource can take the place of another, each is considered a **substitutable resource**. For example, any one of several different prey species (e.g., rabbits, squirrels, *or* mice) may provide a complete diet for a carnivore. A more extensive discussion of resource types may be found in León and Tumpson (1975), Tilman (1982), and Grover (1997). Elser et al. (2007) and Harpole et al. (2011) provide discussions and examples of nutrient co-limitation and other synergistic interactions between multiple limiting resources.

Competition for two essential resources

Let's assume that the environment contains two essential resources, R_1 and R_2, whose abundances we can graph on the x and y axes, respectively (Figure 7.6). Given a consumer with some density-independent mortality rate, we can plot the levels of R_1 and R_2 at which the consumer's per capita birth rate matches its per capita mortality rate and its population is at equilibrium (i.e., $dN_1/dt = 0$). These points define the consumer's **zero net growth isocline** (or in common shorthand, its **ZNGI**; Tilman 1982). For essential resources, the ZNGI is represented by straight lines that run parallel to both axes of the graph, forming an L-shaped curve. At all points along the ZNGI, the consumer population is at equilibrium. At resource levels below the ZNGI, the consumer population goes extinct (birth rate < mortality rate). At resource levels above the ZNGI, the consumer population increases (birth rate > mortality rate).

Note that, for essential resources, there is a level of each resource at which any further increase in abundance will not increase the consumer population's growth rate (i.e., growth rate is limited by the availability of the other essential resource). In Figure 7.6, the consumer population is limited by the availability of R_2 at points along the

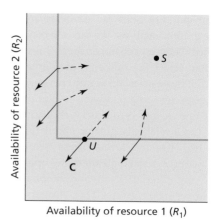

Figure 7.6 The zero net growth isocline (ZNGI) for a consumer species feeding on two essential resources (R_1, R_2). S is the resource supply point, representing the availability of resources R_1 and R_2 in the environment if there were no consumers. The rates at which R_1 and R_2 are consumed are indicated by consumption vectors (C); solid arrows pointing from the ZNGI toward the resource axes; four of many possible consumption vectors are shown here. The slope of each of the consumption vectors is the ratio of R_2:R_1 consumed, and this ratio is determined by the traits of the consumer. The rate of resource supply is indicated by the supply vectors (dashed lines), which point from the ZNGI toward the resource supply point, S. Resource consumption matches resource supply when the consumption and supply vectors are directly opposed (at point U). At this point, consumer and resource populations are in equilibrium. Note that at the equilibrium point U in this example, the consumer population is limited by the availability of resource 2.

ZNGI to the right of the corner of the L-shaped curve; at points along the ZNGI above the corner, the consumer population is limited by the availability of R_1. As mentioned previously, the consumer's ZNGI represents the resource levels at which its population is unchanging. More information is needed, however, to determine whether the system as a whole is at equilibrium (i.e., both consumer and resource populations are unchanging). As Tilman (1982, p. 61) noted,

[F]our pieces of information are needed to predict the equilibrium outcomes of resource competition. These are the reproductive or growth response of each species to the resource, the mortality rate experienced by each species, the supply rate of each resource, and the consumption rate of each resource by each species. An equilibrium occurs when the resource dependent reproduction of each species exactly balances its mortality and when resource supply exactly balances total resource consumption for each resource.

The ZNGI defines the conditions (i.e., abundances of R_1 and R_2) under which the consumer's reproductive rate balances it mortality rate. Thus, we have half of the information we need to determine the outcome of competition. Next, we need to consider the dynamics of the two resources, R_1 and R_2.

Resources will be at equilibrium when the rate of consumption by the consumer matches the rate of resource supply. We can represent consumption at any point on the ZNGI by plotting vectors pointing toward the resource axes; each of these **consumption vectors** (**C** in Figure 7.6) has a slope equal to the relative amounts of R_1 and R_2 consumed. Think of these consumption vectors as indicating the rate and direction in which the consumer is drawing down the two resources. Counterbalancing consumption is the rate at which resources are supplied, which can be represented as a vector pointing toward the **resource supply point** (*S* in Figure 7.6)—the expected standing stock of resources 1 and 2 in the absence of consumption. At one and only one point (*U* in Figure 7.6) along the consumer's ZNGI is the consumption vector in exact opposition to the supply vector. At this point, the system is at equilibrium. Resource consumption matches resource supply, and (because the equilibrium point is on the consumer ZNGI) the consumer's birth rate matches its death rate. For abiotic resources, the equilibrium point is stable (León and Tumpson 1975; Tilman 1980). For biotic resources, stability is not assured; the outcome depends on the resource growth functions and on any interactions between the resources (Grover 1997). For our goal of understanding the potential outcomes of competition for two resources, we will assume that a stable equilibrium between consumer and resources is possible at point *U*.

A second consumer species (N_2) may be added to the system by graphing its ZNGI along with that of consumer species N_1. There are four possible orientations of the consumers' ZNGIs and their consumption vectors, and these orientations dictate the outcome of two-species competition (Tilman 1982; Grover 1997). The descriptions of these outcomes given next follow closely from Tilman (1982).

Species 1 wins

In Figure 7.7A, the ZNGI for species 1 is always "inside" the ZNGI for species 2 (i.e., it is closer to the origin of the graph). Therefore, species 1 requires less of both resources than species 2 in order to maintain its population at equilibrium (birth rate = mortality rate). The positions of the ZNGIs in Figure 7.7A define three regions of potential resource availabilities. Consider what would happen to species 1 and species 2 if they were introduced into an environment where the resource supply points fell within each of these regions. Environments with resource supply points in region 1 would have insufficient resources to support either consumer species (birth rate < mortality rate). Thus, both species would go extinct. In region 2, the supply of resources would be sufficient to maintain the population of species 1, but not that of species 2. Thus, only species 1 would survive in an environment whose resource supply point fell within region 2. Finally, in region 3, there would be enough resources for each species to survive by itself. However, if both species were introduced into an environment with a resource supply point in region 3, the population growth of species 1 would reduce resources to a point on species 1's ZNGI, which is below the ZNGI of species 2. Therefore, species 1 would outcompete and exclude species 2.

Species 2 wins

In Figure 7.7B, the ZNGI for species 2 is always "inside" the ZNGI for species 1. As in Figure 7.6A, the winner of the competition will be the species that can tolerate the lowest level of resource supply. Neither species can survive in environments with resource supply points in region 1, only species 2 can survive in region 2, and species 2 will outcompete species 1 in region 3.

Species 1 and 2 coexist at a stable equilibrium point

In Figure 7.7C, the ZNGIs for the two species cross. The point at which the two ZNGIs cross is a 2-species equilibrium point because at this point, each species' birth rate matches its mortality rate. Whether this equilibrium point is stable or not depends in part on the orientation of the consumption vectors

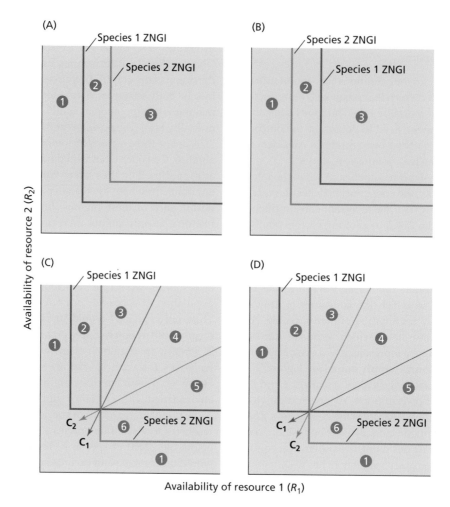

Figure 7.7 Graphic model of two species competing for two essential resources (R_1, R_2), showing four outcomes of resource competition. (A) Species 1's ZNGI lies inside species 2's ZNGI; species 1 outcompetes species 2. (B) Species 2's ZNGI lies inside species 1's ZNGI; species 2 outcompetes species 1. (C,D) The ZNGIs for species 1 and 2 cross. The intersection of the ZNGIs defines an equilibrium point at which the densities of species 1 and species 2 are unchanging (i.e., births = deaths for both species). C_1 and C_2 are the consumption vectors for the two species. (C) The equilibrium point is locally stable because each species consumes more of the resource that most limits its own growth. For environments with resource supply points in region 4, species 1 and species 2 coexist. (D) Consumption vectors for the two species are reversed from those in (C); the equilibrium point is unstable because each species consumes more of the resource that most limits the growth of the other species. For environments with resource supply points in region 4, the outcome of competition depends on initial conditions, and either species 1 or species 2 may win. After Tilman (1982). See http://communityecologybook.org/conres2.html for an online dynamic version of this model.

for each species. (Note that the outcomes of competition in the cases represented by Figure 7.7A,B are unaffected by the orientation of the species' consumption vectors, and therefore we did not include consumption vectors in those graphs.) In Figure 7.7C, the species' ZNGIs and consumption vectors define six regions into which resource supply points may fall. Environments with resource supply points that fall in region 1 have insufficient resources to maintain populations of either species. Supply points that fall in region 2 are sufficient to maintain species 1, but not species 2. Likewise, supply points that fall in region 6 can maintain species 2, but not species 1. Regions 3–5 have supply points that could potentially maintain either species when growing alone.

In competition, however, species 1 will exclude species 2 from environments with resource supply points that fall in region 3, because species 1 will draw resources down to its own ZNGI and to the left of the two-species equilibrium point (a point at which species 2 cannot survive). Likewise, species 2 will exclude species 1 from environments with resource supply points that fall within region 5. Environments characterized by resource supply points that fall in region 4 may support the stable coexistence of both species 1 and species 2 because the joint consumption of resources by both species will draw resource availabilities down to the equilibrium point, where the ZNGIs intersect. *This two-species equilibrium point is locally stable because at this point, each species consumes proportionally more of the resource that most limits its own growth* (León and Tumpson 1975; Tilman 1980). To see this, note that at the equilibrium point in Figure 7.7C, species 1 consumes relatively more of resource 2 (as shown by the slope of its consumption vector), and at the equilibrium point the growth of species 1 is limited by the availability of resource 2. Similar logic can be applied to species 2 at the equilibrium point.

Species 1 and 2 coexist, but the equilibrium point is unstable

Figure 7.7D is similar to Figure 7.7C in that the ZNGIs for the two species cross. However, the consumption vectors for the two species are reversed from those in Figure 7.7C. This reversal causes the two-species equilibrium point to be locally unstable. At equilibrium, species 1 is limited by resource 2 and species 2 is limited by resource 1. However, the slopes of the consumption vectors show that species 1 consumes relatively more of resource 1 and species 2 consumes relatively more of resource 2. Thus, at the equilibrium point, each species consumes relatively more of the resource that most limits the growth of the *other* species, not the resource that most limits its own growth. This situation causes the equilibrium to be unstable; any deviation away from the equilibrium point will be magnified over time until either species 1 or species 2 outcompetes the other. Thus, environments with resource supply points in region 4 will lead to dominance by either

species 1 or species 2 (and the extinction of the other species); the winner depends on the starting conditions. Environments with resource supply points in regions 1, 2, 3, 5, or 6 will lead to competitive outcomes identical to those in Figure 7.7C.

Competition for two substitutable resources

If resources are perfectly substitutable, such that each resource supplies the full complement of substances required for a consumer's growth and one resource can substitute completely for another, then the consumer ZNGIs can be represented by straight lines in resource state space (Figure 7.8). Most heterotrophs (e.g., predators) consume resources that are best viewed as substitutable. León and Tumpson (1975) suggested that a fundamental difference between competition among plants and competition among animals is the difference between essential and substitutable resources. Thus, by considering competition for essential and for substitutable resources, we have covered the two major types of exploitative competition found in natural communities. A number

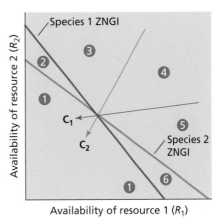

Figure 7.8 Graphic model of two species competing for two perfectly substitutable resources (R_1, R_2). In this example, the ZNGIs for species 1 and 2 cross, and the intersection of the ZNGIs defines an equilibrium point at which the densities of species 1 and species 2 are unchanging (i.e., births = deaths for both species). C_1 and C_2 are the consumption vectors for the two species. The equilibrium point is potentially stable because each species consumes more of the resource that most limits its own growth. In environments with resource supply points in region 4, species 1 and species 2 coexist. See http://communityecologybook.org/conres2.html for an online dynamic version of this model.

(A)

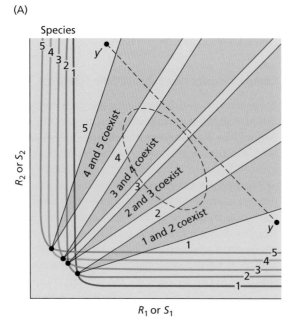

(B)

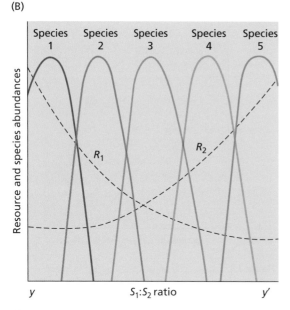

of factors may cause resources to be imperfectly substitutable, resulting in nonlinear ZNGIs. However, as long as the ZNGIs decrease monotonically with a decline in resource abundance, the general conclusions below will apply (Grover 1997).

If the ZNGIs of two consumer species do not intersect, then the species whose ZNGI is closest to the origin will win in competition and exclude the other species (i.e., it will have a lower R^* for both resources). Thus, just as in the case of competition for essential resources, coexistence among two species competing for substitutable resources requires that the species' ZNGIs cross (i.e., there must be a trade-off in the species' abilities to grow on the two resources). The criteria for stable coexistence on two substitutable resources are also similar to those for essential resources, although there are some important distinctions. *For substitutable resources, stable coexistence can occur if each species consumes proportionally more of the resource that most limits its own growth and if equilibrium resource densities are stable and positive* (MacArthur 1972; León and Tumpson 1975; Grover 1997). In the case of biotic resources that are themselves competing, the factors determining stable positive resource populations can be complex (see discussions in Abrams 1988; Grover 1997; Abrams and Nakajima 2007).

Spatial heterogeneity and the coexistence of multiple consumers

The previous examples show how two species may coexist on two resources. Coexistence requires that the species' ZNGIs cross or, in other words, that the species exhibit a trade-off in their abilities to compete for the two resources. In the example shown in Figure 7.7C, species 1 is a better competitor (has a lower R^*) for resource 1, but it is a poorer competitor for resource 2. In general, stable coexistence requires that each species consume more of the resource that

Figure 7.9 Spatial heterogeneity in resources may promote the coexistence of multiple competing species. (A) ZNGIs for five consumer species (1–5) utilizing two resources (R_1, R_2). In this example, there is an interspecific trade-off such that species 1–5 are each progressively better competitors for resource 1 and progressively poorer competitors for resource 2 (i.e., species 1 is the best competitor for R_2 but the worst competitor for R_1). The four dots indicate two-species equilibrium points (where two species' ZNGIs cross). The line from y to y' represents a gradient in resource supply points (S_1, S_2), and the ellipse represents a habitat that has spatial heterogeneity in resource supply points. (B) Exploitative competition causes the five competing species to become separated in space along the resource supply gradient (y to y'). After Tilman (2007b).

most limits its growth. Tilman (1980, 1982) showed how this argument can be extended to explain the coexistence of multiple species on two resources if the species follow the same trade-off curve, such that a species that is better at competing for resource 1 is inferior at competing for resource 2. In this case, there will be regions of the two-resource state space in which different pairs of species may coexist (Figure 7.9A). Moreover, if the environment varies spatially in the supply rates of the two resources, this spatial heterogeneity may allow the coexistence of multiple consumer species in the community. For example, spatial variation in resource supply could be modeled as a gradient in the supply rate of resource 1 and resource 2 (the dashed line from resource supply point y to resource supply point y' in Figure 7.9A). In this case, competitors will be separated in space along the resource supply gradient running from y to y' (Figure 7.9B). If spatial heterogeneity in resource supply is instead modeled as a mixture of resource patches varying in their supply rates of R_1 and R_2 (dashed ellipse in Figure 7.9A), then species would coexist as one- or two-species pairs intermingled in space (Tilman 1982, 2007b).

Testing the predictions of resource competition theory

One of the strengths of resource competition theory is that it provides a mechanistic link between consumers and resources, allowing us to test the theory's predictions using measures of resource use and resource supply. In contrast, more phenomenological competition models (e.g., the Lotka–Volterra model) require estimates of the relative strengths of the interactions within and between species (the competition coefficients in the L–V model) in order to predict the long-term competitive outcomes. In the simple case of multiple species competing for a single limiting resource, resource competition theory predicts that the species with the lowest equilibrium resource requirement (R^*) will exclude all other competing species. The first step in testing this prediction is to measure each species' R^*, by either:

- by growing the species in monoculture on the resource and measuring the resource concentration

when the consumer and resource are at steady state;
- calculating R^* from the species' growth kinetics (Wilson et al. 2007).

When the species are then grown in mixture and allowed to compete, theory predicts that the species with the lowest R^* will win.

A number of laboratory experiments have tested the R^* prediction using two species in competition for a single limiting resource (Figure 7.10 shows one example). Wilson et al. (2007) summarized the results from 43 laboratory competition studies with bacteria, phytoplankton, and zooplankton, and concluded that 41 of the studies were consistent with the R^* prediction, while the two remaining studies were inconclusive. Miller et al. (2005) arrived at a somewhat different conclusion regarding support for the R^* prediction after examining a smaller number of experimental studies, concluding that the results were mixed and that the theory had been insufficiently tested. Wilson et al.'s review demonstrates that "at least in competition experiments with bacteria, phytoplankton, and zooplankton, R^* is almost always a good guide to competitive outcome" (Wilson et al. 2007, p. 704). However, few experiments have tested the R^* prediction for a single resource outside the laboratory. One such study was done by Tilman and Wedin (1991b) and Wedin and Tilman (1993), who grew five perennial grass species in monocultures and measured their R^*s for the limiting soil resource (nitrogen), then tested the outcome of competition among species in pairwise experiments. In all four of the competition pairings, the species with the lowest R^* won, as predicted by the theory.

When two species compete for two limiting resources, resource competition theory predicts four potential competitive outcomes, including stable coexistence. These predictions also have been tested in the laboratory with microorganisms (e.g., Tilman 1977, 1981; Kilham 1986; Sommer 1986; Rothhaupt 1988, 1996; reviewed in Wilson et al. 2007). The results of these laboratory experiments generally support the theory (Wilson et al. 2007; see Figure 7.11 for an example). More recently, Burson et al. (2018) conducted a series of laboratory experiments examining the outcome of competition for two limiting nutrients (nitrogen and phosphorous) and light

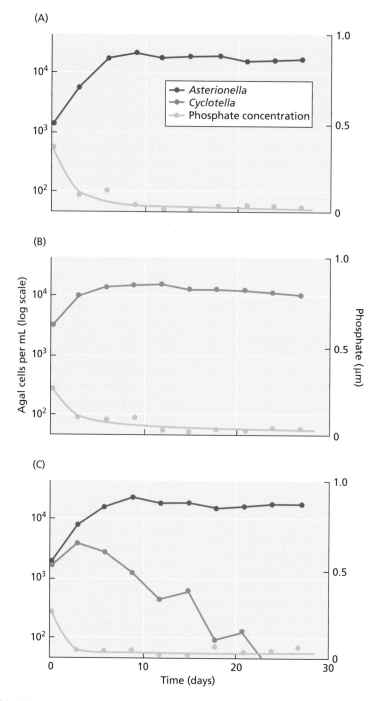

Figure 7.10 A test of the R^* prediction for two species of freshwater phytoplankton (algae) potentially competing for a limiting nutrient. Under the conditions tested in the experiment, both algal species were limited by a single nutrient—phosphorus—and the *Asterionella* species had a lower R^* for phosphorus than did the *Cyclotella* species. (A, B) Each species grew successfully in monoculture. (C) When the two species were grown in competition, *Asterionella* eliminated *Cyclotella*. After Tilman (1977).

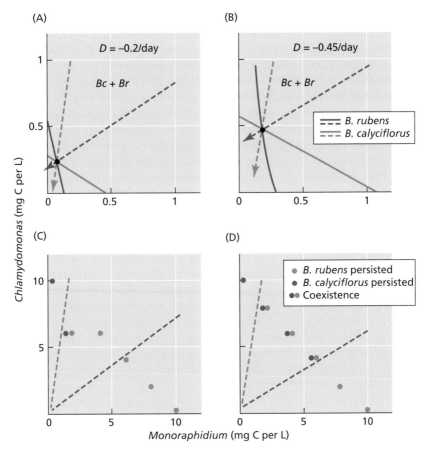

Figure 7.11 Tests of the predictions of resource competition theory for two rotifer species (*Brachionus rubens* and *B. calyciflorus*) competing for two algal resources (*Monoraphidium* and *Chlamydomonas*) in the laboratory. As shown in Figure 7.3B, when growing in monoculture *B. rubens* has the highest growth rate at all densities of *Monoraphidium*, while *B. calyciflorus* has the highest growth rate at all densities of *Chlamydomonas*. (A,B) Predicted outcomes (ZNGIs and consumption vectors) at two different dilution (mortality) rates (*D*). The two rotifers are predicted to coexist at resource concentrations defined by the sector between the dashed lines (*Bc* + *Br*). (C,D) Actual competitive outcomes. Theory correctly predicted the outcome of interspecific competition in 11 of the 12 cases. After Rothhaupt (1988).

using a diverse community of 19 phytoplankton taxa (mostly classified to genus) that they collected from the North Sea and brought into the laboratory. The results of this experiment matched some of the predictions of resource competition theory, but not all. Most notably, four of the 19 taxa coexisted in all the competition treatments, whereas the other (15) taxa were competitively excluded. Different species became dominant under different resource limitations, indicative of trade-offs in competitive abilities for these resources. However, more species coexisted than predicted based on the number of limiting resources.

Dybzinski and Tilman (2007) tested the predictions of resource competition theory in the field using six perennial prairie grasses in an 11-year experiment at the Cedar Creek Reserve in Minnesota looking at competition for soil nitrogen and light. In six of the eight pairwise species combinations, the species with the lowest resource requirements won in competition. Coexistence was predicted and observed in two pairwise combinations of species. However, it was unclear whether species coexistence in these cases was due to a trade-off between competition for nitrogen and competition for light (the limiting resources). Other, more indirect tests of resource competition theory have looked for changes in species composition and in resource concentrations across ecosystems that differ naturally in the supply of limiting nutrients. Tilman (1982) found some support for the theory in the distribution of plant species across nutrient gradients in freshwater and terrestrial ecosystems, but Leibold (1997; see also Chase and Leibold 2003a) found no support for the theory's prediction that nitrogen and phosphorus availability should covary negatively across freshwater ecosystems. Thus, predictions for two species coexisting on two limiting resources are nicely supported by laboratory studies with microorganisms, but there is only limited evidence from field experiments.

It is not surprising that nearly all the direct tests of resource competition theory have come from the laboratory and not the field. In the laboratory, the assumptions of the theory can be met far more easily, and the outcome of competition can be observed over short time scales using microorganisms. It is satisfying to see the predictions of theory generally supported under these controlled conditions. However, the paucity of field tests leaves open the question of whether the theory is useful for predicting the outcome of competitive interactions in the field or the distribution and abundance of organisms in nature (Miller et al. 2005).

It is interesting to note that Dybzinski and Tilman (2007, p. 317), the authors of one of few field tests of resource competition theory, concluded that "while competition for multiple resources might maintain some diversity in natural (plant) communities, other mechanisms are almost certainly operating to maintain the remainder." We will consider other diversity-maintaining mechanisms in the following chapter when we expand on the study of competitive systems to include such factors as environmental variation, variable nutrient requirements, and species interactions in food webs

Apparent competition

We have seen that competition occurs when two or more species share a limiting resource. Species that share a predator also may have negative indirect effects on each other, an interaction that has been called **apparent competition**. The simple mechanism behind this negative interaction is a numerical response by a predator to its prey: an increase in the abundance of one prey species may lead to an increase in predator abundance, which in turn may have a negative effect on the abundance of other prey species that the predator consumes. The predator's numerical response may be behavioral and occur quickly (e.g., predators may aggregate in areas of high prey density), or it may result from slower demographic processes (e.g., predator numbers increase due to higher birth rates or lower death rates in response to increased prey density).

MacArthur (1970, p. 21) noted the potential for shared predation to lead to negative indirect effects on prey species; however, it was Holt (1977) who developed the theory and explored its consequences for species coexistence (see Holt and Bonsall 2017 for a comprehensive review). Holt referred to the negative effect of shared predation as "apparent competition" because the negative interaction between species sharing a predator resembles the negative interaction between species sharing a limiting resource (exploitative competition). Apparent competition has the potential to be a strong interaction in most food webs. In a survey of 23 marine intertidal communities, Menge (1995) found that apparent competition occurred in about 25% of the food webs studied (which made it the second most common indirect effect).

It is possible to understand the action of apparent competition by using a version of the simple predator–prey models developed in Chapter 5. Following Holt et al. (1994), we can define the population growth rates for two prey species i ($i = 1, 2$) as

$$dN_i/dt = r_i N_i - a_i N_i P, \qquad \text{[Eqn 7.7]}$$

where N_i is the abundance of prey species i, r_i is the intrinsic growth rate for prey species i, P is the abundance of the predator, and a_i is the predator's attack rate on prey species i (number of prey eaten per predator, per prey). You should recognize Equation 7.7 as a multispecies version of the familiar L–V model of predator–prey dynamics (see Equation 5.5).

Like the L–V model, Equation 7.7 assumes no self-limitation in the prey population and assumes that the predator has a linear functional response. We can define the population growth rate for the predator as

$$dP/dt = \left(\sum_{i=1}^{2} f_i a_i N_i P - qP \right) + I - EP \qquad \text{[Eqn 7.8]}$$

where f_i is the number of predators born as a result of the consumption of each prey item and q is the predator's density-independent death rate. Equation 7.8 differs from the standard L–V formulation for the predator population (e.g., Equation 5.7) in that it includes terms for the rates of immigration (I) and emigration (E) of predators. Holt et al. (1994) included an immigration rate for the predator to help stabilize the predator's population dynamics. For the present purposes, there is no need to be concerned with these predator migration terms in order to understand the impact of apparent competition on species interactions.

In the simple model described above, the prey species that dominates in apparent competition is the species that withstands (and maintains) the highest density of the shared predator relative to other prey species. This can be seen by noting that the per capita rate of change in the prey population is

$$dN_i/N_i dt = r_i - a_i P \qquad \text{[Eqn 7.9]}$$

Thus, prey species i will have a positive per capita growth rate when $dN_i/N_i dt > 0$, which is true when $r_i > a_i P^*$ (where P^* is the equilibrium density of predators in the system). The prey species that can maintain a positive population growth rate at the highest equilibrium predator density, defined as

$$P^* = r_i/a_i \qquad \text{[Eqn 7.10]}$$

will exclude all other prey species from the system. Holt et al. (1994) labeled this statement the **P^* rule**. The P^* rule is analogous to the R^* rule for interspecific competition, which states that the species able to maintain its population at the lowest resource level (R^*) will dominate all other species when competing for a single resource. Likewise, when two (or more) species share a common predator, the winning species will be the one that can maintain its population at the highest density of predators (P^*). Equation 7.10 emphasizes the two attributes that allow prey species to persist at high predator densities:

- a high intrinsic growth rate for the prey (r_i);
- a low vulnerability to attack by the predator (a_i).

Predator preference and prey vulnerability are two major factors influencing attack rates. Abrams

(1987, 2004a) and others discuss a variety of traits of predators and prey that may affect attack rates.

The impact of apparent competition between prey will be weakened when:

- the predator is not food-limited, but is limited instead by other resources (breeding sites, for example);
- prey species have spatial or temporal refuges from predation; or
- the predator exhibits **prey switching** behavior (i.e., feeding preferentially on the more abundant prey; Murdoch 1969).

If predators concentrate their feeding on the most abundant prey species, then the impact of apparent competition between prey species will be reduced. Such switching behavior may even lead to a short-term indirect mutualism between prey species.

There is a well-developed theory for apparent competition that includes a number of refinements beyond the simple L–V type model presented above (e.g., Holt 1984; Abrams 1987; Holt et al. 1994; Leibold 1996; Noonburg and Byers 2005; Křivan and Eisner 2006). For many years, theory outpaced empirical examples of apparent competition, and the classic study by Schmitt (1987) on apparent competition in subtidal reef communities was often cited as one of the few field experiments showing its effect on species distributions and abundance. However, the impact of apparent competition in natural communities is now well-documented (see reviews by Chaneton and Bonsall 2000; Holt and Bonsall 2017), as is its role in applied ecology (van Veen et al. 2006). It has been applied to predator–prey and host–parasitoid interactions involving insect pests (e.g., Müller and Godfray 1997; Murdoch et al. 2003; Östman and Ives 2003), to the biocontrol of pests through predator introductions (e.g., Willis and Memmott 2005; Carvalheiro et al. 2008), and to studies of the impact of invasive species on native species (e.g., Norbury 2001; Rand and Louda 2004; Orrock et al. 2008; Oliver et al. 2009), including evolutionary and ecological effects of shared predation (e.g., Lau 2012). In both the basic and applied literature, numerous studies have noted that species interactions involving shared enemies often show asymmetrical effects—that is, one species is more strongly impacted by a shared predator than is the other (Chaneton and Bonsall 2000). This asymmetrical impact, predicted by the theory of apparent competition (Holt and Lawton 1994; Brassil and Abrams 2004), may lead to the exclusion of species from areas where they might otherwise exist (e.g., Meiners 2007; Oliver et al. 2009). Finally, species may compete for resources, as well as sharing a predator, leading in the simplest case to a diamond-shaped food web that reflects a mixture of exploitative competition and apparent competition (see Figure 10.13C). Some of the consequences of this "diamond interaction web" will be explored in Chapter 10, when food webs, keystone predation, and ecosystem regulation will be discussed.

Conclusion

This has been quite a long chapter, yet it has only scratched the surface of a vast literature describing the conceptual foundation of interspecific competition. (Interested readers are referred to Grover 1997; Case 2000; and Chesson 2000a for further introduction.) However, despite our limited exposure to competition theory, we have learned a great deal about the nature and outcomes of interspecific competition. We have seen that both the phenomenological L–V model and the more mechanistic consumer–resource model specify general conditions for two-species competition that lead either to competitive exclusion (one species wins and the other is excluded) or to the coexistence of species (with either a stable or an unstable equilibrium point). In both models, we see that *stable coexistence is predicted to be possible when intraspecific competition is stronger than interspecific competition.* The consumer–resource model specifies this condition for stable coexistence mechanistically, based on the species' relative consumption of their limiting resources. This consumer–resource model is sometimes referred to as a "resource-ratio" model of competition because the outcome of competition depends in part on the ratios of resources consumed and supplied (see Figure 7.9B). In the L–V model, when each species has a stronger effect on the growth of its competitor than it does on its own growth, the result is an unstable equilibrium and competitive displacement, with the winner determined by the initial conditions (founder control).

The same outcome is predicted by the consumer–resource competition model when each species consumes more of the resource that limits the other species at the equilibrium point. Thus, *an unstable equilibrium and founder control can occur when interspecific competition is stronger than intraspecific competition.*

Conditions for the competitive displacement of one species by another are most easily visualized with the consumer–resource model. In the case of a single limiting resource, the species with the lowest R^* wins and excludes the other species. In the case of two species competing for two resources, the winner will be the species whose ZNGI falls inside that of the other species (i.e., the species that can maintain a population on a lower level of both resources). Conditions for competitive exclusion in the L–V model are less intuitive, being expressed as the ratios of the species' competition coefficients and carrying capacities.

The rule that coexistence of n competing species requires at least n limiting resources (or n limiting factors) may be relaxed if resources are allowed to vary spatially or temporally (Grover 1997). It is important to note that species coexistence via temporal or spatial resource heterogeneity requires trade-offs in the abilities of different species to use resources. In the case of temporal variation in resource availability (driven by either exogenous or endogenous factors; e.g., limit cycles), species coexistence requires a trade-off such that species that grow relatively well when resources are rare ("gleaners") grow relatively poorly when resources are abundant (and "opportunist" species do the reverse). This trade-off can be expressed as a crossing in the species' nonlinear resource-dependent growth functions (e.g., Figure 7.5A). Chapter 14 will examine how temporal variation in the environment may promote species coexistence by another mechanism, the "storage effect" (Chesson 2000a).

In the case of spatial resource heterogeneity, coexistence of multiple species requires a trade-off in a species' ability to grow on different resources (e.g., Figure 7.9A). As Tilman (2007b, p. 93) notes, "Except for the diversity explanation based on neutrality (Hubbell 2001), all other explanations for the high diversity of life on Earth require tradeoffs." The concept of trade-offs is fundamental to ecology, and ecological trade-offs in a variety of contexts will be considered in later chapters. Although interspecific trade-offs are required for stable, equilibrial species coexistence, we will see in later chapters that equalizing mechanisms can make species coexistence easier to achieve (Chesson 2000a). Equalizing and stabilizing mechanisms of species coexistence will be examined in more detail in Chapter 8, and one particular model of species coexistence via equalizing mechanisms will be considered (Hubbell's neutral theory, alluded to in Tilman's quote above) in detail in Chapter 14.

Summary

1. Interspecific competition is an interaction between species in which an increased abundance of one species causes the population growth of another species to decrease. Exploitative competition (resource competition) occurs when a species consumes a shared resource that limits its own and other species' population growth, thus making that resource less available to other species. Interference competition (contest competition) is when a species restricts another species' access to a limiting resource.

2. The L–V competition model shows that two-species competition has four potential outcomes—species 1 wins; species 2 wins; stable coexistence; or unstable coexistence. The criterion for stable species coexistence in the L-V model is that each species has a greater negative effect on its own per capita growth rate than it does on the per capita growth rate of its competitor. The L–V competition model is a "phenomenological model" in that it does not specify the mechanism by which one species has a negative effect on another.

3. The L–V competition model may be extended to include more than two competing species by using a community matrix approach. Although theorists have made extensive use of the community matrix approach to study multispecies communities, this approach is difficult to apply to natural communities and generally fails to take higher-order interactions into account.

4. Consumer–resource competition models mechanistically link the outcome of interspecific competition to the consumption of resources, specifying how competing consumers affect their resources and how consumer population growth rates are affected by resource abundances.

5. Resources are entities which contribute positively to the growth rate of the consumer population and which are consumed. Essential resources are those required for an individual's growth and survival; one essential resource cannot substitute for another. If one resource can take the place of another, each one is considered a substitutable resource.

6. R^* is the minimum amount of a resource R that consumer species N requires to maintain a stable population at a given mortality rate. At R^*, the consumer's per capita birth rate matches its per capita mortality rate and its population is unchanging.

7. The R^* rule in consumer-resource competition models states that:

   • two species cannot coexist on one limiting resource (at constant densities);
   • the species that can maintain positive population growth at the lowest resource abundance (lowest R^*) will win in competition.

8. If resource abundance fluctuates over time, two species may coexist on a single limiting resource if species 1 has a growth advantage at high resource abundance ("opportunist" species) and species 2 has a growth advantage at low resource abundance ("gleaner" species). Such trade-offs between species types can have important consequences for species coexistence.

9. The abundances of two resources at which a consumer population will be at equilibrium can be plotted as that consumer's zero net growth isocline (ZNGI). At resource levels below the ZNGI, the consumer goes extinct. At resource levels above the ZNGI, the consumer population increases. If a ZNGI for a second consumer species is added to such a plot, we see that the consumer that can survive at the lower level of resources wins in competition.

10. Two consumer species can coexist on two resources at a stable equilibrium if the species' ZNGI's intersect and at the intersection point

each species consumes proportionally more of the resource that most limits its own growth. The equilibrium is unstable, however, if each species consumes relatively more of the resource that most limits the growth of the *other* species.

11. Spatial heterogeneity in the supply of resources within a community can allow more than two species to coexist on two limiting resources.

12. One of the strengths of resource competition theory is that the theory's predictions can be tested using measures of resource supply and species resource use. Laboratory experiments with microorganisms support many of the theory's predictions. However, field tests of the theory are few and their results are more equivocal, suggesting that species coexistence in nature depends on multiple mechanisms.

13. Apparent competition occurs when two (or more) species share a predator and an increase in the abundance of one prey species leads to an increase in shared predator abundance, which may have an indirect negative effect on the abundance of other prey species. The prey species that can maintain its own population at the highest density of the shared predator is favored under apparent competition.

CHAPTER 8

Species coexistence and niche theory

The steady state of a mixed population consisting of two species occupying an identical 'ecological niche' will be the pure population of one of them, of the better adapted for the particular set of conditions   **Georgy Gause 1937**

The Paradox of the Plankton:...The problem that is presented by the phytoplankton is essentially how it is possible for a number of species to coexist in a relatively...unstructured environment all competing for the same sorts of materials.   **G. Evelyn Hutchinson 1961, p. 137**

Stable coexistence...requires important ecological differences between species that we may think of as distinguishing their niches...[along] four axes: resources, predators (and other natural enemies), time, and space.   **Peter Chesson, 2000a, p. 348**

Gause's competitive exclusion principle is encapsulated in the first quote above. As covered in Chapter 7, the theoretical and empirical foundations for this principle are strong. Yet, a freshwater lake may contain 30 or more species of phytoplankton, a coral reef a hundred or more species of fish, and a single hectare of tropical rainforest over 300 species of trees. And as every farmer knows, it takes enormous effort to keep a crop free of weeds. It seems that nature abhors a monoculture as much as it does a vacuum. Ecologists have long puzzled over this wealth of biodiversity in the face of the competitive exclusion principle, especially in situations where the number of available niches seems inadequate to account for the coexistence of so many species (e.g., Hutchinson 1961; Sale 1977; Hubbell and Foster 1986). Hutchinson (1961) was the first to express this dilemma formally. In a paper entitled "The paradox of the plankton," he asked how so many species of freshwater phytoplankton, all of which require the same few limiting resources, coexist in an environment that appears to have little habitat structure (the open water of

a lake). We now know that there are many more opportunities for niche partitioning among phytoplankton species than Hutchinson or others envisioned at the time. Phytoplankton, for example, may require different ratios of nutrients (Tilman 1982; Sommer 1989), may use different wavelengths of light (Stomp et al. 2004), and may suffer different rates of mortality from grazers (McCauley and Briand 1979). But is this enough to explain the coexistence of 30 species of phytoplankton in one lake? Ecologists have devoted considerable attention to these questions in recent decades under the label of "coexistence theory". In this chapter, we will explore the answers modern coexistence theory gives to Hutchinson's "Paradox of the Plankton".

Revisiting the origins of the notion of competitive exclusion

The notion of competitive exclusion as a strong law really took hold in the 1930s due to two separate events, both discussed in detail in Chapter 7:

Community Ecology. Second Edition. Gary G. Mittelbach & Brian J. McGill, Oxford University Press (2019).
© Gary G. Mittelbach & Brian J. McGill (2019). DOI: 10.1093/oso/9780198835851.001.0001

- the development of the Lotka-Volterra model with a prediction of competitive exclusion;
- the demonstration of the correctness of this prediction by Gause in a series of laboratory experiments (Figure 1.1).

Yet everywhere we look we see more than one species, often dozens or even hundreds of species coexisting within a community (i.e., Hutchinson's paradox of the plankton). Does that mean the interspecific competition isn't important? No. There is a wealth of evidence showing that interspecific competition is a strong force in nature (e.g., Adler et al. 2018). Let's begin our search for a resolution of the paradox of the plankton by revisiting the competition models presented in Chapter 7, exploring some aspects of these simple models that can affect their results:

- The Lotka–Volterra (L–V) and consumer-resource models, like all models, make assumptions that may be wrong. If the assumptions do not apply, then the conclusions also might not apply.
- The L–V equations are very phenomenological and, therefore, general, but this makes the model hard to apply to the real world and interpret. The model predicts competitive exclusion if interspecific competition is stronger than intraspecific competition, but the model does not tell us how to measure this. Several attempts, some largely unsuccessful, have been made to expand our understanding of what it means for interspecific competition to be stronger than intraspecific competition.

In the following sections, we explore how both of the points outlined above may help resolve the paradox of the plankton.

How are the assumptions of simple theory violated in nature?

All models make simplifying assumptions, ensuring that they are ultimately wrong or at least incomplete. Box and Draper (1987) remarked "Remember that all models are wrong; the practical question is how wrong do they have to be to not be useful." Today, most theoreticians carefully list the assumptions of their models before proceeding to explore their results. Lotka and Volterra published their models

in a different age and did not do this. In thinking about the assumptions embedded in the L–V model (see Palmer 1994 for a list), it is helpful to know that Lotka originally used very similar equations to model the dynamics of a chemical reaction involving two chemical species in a laboratory test tube (Lotka 1920) before then applying them to competing biological species. It is also helpful to remember that Gause's successful confirmation of the L–V competition model also occurred in a system of laboratory bottles. This suggests a number of assumptions that might apply to chemicals (or microorganisms) in a test tube under controlled laboratory conditions but not to biological species in nature. Namely, both the L–V and consumer–resource models assume:

1. Environmental conditions (e.g. resource or temperature) are constant across the space studied.
2. Environmental conditions are constant through time.
3. The system is closed (no immigration from outside).

Ecologists have shown how violating any of the above assumptions may favor species coexistence. This result will now be examined in more detail.

Spatial variation in the environment can promote species coexistence

Perhaps the most obvious difference between simple competition theory and the real world is the assumption of spatial homogeneity in the environment. **A spatially heterogeneous environment can promote species coexistence**. We have already encountered one example in Figure 7.9, where trade-offs in ability to use multiple nutrients combined with **spatial variation in the resource supply ratio** of multiple nutrients could lead to coexistence. In nature, Lechowicz and Bell (1991) showed that, even in a seemingly homogenous beech–maple temperate forest, the forest floor has substantial variation in the availability of nitrogen and potassium even at the sub-meter scale. To put it another way, they showed there is as much variation in nutrient concentrations in a 2 m × 2 m patch of ground as there is in a 50 m × 50 m patch. Thus, it seems possible that the mechanism in Figure 7.9 might explain coexistence of forest herbs. Spatial variation in even a single resource could also allow species coexistence. For example, the variation in light on a forest floor (so called light specks). Or water consumption and variation in soil moisture that covaries with micro-topography. For coexistence to be explained by variation in the availability of a single resource, trade-offs need to be invoked. For example, there may be a trade-off between tolerance of low resources (or some other environmental stress) and competitive ability, termed the **dominance–tolerance trade-off**. An example of how this trade-off can lead to species coexistence is provided in Box 8.1.

Spatial heterogeneity in non-resource factors can also enhance species coexistence. Animals use vegetation structure as habitat for resting, reproduction, or predator avoidance. Thus, even though vegetation structure is not a consumed resource, its spatial variation can promote species coexistence. Insectivorous birds may reduce interspecific competition by using different habitats (e.g., grassland vs forest) and their habitat use may involve trade-offs. For example, feather coloration for camouflage may differ between birds that live in grasslands vs birds living in a darker forest environment. The wing shapes optimized to fly in an open field and in a forest are also different. These **distinct habitat** type coexistence mechanisms work because each species spends more time in the habitat it is better adapted to and thus encounters more individuals of its own species than the competing species. This means that, even if the species consume the same resources, they are competing mostly against their own species for the resources where they live and this increases the strength of intraspecific competition over interspecific competition. To put this in consumer–resource terms, there are different resource pools in the different habitats, so that even if the two species draw on the same types of resources, they are drawing them down in distinct places (different spatial niches) and not directly engaging in exploitative competition interspecifically. Ironically, although Gause's experiments with the competing species *Paramecium aurelia* and *P. caudatum* were highly influential in cementing the notion of competitive exclusion, at the same time he conducted experiments with *P. caudatum* and *P. busaria* that led to stable coexistence because *P. caudatum* foraged bacteria in open water column, while *P. bursaria* was a benthic forager on the bottom of the jar, a clear case of coexistence through distinct habitats (Gause 1936; Leibold & Chase 2018; Figure 8.1). As the spatial scales coarsen and larger areas are examined, it becomes ever easier to imagine coexistence being facilitated by spatial heterogeneity.

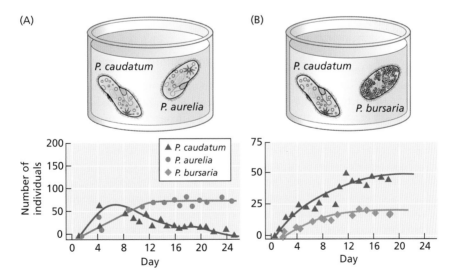

Figure 8.1 Pairs of *Paramecium* spp. may show competitive exclusion (A) or coexistence (B). The competitive exclusion case is more famous. However, Gause also found a case of stable coexistence where *P. bursaria* is a benthic (bottom) habitat specialist and has a distinct niche. After Leibold and Chase (2018).

Temporal variation in the environment can also promote species coexistence

Just as variation in space can lead to coexistence, **temporal heterogeneity of the environment can lead to coexistence**. There are three main ways temporal heterogeneity can lead to coexistence. First, species can partition the annual cycle of seasons or the daily cycle of night and day, living or being active at different parts of these cycles; but care is required. Such **temporal niche partitioning** does eliminate interference competition, but temporal niche partitioning does not automatically guarantee reduced resource competition. This is because species active at different times but living in the same space, may well share the same resource pool. For example, spring annual plants and fall annual plants may well access the same soil nutrient pool. Temporal niche partitioning can only eliminate resource competition if the resources available at different times are different (e.g., diurnal swallows and nocturnal bats feed on different insect species), or if the resources are replenished faster than the temporal niche partitioning time scale. For example, herbivorous insects that live in the spring probably access different plants than herbivorous insects in the fall. Of course, spring annual plants and fall annual plants may still compete for nutrients, but they do not compete with each other for light or probably water, and thus meet the requirement for accessing different resource pools through their temporal niche partitioning.

The second major way temporal heterogeneity can lead to coexistence in resource competition is when the resource itself fluctuates in availability over time and the consumer species have trade-offs in their relative ability to exploit low and high resource densities. This was discussed in Chapter 7, when resource competition theory was introduced. Consider again Figure 7.5. In this example, the species' birth rate curves cross, such that one species has a relatively greater population growth rate at low resource levels and the other species has a relatively greater population growth rate at high resource levels. Note that the absolute ability is always lower for both species on a low density resource—it is the relative ability that changes. These two consumer types have been termed **gleaners and opportunists**,

respectively (Frederickson and Stephanopoulos 1981; Grover 1990). Species 2 is a gleaner because it grows relatively well when the resource is scarce. Species 1 is an opportunist—it grows relatively poorly at low resource levels, but grows well when resources are abundant. In a constant environment, we would expect species 2 to win in competition because it has the lower R^*. However, as we saw in Chapter 7, if resource levels fluctuate between high and low abundance then the two species may coexist because species 1 (opportunist) has a growth advantage at high resource abundance and species 2 (gleaner) has a growth advantage at low resource abundance. Note that this coexistence is possible because of non-linear resource use curves - this allows them to cross multiple times and create multiple domains with different outcomes. Chesson (2000a) calls this mechanism involving non-linear resource use combined with fluctuation in resources **relative non-linearity.**

A third major way environmental fluctuation can lead to coexistence occurs when organisms have long-lived adults that are relatively insensitive to environmental variation (e.g. trees), but the juveniles of different species are sensitive to the environment and do best under different environmental conditions. This is called the **storage effect** (Chesson & Huntly 1997) and is covered in more detail in Chapter 15.

Ways in which dispersal and immigration can promote species coexistence

An open system with immigration can also lead to coexistence. Imagine a system with two species and two patches. Suppose species A is better adapted to patch a, and species B is better adapted to patch b. If there was no dispersal, then one would expect A to competitively exclude species B in patch a, and vice versa. Now imagine that a small fraction of the individuals born each generation disperse to the other patch. This would prevent competitive exclusion from happening because there is a constant supply of new individuals who have not yet experienced competition. Such a system is an example of source–sink dynamics or mass-effects in metacommunities, both of which will be discussed in more detail in Chapters 13 and 14. Levin (1974) analyzed an even more extreme case where the two patches were identical and the two species were such that the contingent exclusion case of L–V applies (whichever species starts with more individuals competitively excludes the others). If one patch starts with more of species A and one patch starts with more of species B, then coexistence occurs by dispersal maintaining species in patches from which it would otherwise be excluded. If dispersal is non-existent then competitive exclusion will occur in each patch—species A & B will not coexist within a single patch. If dispersal is too high then the system functions like one patch and one species will competitively exclude the other in both patches. However, if dispersal is intermediate, then coexistence within a patch will occur. Note that this coexistence occurs even when both patches are identical so no spatial heterogeneity is required. Multiple patches and dispersal is all that is needed for coexistence in this **source-sink coexistence**. Another way dispersal can potentially promote the coexistence of a large number of species in the **competition-colonization** trade-off model (Hastings 1980; Nee & May 1992; Tilman 1994). In this model, there is a trade-off between competitive ability (e.g., lower R*) and dispersal ability (specifically the ability to get to new patches). Competition–colonization trade-offs are discussed in more detail in Chapter 13. Bolker and Pacala (1999) identified a third coexistence scenario that they call the **exploiter strategy**. The exploiter strategy is similar to the competition–colonization trade-off, but instead of specializing in getting to newly open patches quickly, exploiters specialize in very quickly extracting resources from new patches before competitively superior species arrive and then disperse locally to maximize their ability to exploit new patches.

Can dispersal in the absence of a trade-off lead to coexistence? A number of authors (Atkinson and Shorrocks 1981; Ives and May 1985; Silvertown et al. 1992; Bolker and Pacala 1997) have pointed out that if species disperse locally, then each species is more likely to come into contact with individuals of its own species than its competitor, increasing the strength of intraspecific competition vs interspecific. This is especially true when the aggregation is enhanced by a patchy environment (Atkinson and Shorrocks 1981; Shorrocks and Svenster 1995). This

mechanism cannot prevent competitive exclusion in the long run (Durrett & Levin 1994; Bolker et al. 2003), but it can slow exclusion down for very long periods of time, a conclusion with empirical support (Stoll and Prati 2001). Whether this mechanism leads to coexistence is partly a matter of definition of coexistence. It cannot lead to what we will call stable coexistence in a few sections hence, but it could be a form of unstable coexistence that may be an important explanation for the coexistence we observe in the natural world.

This distinction between permanent coexistence vs coexisting for a long time before stochastic drift causes a stochastic extinction that cannot be reversed, came to the fore during ecology's discussion about neutral models in the 2000s (discussed in detail in Chapter 14). Before that time, few bothered to draw a distinction. However, neutral theory caused many people to recognize the potential importance of this distinction, which became one of the motivations for advancing the study of coexistence theory. In the next section we will give a precise definition of stable coexistence. To do so we need to return to the models of Chapter 7, which have an attractor (the system returns to an equilibrium with coexistence whenever stochastically perturbed off of it). These models specify the conditions under which a stable coexisting equilibrium exists.

How do we tell if the strength of intraspecific competition is greater than interspecific competition ($\alpha_{ii} > \alpha_{ji}$)?

Two major types of equilibrium models were presented in Chapter 7, which presents ecologists studying competition with a dilemma—whether to use the L–V model or a consumer–resource model. On the one hand, L–V models are quite generic in the type of competition they can represent. Indeed, they simply implement the definition of competition by having each species have a phenomenological parameter for their negative impact on the growth rate of the other species (the α_{12} and α_{21} of Equation 7.5). Unfortunately, interpreting these phenomenological parameters then becomes pivotal in determining whether species coexistence is possible. The qualitative claim that intraspecific competition has to be stronger than interspecific competition to allow two competing species to coexist has a precise mathematical statement, $\alpha_{jj} > \alpha_{ji}$ (switching now to the formulation of Equation 7.5). How do we look at a real world situation and know if $\alpha_{jj} > \alpha_{ji}$? Because the α's apply broadly across diverse scenarios of competition, such as resource competition, niche partitioning, interference competition, etc., the answer is not obvious. On the other hand, consumer–resource models are explicitly mechanistic and make predictions about R* that are directly interpretable and testable (e.g., Figure 7.11) As shown in Chapter 7, R* can be measured in the real world and used to make predictions about coexistence. On the down-side, consumer–resource models poorly represent some common types of competition (e.g., interference competition and apparent competition discussed in Chapter 7). Even for exploitation competition, consumer resource models depend on the notion of distinct resource types. However, a common form of exploitation competition involves only a single type of resource that is differentiated along a continuous gradient.

Let us briefly examine a well-studied example of exploitation competition along a continuous resource gradient involving Darwin's Finches (*Geospiza*) and their consumption of seeds of different sizes. In Darwin's finches, both the birds (specifically their beaks) and the seeds show continuous variation in size. While most birds can eat most seeds, species with larger beaks are more efficient with and prefer larger seeds and vice versa (Schluter and Grant 1984) (Figure 8.2). When two species of Darwin's Finches, *G. fortis* and *G. fulginosa*, occur alone on different islands they show substantial overlap in beak size (and, hence, seed size preference). However, when the two species compete with each other on the same island, their beak sizes evolve such that there is almost no overlap in size and hence little overlap in resource use (Figure 8.3). This is a classic example of **character displacement** (Lack 1947; Schluter et al. 1985; Grant 1986), which is "the situation in which, when two species of animals overlap geographically, the differences between them are accentuated in the zone of sympatry and weakened or lost entirely in the parts of their ranges outside this zone" (Brown & Wilson 1956). Character displacement is an

evolutionary response to reduce the intensity of competition by reducing niche overlap. Competition is clearly an important force in these birds, strong enough to drive evolution, but this scenario does not map well to a consumer–resource model. If we use a L–V model, can we say whether intraspecific competition is stronger than interspecific in the allopatric case? How about the sympatric case? How can we map this very detailed and persuasive example to the α coefficients in a L–V model?

Early empirical efforts to estimate the α's in the L–V model focused on measuring overlaps in resource use between species competing for resources that could be arrayed on a continuous axis (see Figure 1.4 for a conceptual version and Figure 8.4 for a real-world example). This approach has some grounding in reality. As already discussed, birds within a guild often divide up resources with bigger birds eating bigger seeds (Pulliam 1985; Schluter and

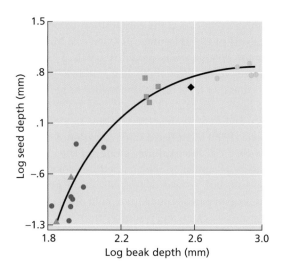

Figure 8.2 Preferred food size is related to beak size in five species of Darwin's finches (Geopsiza). These bird species feed mostly on seeds and a bird's beak is its chief instrument for acquiring, manipulating, and crushing seeds. The width of the smallest preferred seed increases with the size of the beak. This means that preferences for different seed sizes is an important way in which these species differ in their niche, causing interspecific competition to be weaker than intraspecific competition. Symbols correspond to finch species; *Geospiza fuliginosa* (red circles), *G. difficilis* (blue triangles), *G. fortis* (green squares), *G. magnirostris* (yellow circles), and *G. conirostris* (black diamond). Line fit by non-linear regression, with one outlier data point removed. After Schluter and Grant (1984).

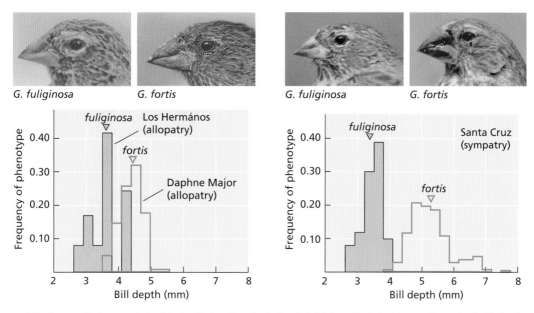

Figure 8.3 Character displacement in two Darwin's finches, *Geospiza fortis* and *G. fuliginosa*. Beak depth is strongly correlated with the sizes of seeds the birds are able to eat; mean beak depth is indicated by triangles. Note that on Santa Cruz, where the two species are sympatric (co-occur), beak depth distributions are displaced relative to their distributions on Daphne Major and Los Hermános, where each species is found alone (allopatry). Such enhanced difference in a character trait (here, beak depth) between sympatric species is one response to competition for a resource. From Grant (1986).

Grant 1984) and similarly for seed-eating mammals (Figure 8.4), and for insectivores and carnivores. Likewise, plants often show similar responses in, for example, the rate of photosynthesis vs temperature (Mooney and Ehleringer 2003).

Thus, ecologists first looked to quantify the α-coefficients in the L–V competition model from measures of overlap in species' resource use. Some of these attempts were quite successful. For example, MacArthur (1968) calculated measures of niche overlap for four species of warblers based on how much time they spent foraging in different parts of a spruce tree. He then converted niche-overlaps into α-coefficients and parameterized a multi-species L–V model that closely matched observed population dynamics. This seemed to suggest that niche overlap was all that was needed to estimate α-coefficients and the race was on, rapidly proliferating into hundreds of papers and a multitude of different metrics for niche overlap (Colwell and Futuyma 1971;

Hurlbert 1978). Unfortunately, few of these efforts proved as successful as MacArthur's. In hindsight, we now know that there is no reason they should. The α-coefficients in the L–V model are measures of how much an individual of one species suppresses the per capita population growth rate of another species. This may have little to do with how much their diet niches overlap, or, at best, the relationship between overlap in resource use and population growth rate impacts might be highly non-linear.

Using resource overlap to estimate the strength of competition also led to the historically important idea of **limiting similarity** (briefly introduced in Chapter 1). To address the question of how similar species can be in their niches and still coexist, MacArthur and Levins (1967) started with the assumption that overlap in resource utilization (niche) could be equated to the α-coefficients in the L–V model. They then added in the assumption that

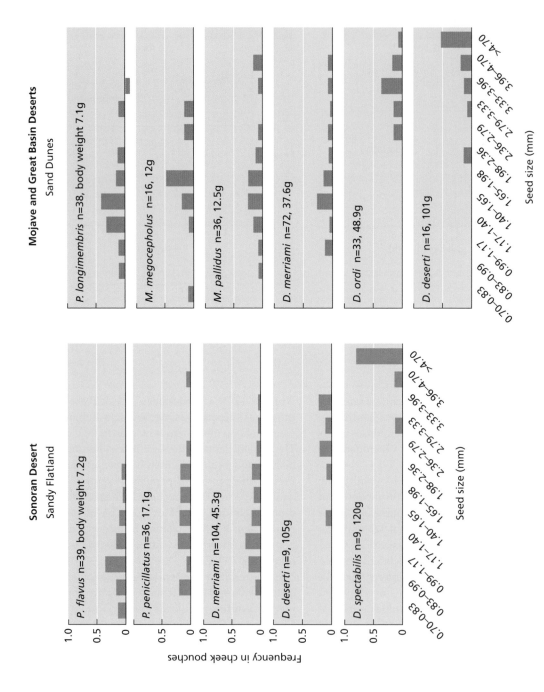

Figure 8.4 An example of resource utilization curves for nine seed-eating rodent species found in three NA deserts (Sonoran, Mojave and Great Basin). Note that the size of seed harvested is generally correlated with the size of the rodent, but there is considerable overlap between species. After Brown (1975).

resource utilization curves were normal (bell curve) shaped and used a particular metric for calculating niche overlap. They compared the distance between the peaks of the resource utilization curves, which they called d, and the width of the resource utilization curves, w, which was defined as the standard deviation of the normal-shaped resource utilization curve (Figure 1.4). They found that for coexistence to occur required $d \geq w$. This was an intuitively appealing and mathematically precise result. Here, at last, was a limit on how similar species could be to meet the vague L–V requirement that intraspecific competition be greater than interspecific. May (1973a) and May and MacArthur (1972) extended this result to show that in a stochastically variable environment one needed $d \geq w\sigma$ where σ was the standard deviation of environmental variation. Unfortunately, the notion of a universal limit to similarity quickly began to unwind (reviewed in Abrams 1983b). As already noted, resource overlap does not equate to impacts on species' demographics and the theoretical result turned out not to be very mathematically robust. If you change the method of calculating overlap, limiting similarity falls apart (Abrams 1975) as it also does if you use a resource utilization curve shape other than normal (Roughgarden 1974). We now know there are no universal (or "hard") limits to similarity as originally envisioned by MacArthur and Levins (1967). However, recent work has shown that, while there are no formal limits to similarity, the more tightly packed a community is in terms of niche space the more fragile is species coexistence to any perturbation in model parameters (Meszéna et al. 2006; Barabás et al. 2012, 2013). The extreme fragility of tightly packed communities suggests a reinterpretation of the limiting similarity principle, rather than its abandonment.

The current best attack on the question of how to interpret the meaning of intraspecific competition> interspecific competition (or $\alpha_{jj} > \alpha_{ji}$) starts by abandoning the L–V model altogether. It was begun by Chesson and Huntly (1997) and expanded by Chesson (2000a, 2018), Adler (2007, 2010) Levine and HilleRisLambers (2009) and Letten, Ke and Fukami (2017) among others. Chesson and Huntly (1997) boiled competition down to its bare

essence and developed a very simple, but very general model

$$\frac{1}{N_s}\frac{dN_s}{dt} = r_{t,s} - b_s F_t \text{ for } s = 1 \dots S \qquad [Eqn\ 8.1]$$

Here, s is the subscript for the S different species in the system. In this model, the per capita growth rate of a species in competition was separated into two components. A density-independent component, r_t captures the fit of the organism to its environment and probably fluctuates over time as the environment changes (hence, the t subscript). The second component consists of one density-dependent factor (F_t), be it resource competition, interference competition, predators, etc., where the effect on the per capita growth rate depends on the population size, N (i.e. density). The parameter b is simply the sensitivity of the species to those density-dependent factors. Expansions of the model having multiple density-dependent limiting factors (F_t) simply reconfirm that multiple limiting factors are needed for multiple species to coexist and are not explored further here (but see Chesson & Huntly 1997).

Note that F_t probably depends on the density of the other species, but we do not specify a specific form. All that is captured is that it, too, varies over time. This means we cannot explicitly solve for an equilibrium like we did with the models in Chapter 7, but we do not have to! In fact, this is one of the strengths of the Chesson–Huntly approach. It replaces strong assumptions about linear density dependence that allow for equilibrium analysis with more general assumptions about density-dependence that no longer allow equilibrium analysis. Instead this approach analyzes invasion ability. Invasion analysis looks at whether a species has a positive growth rate when it is rare and competing against another species (or set of species) that is fully established (i.e. at its carrying capacity for when no other species are present). A species that can invade under such conditions cannot be competitively excluded, while a species that cannot invade can be competitively excluded. It may be possible for two or more species that cannot mutually invade each other to coexist for a long time. Chesson (2000a) calls this **unstable coexistence.** The neutral models discussed in Chapter 14 are an example, as are the examples discussed earlier in this chapter, where limited dispersal kept individuals more in competition with their own species. Ultimately, however, stochastic drift makes species with unstable coexistence go extinct and, then, because they cannot invade, they can no longer coexist. The example discussed earlier where short-distance dispersal that causes a species to come in contact with and compete more with its own species is another example of unstable coexistence. It can prolong coexistence, but it does not have a recovery mechanism when a chance event eliminates one species. Chesson and Huntly instead focused on **stable coexistence.** This occurs when each species is able to invade the system when rare. Here, even if a stochastic event causes a species to go extinct, it can re-invade and re-establish coexistence.

Using this very general model of competition and invasion analysis, Chesson and Huntly were able to reason in general about competition. A number of their key insights relate to fluctuating environments and will be covered in Chapter 15. However, one can map this general equation (Equation 8.1) back to a number of specific models including consumer-resource models, L–V models (Chesson and Huntly 2007) and a plant competition model (Adler et al. 2007). The details of the exact mappings vary, but the results all share several common features. We start with the example of mapping the Chesson–Huntly model (Equation 8.1) back to the consumer resource model (Chesson 2000a). This gives the per-capita-growth rate for species i trying to invade against an established species, s, as:

$$\frac{1}{N_i}\frac{dN_i}{dt} = b_i\left(\frac{r_i}{b_i} - \frac{r_s}{b_s}\right) + b_i(1-\rho)\frac{r_s}{b_s} \qquad \text{[Eqn 8.2]}$$

Here the r_i/b_i terms capture our original notion that fitness has a density independent component (r_i) decreased or rescaled by sensitivity to density dependence (b_i). Together these terms give what works as the notion of invasion ability or invasion fitness, $f_i = r_i/b_i$, so we can rewrite (2) as

$$\frac{1}{N_i}\frac{dN_i}{dt} = b_i(f_i - f_s) + b_i(1-\rho)f_s \qquad \text{[Eqn 8.3]}$$

The only new variable is ρ, which stands for resource overlap. Here, at last, we have a model that actually incorporates resource overlap. Note that, when

overlap is 100% (i.e. $\rho=1$), the red term disappears and the model simplifies to the blue term

$$\frac{1}{N_i}\frac{dN_i}{dt} = b_i(f_i - f_s) \qquad \text{[Eqn 8.4]}$$

These equations lead to the following points:

1. The key measure of competitive ability that determines winners of competition is invasion fitness, f_i, which is an increasing function of the density-independent responses to the environment (r_i) and a decreasing function of sensitivity to competition (b_i). The latter can also be phrased as an increasing function of tolerance ($1/b_i$) to competition. These are the two main factors which determine superior competitors.

2. Without niche separation (i.e. Equation 8.4), when invasion fitnesses differ ($f_i \neq f_s$), then only the species with the largest f_i will survive. Every other species will have a negative per-capita growth rate in invasion conditions (since $f_i < f_s$ for all $i \neq s$). This is the competitive exclusion scenario (only one species remains).

3. Without niche separation (Equation 8.4), if fitnesses (f_i) are all equal, they will neither grow nor shrink deterministically, but will change only through stochastic variation (i.e. drift). However, once a species drifts to extinction it cannot reinvade (invasion per capita growth rate is zero). This is the unstable coexistence scenario discussed above.

4. With niche separation (Equation 8.3), the blue portion of Equation 8.3 is negative for all but the most invasion capable species (highest-ranked f_i). However, the red term is positive and added to the blue part. Thus, it is possible that if the red term is positive enough the per capita growth rate can be positive even when the fitness difference is negative (i.e., it is not the top-ranked species).

These observations lead to the following terminology. The red part of Equation 8.3 is called the **stabilizing effect.** When it is larger (i.e., when niche overlap, ρ, is smaller) it makes it easier for species to coexist. When the differences in invasion fitness is large (the blue part of Equation 8.3), it is very hard for the niche overlap (red) part of the equation to compensate, making coexistence unlikely. Conversely, when the differences in invasion fitness between species (f_i–f_s)

Box 8.1 Shared vs distinct preference niche structuring

Characterizing the resource use of species along a continuous axis of resource availability (e.g., seed size for birds and mammals, soil moisture for plants) has played a central role in how ecologists have thought about niche partitioning and niche overlap. But it has been recognized for decades that even when the resource is continuous there are at least two ways species can differentiate along the axis (Colwell and Fuentes 1975; Keddy 1989; Wisheu 1998; see Figure 8.5). The classic scenario already discussed (Figures 1.4, 8.2, 8.5A) is the **distinct preference** scenario. Each species is optimal at a different location along the continuous axis and would be found to be most abundant at that optimum, regardless of the presence or absence of other species. Here, the trade-off is in where the optimum goes. We saw such a trade-off in Darwin's finches where different beak sizes in birds are optimal at handling different seed sizes (Figure 8.2). In the **shared-preference** scenario, every species prefers to be at the same optimum environment (in Figure 8.5B high soil moisture) and does best there. In other words, they *share* the same optimum. However, the species differ in two other

aspects. First, some species are able to tolerate poorer environments (here, low soil moisture; in Figure 8.5B black can tolerate the worst environment). Secondly, some species do better in the shared optimal environment than others (in Figure 8.5B yellow can perform the best in the optimal environment. We call this species dominant over the other species. The dominant species would competitively exclude (or at least greatly lower the abundance of the other species) in the optimal environment. This often happens by having a superior R* for exploitation competition or by being physical larger or more aggressive and therefore able to win in interference competition. These two factors are part of a trade-off such that the most tolerant species perform the worst in the optimal environment and vice-versa. This is called dominance-tolerance trade-off (Grime 1977; Rosenzweig 1991).

One of the first experimental field studies of competition provides a good example of shared preferences (even though it is sometimes mistakenly described as an example of distinct preference). Connell (1961a,b) studied two species of barnacles, *Chthamalus* and *Balanus*, that grow at the same sites on the Scottish coast. The two species show strong zonation along the vertical tidal gradient. *Chthalamus* is found higher up in the intertidal zone, while *Balanus* is found lower down. Through a series of careful experiments and observations, Connell showed that *Balanus* and *Chthamalus* larvae can both settle in the high intertidal (which is out of the water much of the time), but that *Chthamalus* survives better than *Balanus* because it tolerates desiccation better. Removal of *Balanus* from the high intertidal had little effect on *Chthamlus* survival. *Chthamalus* and *Balanus* larvae also settle in the middle intertidal, where both species survive well initially. However, in the middle and lower intertidal, *Chthamlus* eventually died. Connell showed that its death was due to interference competition from *Balanus*. *Balanus* overgrew, crushed and popped loose the *Chthamlus* individuals, killing them, but when *Balanus* was experimentally removed, *Chthamalus* survived in this zone for the duration of the study. Thus, both species did well in the middle intertidal zone, but only *Balanus* was found living there because it was competitively dominant over *Chthamalus*. In the high intertidal, only *Chthamlus* could survive the environmental conditions and thus was the environmentally tolerant species. In Figure 8.5B if we change the horizontal axis to distance below high tide, *Chthamalus* is the red species and *Balanus* is the blue species.

Dominance–tolerance trade-offs are also found in plants. In these scenarios, the dominance (or interference competition)

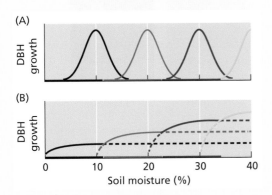

Figure 8.5 Comparing distinct preference niches (A) vs shared preference niches (B). In (A) each species (differently colored lines) has optimal performance at a distinct part of the gradient. This leads directly to the species existing (or at least being most abundant) at a particular part of the gradient (shown by the colored horizontal lines along the x-axis). In (B) each species performs optimally in the same part of the gradient (high soil moisture), but there is a trade-off where some species are better able to tolerate poor environments (e.g., the blue species), but are not strong competitors while other species are strong competitors at the cost of low tolerance for poor conditions (e.g., yellow species). Note that the pattern of where species occur (or are most abundant), again shown by horizontal lines on the x-axis, is almost identical between (A) and (B); only removal or density manipulation experiments can distinguish these two scenarios.

continued

Box 8.1 *Continued*

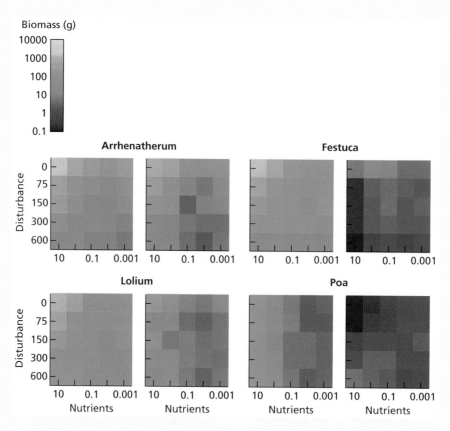

Figure 8.6 Some of the results from the Campbell and Grime (1992) experiment discussed in the text. A logarithmic gradient of nutrients was created running from high on the left to low on the right. A disturbance gradient was created from none on the top to high on the bottom. Each of seven species was planted in a monoculture (left graph of each pair). All seven species were also grown together (equal seed frequencies) in a mixed plot (right graph of each pair).

portion often occurs by growing bigger and especially taller to block the other species from receiving light, although it can include allelopathy and a variety of other mechanisms. A nice demonstration of shared preference can be found in an experiment by Campbell and Grime (1992). They constructed a series of garden beds with a 2-dimensional experimental design. In one direction ran a gradient of low to high nutrients. In the other direction ran a gradient of low to high disturbance (clipping) so that all possible combinations could be found in one planting bed. They then grew seven species of grasses in monoculture, and all seven species together (with an initial seed density of 1/7 of the mix for each species). Water and light were controlled and equal across all treatments. After 1 year, the biomass of each

species in each treatment was measured. Each species' biomass was greatest in the low disturbance, high nutrient treatment (i.e. shared preference, see left columns of Figure 8.6). What was interesting was how the different species responded in the seven-species competition experiment. Some competitively dominant species (*Arrhenatherum* in Figure 8.6) continued to be most abundant in the optimal conditions, even though their abundance was reduced by competition. Other species shifted to show a maximum abundance in suboptimal conditions once competition was present (*Festuca* was most abundant in low disturbance, intermediate nutrient regimes and *Lolium* was most abundant in high nutrient, intermediate disturbance regimes). *Poa* was strongly affected by competition and its highest biomass under competition was found in high-nutrient, high-disturbance regimes, its least preferred regime without competition.

This experiment also shows how shared preference can be extended when there are multiple gradients of stress to what is called **centrifugal organization** (Rosenzweig and Abramsky 1986; Keddy 1989; Rosenzweig 1991; Wisheu 1998). The word "centrifugal" means "fleeing the center". In community organization it means every species prefers to grow in the best conditions of each gradient (e.g., no disturbance, high nutrients). However, only the competitive dominants get to grow abundantly there. The higher environmental tolerance, competitive subordinates flee those conditions under competition, but because there are multiple

gradients, they can flee in different "directions" (*Lolium* to more disturbance, but still high nutrients and *Festuca* the opposite). An important point is that you cannot distinguish a distinct preference from a shared preference scenario simply by observation in the field; in both shared and distinct preference species are found to serially replace each other along a gradient. However, the underlying mechanisms are very different. Only removal or density manipulation experiments can determine which scenario is operating. In a meta-analysis of field competition experiments that performed these manipulations, shared preference organization was twice as likely as distinct preference (34 vs 17 cases; Wisheu 1998).

is small, this is called an **equalizing effect**. With more equalizing effects, relatively weaker stabilizing effects are needed to make the per capita growth rate positive (leading to the ability to invade and create stable coexistence). Specifically, when the stabilizing component (red) is larger than the fitness differences (blue component), then the per capita invasion growth rate becomes positive and stable coexistence occurs.

Again, the details of Equations 8.2–8.4 depend on the exact model used to derive them. Recall Equations 8.2–8.4 came from a consumer resource model (also see Letten et al. 2017). However, the general principals of fitness being a function of both density independent growth rates and tolerance of competition, and fitness differences balancing with stabilizing effects (niche differences) apply across a wide suite of models that have been put into this Chesson–Huntly decomposition framework including L–V models (Chesson and Huntly 1997, Appendix A) and a plant competition model (Adler et al. 2007; Levine & HilleRisLambers 2009). Two up-to-date, pedagogically-oriented derivations of coexistence theory may be found in Letten et al (2017) and Barabás et al. (2018). Letten et al. (2017) emphasize the consumer–resource model derivation and build links to niche theory, pointing out that fluctuating resource supply vectors are a form of equalizing mechanism, while varying impact (consumption) vectors are a form of stabilizing mechanism. Barabás et al. (2018) develop a very general derivation of coexistence theory based on approximation methods and suggest that caution is needed in drawing a sharp contrast

between equalizing and stabilizing mechanisms. Chesson has also provided his own recent overview and synthesis (2018).

Predation and coexistence

For almost 100 years, coexistence theory was framed in the context of competition. L–V theory and Gause developed the concept of competitive exclusion. The notions of limiting similarity (MacArthur and Levins 1967) consumer–resource models (Tilman 1982) and coexistence theory (Chesson and Huntly 1997; Chesson 2000a) are all framed in terms of competition. Even neutral theory which in some senses is the opposite of stable coexistence is developed for a single trophic level.

However, Holt's development of apparent competition (1977, reviewed in Holt and Bonsall 2017 and Chapter 7) began to suggest that coexistence might require thinking beyond competition to multiple trophic levels. Consumer-resource models focus on R* as the key predictor between competitive consumers. However, there is an analogous P* notion for the prey that can withstand the highest predator densities (Holt et al. 1994). Recently, Chesson and Kuang (Kuang and Chesson 2008; Chesson and Kuang 2008) have combined Holt's notion of apparent competition via predation with the Chesson and Huntly's (1997) approach to coexistence based on analyzing invasion capabilities when rare. Just as in formulas 1–4 we can break a two-predator, two-prey system into notions of fitness (with fitness broken into a density independent component and a density-dependent tolerance of predation component) and niche overlap. Stable coexistence occurs when niche overlap is small or fitness differences are small. As a result, species can coexist even when their resource use is identical if their differences in sensitivity to predators is small or their "predator niche" overlap is low. The prey just need to differ in the degree to which they are limited by different natural enemies (see discussion of apparent competition in Chapter 7 for more details). The notions of temporal coexistence mechanisms such as relative non-linearity (Chapter 7) and fluctuating coexistence like the storage effect (Chapter 14) also extend into predator-prey systems (Kuang and Chesson 2008). Thus, it is not a reach to say that all of the theory of coexistence in competitive systems has parallel concepts and effects in predator–prey systems. It is a satisfying result that instead of debating the relative importance of competition and predation, we can now say that they are symmetric and in a fundamental sense equivalent.

Conclusion

Given the propensity for competition models to predict competitive exclusion, how can we explain the coexistence of so many species everywhere we look? Especially in organisms like trees or phytoplankton where the basic requirements for light, water, and nutrients are very similar and the number of possible dimensions for niche separation seem small. As Hutchinson (1961), in his paradox of the plankton paper, suggested much of the answer lies in spatial and temporal heterogeneity. While Hutchinson's basic intuition was right, we now better understand that not all spatial and temporal heterogeneity produces coexistence (see Chapter 15, Section "Why resetting competition does not enable coexistence"). Thus, ecologists have made significant advances in understanding when spatial and temporal heterogeneity can or cannot lead to coexistence. Hutchinson also missed the notion that dispersal and spatial structure alone (even devoid of heterogeneity) can lead to coexistence. To summarize, species can coexist by partitioning food and resources, habitats, and predators along multiple axes of variation, and in time and space, and also by dispersal (see Chesson quote at

the start of this chapter). In short, the state of modern coexistence theory (Figure 8.7) makes it clear that there is no paradox of plankton. Rather, the paths to coexistence are many. Figure 8.7 presents the current state of coexistence theory as a branching tree of choices. Of the 16 tips of the tree, 12 involve coexistence and only four involve competitive exclusion (although this should not be taken as an estimate of the relative frequency of coexistence mechanisms). Thus, we see that Hutchinson's paradox exists only in very simplified models and scenarios. Coexistence should be and is common.

The main lesson from this chapter is not that coexistence is hard or surprising, but that ecologists have not yet constructed a clear theory of the niche that maps onto coexistence predictions. One such mapping would be to go from an understanding of the niche to calculating the α-coefficients of the L–V model. This would allow us to understand when intraspecific competition was stronger than interspecific in terms of niches.

However, there are some general outlines of what such a general theory of niches and coexistence might look like. First and foremost, trade-offs play a central role. As Tilman (2007b, p. 93) notes, "Except for the diversity explanation based on neutrality (Hubbell 2001), all other explanations for the high diversity of life on Earth require tradeoffs." The concept of trade-offs is fundamental to ecology and ecological trade-offs in a variety of contexts will be considered in later chapters. Just in the context of coexistence we have seen many important trade-offs including:

1. Different efficiencies in using essential resources like nutrients.
2. Different efficiencies along continuous gradients of resources like seed size, and habitats like temperature and vegetation.
3. Different efficiencies in avoiding particular predator species.
4. Different efficiencies in using day/night or seasonality (temporal niches).
5. A trade-off between competitive dominance and environmental tolerance.
6. A trade-off between competitive dominance vs colonization ability.
7. A trade-off in relative efficiencies in using low and high resource levels (gleaner/opportunist).

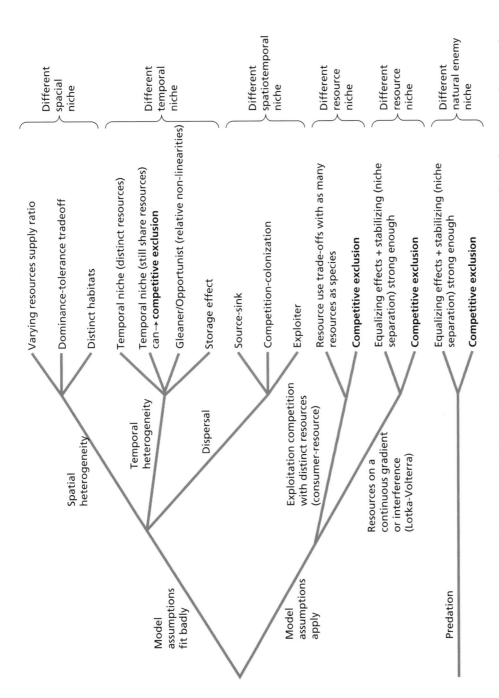

Figure 8.7 A graphical summary of the current state of understanding of coexistence. Each fork represents a choice of types of coexistence mechanisms, and the tips of the tree represent 12 different types of theories leading to coexistence and four cases where choices predict competitive exclusion. The right hand side of the graph represents that there are five general aspects of niche on which species can be differentiated to coexist.

Thus, while all that we need to model for predicting coexistence is the resulting niche differentiation, it seems likely that a full theory of niches and coexistence will need to incorporate the evolutionary and physiological origins of trade-offs.

Secondly, the body of coexistence theory is incredibly complex and contains a long list of specific cases with seemingly little connection. Broad synthetic theory unifying coexistence mechanisms is needed. Such a theory appears to be emerging. Chesson (2000b, also see Amerasakare 2003) point out that there are exact parallels between coexistence mechanisms in varying space and time. The storage effect of time is analogous to what we called the source–sink coexistence model of space, and the partitioning mechanisms in space and time have parallels with partitioning of resources (Chesson 2000a,b), all three of which also have parallels with partitioning of predators (Chesson & Kuang 2008). We are close to a fully synthetic theory of at least the coexistence side of the story across the four dimensions of resources, predators, space, and time, but this work still needs to be linked into the theory of niches.

Finally, we need to build a better, more synthetic understanding of the empirical patterns of niche differentiation. Although we now have hundreds of examples of niche differentiation, they have not really been pulled together into a general understanding of what types of niche differentiation are more common or rare in nature, or what types of niche differentiation can or cannot co-occur within a single system. Early papers by Schoener (1974) and Wisheu (1998) give some interesting hints, but they have been poorly followed up on. We noted earlier that Wisheu found that shared-preference niches were twice as common as distinct preference, but she also found more detailed patterns. Autotrophs and heterotroph subgroups within a single species (e.g. ontogenetic or sexual subgroups), almost always partitioned by shared niches, while herbivore or parasite preferences for hosts almost always showed distinct preferences. Only interspecific heterotrophs truly showed a mix of shared and distinct preferences. Schoener (1974) found that habitat differences were more common than food differences that were more common than temporal partitioning. Temporal partitioning was rare compared with food and habitat (although he did not look at the

temporally fluctuating mechanisms), but daily niche separation was more common among predators and terrestrial poikilotherms, while seasonal niche separation was rare in vertebrates. Food partitioning is more common when the food is larger than usual relative to the body size of the consumer. Schoener also found that all combinations of differentiation along multiple axes (e.g., habitat and food) could be found. All of these patterns have obvious ecological explanations, but there has been little to no follow up on any of these hypotheses in the last four decades and it is unknown how robust they are.

In summary, both niche theory (Chapter 1) and competition theory (Chapter 7) have played central roles in community ecology for over 100 years and they naturally fit together. Grinnell linked competitive exclusion and niche theory as far back as 1917: "It is, of course, axiomatic that no two species regularly established in a single fauna have precisely the same niche relationships". However, this unification of the two theories has proven to be harder than it might seem with many false turns in the road. We now have a rich understanding of how species can coexist and now need to return to the understanding of the complexities of the niche and how they map onto coexistence.

Summary

1. The prevalence of competitive exclusion in models of competition stands in contrast to the rarity of monocultures in nature.
2. Consumer-resource models, L–V models and Gause's bottle experiments all remove spatial heterogeneity, temporal heterogeneity, and dispersal and migration. Each of these factors can lead to coexistence in several ways.
3. L–V models are the most general and apply to all forms of competition, but they are phenomenological and difficult to interpret in the real world. Equating the α coefficients to niche overlap and the idea of limiting similarity were early efforts that did not survive the test of time.
4. A more effective approach to understand coexistence involves analyzing invasion capabilities. Unstable coexistence mechanisms can occur (such as neutral theory or limited dispersal), but stable coexistence only occurs when each species can

invade the other when rare. Models point to the importance of fitness differences based on density-independent growth rates and tolerance to competition. Coexistence occurs by a combination of equalizing effects (minimizing fitness differences) and stabilizing effects (minimizing niche overlap).

5. All of the features of coexistence theory that exist for competition can also be found in predation.

6. Niche separation can be broadly classified as shared or distinct preferences. They provide the same end result, serial replacement of species along an environmental gradient. In shared preferences every species prefers the same portion of the resource axis. This difference from distinct preference can only be detected by removal or similar experiments. Shared preferences often involve a trade-off between environmental tolerance and competitive dominance.

7. There are many different coexistence mechanisms. They typically involve either shared or distinct niche separation, which invariably involves some type of trade-off. A promising research direction is to link coexistence mechanisms to niche theory and trade-off theory.

CHAPTER 9

Beneficial interactions in communities: mutualism and facilitation

From the algae that help power reef-building corals, to the diverse array of pollinators that mediate sexual reproduction in many plant species, to the myriad nutritional symbionts that fix nitrogen and aid digestion, and even down to the mitochondria found in nearly all eukaryotes, mutualisms are ubiquitous, often ecologically dominant, and profoundly influential at all levels of biological organization.
Edward Allen Herre et al., 1999, p. 49

As we expand the scope of our studies to include broader ecological contexts, mutualisms will continue to claim an overdue and more prominent role in community ecology.
Palmer et al., 2015, p. 160

What role do beneficial interactions—notably facilitation and mutualism—play in community ecology? Are they as important in determining species diversity or regulating the abundances of individuals or species as are the negative interactions of competition and predation? The answer is, we don't know. Mutualism has long been the "ugly duckling" of community ecology. In contrast to competition and predation, until recently there was little general theory on how mutualisms affected patterns of species distribution and abundance. Why was this so? It was not because mutualisms are uncommon; as the above quote from Herre et al. (1999) notes, mutualisms are everywhere, and most organisms are involved, either directly or indirectly, in a variety of positive interactions. Nor was it because mutualisms are uninteresting; some of the most fascinating examples in natural history involve mutualistic interactions, especially those found in the species-rich

tropics. This was not because mutualisms were only recently recognized; Aristotle described mutualisms in the fourth century BCE.

Some have argued (e.g., Bruno et al. 2003) that the failure to include beneficial interactions in modern ecological theory was the result of the dominant role that interspecific competition (and to a lesser extent predation) played in the conceptual development of community ecology, from the formulation of Gause's competitive exclusion principle and Lotka's and Volterra's competition models to Hutchinson's and MacArthur's work on the niche and limiting similarity. However, while interspecific competition and predation have indeed provided much of the foundation for the development of community ecology (as described in Chapter 1), we disagree with the idea that ecology's early focus on negative species interactions somehow biased the thinking of current researchers, causing them to ignore the importance

Community Ecology. Second Edition. Gary G. Mittelbach & Brian J. McGill, Oxford University Press (2019).
© Gary G. Mittelbach & Brian J. McGill (2019). DOI: 10.1093/oso/9780198835851.001.0001

of mutualism and facilitation. Rather, there are fundamental reasons why ecologists have been slow to incorporate positive species interactions into general ecological theory. This is changing. Before we explore the role of beneficial interactions in communities, however, we need to consider some definitions.

Mutualism and facilitation: definitions

Most ecologists would agree that mutualism can be defined as a reciprocally positive interaction between species (Bronstein 2009), but the definition of facilitation is less clear. Callaway (2007) suggests that facilitation can include any pairwise interaction in which at least one species benefits. Under this definition, mutualism would be a subset of facilitation. However, there are important distinctions between mutualism and facilitation (Bronstein 2009). For example, facilitation generally requires close proximity between the interacting species, whereas mutualisms may link species over long distances (e.g., seed dispersal and pollination mutualisms). In addition, most mutualisms involve multiple trophic levels, whereas facilitative interactions generally occur within a trophic level. Thus, Bronstein (2009, p. 1160) suggests that facilitation be defined as "an interaction in which the presence of one species alters the environment in a way that enhances growth, survival, or reproduction of a second, *neighbouring* species" (authors' italics). Examples of facilitation include nurse plants that aid the establishment of seedlings of other species by reducing thermal or evaporative stress (Went 1942; Bertness and Callaway 1994), nitrogen-fixing plants that benefit non-nitrogen-fixing neighbors (Vandermeer 1989; Loreau and Hector 2001; Temperton et al. 2007), and salt marsh plants that provide shade and reduce evapotranspiration, thus reducing soil salinity and allowing less salt-tolerant species to survive (Figure 9.1; Bertness and Hacker 1994; Hacker and Bertness 1999).

The most pervasive examples of facilitation involve species whose very presence so modifies the environment that whole communities of organisms can develop there. Some species provide structural habitat for other species; these are often termed foundation species (e.g., corals, kelp, mangroves, and trees; Dayton 1975). Species that modify the environment so as to make it more or less suitable for other species are often termed ecosystem engineers (Jones et al. 1994; Wright and Jones 2006). The construction of ponds by beavers is the quintessential example of ecosystem engineering. In this case, it is clear how beavers facilitate the coexistence of some species, but not others. Because all organisms affect and are affected by their physical environment, ecologists have debated the use and relevance of the terms "foundation species" and "ecosystem engineers," questioning when and how to apply these concepts, and how to distinguish organisms that do and do not fall into these categories (e.g., Jones et al. 1997; Power 1997; Reichman and Seabloom 2002). There is no doubt, however, that facilitation by means of habitat modification plays an important role in most communities.

Some ecologists have suggested that the definition of facilitation should be broadened to include positive effects between two species that act through a third species—that is, indirect effects. "After all, although rarely recognized as such, a trophic cascade is simply an indirect facilitation" (Bruno et al. 2003, p. 124). However, while indirect effects may often lead to positive interactions between nonadjacent species, we gain little by relabeling coupled consumer–resource interactions that produce positive, but indirect effects as facilitation. We would prefer to limit use of the term "facilitation" to describing a particular type of pairwise interaction, just as we use the terms "competition" and "predation." Thus, we will adopt Bronstein's definition of facilitation throughout this chapter: *facilitation is an interaction in which the presence of one species alters the environment in a way that enhances the growth, survival, or reproduction of a second, neighboring species.* We recognize that even this definition permits ambiguity with regard to indirect effects. For example, if we include the biotic, as well as the abiotic environment in the above definition, then an interaction between two plant species in which the presence of an unpalatable species reduces herbivory on a nearby palatable species (Hay 1986) could be classified as facilitation under Bronstein's definition, even though the interaction is an indirect effect involving another trophic level (the herbivore). The indirect effects (positive and negative) of species interactions will be discussed in detail in Chapter 10. For now,

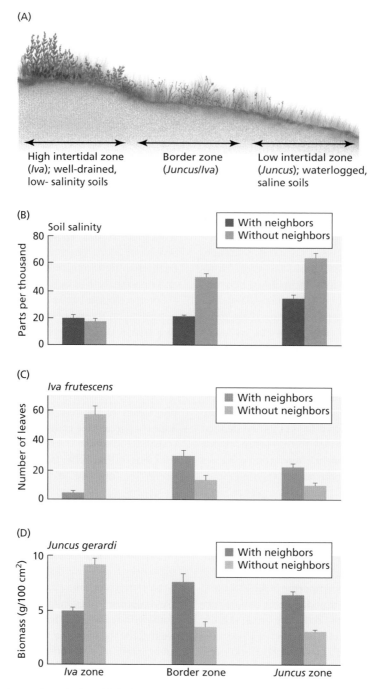

Figure 9.1 Facilitation in two salt marsh plants. (A) The three major vegetation zones, along with their physical environments, in a typical New England salt marsh. Notice the separation in distribution of *Iva* and *Juncus* species, and their zone of overlap. (B–D) Results of transplant experiments in which adult *Iva* and *Juncus* plants were transplanted into each of the three vegetation zones with and without the other species. *Iva* performed better in the low intertidal and border zones when transplanted with *Juncus*; *Juncus* presumably provides shade and reduces evapotranspiration, thus lowering soil salinity and allowing the less salt-tolerant *Iva* to survive. At higher marsh elevations, where soil salinity is low and environmental conditions are less harsh, the interaction between *Iva* and *Juncus* shifts from facilitation to competition. After Bertness and Hacker (1994).

it seems we will simply have to accept some "wiggle room" in our definition of facilitation.

Obligate mutualisms occur when a species must have its mutualistic partner to survive or reproduce. The opposite of obligate is facultative. An example of a facultative mutualism is the fish-cleaning mutualism. Fish that have their parasites removed by another (cleaner) fish are healthier, but not receiving cleaning services is not inherently fatal. Closely related concepts are specific (only one particular species can meet the need of the mutualist) and general (many different taxa can meet the need). Obligate mutualisms are more typically specific (Bronstein 1994), but all combinations can occur and the relation can be asymmetric. For example, some orchid species can only be pollinated by a single species of bee (obligate and specific), but that bee species can pollinate (and receive food from) many different species of orchids and other plant taxa (facultative and general; Bronstein 1994).

Symbiosis refers to an intimate, persistent interaction between species. Mutualisms often involve some form of symbiosis, such as endosymbiosis, where one species lives inside another (e.g., *Rhizobium* bacteria living within the root nodules of leguminous plants) or ectosymbiosis, where one species lives on a body surface of another species (e.g., chemosynthetic bacteria living on the surfaces of shrimp and crabs in deep-sea hydrothermal vent communities). Examples of mutualistic endo- and ectosymbiotic relationships will be discussed in the next section. The definition of symbiosis, like that of facilitation, is somewhat controversial, and biologists debate whether a definition of symbiosis should include both positive and negative (e.g., parasitic) interactions. In reality, symbiotic interactions include a continuum of positive and negative effects (Douglas 2008). This chapter, however, uses the term symbiosis primarily to refer to a positive, persistent, and intimate association between species.

A brief look at the evolution of mutualism and facilitation

How mutualisms evolve and how they are maintained has been an active area of study for decades, dating back to Darwin (1859). In contrast, the study of facilitation, and particularly the evolution of facilitative interactions, began only recently (Bronstein 2009). The evolution of mutualisms appears to depend on the relative costs and benefits of the interaction to the participating species. The current thought is that mutualistic interactions are best viewed as reciprocal exploitation that, nonetheless, provides net benefits to each partner (e.g., Leigh and Rowell 1995; Doebeli and Knowlton 1998; Sachs et al. 2004; Bronstein 2015). Mutualisms can therefore be thought of as biological markets, where members of each species exchange resources or services (Noë and Hammerstein 1994; Schwartz and Hoeksema 1998; McGill 2005; Akcay and Roughgarden 2007). Markets are a particularly useful framework for studying the evolution of mutualism because they allow us to incorporate the notion of comparative advantage. For example, suppose species 1 needs A and B, but is relatively more efficient (compared with species 2) at producing A, whereas species 2 (who also needs A and B) is relatively more efficient at producing B. Under these conditions, both species benefit from trading in the "marketplace" (species 1 trades A for B from species 2). Thus, it is not surprising that a great many mutualisms involve plants that are efficient at producing carbon energy and that trade this carbon energy for services from animals that can move (e.g., pollination and seed dispersal) or energy for nutrients (e.g. nitrogen fixing bacteria). However, if a species can take what it wants, rather than trade for what it wants, it will. This makes mutualism markets vulnerable to exploitation and cheating.

What factors help align mutual interests and limit cheating? **Partner fidelity** and **partner choice** are two important mechanisms that can constrain exploitation. Partner fidelity is enhanced by the vertical transmission of symbionts from parent to offspring (Fine 1975; Axelrod and Hamilton 1981) and it provides a mechanism for fitness benefits to become aligned. That is, benefits to one partner result in increased benefits to the other partner. There are, however, many examples of the movement of mutualistic symbionts (e.g., pollinators, marine algae, mycorrhizal fungi) between unrelated host individuals (horizontal transmission). For some cases of horizontal transmission, there is evidence that hosts preferentially select among strains of symbionts, benefiting those strains that also benefit

(A)

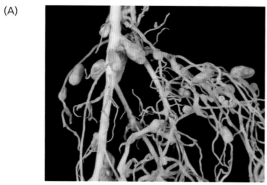

(B)

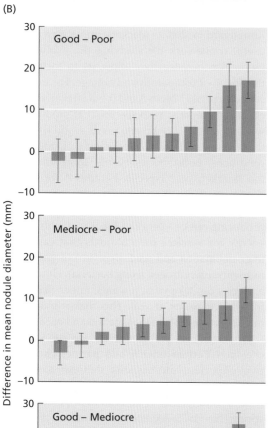

Difference in mean nodule diameter (mm)

Good – Poor

Mediocre – Poor

Good – Mediocre

Individual plants

the host. Thus, partner choice may play an important role in reducing the impact of "cheating" in mutualistic interactions (Sachs et al. 2004).

Simms et al. (2006) provide a particularly nice example of partner choice in the yellow bush lupine (*Lupinus arboreus*). Lupines and other leguminous plants maintain a symbiotic association with bacteria of the genus *Rhizobium* that may benefit both species. Rhizobia occupy nodules in the plant's root system (Figure 9.2A). Within these nodules, the rhizobia convert atmospheric dinitrogen (N_2) into biologically available ammonium (NH_4^+), which the rhizobia exchange for photosynthates (carbon compounds) produced by the plant. However, this interaction is not always cooperative and there is clearly the potential for cheating on the part of the rhizobia. A number of authors have hypothesized that rhizobial cooperation could be promoted by plant traits that would differentially allocate resources to those nodules harboring the most cooperative strains of rhizobia (e.g., Denison 2000; Simms and Taylor 2002; West et al. 2002). Simms et al. (2006) showed that yellow bush lupine plants did exactly that. When inoculated with different strains of *Rhizobium*, the plants allocated more resources (provided larger nodules) to those strains that provided the greatest benefits to the plants (Figure 9.2B). These results suggest that post-infection partner choice may be an important mechanism constraining exploitation in this mutualistic interaction. Other examples of partner choice in mutualisms can be found in Sachs et al. (2004).

The study of the evolution of facilitation lags well behind the study of evolutionary relationships in mutualisms. Bronstein (2009, p. 1166) explains,

Figure 9.2 Partner choice in a legume–*Rhizobium* mutualism. (A) Legumes produce root nodules that house nitrogen-fixing *Rhizobium* bacteria. (B) The yellow bush lupine (*Lupinus arboreus*) exhibits partner choice by providing larger root nodules to more beneficial strains of rhizobia. Each histogram represents a single plant that was inoculated with two different strains of *Rhizobium*. The average difference in quality between the inoculated *Rhizobium* strains was: good to poor (top panel), mediocre to poor (middle panel), and good to mediocre (bottom panel). The size of the histogram bar indicates the difference between the average sizes of nodules housing the best and the worst strain in a single plant. In general, plants allocated resources to producing larger nodules for the more beneficial *Rhizobium* strain. Photo by David McIntyre; (B) from Simms et al. (2006).

"There has been minimal consideration to date of how facilitative interactions arise either de novo or in transition from other forms of interaction. Nor, to my knowledge, have the conditions that favor their evolutionary persistence or that lead them to break down been considered." Like the evolution of mutualism, the evolution of facilitation appears to depend on the relative costs and benefits to the participating species. Also like mutualism, facilitation ranges from relatively specific to highly generalized interactions (Callaway 1998; Bronstein 2009). Progress in understanding the evolution of facilitative relationships, and in determining whether these relationships have evolved via changes in just one of the partners (e.g., the facilitated species have evolved traits that promote physical proximity to the facilitator species) or in both partners (co-evolution of traits), will require experimental work on the costs and benefits to the associated species, ideally within a known phylogenetic context. Such studies are currently in their infancy (Bronstein 2009).

Incorporating beneficial interactions into community theory

As mentioned at the start of this chapter, beneficial interactions have long received short shrift in ecology textbooks. Most of the focus on pairwise interactions and, in particular, on the presentation of theory centers on interspecific competition and predator–prey interactions. Bruno et al. (2003) suggest that this focus reflects historical bias because models of competition and predation were developed first. However, over 70 years ago, Gause and Witt (1935) explored adding positive interactions to the ecological theory of their time: the Lotka–Volterra model of two-species interactions. According to Bruno et al. (2003, p. 121), "As Gause and Witt demonstrated in 1935, changing the Lotka–Volterra competition model into a mutualism model by switching the sign of the interaction coefficients predicts that mutualisms can result in a stable equilibrium, where the densities of both species can be greater when they co-occur." However, this statement glosses over a major problem with simple pairwise models of mutualism or facilitation (May 1981).

The equations below show how the Lotka–Volterra competition model can be rewritten as a model for positive species interactions simply by changing the sign of the interspecific effect from negative to positive

Lotka – Volterra competition   Lotka – Volterra mutualism

$$dN_1/dt = rN_1\left(\frac{K_1 - N_1 - \alpha_{12}N_2}{K_1}\right) \qquad dN_1/dt = rN_1\left(\frac{K_1 - N_1 + \alpha_{12}N_2}{K_1}\right)$$

$$dN_2/dt = rN_2\left(\frac{K_2 - N_2 - \alpha_{21}N_1}{K_2}\right) \qquad dN_2/dt = rN_2\left(\frac{K_2 - N_2 + \alpha_{21}N_1}{K_2}\right)$$

In this Lotka–Volterra type model of mutualism, species coexistence is stable only if the product of the mutualism coefficients is less than unity (i.e., $\alpha_{12}\alpha_{21} < 1$ or is less than the product of the intraspecific competition coefficients (May 1975; Holland et al. 2006; Holland 2015). To see this, consider the situation in which the two species populations are at some density and one individual is added to each species' population (N_1 and N_2). If $\alpha_{12} = \alpha_{21} = 1$, then the *intraspecific* negative density-dependent effect of adding one individual is exactly countered by the *interspecific* positive density-dependent effect of adding one individual. Therefore, both species populations continue to grow to infinity (Gause and Witt 1935; Figure 9.3). Thus, May (1981, p. 95) concluded that modeling positive interactions using the Lotka–Volterra equations is unrealistic:

[S]imple, quadratically nonlinear Lotka–Volterra models that capture some of the essential dynamical features of prey–predator (namely, a propensity to oscillation) or competition (namely, a propensity to exclusion), are inadequate for even a first discussion of mutualism, as they tend to lead to silly solutions in which both populations undergo unbounded exponential growth, in an orgy of mutual benefaction. Minimally realistic models for two mutualists must allow for saturation in the magnitude of at least one of the reciprocal benefits.

Thus, a simple exploration of the Lotka–Volterra model demonstrates that for stable species coexistence to occur, the positive effects of mutualism need to be limited. This limitation can occur via negative density dependence resulting from strong intraspecific competition or via a saturating functional response for the positive feedback effects of mutualism on the reciprocal species (Holland 2015). Alternatively, negative density dependence could occur via feedbacks from other life stages (e.g., nurse plants benefit seedlings, but not adults), or from

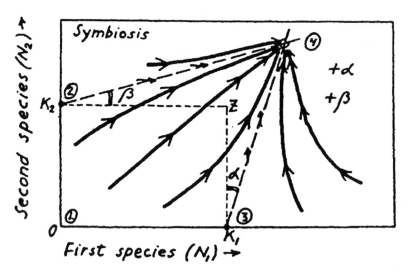

THE AMERICAN NATURALIST   [Vol. LXIX

Figure 9.3 Gause and Witt (1935) modified the Lotka–Volterra (L–V) competition model to consider conditions in which two species benefit each other ("symbiosis") by changing the signs of the L–V interaction coefficients (α and β in this figure, which correspond to α_{12} and α_{21} in the text) from negative (competition) to positive (mutualism). The results are illustrated in this phase-plane diagram taken from Gause and Witt's 1935 paper. As the authors stated, when the interaction coefficients are positive, both species together attain larger biomasses than separately, and a "stable knot" (equilibrium point) occurs when $\alpha\beta < 1$: "As coefficients of mutual aid α and β increase the 'knot' continuously moves up and rightwards and finally with $\alpha\beta = 1$ passes into infinity" (Gause and Witt 1935, pp. 603–4). Thus, Gause and Witt showed that considering mutualisms within the simple framework of two-species Lotka–Volterra interactions is unrealistic.

fluctuating environments (sometimes good, sometimes bad for participants in a mutualism), or from other members of the food web.

The above analysis provides some insight into why beneficial interactions may have been "ignored" in the development of early ecological theory (Bruno et al. 2003). On the other hand, maybe this whole question simply amounts to barking up the wrong tree. Asking how mutualisms limit populations is a poor question because they typically do not—they just lessen the impact of other pre-existing limitations. Instead, the most interesting questions about positive interactions are the co-evolutionary questions of how they evolve and cheaters are prevented. Recent models have shown how mutualisms could evolve from parasitic interactions (Roughgarden 1975; de Mazancourt et al. 2005) and even from competition (de Mazancourt and Schwartz 2010). In the latter case, pairs of coexisting competitors could evolve toward a mutualistic "trading partnership" if the two competing species require the same resources, and if each species consumes in

excess the resource that least limits its growth, and then trades this excess resource to a partner whose growth is more limited by that resource. This mechanism works when the cost of excess resource acquisition and resource transfer between species is negligible (de Mazancourt and Schwartz 2010).

A productive conceptual model for the evolution of mutualism and facilitation asks about the balance of the benefits (b) and costs (c) of interacting (Bronstein 1994). Typically, having a mutualistic (or facilitation) partner involves both benefits and costs to the focal species. For example, in the legume-*Rhizobium* mutualism described above the legume gains nitrogen in the form of ammonia (receives benefit b) and loses carbon energy (products of phytosynthesis) that are consumed by the bacteria as well as allocating carbon energy to build nodules to house the bacteria (costs c). The costs and benefits are roughly reciprocal for the bacteria. Thus, for this to be a feasible mutualism requires only that b > c for each species, not that b > 0 and c = 0. By framing species interactions as having both benefits and costs, we can more clearly focus on the contexts in which b > c for each species and where beneficial interactions should evolve. Such "context-dependency" may involve the presence of other species, the nature of the abiotic environment, and even the age or life-stage of the species involved. Thus, b > c may be possible in some contexts and not in others.

Mutualisms may be context-dependent

In many cases, understanding the evolution and maintenance of positive interactions in communities requires that we consider the broader web of interactions and abiotic conditions in which mutualisms are embedded (Frederickson 2013; Hoeksema and Bruna 2015; Palmer et al. 2015). The legume–*Rhizobium* mutualism is a good example. The existence of this positive interaction between leguminous plants and bacteria appears to be context dependent. In low-nutrient environments, the ability to get nitrogen from their rhizobial partners should make legumes better competitors against non-nitrogen-fixing plant species. However, in high-nutrient environments the competitive advantage of legumes should be diminished and we might also expect the resource trading mutualism between rhizobia and legumes to breakdown.

A long-term study by Weese et al. (2015) confirms these context-dependent aspects of the legume-*Rhizobium* mutualism. After 21 years of nitrogen addition to field plots, the abundance of three species of *Trifolium* (clovers, a legume) was 10 times lower in fertilized plots than in control (unfertilized) plots, suggesting that nitrogen addition reduced the competitive advantage of legumes over non-legumes. Even more interesting, Weese et al. (2015) found that 21 years of nitrogen addition led to the evolution of less-mutualistic bacteria. In a greenhouse experiment, plants inoculated with rhizobium strains isolated from fertilized plots were significantly smaller and had reduced chlorophyll content (a trait correlated with plant nitrogen content) compared with plants inoculated with rhizobium strains from unfertilized control plots. Thus, in a high nitrogen environment the bacteria evolved to supply less nitrogen to their resource-trading partner. Another widespread mutualism, that between plants

and their mycorrhizal fungi, also appears to be context-dependent (Johnson 2010; Johnson et al. 2015).

The mutualism between plants and mycorrhizal fungi is probably as old as the evolution of land plants themselves, as structures resembling arbuscular mycorrhizae have been found on the fossil remains of plant roots from hundreds of millions of years ago. In general, plants deliver carbon compounds to their associated fungi and, in return, receive mineral nutrients—most commonly phosphorus and in some circumstances nitrogen. In nutrient-poor ecosystems, plants may obtain up to 80% of their requirements for nitrogen and up to 90% of their phosphorus from mycorrhizal fungi (van der Heijden et al. 2008). However, this positive interaction between plants and fungi may shift to an antagonistic interaction under high-nutrient conditions, in which there is often no benefit to the association and plant growth may even be slightly reduced due to the carbon demand of the fungus (Johnson et al. 1997; Smith et al. 2009).

Johnson (2010) presents an excellent review of the ecology and evolution of mutualistic interactions between plants and arbuscular mycorrhizae, casting this relationship in the context of nutrient stoichiometry (Sterner and Elser 2002) and trading partnerships (e.g., Schwartz and Hoeksema 1998; Hoeksema and Schwartz 2003; Grman et al. 2012). Johnson (2010, p. 2) suggests that the dynamics of this mutualistic relationship can best be understood by "defining plants and fungi as chemical entities which must obey the laws of definite proportion and the conservation of mass and energy" (i.e., the stoichiometric perspective), and by assuming that "the symbiotic dynamics of mycorrhizal trading partnerships is defined by two factors: resource requirements and the ability to acquire resources." Figure 9.4 presents a conceptual trade balance model for the sharing of carbon, nitrogen, and phosphorus between plant and fungal trading partners. The model shows how the relative abundances of nitrogen and phosphorus in the soil may shift the plant–fungus interaction from mutualism to commensalism to parasitism. Following these scenarios, global climate change and the concomitant enrichment of above- and below-ground resources (e.g., nitrogen deposition and fertilization; CO_2 enrichment) is expected to affect the nature of the beneficial relationships between plants

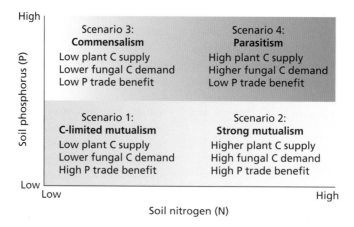

Figure 9.4 A trade balance model showing how the interaction between plants and mycorrhizal fungi is expected to shift from mutualism to commensalism to parasitism as supplies of nitrogen (N) and phosphorus (P) in the soil vary. Four scenarios are predicted based on the relative abundance of N and P. (Scenario 1) A carbon-limited mutualism is predicted when levels of both N and P are low. Even though plant growth is limited by low nutrient levels and little carbon (C) is available for trading, a limited trade of C for P is still favorable to both partners. (Scenario 2) A strong mutualism is predicted when N is abundant, but P is rare. The plant has abundant carbon to trade, making the C-for-P trade extremely beneficial to both partners. (Scenario 3) Commensalism is predicted when P is abundant and plants have little to gain from the C-for-P trade. In this case, mycorrhizal fungi gain by receiving carbon from plants, but fungal demand for C is low because fungal growth is limited by low N. (Scenario 4) Parasitism is predicted when both N and P are abundant and plants gain no benefit from a C-for-P trade. Fungal demand for carbon is high, however, because N is abundant; thus, the fungus is predicted to parasitize the plant. After Johnson (2010).

and mycorrhizae (Johnson 2010). The model of partner trade between plants and mycorrhizal fungi provides an excellent example of how the ecological concepts of stoichiometry, optimal foraging theory, and shared costs and benefits may be applied to understanding the factors controlling the nature of beneficial interactions under different environmental conditions (Johnson 2010; Johnson et al. 2015; Grman et al., 2012).

Ant–plant mutualisms provide another example of the role played by the broader environment (this time the biotic environment) in driving the evolution and maintenance of a mutualism. Mutualisms among a variety of species of ants and acacia trees throughout the tropics are a textbook example of a beneficial interaction (Janzen 1966; Heil and McKey 2003). The trees provide the ants with shelter (e.g., hollow thorns) and food (extrafloral nectaries and Beltian bodies; Figure 9.5A). The ants, in turn, protect the trees from herbivores and may also reduce competition with other trees (by clipping plants near the acacia). This well-studied example shows clear benefits that each species in the mutualism receives from the other. However,

the positive nature of this interaction depends critically on the presence of other, antagonistic species (predators, competitors) in the community. Palmer et al. (2008) showed that the experimental exclusion of large mammalian herbivores from an African savanna shifted the balance away from a strong mutualism and toward a suite of less-beneficial behaviors by the interacting plant and ant species (Figure 9.5B,C). Ten years of herbivore exclusion lead to a reduction in the nectar and housing provided by the trees to the ants, increased antagonistic behavior by a mutualistic ant species, and shifted competitive dominance from a nectar-dependent mutualistic ant species to a less-beneficial ant species. Trees occupied by the less-beneficial ant species grew more slowly and experienced higher mortality than trees occupied by the more mutualistic ant species. Thus, a change in the herbivore community resulted in a rapid breakdown of the ant–plant mutualism. Again, we can see from this example that realistic models of mutualistic interactions often depend on the inclusion of other species and the recognition of a balance between positive and negative effects (Chamberlain and

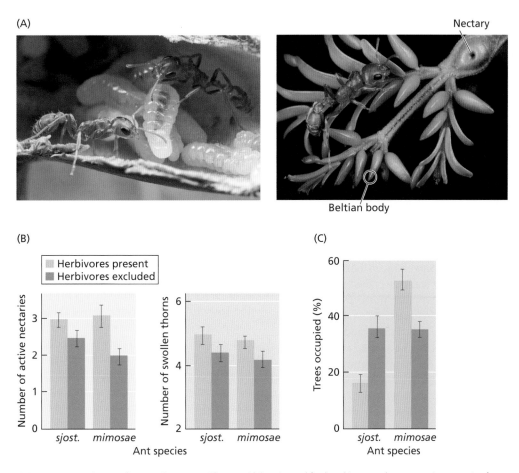

Figure 9.5 In an ant–acacia mutualism, acacia trees provide ants with housing and food, and in return the trees receive protection from herbivores. (A) Ants may live and raise their broods within the tree's specialized swollen thorns (domatia), and often are nourished by the tree's extrafloral nectaries and Beltian bodies. (B) Studies done in East Africa showed that excluding large herbivores caused trees of the whistling-thorn acacia (*Acacia drepanolobium*) to reduce the level of rewards offered to ants. Both the number of nectaries and the number of swollen thorns declined when herbivores were excluded (orange histograms) compared with when herbivores were present (gold histograms). (C) In the same study, most trees were occupied by ants of either of the species *Crematogaster mimosae* or *C. sjostedti*. *C. mimosae* aggressively defends host acacias from herbivores and relies on domatia to raise its broods; *C. sjostedti* is a less aggressive defender and does not live within domatia. Under natural conditions, *C. mimosae* is the most abundant ant symbiont. However, when herbivores were excluded, the better mutualist (*mimosae*) decreased in abundance and the less-beneficial ant species (*sjostedti*) increased. (A) © A. Wild, Visuals Unlimited and M. Moffett, Minden Pictures; (B,C) after Palmer et al. (2008).

Holland 2009; Palmer et al. 2015). Indeed, a recent study by Prior and Palmer (2018) found that the presence of a third trading partner (scale insects) strengths the ant–plant mutualism. In dry periods, when the production of nectar from acacia extrafloral nectaries is reduced, honeydew produced by scale insects provides ants with supplemental carbohydrates, allowing them to remain active and protect trees from elephant grazing.

Interactions may change from positive to negative across life stages

As we have seen above, both the physical and the biotic environment can affect the nature of the interaction between species, and may change the overall sign of the interaction from negative to positive. In addition, the signs and strengths of interspecific interactions can change across life stages, as illustrated by the following examples.

Nurse Plants

Plant recruitment in harsh environments is often facilitated by the presence of nurse plants—species whose canopies provide a protective microclimate promoting the germination and growth of other species. Particularly dramatic examples of the "nurse plant syndrome" can be seen in deserts, where trees such as paloverde (*Cercidium* spp.), mesquite (*Prosopis* spp.), and ironwood (*Olneya tesota*) shade the substrate, reduce water loss by evaporation, and/or increase soil fertility, thereby increasing seed germination and seedling survival rates for other plant species (see Shreve 1951; Hutto et al. 1986; Tewksbury and Lloyd 2001). However, the net benefit of association with a nurse plant may be positive or negative depending on the environmental context as well as the life stage of the "nursed" plant.

Studies have shown that nurse plants provide greater benefits to individuals in more water-stressed (xeric) environments (e.g., Franco and Nobel 1989). In addition, nurse plants have been shown to have the greatest impact on plant community structure and species richness when water stress is high (Tewksbury and Lloyd 2001). Other studies, however, have shown that temporal variation in the extent of drought stress did not change the beneficial impacts of nurse plants (e.g., Tielbörger and Kadmon 2000). Finally, the sign of the interaction may change over an individual's lifetime (Schupp 1995; Callaway and Walker 1997; Miriti 2006). Individual plants may benefit from their interaction with nurse plants when they are young and small, but face increased competition with neighbors of the nurse plant species when older and larger; in other words, "the best place for a seedling to grow may not be the best place for a sapling to reach maturity" (Nuñez et al. 2009, p. 1067). Such ontogenetic shifts in the type and strength of species interactions are particularly common in size-structured or stage-structured species (Yang and Rudolf 2010).

Mycorrhizal Associations

Earlier in this chapter, we discussed the fact that most plant species are engaged in symbiotic relationships with mycorrhizal fungi. However, not all plant–mycorrhizal interactions are positive, and the sign of the interaction may become neutral or negative under high-nutrient conditions (Johnson et al. 1997; Smith et al. 2009). Thus, the plant–mycorrhizal interaction provides another important illustration of the context dependence of facilitative interactions. Moreover, many mycorrhizal fungi are not host-specific, and these fungi may link the root systems of multiple plants. Thus, mycorrhizal networks have the potential to distribute resources among different plants irrespective of their species or size (van der Heijden and Horton 2009). It has been shown, for example, that seedlings are often inoculated with mycorrhizal fungi through their association with adult plants (Read et al. 1976; Dickie et al. 2002). When this occurs, the survival or growth of seedlings may be facilitated by their proximity to adults (e.g., Horton et al. 1999; Dickie et al. 2002). However, adult plants may also reduce the level of light available to seedlings and may compete with them for resources. Thus, the net outcome of the interaction between seedlings, adults, and their shared mycorrhizal fungi depends on the balance of positive and negative effects, and this balance may change spatially and with ontogeny (Dickie et al. 2005; van der Heijden and Horton 2009; Booth and Hoeksema 2010). In general, positive seedling–adult interactions shift to negative effects as seedlings mature, grow and compete with their neighbors.

Pollinator Facilitation

It has been estimated that about 80% of all plant species are pollinated by animals. Plants clearly benefit from this mutualism. Less clear, however, is whether plants might facilitate each other's reproductive success by increasing pollinator visitation. For example, if one plant species attracts pollinators to the neighborhood of a second plant species, one or both of the species may benefit through increased seed set and higher reproductive rates (Schemske 1981; Rathcke 1988). Of course, neighboring plant species may also compete for pollinators or other resources, so the balance of their interaction may be positive or negative.

Population models show that pollinator facilitation is likely to occur only if a pollinator's visitation rate accelerates (at least initially) as the total number of flowers within a patch increases (i.e., the pollinator has a sigmoidal or type III functional response; see Chapter 5). Facilitation is unlikely when pollinators display a saturating (type II) functional response (Feldman et al. 2004). Thus, while facilitation between plant species is possible via pollinator sharing, the conditions for this to occur seem rather restrictive. Empirical studies looking at the effects of hetero-specific floral density on pollinator visitation of a focal species have found positive effects (e.g., Laverty 1992; Ghazoul 2006; Hegland et al. 2009), negative effects (e.g., Thomson 1978; Campbell and Motten 1985), and no effects (e.g., Campbell and Motten 1985; Caruso 2002).

In a meta-analysis, Morales and Traveset (2009) examined the evidence for pollinator facilitation and competition for pollinators between native plant species, and between native and exotic plant species in a total of 40 studies involving 57 focal species. They found that most shared pollinator interactions between native species had no effect on rates of pollinator visitation or plant reproductive success. However, exotic species had significant negative effects on pollinator visits to native species and on native species' reproductive success (Figure 9.6). Morales and Traveset (2009) suggest that exotic species are superior and more attractive competitors for pollinators than native co-flowering species on a per capita basis.

Combining positive and negative effects

Although the consequences of positive interactions for species diversity and community structure are still poorly understood, there have been some recent steps toward developing a general theory of species interactions that includes a combination of positive and negative interactions. In a notable example, Gross (2008) modified the standard consumer–resource competition model developed to study interspecific competition (see Chapter 7; Tilman 1982; Grover 1997) to include positive interspecific effects. In Gross's model, per capita mortality or the maintenance requirements of a resource competitor may be reduced by the presence of another competing species. Such positive interactions incorporate the general features of facilitation, in which one species ameliorates a stressful environment for a second species. In a two-species model, Gross (2008) found that both species could coexist on a single resource if facilitation by the superior resource competitor reduced the mortality rate of the inferior resource competitor enough to allow the "inferior" species to invade a community consisting solely of the superior competitor. Positive and negative effects might also be incorporated into the growth rate terms. The important point is that species may have both positive and negative effects on each other, and that considering these effects separately provides additional mechanisms for coexistence that are absent in models of the Lotka–Volterra type (in which the average effect of an interacting species is summarized in a single term, the interaction coefficient, α).

Gross's theoretical analysis shows how positive interactions could allow species coexistence and

(A) Pollinator visitation frequency

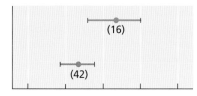

(B) Reproductive success

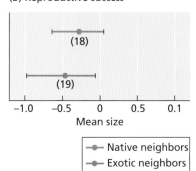

Figure 9.6 Results of a meta-analysis showed that the identity of a neighboring plant species (exotic vs native) affects the pollination and reproductive success of a focal plant species. The graphs show the mean effect size (Hedges' *d*) and 95% confidence intervals for the effect of exotic or native neighboring plant species on pollinator visitation frequency (A) and reproductive success (B) of focal plant species. Exotic neighbors had a significant negative effect on both pollinator visitation and reproductive success (95% confidence intervals do not include zero mean effect size), whereas native neighbors had no significant overall effect (95% confidence intervals include zero mean effect size). Sample sizes are shown in parentheses. From Morales and Traveset (2009).

promote diversity even when they only partially counteract exploitative competition. Two empirical studies illustrate additional pathways by which positive interactions may mediate the effects of interspecific competition and thereby increase the probability that competing species will coexist. Bever (1999, 2002) showed that two competing plant species (*Plantago lanceolata* and *Panicum sphaerocarpon*) may coexist if the mutualistic mycorrhizal fungal species that confers the greatest growth benefit to these plants associates preferentially with *Panicum*—the otherwise inferior competitor. This result has interesting parallels to how shared predation may promote species coexistence if the predator selectively preys on the superior resource competitor (see Figure 6.6).

In the second study, Schmitt and Holbrook (2003) found an interesting wrinkle in the well-known mutualism between sea anemones and anemonefish, showing how this interaction could promote coexistence between competing fishes. Anemonefish of the genera *Amphiprion* and *Premnas* find shelter in the stinging tentacles of sea anemones (Figure 9.7);

in return, the anemonefish drive off specialized anemone predators. Schmitt and Holbrook found that sea anemones grew faster and reached larger sizes when anemonefish were present. This increase in anemone size allowed a second fish species, the damselfish (*Dascyllus trimaculatus*), to occupy an anemone together with its superior competitor, the more aggressive anemonefish. Coexistence occurred in large anemones, but not in small anemones, because social interactions between anemonefish limit them to two individuals per anemone, regardless of anemone size. Thus, the intensity of intraspecific competition between anemonefish remains constant regardless of anemone size; however, the intensity of interspecific competition between damselfish and anemonefish decreases with anemone size. Recall from Chapter 8 that species coexistence is promoted when the ratio of per capita intraspecific to interspecific competitive effects increases. Schmitt and Holbrook suggested that this mechanism may be prevalent in a variety of systems and thus adds another pathway by which mutualism may enhance diversity.

Figure 9.7 Sea anemones grow faster and reach larger sizes when anemonefish—their mutualistic partners—are present. Photograph © Jodi Jacobson/Shutterstock.

The stress gradient hypothesis for plant facilitation

If we assume that species interactions often represent a balance between positive and negative effects, we can ask what factors might tip the balance, so that we see a preponderance of either net positive or net negative effects within a community. In plants, facilitation appears to be more important, or to occur more frequently, in harsh environments than in mild environments. Primary succession—the colonization of newly exposed substrates such as may be made available by receding glaciers or shoreline, or by volcanic activity—has long been viewed by plant ecologists as a facilitative process (Chapter 14; Clements 1916; Connell and Slatyer 1977). During primary succession, early colonizing plant species may ameliorate the harsh physical conditions of the new habitat by reducing water stress or wind intensity, or by promoting the accumulation of organic material, thereby allowing less hardy plant species to colonize (Chapin et al. 1994; Walker and del Moral 2003). Facilitation has also been frequently observed in harsh environments that are not undergoing succession, such as arid, alpine, and salt marsh environments. This general observation has led to a variety of conceptual models of facilitation in plant communities that can be collectively termed the stress gradient hypothesis (SGH) (Bertness and Callaway 1994; Callaway and Walker 1997; Brooker and Callaghan 1998; Callaway 2007; Butterfield 2009).

Simply put, the stress gradient hypothesis postulates that the relative importance of competitive versus facilitative interactions among plant species changes along a gradient of environmental harshness (Figure 9.8). Competitive interactions should dominate in benign environments, whereas facilitative interactions should become more important in stressful environments. One of the best empirical examples supporting the predictions of the SGH is the work of Callaway et al. (2002), in which a team of researchers from around the world compared the responses of alpine plant species to the removal of neighboring plant species at low elevations (subalpine vegetation) and at high elevations (alpine vegetation) at eleven different mountainous sites across the globe (Figure 9.9). High-elevation sites were harsher environments than low-elevation sites. In addition, the shift from competition to

(A)

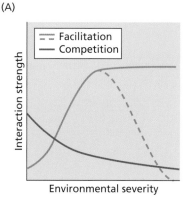

(B)

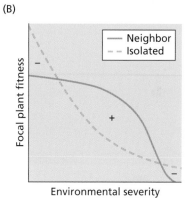

(C)

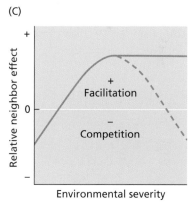

Figure 9.8 Predictions of the stress gradient hypothesis. (A) According to the SGH, the strength of interspecific competition decreases and the strength of facilitation increases with the severity of the physical environment. In very harsh environments, however, the strength of facilitation may decline (dashed line). (B) As a result of these relationships, the fitness of the focal plant species is expected to decline less in the presence of neighboring plants as environmental severity increases (C) The relative neighbor effect is expected to switch from negative (indicating competition) to positive (indicating facilitation) with increased environmental severity. After Butterfield (2009).

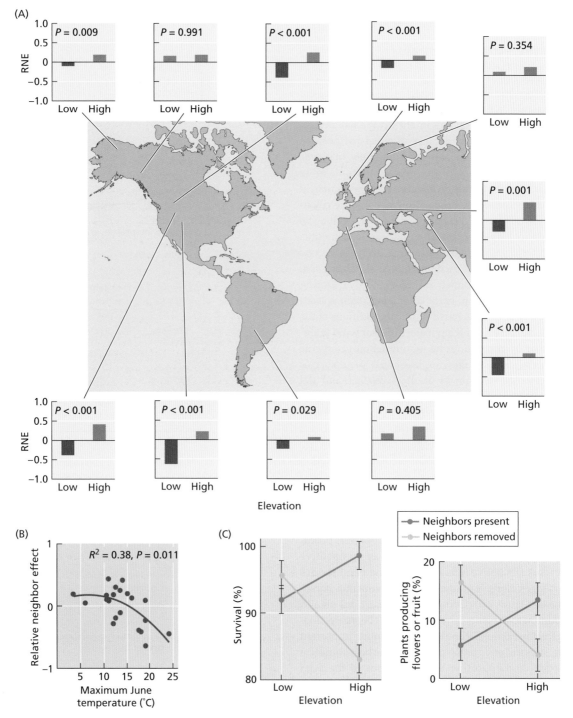

Figure 9.9 Empirical support for the stress gradient hypothesis. Removing neighboring plant species has a detrimental effect on alpine plant species located in harsh alpine habitats, but a positive effect on the same species located in milder, subalpine habitats. (A) Worldwide distribution of the experimental sites. At each experimental site, all neighboring plants within 10 cm of a focal plant were removed and the change in biomass of the focal individuals was measured. RNE is the relative neighbor effect, where positive responses indicate facilitation (blue histograms) and negative responses indicate competition (red histograms). (B) The sign of the RNE changed from facilitation (positive RNE) to competition (negative RNE) as the harshness of the environment decreased (i.e., as June temperatures became warmer). (C) Neighbor removal had a negative effect on focal plant survival and reproduction (i.e., flower/fruit production) in high-elevation plots, but a positive effect at lower elevations. After Callaway et al. (2002).

facilitation was inversely correlated with one measure of environmental harshness, maximum summer temperature (Figure 9.9B). The pattern of response was very consistent at all eleven experimental sites: removing neighbors at low elevations resulted in an increase in target plant survival and in the proportion of target plants that flowered (evidence for interspecific competition), whereas removing neighbors at high elevations generally led to a decrease in target plant survival and flowering (evidence for facilitation; Figure 9.9C).

Although Callaway et al. (2002) and others have provided experimental evidence in support of the stress gradient hypothesis, a number of studies have found contradictory results (e.g., Tielbörger and Kadmon 2000; Pennings et al. 2003; Maestre and Cortina 2004). Moreover, a meta-analysis of field and common garden studies (Maestre et al. 2005) concluded that facilitation does not appear to increase steadily in importance with abiotic stress. Maestre and Cortina (2004) and Butterfield (2009), working in arid systems, have proposed that the facilitative effects of neighbors are likely to be a peaked function of environmental harshness when examined along the full gradient of environmental stress (see Figure 9.8). Butterfield (2009, p. 1194) notes that:

[R]egardless of the mechanism by which plants ameliorate environmental severity, there has to be some set of environmental conditions in which that amelioration is optimal, with reduced ameliorative effects in more or less severe environments. Evidence suggests that plants in many of the most extreme alpine environments approach or reach just beyond this optimal buffering ability (Callaway et al. 2002), whereas perennial plants in the Sonoran Desert lie beyond this optimum (Butterfield et al. 2010).

It remains to be seen whether the stress gradient hypothesis (including its proposed modifications; Maestre et al. 2009) describes a general feature of plant communities or not. In a similar vein, Thrall et al. (2007) proposed that low-productivity environments should favor the evolution of symbiotic mutualisms (positive interactions), whereas high productivity environments should favor symbiotic interactions that are predominantly parasitic (negative interactions). As with the stress-gradient hypothesis, the evidence in support of this hypothesis is mixed and it does not appear that increasing

environmental quality generally favors the evolution of antagonisms over mutualism (Hoeksema and Bruna 2015).

As theoretical models have shown, determining whether the net effect of a species interaction is positive or negative tells only part of the story (Gross 2008). What may be most important to species coexistence and species diversity is the joint action of positive and negative effects. As Gross (2008, p. 935) notes,

Positive interactions can also drive coexistence when they only partially counteract exploitative competition. This latter effect is more subtle, but may be more common in nature. The fact that these more subtle positive interactions go undetected if interactions are classified only by their net effect emphasizes the need to understand all the components of an interaction between species (Callaway and Walker 1997). Even in communities that appear to be predominantly structured by resource competition, positive interactions may provide a key to explaining how many different species coexist.

Looking ahead

Community ecology has made significant strides in recent years in understanding the consequences of beneficial interactions for species diversity and community structure (Bronstein 2015). However, most of these advances have taken place in the context of studying local communities. The role of beneficial interactions in determining macroecological patterns, regional diversity, and species dynamics in metacommunities is essentially unknown (see Gouhier et al. 2011; Menge et al. 2011, for examples of how recruitment facilitation may affect local and regional species diversity in marine systems). Many studies suggest intriguing relationships between beneficial interactions and macroecological patterns. For example, ecologists since the time of Darwin have noted what seems to be a preponderance of mutualistic interactions in the tropics. May (1981, p. 98), based on his mathematical examination of the stability of mutualisms in different environments, speculated that obligate mutualisms are more likely to be favored in relatively stable environments (e.g., the tropics):

These dynamical considerations [stability analyses] may go some way towards explaining why obligate mutualisms of this sort are less prominent in temperate and boreal ecosystems than in tropical ones. A good survey of empirical evidence bearing on this point is given in Farnworth and Golley (1974, pp. 29–31), where it is observed that not only are there no obligate ant–plant mutualisms north of 24 degrees, no nectarivorous or frugivorous bats north of 33 degrees, no orchid bees north of 24 degrees in America, but also within the tropics mutualistic interactions are more prevalent in the warm, wet evergreen forests than in the cooler and more seasonal habitats.

Whether mutualisms are more prevalent in the tropics and whether beneficial interactions play a significant role in the development of the latitudinal diversity gradient are fascinating questions that remain to be explored (Schemske et al. 2009).

As we have seen in this and preceding chapters, interactions between species across trophic levels can result in positive indirect effects (e.g., predators may indirectly benefit plants by reducing the abundance or feeding rate of herbivores). The following chapters will explore the consequences of indirect effects (positive and negative) in food chains and food webs, addressing questions of species diversity, stability, cascading effects, and the regulation of trophic level biomass. In addition to studying indirect effects in traditional food webs (which describe who eats whom within a community), ecologists are making significant progress in incorporating beneficial species interactions (mutualisms) directly into ecological networks (e.g., plant–pollinator networks, plant–seed disperser networks). In the next chapter, we will examine the properties of mutualistic networks and compare the properties of these networks of beneficial species interactions to ecological networks based upon feeding relationships (food webs).

Summary

1. Mutualism may be defined as a reciprocally positive interaction between species. Facilitation may be defined as an interaction in which the presence of one species alters the environment in a way that enhances the growth, survival, or reproduction of a second, neighboring species.
2. The evolution of mutualism and facilitation depends on the relative costs and benefits of these interactions to the participating species. Mutualistic interactions may best be viewed as reciprocal exploitation between species that nonetheless provides net benefits to each partner.
3. The consequences of beneficial interactions for the diversity and functioning of communities are still poorly understood. This failure to incorporate positive interactions into community theory is not simply the result of a historical bias in favor of studying competition and predation, but also stems from the complexity of these interactions.
4. The Lotka–Volterra competition model can be written as a model for positive species interactions simply by changing the sign of the interspecific effect from negative to positive; however, this model is not realistic because it can result in both species' populations growing without bound.
5. Recent models have shown the mutualisms may evolve from parasitic or competitive interactions. Pairs of coexisting competitors may evolve toward a mutualistic "trading partnership" if the two competing species require the same resources, and if each species consumes in excess the resource that least limits its growth, then trades this excess resource to a partner whose growth is more limited by that resource.
6. A change in the environment may affect the costs and benefits of an interspecific interaction. Realistic modeling of mutualistic interactions often depends on the inclusion of other species and on the recognition of a balance between positive and negative effects.
7. The type and strength of an interaction between species may be context-dependent. The extent to which it is beneficial versus antagonistic may change with the harshness of the abiotic environment, the presence of other species (competitors or predators), and the ontogeny or life stage of the interacting individuals.
8. Although the consequences of positive interactions for species diversity and community structure are still poorly understood, there have been some recent steps toward developing a general theory of species interactions that includes a combination of positive and negative interactions. One model shows that two species could coexist on a single resource if facilitation by the superior competitor reduced the mortality rate of the

inferior competitor enough to allow it to invade a community consisting solely of the superior competitor.

9. The stress gradient hypothesis postulates that the relative importance of competitive and facilitative interactions between plant species changes along a gradient of environmental harshness: competitive interactions dominate in benign environments, whereas facilitative interactions become more important in stressful environments.

PART III

Putting the Pieces Together: Food Webs, Ecological Networks and Community Assembly

CHAPTER 10

Species interactions in ecological networks

It is interesting to contemplate a tangled bank, clothed with many plants of many kinds, with birds singing on the bushes, with various insects flitting about, and with worms crawling through the damp earth, and to reflect that these elaborately constructed forms, so different from each other, and dependent upon each other in so complex a manner, have all been produced by laws acting around us. There is grandeur in this view of life...from so simple a beginning endless forms most beautiful and most wonderful have been, and are being evolved.

Charles Darwin, 1859, pp. 489–90

Ever since Darwin contemplated the interdependence of the denizens of his "entangled bank," ecologists have sought to understand how this seemingly bewildering complexity can persist in nature.

Thomas Ings et al., 2009, p. 254

Ecological networks attempt to summarize the multitude of potential interactions between species within a community by representing those species as nodes in the network and using links between nodes to represent interactions between species (Figure 10.1). Traditionally, ecologists have approached the description of ecological networks from the perspective of consumers and their resources—basically, by constructing a diagram showing who eats whom within a community, which is known as a food web. More recently, ecologists have expanded their studies of ecological networks to include other types of species interactions, such as mutualistic webs (e.g., Pascual and Dunne 2006; Ings et al. 2009; Thébault and Fontaine 2010; Schleuning et al. 2014) and host–parasitoid webs (e.g., Hawkins 1992; Van Veen et al. 2008; Condon et al. 2014). Food webs and host–parasitoid webs describe antagonistic interactions, in which one species benefits and another

loses. Mutualistic webs describe interactions that benefit both species (e.g., plant–pollinator networks, plant–seed disperser networks).

Food webs focus on "typical" predator–prey interactions, where consumers are generally larger than their prey. Host–parasitoid webs concentrate on a special type of "predator–prey" feeding relationship, in which parasitoids lay eggs within their host; the larvae that hatch from those eggs then consume and kill the host. Parasites, which feed on, but may not kill their hosts, are conspicuously absent from most studies of ecological networks, although the need to better incorporate parasites more fully into food webs is well recognized (Marcogliese and Cone 1997; Lafferty et al. 2006, 2008; Dunne et al. 2013). Similarly, interactions between plants and herbivores (which often result in the partial consumption, but not the death of an individual plant) do not fit neatly within the traditional food web approach, and the systematic

Community Ecology. Second Edition. Gary G. Mittelbach & Brian J. McGill, Oxford University Press (2019).
© Gary G. Mittelbach & Brian J. McGill (2019). DOI: 10.1093/oso/9780198835851.001.0001

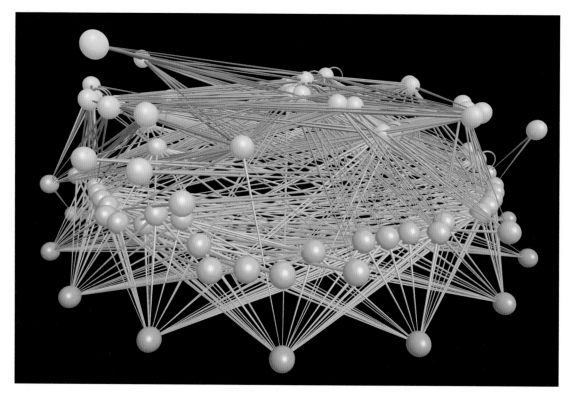

Figure 10.1 An ecological network, or food web, for species in the East River Valley, near Crested Butte, Colorado (based on the research of Neo Martinez and Brett Harvey). Node colors represent trophic levels: red nodes represent basal species, such as plants and detritus; orange nodes represent intermediate consumer species; yellow nodes represent top consumers (predators). Links characterize the interaction between nodes, with the link being thicker at the consumer end and thinner at the resource end. Image produced with FoodWeb3D, written by R. J. Williams and provided by the Pacific Ecoinformatics and Computational Ecology Lab: www.foodwebs.org.; Yoon et al. (2004).

study of herbivory networks is relatively recent (e.g., Melián et al. 2009; Thébault and Fontaine 2010; Fontaine et al. 2011).

The study of ecological networks is an active and fast-moving field, fueled in part by analytical techniques developed in the larger field of "network science," which includes social, biological, and technological networks (e.g., the Internet), and by the ability to access information on ecological networks through online databases (e.g., the Interaction Web Database at http://www.nceas.ucsb.edu/interactionweb/index.html). This chapter will explore a variety of different ecological networks and their properties, beginning with the best-studied networks—food webs.

Food webs

Although simple food web diagrams appeared as early as the late 1800s (e.g., Camerano 1880; Shelford 1913), it was Elton (1927) who introduced the concept of the food web to the mainstream of ecology (see discussion in Leibold and Wootton 2001). Figure 10.2 is a typical food web diagram—in this case, for the Benguela marine ecosystem off the southwestern coast of Africa (Yodzis 2001). This diagram is an example of a connectedness web (Paine 1980) or structural web, in which the trophic links between resources and consumers (i.e., who eats whom) are illustrated by arrows pointing toward the consumer.

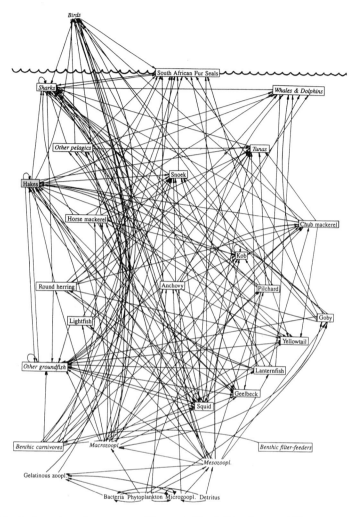

Figure 10.2 A food web for the Benguela marine ecosystem off the coast of South Africa. Arrows point from a resource to a consumer. Hake, the commercial fish of interest in this food web, are located near the upper left. From Yodzis (2001).

Connectedness webs

Connectedness webs show the presence of an interaction between species, but they do not specify the strength of the interaction. As in most ecosystems, the predators in Figure 10.2 feed on multiple species of prey, and the prey species likewise have many predators. Note, too, that consumers near the top of the food web are identified by species or feeding guilds, whereas consumers and prey lower in the food web tend to be aggregated into broad functional groups (e.g., "benthic filter-feeders," or

even "bacteria" and "phytoplankton" at the base of the food web in Figure 10.2). This is typical of most food webs, which tend to be better resolved at the top than at the bottom, in part because species richness is greater at lower trophic levels and in part because species at lower trophic levels tend to be small, difficult to identify, and have feeding relationships that are hard to quantify (Dunne 2006). Chapter 11 will carry this approach of horizontally collapsing species within a trophic level into compartments to its logical extreme, and look at food chains that just have primary producer, herbivore, and carnivore boxes, and that focus on vertical interactions between trophic levels. We will use simplified food chains to address such questions as: "Why is the world green, how important is top-down vs bottom-up regulation, and what determines food chain length?" This chapter, however, will consider full networks that include within-trophic level (horizontal) complexity, and which mix vertical and horizontal processes.

Ecologists often separate food webs into those in which the basal trophic level is composed of primary producers ("green food webs") and those in which the basal trophic level is composed of detritus ("brown food webs"). Both primary production and detritus are sources of energy in communities and there are increasing efforts to link the "green" and "brown" portions of food webs (e.g., Rooney et al. 2006; Butler et al. 2008; Ward et al. 2015). Rooney and McCann (2012) suggest that the "green" and "brown" portions of food webs may represent "fast" and "slow" energy channels, respectively, and that these different energy channels play an important role in determining the stability of food webs. We will return to the question of network structure and stability in more detail later in this chapter.

The Benguela food web shown in Figure 10.2 was analyzed extensively by Peter Yodzis, a theoretical ecologist at Guelph University. Yodzis was asked by the South African government to address whether culling (i.e., killing) Cape fur seals (*Arctocephalus pusillus*), which can consume up to their own body weight daily in hake (*Merluccius capensis*; a delicacy in Europe), would increase the fish harvest for humans. The annual consumption of commercial fish by seals is about 2 million tons, which equals the annual catch by fishermen. When we look at the food web in Figure 10.2, it is clear that seals have a direct, negative effect on the abundance of hake. However, seals also eat hake predators and hake competitors. Therefore, as Yodzis (2001) noted, the combined impact of the direct and indirect effects of seals on hake is not obvious. Yodzis (1998) calculated that there are over 28 million potential interaction pathways between seals and hake. His tentative conclusion was that culling seals was unlikely to benefit the hake fishery because of the multitude of indirect pathways between seals and hake—some of which are positive and some

negative. We will return to the question of compounding indirect effects later in this chapter.

Ecologists have many examples of connectedness webs because the data needed to generate them (i.e., presence or absence of interactions) are relatively easy to collect (although there are significant differences in the quality of published food webs and in whether they are based on observations of individuals or species; Woodward et al. 2010). Thus, ecologists have spent considerable time and effort in analyzing the properties of connectedness webs (also referred to as "binary webs" because they record only the presence or absence of an interaction and not the strength of the interaction). Initial work on connectedness food webs suggested some general topological patterns, and the structural properties of networks are known as **network topology**. (For reviews of the early literature, see Cohen et al. 1990; Pimm et al. 1991; Martinez 1992.) For example, it was once thought that the proportions of species found at different trophic levels (i.e., top predators, consumers, producers) remained relatively constant across webs of different species richnesses, and that the ratio of the total number of links (L) to the total number of species (S) was roughly constant at a value of about 2. Hence, the average number of species with which a given species interacts, $n = 2L/S$, was estimated to be about 4 and appeared to be independent of the total number of species in the web (Cohen et al. 1990; Polis 1991). More recent work, however, using more fully resolved food webs, suggests that the above generalizations are not all that general (Pascual and Dunne 2006; Ings et al. 2009). Other topological properties of food webs (and of other ecological networks), however, appear more robust. We will discuss three of these properties—connectance, nestedness, and modularity—in detail when we compare the structures of trophic and mutualistic networks.

Current interest in network topology is focused on the *mechanisms that produce the structural patterns* of food webs, as well as on the patterns themselves (Pascual and Dunne 2006). Williams and Martinez (2000), for example, showed that many of the structural properties of empirical food webs (i.e., those based on actual observations)—including generality (the number of species that are prey for a particular species), vulnerability (the number of predators that prey on a particular species), and average food chain length—can be predicted from a simple model specifying that predators consume prey species falling within a contiguous range of "niche values" (e.g., body sizes). In other words, large species consume smaller species that are within some particular size range. Although this "niche model" and related approaches can predict many of the structural properties of real food webs, these models are still largely phenomenological because the rules specifying consumer diets are not based on explicit ecological or evolutionary processes (Stouffer 2010). More mechanistic approaches to understanding food web structure have been proposed. For example, Loeuille and Loreau (2005) and Rossberg et al. (2006) examined the evolutionary development of food webs,

and Beckerman et al. (2006) and Petchey et al. (2008) used optimal foraging theory to predict predator diets and trophic links. Despite their differences, the phenomenological and mechanistic approaches to understanding food web structure share some common features, most notably a strong focus on body size (mass) as a key species trait (Stouffer 2010).

Although relatively easy to construct, connectedness webs miss a lot of biological reality. In particular, they ignore the amount of energy (or biomass) transferred between nodes within the food web, and they are mute about the strength of interactions between species. Paine (1980) suggested two other ways to construct food webs that addressed these deficiencies: energy flow webs and functional webs (Figure 10.3).

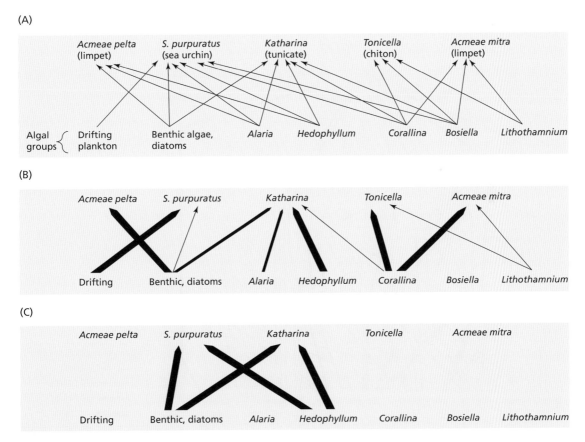

Figure 10.3 Three distinct approaches to constructing food webs, illustrated for the same set of species/functional groups found in the marine intertidal zone of Tatoosh Island, off the coast of Washington State. Arrows point from algal groups to consumers of algae (grazers). (A) A connectedness web, based on observations of who eats whom. (B) An energy flow web, based on estimates of biomass consumption (plus values from the literature). Arrow thicknesses correspond to different amounts of energy flow. (C) A functional web, based on species removal experiments. Arrows connect strongly interacting species. From Paine (1980).

Energy flow webs

An energy flow web measures the amount of energy (biomass) moving between species within a food web (Figure 10.4). Constructing an energy flow web for a community involves much more effort than building a connectedness web and, as a consequence, high-quality energy flow webs for complex communities are relatively rare (see Baird and Milne 1981; Baird and Ulanowicz 1989; Raffaelli and Hall 1996; and Benke et al. 2001 for examples). Implicit in the energy flow approach to food webs is the idea that there is a relationship between the amount of energy flowing through a pathway and the importance of that pathway to community dynamics. This is true for some community properties, such as the bioaccumulation of pollutants like PCBs or mercury, which closely mirrors the flow of energy between species or trophic levels (Rasmussen et al.

1990; Vander Zanden and Rasmussen 1996). However, energy flow has been shown to be a surprisingly poor predictor of the strength of interactions between species or of the impact of removing a particular species from a community (e.g., Paine 1980; Berlow 1999).

When one of us (Mittelbach) was a graduate student in Earl Werner's lab in the late 1970s, the limnologists in Robert Wetzel's laboratory (located one floor below at the Kellogg Biological Station) would needle us about the relevance of our research, suggesting that the fishes we studied (bluegill and bass) might be interesting, but probably didn't contribute much to the functioning of lake ecosystems because only 1% of the primary production of lakes ended up in the bass population. Ironically, less than 10 years later, pioneering studies by Power et al. (1985) and Carpenter et al. (1985) showed that

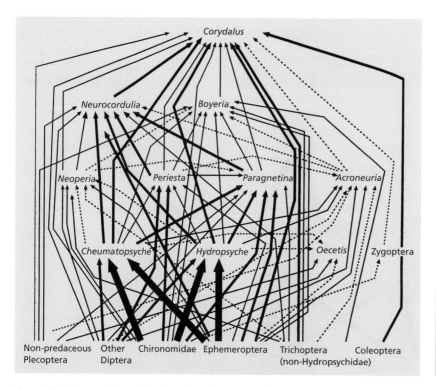
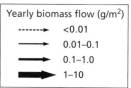

Figure 10.4 An energy flow web for the major predaceous invertebrates living on woody debris in the Ogeechee River, Georgia. Arrows point from prey to predators; arrow thicknesses indicate the amount of energy (biomass) flow. Primary consumers (non-predators) are in the bottom row. Predators are arranged in a hierarchy according to their distance, in food chain links, from the primary consumers. *Cheumatopsyche* and *Hydropsyche* are omnivores, and their ingestion of basal food resources is not shown. From Benke et al. (2001).

bass may function as keystone species in streams and lakes, and that the strong cascading effects of these top predators on species at lower trophic levels could have major impacts on community composition, primary production, and even water quality (see also Mittelbach et al. 1995, 2006). In Chapter 11, we will talk much more about keystone species and cascading effects, but for now the take-home message is this—energy flow is often a poor predictor of the impact that a particular species or group of species may have within a food web. This recognition leads us to functional webs, the third type of food web suggested by Paine (1980).

Functional webs

A functional web (or interaction web) shows the strength of the interactions between species within a community, implicitly recognizing that not all species and interactions are equally important. Paine (1980) proposed that we measure interaction strengths by experimentally removing species from the community and looking at the responses of the remaining species. He did this for a portion of the algal/grazer community found in the rocky intertidal of Tatoosh Island, on the outer coast of Washington State, and found "few strong interactions embedded in a majority of negligible effects" (Paine 1992; see also Figure 10.3).

Since Paine's pioneering work, ecologists have developed a variety of ways to measure interaction strengths in theoretical and real food webs (Berlow et al. 2004). Measures of interaction strength in

model food webs tend to focus on the individual interactions between species pairs within a web (i.e., elements of the community or interaction matrix; see Levins 1968 and Chapter 7), whereas measures of interaction strength in empirical food webs tend to focus on the impact of one species on the rest of the web as measured by removal experiments (Laska and Wootton 1998; O'Gorman and Emmerson 2009). As noted by Berlow et al. (2004: 587), "In the first category individual interaction strengths are independent of the network, in the second category they are in theory inseparable from their network context" (see also Emmerson and Raffaelli 2004).

Table 10.1 lists four indices commonly used to measure interaction strength in empirical food webs, including the metric used by Paine (1992). All of these metrics require that responses be measured over relatively short time intervals; otherwise, indirect effects and density-dependent feedbacks may occur and influence the estimates of interaction strength between directly interacting species (Bender et al. 1984; Laska and Wootton 1998). One of the measures listed in Table 10.1, the "dynamic index" (also called the "log response ratio" or "log-ratio metric"), compares the log of the ratio of prey abundance "with" vs "without" predators; it is equivalent to the coefficients of interaction between species in the Lotka–Volterra predator–prey model (Navarrete and Menge 1996; Osenberg et al. 1997; Laska and Wootton 1998). This metric provides a useful link between theoretical and empirical estimates of interaction strength in food webs; however,

Table 10.1 Indices used to calculate interaction strengths in experimentally manipulated food webs

Index	Formula[a]	Original intent of index	References
Raw difference	$(N - D)/Y$	Commonly used in studies to show absolute, untransformed treatment effects	
Paine's Index (PI)	$(N - D)/DY$	Quantify effect of consumer on competitive dominant resource that has potential to form monoculture	Paine 1992
Community importance (CI)	$(N - D)/Np_y$	Quantify effect of a species relative to its abundance (i.e., distinguish "keystones" from "dominants"	Power et al. 1996
Dynamic Index	$ln(N/D)/Y_t$	Quantify an effect size that is theoretically equivalent to the coefficient of interaction strength in the discrete-time version of the Lotka–Volterra model	Osenberg & Mittelbach 1996; Wootton 1997

[a] Where N is the abundance of prey in the treatment with predators present (i.e., Normal condition); D is the abundance of the prey in the treatment where predators are "Deleted" (i.e., experimentally removed); Y is the abundance of the predator; p_y is the proportional abundance of the predator; and t is time.

Source: From Berlow et al. 1999.

it does so only for the simplest predator–prey model, assuming a linear functional response and no predator interference (Berlow et al. 2004). Bascompte et al. (2005) proposed an additional measure of interaction strength in food webs using observational rather than experimental data, where interaction strength is measured as

$$IS = \frac{(Q/B)_j \times DC_{ij}}{B_i} \qquad \text{[Eqn 10.1]}$$

where $(Q/B)_j$ is the number of times a population of predator j consumes its own weight per day, DC_{ij} is the proportion of prey i in the diet of predator j, and B_i is biomass of prey i.

Despite the challenges of measuring interaction strengths, a consistent feature has emerged from many different studies of both theoretical and empirical food webs: *most food webs contain a few strong and many weak links*. For example, Wootton and Emmerson (2005) analyzed the frequency distributions of the absolute value of per capita effects of consumers on their prey (using a form of the dynamic index; see Table 10.1) using data collected from experimental manipulations of a number of different food webs. They found that interaction strengths in the communities studied were not normally distributed; rather, strong interactions occurred between relatively few species and weak interactions occurred between most species (Figure 10.5). A very similar result was found by Bascompte et al. (2005; Figure 10.6) and Rooney and McCann (2012). Theoretical studies show that weak interactions, which are overrepresented in nature, can be a powerful stabilizing force in food

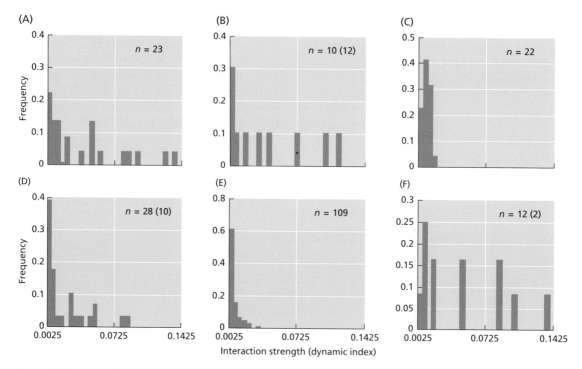

Figure 10.5 Frequency distributions of the absolute values of per capita effects of consumers on their prey. Interaction strengths between consumers and prey were calculated using a version of the dynamic index in Table 10.1 (Osenberg et al. 1997; Wootton 1997). The distributions were calculated using raw data obtained from the studies detailed in (A) Paine 1992, (B) Raffaelli and Hall 1996, (C) Sala and Graham 2002, (D) Wootton 1997, (E) Fagan and Hurd 1994, and (F) Levitan 1987. All distributions are skewed toward many weak interactions and few strong interactions, a pattern that is consistent across all systems studied. Numbers on each graph represent sample size (number of interactions); numbers in parentheses indicate the number of interaction strengths found above the scale of the graphs. After Wootton and Emmerson (2005).

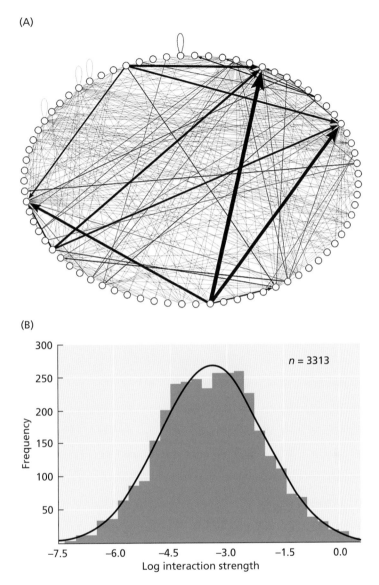

Figure 10.6 Variation of interaction strengths in a real food web. (A) A random sample of 30% of the species and 11% of the interactions in a large Caribbean marine food web (249 total species/trophic groups). Arrows connect predators and their prey; arrow thickness is proportional to interaction strength. Loops represent cannibalism. (B) The frequency distribution of interaction strengths in the food web in A, as calculated by Equation 10.1. The solid line represents the best fit to a lognormal distribution. After Bascompte et al. (2005).

webs (McCann et al. 1998; Emmerson and Yearsley 2004; Valdovinos et al. 2010; Rooney and McCann 2012). We will explore this idea further in our discussion of complexity and stability later in this chapter.

Keystone species

The skewed distribution of interaction strengths in real food webs is also consistent with the observation that some species play extraordinarily large roles within communities. Paine (1969) called these

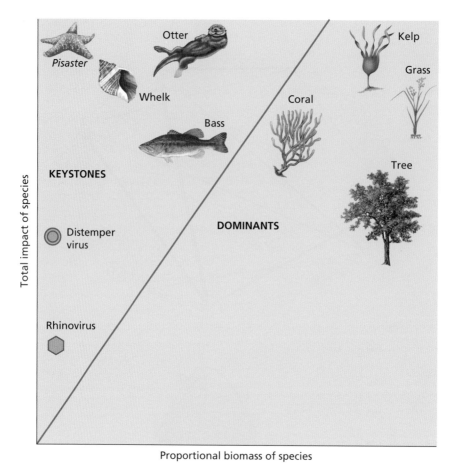

Figure 10.7 Keystone species (upper left) are defined as species whose impacts on the community are large relative to their biomass. Dominant species (upper right) are those that constitute a large fraction of a community's biomass and whose impacts are large but not disproportionate to their abundance. After Power et al. (1996).

keystone species—species whose effect on the community is disproportionately large relative to their abundance (Figure 10.7; Power et al. 1996). Classic examples of keystone species and their impacts include:

- Predation by *Pisaster* starfish increases species diversity by preventing the monopolization of space by mussels in the Pacific intertidal (Paine 1966, 1969).
- Sea otters limit the abundance of grazing sea urchins, thereby allowing kelp forests and their associated species to flourish (Estes and Palmisano 1974; also see Chapter 11).

- Piscivorous bass (*Micropterus*) control the abundance of small fishes in lakes (Figure 10.8), affecting a trophic cascade down through the grazer and algal trophic levels, and ultimately affecting water clarity (Carpenter et al. 1985; Power et al. 1985; Mittelbach et al. 1995, 2006).

A longer list of likely keystone species and their mechanisms of action can be found in Power et al. (1996).

Identifying a keystone species is relatively easy after the species is lost or purposefully removed from an ecosystem; identifying keystones *a priori* is much harder. Ecologists have struggled to list the

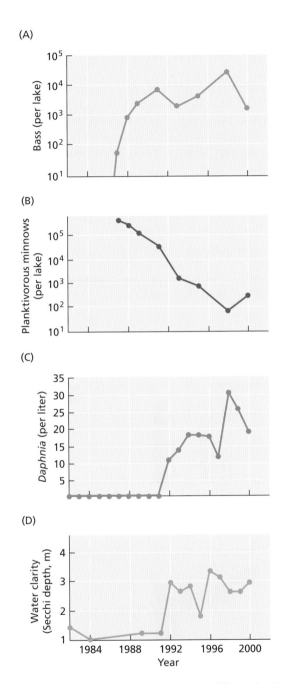

traits that might characterize keystone species. For example, a high feeding rate and a preference for consuming prey that are competitive dominants typify some keystone predators (Power et al. 1996). However, in many cases, whether a species acts as a keystone species or not depends on the context of the interaction ("context" here includes community composition, geographic location, environmental productivity, etc.; Menge et al. 1994; Power et al. 1996).

Similarly, ecologists are searching for general properties or traits of species that might predict strong and weak interactions in food webs. As Berlow et al. (2004, p. 590) note, "Identifying consistencies in a community-level pattern of interaction strengths does not necessarily provide information about the identity of key species or particularly vulnerable species. Similar community-level patterns of a few strong and many weak interactions for two different interaction strength metrics do not necessarily mean that the same species will be identified as strong players in each case." One trait that holds promise for characterizing strong and weak interactions in food webs is body size.

Body size, foraging models, and food web structure

Not surprisingly, Charles Elton (1927, p. 59) was one of the first to recognize the importance of predator and prey body sizes in structuring food web interactions: "A little consideration will show that size is the main reason underlying the existence of these food chains, and that it [body size] explains many of the phenomena connected with the food-cycle [food web]." Species attributes that influence their interactions with other species often scale with body size; examples include metabolic rate, movement speed, rate of encounter, handling time, and feeding rate (Peters 1983; Brown et al. 2004; Woodward et al. 2005; McGill and Mittelbach 2006; Pawar et al.

Figure 10.8 Cascading effects across trophic levels following the reintroduction of a top predator, the largemouth bass, to a small Michigan lake. About 600 fingerling bass were reintroduced to Wintergreen Lake in 1987, after a winterkill event had eliminated all bass from the lake 10 years earlier. (A) The bass population increased rapidly following introduction. (B) As the bass increased in numbers, they decimated the population of planktivorous minnows (golden shiner, *Notemigonus crysoleucas*). (C) Loss of the minnows allowed for the return of large-bodied herbivorous zooplankton (*Daphnia* spp.). (D) The return of *Daphnia* resulted in a dramatic increase in water clarity, as measured by Secchi depth. After Symstad et al. (2003); see also Mittelbach et al. (1995).

2012; DeLong et al. 2015, see also Chapter 6). Thus, we might expect the relative sizes of predator and prey to go a long way toward determining who eats whom within food webs (the trophic links) and, thus, potentially, the strength of ecological interactions.

Beckerman et al. (2006) and Petchey et al. (2008) incorporated size-specific handling times for predators, along with the size-based energy content of prey, into a foraging model to predict trophic links within a food web. By modeling handling time as an increasing function of the ratio of prey size to predator size (assuming a minimum handling time; see Chapter 6), and by modeling prey energy content as a positive function of prey size, they correctly predicted up to 65% of the trophic links in four real-world food webs, based on the assumption that predators prefer to eat the most energetically rewarding prey (Figure 10.9). Their study thus

makes an important stride toward linking foraging theory and species interactions. The link, however, is still incomplete, as Petchey et al. (2008) were not able to parameterize all components of the optimal foraging model independent of the data used to test the model, and Allesina (2011) showed that a phenomenological food web model performed as well as Petchey et al.'s more mechanistic model. However, this does not diminish the value of applying optimal foraging theory to the study of food webs (Petchey et al. 2011) and Valdovinos et al. (2010) show how adaptive foraging behavior can stabilize food web dynamics. More recent studies also have focused on predator and prey body sizes as key traits to predict the realized interactions in food webs (e.g., Laigle et al. 2018; Brousseau et al. 2018), although unlike Petchey et al. (2008) these studies do not attempt to use optimal foraging theory to make these predictions.

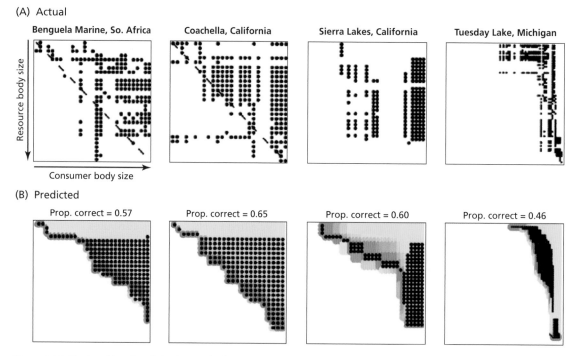

Figure 10.9 Size-based foraging relationships predict food web structure. (A) Trophic links from four real food webs in matrix format, with resources in rows and consumers in columns. Body size increases from left to right and top to bottom. A black dot indicates that the consumer in that column feeds on the resource in that row. (B) Trophic links predicted by a model of optimal foraging with parameters tuned to the data in A. Yellow to red indicates low to high resource profitability. Consumer diets always include the darker red (most profitable) resources and extend by different amounts into the yellow (less profitable) resources. From Petchey et al. (2008).

We might wonder, then, is there a connection between the relative sizes of predators and prey, and the strength of their interactions as well? Perhaps. In a preliminary analysis, Wootton and Emmerson (2005) showed that per capita inter-action strength (measured by the dynamic index) was positively related to the ratio of prey weight to predator weight in a collection of four food web studies (Figure 10.10). They noted, however, that this relationship showed a great deal of variation between studies, and they suggested that the underlying relationship might be unimodal rather than linear. Emmerson and Raffaelli (2004) also found that the predator/prey size ratio was cor-related with the strength of trophic interactions in experimental food webs, but that the form of this relationship varied with predator species. Finally, a unimodal relationship between predation rate (a measure of energy flux) and the predator–prey body mass ratio was found in a laboratory study using predaceous ground-dwelling beetles and spiders feeding on collembolans, crickets, or fruit flies (Brose et al. 2008). In this experiment, the observed predation rate for predators and their different sizes of prey was accurately predicted by a foraging model predicated on the size-specific scaling of handling times, prey detection, and prey escape, but did not conform well to the predictions of a model based on the metabolic scaling of pre-dation rates Figure 10.11). On the great plains of Africa, adult body size has been shown to be a strong predictor of predator-driven mortality rates (Figure 10.12; Sinclair et al. 2003) and larger-bodied predators induce stronger trophic cascades (DeLong et al. 2015), suggesting that the loss of large preda-tors should have greater consequences for ecosys-tems than the loss of smaller predators (Brose et al. 2016).

To summarize, it appears that Elton's intuition was correct; body size relationships play a major role in determining the pattern and strength of trophic interactions within food webs (e.g., Williams and Martinez 2000; Gravel et al. 2013). In addition, by developing predator–prey models in which important parameters such as encounter rates and handling times are allometric functions body size (e.g., McGill and Mittelbach 2006; Pawar et al. 2012), ecologists have begun to forge important links between individual behaviors and the structure and dynamics of ecological networks (Petchey et al. 2008; Stouffer 2010; Pawar et al. 2012; Brose et al. 2016; Costa-Pereira et al. 2018).

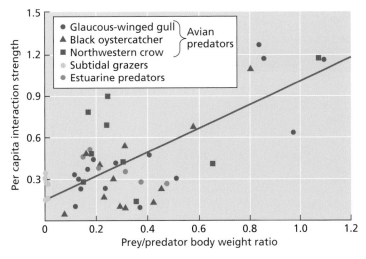

Figure 10.10 Per capita interaction strength increases with the ratio of prey to predator body weight (fourth-root transformed data). This conclusion is based on data from avian rocky intertidal predators, subtidal grazers, and estuarine predators. The regression line shown is for avian intertidal predators only. After Wootton and Emmerson (2005).

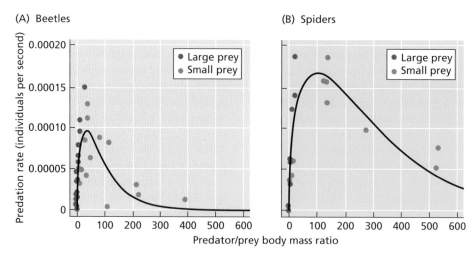

(A) Beetles

(B) Spiders

Predator/prey body mass ratio

Figure 10.11 Predatory beetles (A) and spiders (B) feeding on large and small prey items in the laboratory show feeding rates that are a hump-shaped function of the ratio of predator mass to prey mass. Observations suggest that predation was limited by the relatively long time needed to subdue and handle prey at low predator–prey body mass ratios. At high predator–prey body mass ratios, predation was limited by the high escape efficiencies of prey species due to fast reaction times and their use of small refuges. After Brose et al. (2008).

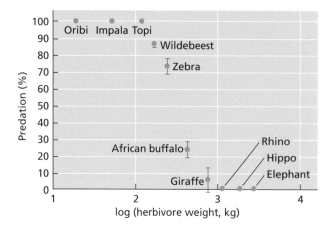

Figure 10.12 Predator-related adult mortality in nonmigratory African ungulates drops off sharply at a body size of about 150 kg. Error bars are 95% confidence limits. After Sinclair et al. (2003); photo © Images of Africa/Alamy.

Species traits and the structure of ecological networks

We have seen the body size is a key trait predicting the links between predators and prey in a food web. Are there other organismal traits that are important in predicting species interactions within ecological networks and can we identify which (and how many) traits contribute to explaining network structure? Eklof et al. (2013) tackled this question by examining a set of 200 ecological networks, including food webs, and mutualistic and antagonistic networks. For each network, they determined the number of trait-axes (dimensions) needed to specify the interactions between species within the network. Figure 10.13 illustrates this process for two traits— body size and habitat (water depth in this example). In panel (a), the focal species (red) interacts with all species in the body size range ($b_1 B_1$). In panel (b) the focal species interacts with the species in the body size range ($b_1 B_1$) but only if they are also found in depth range ($d_1 D_1$). Eklof et al. (2013) found that the number of dimensions (species traits) needed to specify all the interactions in an ecological network was always <10. Moreover, by using only 3 traits they could often rule out many potential interactions within a food web. For example, in a food web for the Weddell Sea composed of 488 species and 238,144 possible connections, by using three traits (body mass of the consumer, body mass of the resource, and mobility of the resource) they were able to eliminate almost 200,000 connections as "forbidden links", leaving about 40,000 possible interactions (of which 15,580 were realized). Thus, we can tentatively conclude that relatively few niche dimensions may structure most ecological networks.

In a similar vein, Kéfi et al. (2016) examined the organization of a complex network of 106 interacting species from the Chilean intertidal zone. In this "multiplex" ecological network, the species interactions include a mix of predation, competition, and facilitation (the networks examined by Eklof et al. 2013 were all single-interaction type). Kéfi et al. (2016) found that the Chilean network of multiple interaction types was organized into functional groups, with each group gathering closely related species. Moreover, these functional groups

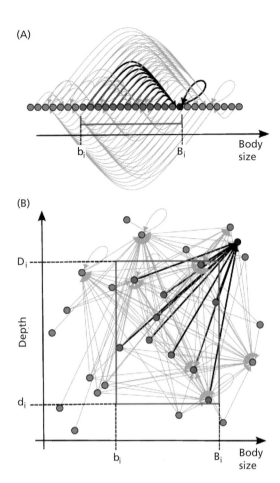

Figure 10.13 Schematic description of species participation in an ecological network based on their traits. In (a) the focal species (red) interacts with all species in the body size range [$b_i B_i$]. In (b) the focal species interacts with the species in the body size range [$b_i B_i$] but only if they are also present at depth [$d_i D_i$]. After Eklof et al. (2013).

were well-predicted by simple traits, such as trophic level, mobility, and habitat. The authors suggest that the combinations of interactions actually realized in this community are constrained to be far fewer than possible, which may "… open up a pathway towards simplifying ecosystem complexity into basic building blocks" (Kéfi et al. 2016, p. 9).

Indirect effects

Looking at a food web, or any other diagram of an ecological network, emphasizes the potential importance of indirect effects in communities. An indirect effect occurs when one species affects a second species through a "change" in an intermediate (third) species (Abrams et al. 1996). This "change" may be an alteration in the abundance of a species, referred to as a density-mediated effect, or an alteration in a species' phenotype (e.g., behavior, habitat use, morphology), referred to as a trait-mediated effect.

There are many types of indirect effects in food webs; a few are modeled in Figure 10.14. While all are potentially important, four types of indirect effects stand out as being particularly common and significant: exploitative competition, apparent competition, trophic cascades, and keystone predation. We discussed exploitative competition and apparent competition in Chapters 7 and 8. In Chapter 11, we will examine trophic cascades and keystone predation in relation to top-down versus bottom-up control of trophic-level biomass and species richness.

Because indirect effects involve the influence of one species on another via an intermediate species, the pathway for an indirect effect is necessarily longer than that for a direct effect. Because of this, we might expect that:

- indirect effects will take longer to develop than direct effects;
- indirect effects will be weaker than direct effects (because of attenuation of effect size with each trophic link involved).

An early literature review by Schoener (1993) found some support for these predictions. Schoener summarized the detailed findings of six experimental studies of community regulation in terrestrial and aquatic habitats. He concluded that:

- direct effects were usually stronger than indirect effects;
- short-chain indirect effects were stronger than long-chain indirect effects.

Schoener's analysis was insightful, but it was limited in scope because of the few studies available at the time.

In a later analysis, Menge (1995) examined the results of perturbation experiments in 23 marine rocky intertidal habitats around the globe. Although the mean change in species densities due to direct effects was greater than that due to indirect effects in 12 of 18 food webs, this difference was not significant (Figure 10.15A,B). On average, 40–50% of the change

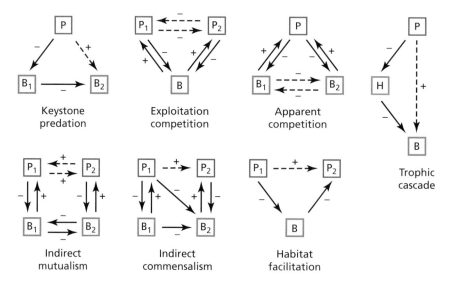

Figure 10.14 Models of seven types of indirect effect sequences. Solid arrows represent direct effects; indirect effects are shown by dashed arrows. Plus and minus signs indicate positive and negative effects, respectively. B, basal species; P, predator; H, herbivore. After Menge (1995).

in species densities after an experimental perturbation was due to indirect effects, with the remainder due to direct effects. Thus, direct and indirect effects were comparable in magnitude in the communities studied. Menge (1995) also found no difference in the time it took to observe significant direct and indirect effects (Figure 10.15C,D). Thus, based on Menge's review, we may conclude that indirect effects are comparable in magnitude to direct effects, and that direct and indirect effects take place at similar rates. However, we need to temper these conclusions somewhat because the results are based on a single well-studied habitat—the marine intertidal zone.

To our knowledge, no other comprehensive reviews comparing the strength of indirect and direct interactions in field experiments have been published since Schoener (1993) and Menge (1995) conducted their studies, which is surprising. Montoya et al. (2009) evaluated the relative importance of direct and indirect effects in nine well-studied empirical food webs using a non-experimental approach. They estimated per capita interaction strengths between all the species in each of the nine food webs from a combination of field measurements and prey/predator body size ratios, then used the inverse of the community matrix (Bender et al.

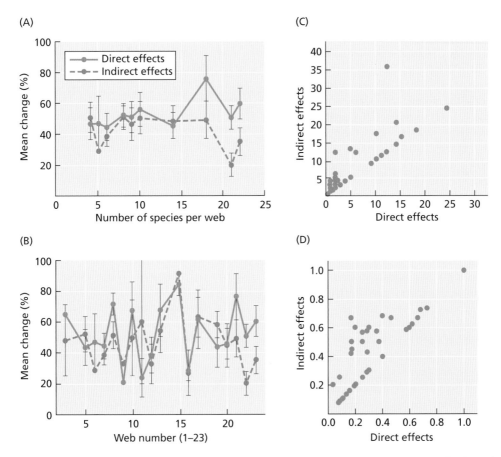

Figure 10.15 Importance of direct and indirect effects in marine intertidal food webs. (A) Mean percentage change in the abundance of organisms (absolute value) caused by experimentally manipulated direct and indirect effects plotted against the number of species in each food web. Webs with equal numbers of species are lumped. (B) The same data plotted for each food web individually (webs are not lumped). Error bars are ±1 SE. (C) The number of months required to demonstrate significant direct and indirect effects. (D) The same data expressed as the proportion of the duration of each experiment, in order to adjust for variation in experiment duration. After Abrams et al. (1996); data from Menge (1995).

1984; Yodzis 1988; see Chapter 7) to explore how each food web should respond to a theoretical perturbation that alters species' abundances. Like Menge (1995), they found that direct and indirect effects played similarly important roles in determining a response to perturbation, and that the combined effects of indirect interactions could often counteract the direct effects. For example, they found that for 40% of predator–prey links, predators had a positive net effect on the abundance of their prey due to a predominance of indirect effects. This counterintuitive result harkens back to a question raised earlier in this chapter—what is the predicted effect of culling a top predator (the Cape fur seal) on the abundance of one of its prey (hake). Recall that the scientist asked to answer this question, Peter Yodzis, replied that because of the multitude of possible indirect effects, the answer was unknown.

Non-consumptive (trait-mediated) effects of predators on their prey (e.g., changes in prey behavior, habitat use, morphology, etc.; see Chapter 6) can also modify the trophic interactions in food webs and lead to a variety of indirect effects (Abrams 1995). In Chapter 11 we will show how the non-consumptive effects of predators can induce or modify the strength of a trophic cascade (one of the indirect effects in Figure 10.14). We know much less about the impact of non-consumptive effects on other types of indirect interactions in food webs, although recent work suggests that they play a significant role (Krenek and Rudolf 2014; Terry et al. 2017; Trussell et al. 2017). Jonsson et al. (2018) provide a particularly nice example of this from an experimental system containing one to four predatory insect species and their aphid prey. When each predator species was alone with its prey, the interaction strength between predator and prey was well-predicted by a model where attack rates and handling times were based on predator-prey body mass ratios (panel (A) in Figure 10.16). However, in food webs containing multiple predator species (2–4 species), the predictive ability of the model was greatly decreased (panel (B) in Figure 10.16; $R^2 = 0.43$ compared with $R^2 = 0.90$). Jonsson et al. (2018) attributed the model's inability to accurately predict interaction strengths when multiple predator species were present to behaviorally-mediated effects that were not included in the model.

Other types of ecological networks

In contrast to food webs, which have been recognized and studied for over a century, ecologists have only recently begun to examine networks of other types of species interactions. Mutualisms, in particular, have become the focus of much recent work on ecological networks.

Mutualistic networks

Two of the most common and important mutualistic interactions occur between plants and animals: pollination and seed dispersal. Over 90% of angiosperm species in the tropics, for example, are thought to depend on animals for pollination (Bawa 1990) and dispersal (Jordano 2000). Many examples of plant–animal mutualisms were discussed in Chapter 9. Here, we will consider some of the patterns of interaction observed between plant and animal mutualists within communities and their potential consequences.

Like trophic interactions, mutualistic interactions can be diagrammed as webs of links between species (or functional groups). However, unlike food webs, in which species are often represented by nodes of a single type (one-mode networks), plant–animal mutualisms have two well-defined types of nodes (e.g., plants and pollinators), and interactions occur between, but not within, node types (Bascompte and Jordano 2007). These two-mode networks may be represented by bipartite interaction webs. Figure 10.17 shows an example of such an interaction web for plants and their bee pollinators (Bezerra et al. 2009). In this example, the lines connect plant species (circles) to the bee species (squares) that pollinate them. The strength of the interaction is represented by the thickness of the connecting line, which indicates the relative number of visits by a bee species to a plant species.

As in food webs, we can very often predict the likelihood of species interactions in mutualist networks by matching the traits of the potential interactors (plants and their animal mutualists). For example, in hummingbird pollination networks the most important traits predicting the structure of species interactions are bill curvature, bill length, and body mass for the hummingbirds and corolla

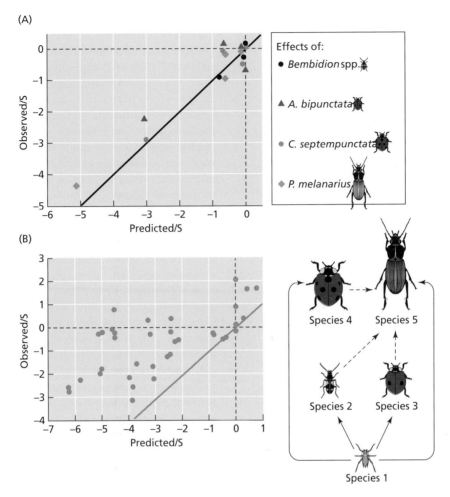

Figure 10.16 Trophic interactions between aphids and their predators (solid lines) and intraguild predation between predator species (dashed lines) in an experimental food web. Sp. 1: the aphid *Rhopalosiphum padi*, Sp. 2 *Bembidion* spp., Sp. 3 *Adalia bipunctata*, Sp. 4 *Coccinella septempunctata*, and Sp. 5 *Pterostichus melanarius*. Body size predicts observed interaction strengths (*IS*) well in the absence (A) but not in the presence (B) of other species. Dashed lines in (A) and (B) are 1:1 diagonals and not regression lines. After Jonsson et al. (2018).

length, curvature, and volume for the plants (Maglianesi et al. 2014). These traits were also strong and significant predictors of interaction strength in the network. For frugivory (seed dispersal) networks the most important traits are often beak size and fruit size, with body mass and crop mass also being significant (Dehling et al. 2014).

Some important topological properties of mutualistic networks are:

- a high degree of nestedness;
- a moderate level of modularity;

- relatively low connectance (Thébault and Fontaine 2010; Schleuning et al. 2014; Almeida and Mikich 2017).

Connectance refers to the observed number of links in the network, expressed as a proportion of the total number of possible links in the network. Nestedness refers to a specific type of interaction structure in which species with many interactions (generalists) form a core of interacting species and species with few interactions (specialists) interact mostly with generalists (Bascompte et al. 2003;

Plant
species

Bee
species

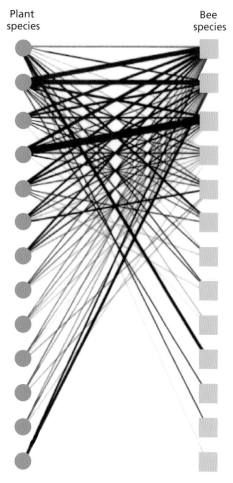

Figure 10.17 A plant–pollinator interaction network. Boxes represent pollinating bee species; circles represent plant species. The thickness of the line connecting bee and plant indicates the number of visits by a bee species to a plant species (thicker lines indicate more visits). There are an equal number of bee and plant species in this example simply due to chance. From Bezerra et al. (2009).

Melián and Bascompte 2004; Ings et al. 2009). Modularity exists in a network when groups of species interact more among themselves than with species from other groups (**Figure 10.18**).

Non-random patterns of nestedness have been found in ant–plant networks (Guimarães et al. 2007), pollination and seed dispersal networks (Bascompte et al. 2003; Bezerra et al. 2009; Bascompte et al. 2013; Almeida and Mikich 2017), and marine cleaning networks (Guimarães et al. 2006). These observations suggest that mutualisms may evolve

into a predictable community structure (Lewinsohn et al. 2006). Thébault and Fontaine (2010) compared the structures of 34 mutualistic (pollination) networks with those of 23 trophic (herbivory) networks. They also analyzed models of population dynamics to assess how differences in network topology (nestedness, modularity, and connectance) affected the persistence and resilience of mutualistic and trophic networks. They found that increased nestedness and connectance promoted stability in model mutualistic networks, whereas increased modularity promoted stability in model trophic networks (Figure 10.19). These same properties were found to differ between real-world mutualistic and trophic networks: mutualistic networks had higher connectance and higher nestedness than did trophic networks, whereas trophic networks tended to have higher modularity (Figure 10.20). Thus, their analyses suggest that food webs and mutualistic webs differ systematically in their topologies, and that these differences differentially affect network stability such that each type of network develops a structure that tends to stabilize that network.

The study of mutualistic networks to date has focused largely on describing patterns of interaction and predicting their consequences for network functioning. Future steps toward understanding these networks should include experiments (Vázquez et al. 2015). Just as the study of food webs has moved from the analysis of patterns in connectedness webs to the construction of functional webs, experimental manipulations of mutualistic webs (e.g., species removals) are needed to judge whether the patterns observed in these networks reflect their functional behaviors. For example, interaction strength in pollination webs has been estimated by the observed number of visits by a pollinator species to a plant species (e.g., Bezerra et al. 2009), yet we do not know whether the number of pollinator visits in fact reflects the degree to which the reproductive success of the plant species depends on a particular pollinator species (but see Vázquez et al. 2005). Likewise, we do not know how the extinction of a particular pollinator (or plant species) would influence the persistence of other pollinator or plant species within the network if pollinators are flexible in their behaviors and, therefore, shift the structure of the interaction web in response to the extinction

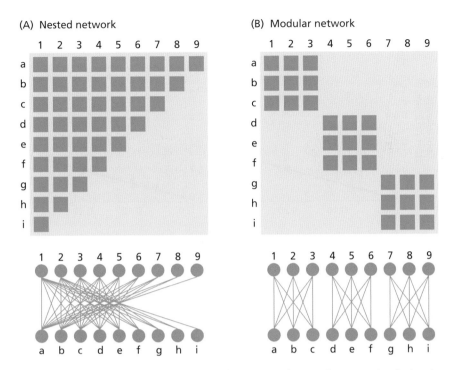

Figure 10.18 Schematic representation of nested (A) and modular (B) bipartite networks. In matrix representations (top), each row and column corresponds to a species; squares represent species interactions. In web representations (bottom), each node represents a species, and interacting species are connected by lines. From Fontaine et al. (2011).

(or removal) of a particular plant or pollinator species (Rezende et al. 2007). Quantifying interaction strength becomes especially challenging when the network is made up of different interaction types (e.g., pollination and herbivory) because we need to find a common currency (Kéfi et al. 2015; Sauve et al. 2016).

Parasites and parasitoids

Most of the world's species are probably parasitic (Price 1980), especially if we take a broad view of parasites that includes bacteria, viruses, fungi, and other symbiotic organisms, as well as protozoans and metazoans that feed on their hosts without killing them. Indeed, Lafferty et al. (2008) argue that parasitism is the most common consumer strategy among organisms. Yet parasites are rarely included in the study of food webs (insect parasitoids, which *do* kill their hosts and thus act like "standard" predators, are an exception; see van Veen et al. 2008

for examples of parasitoid food webs). Why have parasites been "ignored" in the study of ecological networks, despite the fact that in one of the earliest discussions of food webs in ecology, Elton (1927) spent nearly as much time talking about parasite–host webs as about predator–prey webs? There are many possibilities. For one, parasites are often small and cryptic (living within larger organisms), and they require a level of taxonomic expertise to identify that is outside the range of most ecologists. Thus, Lafferty et al. (2006, 2008) suggest that parasites are absent from most food webs because they are difficult to quantify by standard ecological methods. No doubt this is part of the answer. There are, however, a number of recent examples in which researchers have included parasites in highly resolved food webs (e.g., the seven food webs listed in Dunne et al. 2013). Analyses of food webs containing parasites reveal several interesting patterns. For example, while free-living species are involved in more food web links than are parasites (by necessity), parasite

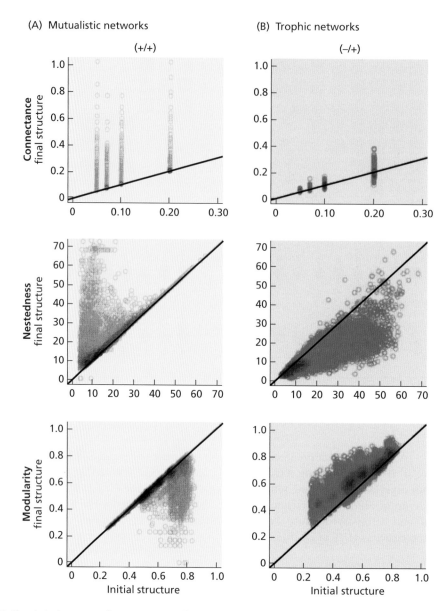

Figure 10.19 The relative importance of connectance, nestedness, and modularity in the stability and architecture of mutualistic (plant–pollinator in this study) and trophic (plant–herbivore) model networks. Final network structure is plotted against initial network structure for a series of model simulations in which mutualistic (A) and trophic (B) networks were allowed to develop over time. During the simulations, some species became extinct before equilibrium was reached, thus altering network structure. These extinctions caused mutualistic networks to become more connected and more nested over time, whereas trophic networks became more modular over time. After Thébault and Fontaine (2010).

(A)

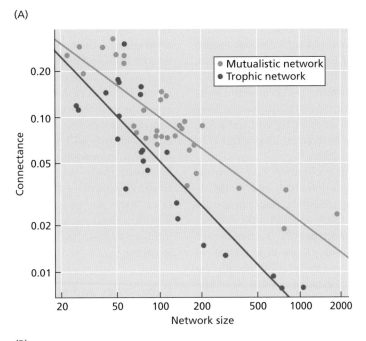

(B)

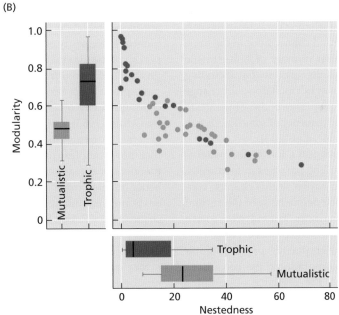

Figure 10.20 The degrees of connectance, nestedness, and modularity differ among real mutualistic and trophic networks. Each dot represents an empirical network involving either pollination (green) or herbivory (red). (A) Connectance plotted as a function of network size. Connectance is greater in pollination networks than in herbivory networks. (B) Relationship between network nestedness and modularity. Box plots of relative nestedness (below) and relative modularity (left) show that nestedness is greater in pollination (mutualisitic) networks and modularity is greater in herbivory (trophic) networks. After Thébault and Fontaine (2010).

interactions may contribute up to 50 percent of the links in some food webs (Dunne et al. 2013) and the total biomass of parasites in food webs may exceed that of top predators (Kuris et al. 2008). In addition, including parasites in food webs reduces the robustness of food webs to species loss because the webs are more sensitive to secondary extinctions (Lafferty and Kuris 2009; Chen et al. 2011) and including parasites can increase food chain length and connectance (Lafferty et al. 2006).

Species richness will necessarily increase when parasites are added to a food web and ecologists know that simply increasing the number of species in a web will increase the number of links per species and may alter connectance (the fraction of possible links realized). Thus, Dunne et al. (2013) set out to test whether parasites uniquely change the properties of food webs relative to what would be expected from the generic effects of adding any type of species and links to a web. What they found was very interesting. By analyzing seven highly resolved food webs that included metazoan parasites, they showed that including parasites in the web increased link density and connectance in the same way as adding a similar number of non-parasite species. Thus, their results suggest that including parasites in food webs may not fundamentally change many of the structural properties of these interaction networks (although the frequency of some interaction types will change; Dunne et al. 2013).

The above studies suggest a few of the patterns that emerge from including parasites in food webs. However, as we noted earlier in this chapter, connectedness webs (which define most of the current attempts to include parasites in food webs) have their limitations. Moreover, other traits of parasites—beyond their size and crypticity—present significant challenges to incorporating these consumers into food webs. First, the presence of complex life cycles and multiple hosts makes it difficult to position a parasite species within a food web. (This problem is similar to the one raised by ontogenetic niche shifts in free-living organisms that change their diet as they grow; see, for example, Rudolf and Lafferty 2011.) Secondly, parasites feed on, but rarely kill, their hosts. This trait makes them similar to many herbivores, but measuring the energy transfer from hosts to parasites is more difficult than measuring energy transfer from plants to herbivores. Finally, parasites affect the behaviors of their hosts, often making them more susceptible to predators (Moore 2002) and thereby influencing host mortality rates and the rate of energy transfer between trophic levels. This trait makes their effects similar to (but opposite in direction from) the non-lethal (non-consumptive) or trait-mediated effects of predators on their prey (see Chapter 6). Thus, while the challenges of incorporating parasites into food webs are qualitatively no different from the challenges incurred with other types of consumers (Lafferty et al. 2008), the sum total of these challenges is significant and explains why so few webs to date include parasites, despite their recognized importance in nature.

Complexity and stability

From their very beginning, studies of food webs and other ecological networks have wrestled with the question of whether more diverse communities are more stable (McCann 2000). Charles Elton, a pioneer in the study of food webs, was a strong proponent of the idea that simple communities showed greater fluctuations in species abundance, and were more prone to species invasions, than were diverse communities (Elton 1927). Elton's ideas were echoed by Odum (1953) and MacArthur (1955), and the general consensus among ecologists at the time was that more complex communities, with their greater numbers of consumer and resource species, were better buffered from the impacts of species loss and environmental fluctuations. This idea was severely challenged by Robert May (1973a), whose theoretical analysis of model ecosystems showed that more diverse communities tended to be less stable.

May constructed his theoretical communities by randomly assigning interaction strengths to the different species. We now know that the random assignment of interaction strengths in May's model was critical to his finding that diversity tended to destabilize community dynamics and that the distribution of interaction strengths in real communities is distinctly nonrandom. May was well aware of the limitations of his analyses and he suggested that "theoretical effort should concentrate on elucidating the very special and mathematically atypical sorts of complexity which could enhance stability, rather than seeking some (false) 'complexity-implies stability' general theorem" (May 1973a p. 77). This search for the ways in which interaction networks found in the real biological world are a special subset of all possible networks is a theme running throughout the history of the study of interaction networks.

As discussed earlier in this chapter, natural communities show a pronounced skew in the distribution of interaction strengths: interactions between a few species are very strong, but most species interact only weakly (Bascompte et al. 2005; Wootton and Emmerson 2005). Recent theoretical studies of food webs show that this skewed distribution of interaction strengths strongly promotes stability. McCann et al. (1998) and McCann (2000) showed how weak interaction strengths promote stability in food webs by damping population oscillations associated with strongly interacting species. Weak interactions can limit runaway consumption along strong consumer–resource pathways, and they can generate negative covariances between resources that share a consumer, thus helping to ensure that consumers have weak effects on a resource when that resource is at low density. Adaptive foraging behavior also can cause predators to reduce their foraging efforts on prey species that become rare, thus helping to stabilize food-web dynamics (Valdovinos et al. 2010). With regard to mutualistic webs, an increase in nestedness has been shown to increase the range of environmental conditions compatible with species coexistence (Rohr et al. 2014; Saavedra et al. 2016). Thus, we might expect to find more nested mutualistic webs in more unpredictable environments (Song et al. 2017).

Empirical food webs differ from the randomly constructed webs studied by May (1973a) by being significantly more modular or compartmentalized (e.g., subsets of species interact more often amongst themselves than with other species; Krause et al. 2003; Rezende et al. 2009; Guimerà et al. 2010; Thébault and Fontaine 2010; but see Kondoh et al. 2010). Modularity has been shown to increase food web persistence by buffering the propagation of species extinctions through the community (Melián and Bascompte 2002; Thébault and Fontaine 2010; Stouffer and Bascompte 2011). The stabilizing effects of modularity and weak interactions can be promoted in food webs by the segregation of energy flow into "fast" and "slow" channels–energy channels defined as groups of highly interacting species at lower trophic levels that derive their energy largely from the same basal resource. Fast energy channels tend to have smaller, faster-growing populations with higher biomass turnover rates than slow energy channels (Rooney and McCann 2012). In marine food webs, for example, the flow of energy originating from phytoplankton production moves through a "fast channel" and the energy flow originating from detritus moves through a "slow channel." Mobile consumers at the top of the food web couple the energy flow from these two channels (Ward et al. 2015). Species diversity tends to be greater in the slow, detritus-based channel than in the fast, plankton-based channel, and Rooney and McCann argue that the many weak interactions observed within the slow channel are critical to stabilizing the overall food web. Thus, skewed distributions of interaction strengths and modularity of interaction webs appear to be two of the answers to May's challenge.

Conclusion

The boom in network analysis in the natural and social sciences during the past decade has spawned a variety of analytical and graphic tools that are helping to power the next generation of studies of ecological networks (Watts and Strogatz 1998; Barabasi 2009; Bascompte 2009; Fontaine et al. 2011). These tools make it possible to better visualize the structure of complex ecological networks, characterize their properties, and develop testable models. Food webs continue to be the most studied and best understood of all ecological networks. Our current knowledge of food webs shows that many of the patterns observed in the early analysis of connectedness webs are less general than once thought, but that some properties, such as modularity, are common to most food webs. Studies of interaction webs have increased our understanding of the roles of individual species and of particular types of interactions within food webs. While species remain the principal focus in most studies of ecological networks, ecologists are beginning to broaden their scope to focus on more general properties of organisms (e.g., body size) that determine metabolic rates or rates of consumer–resource interactions. Focusing on the traits of organisms may allow greater insight into the mechanisms underlying the structure and functioning of food webs (Beckerman et al. 2010; Eklof et al. 2013).

Although the new tools of network analysis have greatly increased the ability of ecologists to visualize ecological networks and analyze their structural properties, fundamental challenges remain. Perhaps the greatest of these challenges is to incorporate multiple types of species interactions into the same ecological network (Melián et al. 2009; Fontaine et al. 2011; Kéfi et al. 2015, 2016; García-Callejas et al. 2018). Ecologists have significantly broadened their focus to include networks that include species interactions beyond the traditional predator–prey relationships considered in food webs. However, analyses of these different types of ecological networks are mostly proceeding in parallel (i.e., ecologists study either pollination networks or plant–herbivore networks or host–parasite networks), whereas real communities are composed of multiple species interacting in multiple ways (e.g., mutualism, facilitation, predation; see Kéfi et al. 2015 for an example). Multilayer network theory is one possible framework for including multiple interactions types within the same ecological network (Pilosof et al. 2017; García-Callejas et al. 2018). Experimental studies of interaction webs in nature have demonstrated the importance of multiple modes of interaction (e.g., Menge 1995). Thus, we should expect that new insights and a far greater understanding of how network structure affects the diversity and stability of communities will result from the integration of multiple interaction types within a single network.

Summary

1. Interactions among the species in a community can be diagrammed as ecological networks. Food webs focus on "typical" predator–prey interactions, in which consumers are larger than their prey. Less commonly studied are mutualistic webs, host–parasitoid webs, and the roles of parasites and herbivores.
2. Food webs tend to be better resolved at the top than at the bottom, in part because species richness is greater at lower trophic levels and in part because species at lower trophic levels tend to be small, difficult to identify, and have feeding relationships that are hard to quantify. Ecologists often separate food webs into those in which the basal trophic level is made up of primary producers ("green food webs") and those in which the basal trophic level is detritus ("brown food webs").
3. Connectedness or structural webs show the presence of an interaction between species but do not specify the strength of that interaction.
4. Energy flow webs measure the amount of energy (biomass) moving between species in a food web. Implicit in this approach is the idea that there is a relationship between the amount of energy flowing through a pathway and the importance of that pathway to community dynamics. However, energy flow has been shown to be a surprisingly poor predictor of the strength of interactions between species or of the impact of removing a particular species from a community.
5. Functional, or interaction, webs measure the strength of the interactions between species within a community, implicitly recognizing that not all species and interactions are equally important. Measures of interaction strength in model food webs tend to focus on the individual interactions between species pairs, whereas measures of interaction strength in empirical food webs tend to focus on the impact of one species on the rest of the web, as measured by removal experiments.
6. Studies of theoretical and empirical food webs have shown that most food webs contain a few strong links and many weak links. A keystone species is one whose effect on the community is disproportionately large relative to its abundance. Body size relationships play a major role in determining the pattern and strength of trophic interactions within food webs.
7. Evidence suggests that the net effects of indirect interactions—when the actions of one species influences a second species via a third species—are important in food webs. Such indirect interactions include exploitative competition, apparent competition, cascading effects, and keystone predation.
8. Mutualistic interactions can be diagrammed as webs of links between two well-defined types of nodes, in which interactions occur between,

but not within, node types. Some important properties of mutualistic webs are a high level of connectance, a high degree of nestedness, and relatively low modularity.

9. Parasites are left out of most food webs because their interactions and impacts on other species are difficult to quantify by standard ecological methods. Ecologists need to find ways to better incorporate parasites and other infective agents into ecological networks.

10. Food webs that are more diverse tend to have more weak interactions and greater modularity than simpler communities, and thus tend to be more stable.

CHAPTER 11

Food chains and food webs: controlling factors and cascading effects

Why the sky is blue is a matter of basic physics, but why land is green is a much trickier question.
Shahid Naeem, 2008, p. 913

[P]opulations in different trophic levels are expected to differ in their method of control.
N. Hairston, F. E. Smith, and L. B. Slobodkin, 1960, p. 421

Everything should be made as simple as possible, but not simpler.   **Albert Einstein**

The previous chapter showed how the multitude of links between species in a food web can generate a complex network of potential interactions and indirect effects (see, for example, Figure 10.2). This chapter shows how these complex networks can be simplified into chains of linked consumer–resource populations in order to address the fundamental question of what controls the abundances of organisms at different trophic levels. Are these abundances controlled from the bottom up, by resource availability, or from the top down, by predation? Are simple food chain models, which lump together all the organisms at each trophic level, too simple? We'll begin here by looking at two classic papers that launched a vigorous exploration of the factors controlling the abundances of producers, herbivores, and carnivores in ecosystems.

Why is the World Green?

Given the complexity of natural food webs, it is surprising that one of the most influential and highly

cited papers in community ecology (over 2000 citations in ISI Web of Science as of June 2018) takes as its premise that communities can be simplified into just three interacting trophic levels: producers, herbivores, and carnivores. In 1960, three ecologists from the University of Michigan, Nelson Hairston, Frederick E. Smith, and Lawrence B. Slobodkin, stirred up a hornet's nest when they published a short note to the *American Naturalist* entitled "Community structure, population control, and competition." Hairston et al. (1960; hereafter referred to as HSS) were interested in what controlled the abundance of populations and by extension, the total abundances of organisms at different trophic levels. They reasoned as follows:

1. In the absence of higher-level predation, carnivores should be limited by competition for their food (herbivores).
2. Consequently, herbivore populations should be held below their carrying capacity by carnivores and have little impact on their food (plants).

Community Ecology. Second Edition. Gary G. Mittelbach & Brian J. McGill, Oxford University Press (2019).
© Gary G. Mittelbach & Brian J. McGill (2019). DOI: 10.1093/oso/9780198835851.001.0001

3. In the absence of control by herbivores, plants should be dense and limited by competition.

Thus, they concluded that "populations in different trophic levels are expected to differ in their methods of control" (Hairston et al. 1960, p. 421)—in other words, that carnivores and plants are limited by resource competition and herbivores are limited by predation. Their argument quickly became known as the "world is green hypothesis" because the mechanisms it proposed would lead to a world rich in plants; they cited as evidence for their reasoning that "obvious depletion of green plants by herbivores are exceptions to the general picture" (Hairston et al. 1960, pp. 423–4). However, HSS's logic did not sit well with a number of ecologists, who pointed out that many plants had chemical and other defenses to prevent them from being eaten, that populations were limited by factors other than resources and predators, and that simple deductive arguments are not proof of underlying processes (Murdoch 1966; Ehrlich and Birch 1967). After some spirited debate, the ideas of HSS faded from view, only to be resurrected 20 years later in a landmark study by Oksanen et al. (1981).

The papers by HSS and Oksanen et al. launched a vigorous exploration of the factors controlling the abundances of plants, herbivores, and carnivores in ecosystems. Studies of trophic-level regulation and related topics (e.g., top-down vs bottom-up control, trophic cascades, and determinants of food chain length) have played a major role in the study of ecological networks. Summarizing this wealth of information is a challenge. We have chosen to follow (albeit roughly) the historical development of these ideas and their empirical tests. The older papers (e.g., Hairston et al. 1960; Rosenzweig 1977; Oksanen et al. 1981) provide a conceptual foundation for much that follows, and this historical thread illustrates the progress that can be made when theory gives rise to testable predictions and empirical tests, in turn, guide the development of theory. As we will see, viewing food chains and food webs as coupled consumer–resource models yields some very interesting and testable predictions, and the outcome of testing these predictions has greatly increased our understanding of the factors controlling the abundances of producers and consumers in ecosystems.

What determines abundance at different trophic levels?

Oksanen et al. (1981), building on the mathematical framework of Rosenzweig (1973), formalized and extended the ideas of HSS. Their approach was to apply consumer–resource (predator–prey) equations to interacting trophic levels. Like HSS, Oksanen and colleagues considered trophic levels to be ecological units that could be modeled as linked consumer or resource populations, and they treated each trophic level as a single, homogeneous population. This approach simplifies nature to an extreme, but it turns out to be quite insightful.

Oksanen et al. focused on how the biomass at each trophic level should change with a change in the potential primary productivity of the ecosystem. "Potential primary productivity" (G) can be thought of as the maximum gross primary productivity allowed by the environment if there were no consumers present; for simplicity, we will sometimes refer to this as "potential productivity." In many systems, G is correlated with the supply of a limiting factor, such as phosphorus or nitrogen in lakes (Elser et al. 2007) or precipitation in grasslands (Scurlock and Olson 2002). The Oksanen et al. model predicts that ecosystems with low potential primary productivity will support fewer trophic levels than ecosystems with high potential productivity. We will examine whether food chain length increases with productivity in natural systems later in this chapter. For now, we will take as a given that more productive systems can support more trophic levels, and we will use a simplified version of Oksanen et al.'s analysis to examine HSS's prediction that "different trophic levels are expected to differ in their methods of control."

Using three-dimensional isocline plots (such as Figure 11.1) based on the Rosenzweig and MacArthur predator–prey model (see Chapter 5), Oksanen et al. showed how the abundances of producers, herbivores, and carnivores should change with potential primary productivity. Recall that in Chapter 5 we used two-dimensional isocline plots to visualize

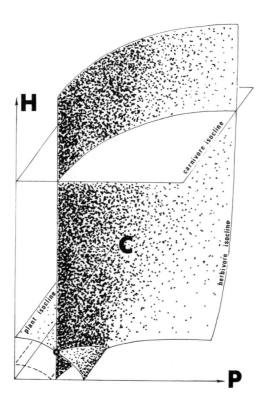

Figure 11.1 An example of a three-dimensional isocline plot showing the equilibrium biomass of three trophic levels (plant, herbivore, and carnivore abundances) in a system of linked consumer–resource populations. From Oksanen et al. (1981). See http://communityecologybook.org/tritrophic.html for an online dynamic version of this model.

the responses of predator and prey populations to an increase in the prey's carrying capacity (see Figure 5.9). Rather than using complex three-dimensional isocline plots here, however, we will focus instead on a simplified picture of Oksanen et al.'s main result.

Figure 11.2 shows how the number of trophic levels is predicted to increase as potential productivity (G) increases and how the equilibrium biomass of a trophic level is expected to change with an increase in potential primary productivity. Note the interesting stair-stepped pattern. Whether a trophic level responds to an increase in potential primary productivity depends on the number of trophic levels in the system, with adjacent trophic levels showing an alternating pattern of response. That is, in food chains with an odd number of trophic levels (e.g., three), the abundances of producers and of primary carnivores

are predicted to increase with potential primary productivity and the abundance of herbivores is predicted to stay constant. In food chains with an even number of trophic levels (e.g., four), producer and primary carnivore abundances stay constant, and herbivore and secondary carnivore abundances increase with potential productivity.

To better understand what drives this stepwise response to potential productivity, let's focus on a system with three trophic levels (as did HSS). It is important to remember that all trophic-level abundances in Figure 11.2 represent equilibrium biomasses. Therefore, one can view Figure 11.2 as representing different ecosystems with different G's or, alternatively, as a single ecosystem that has undergone a change in G in response to which the various trophic levels have reached a new equilibrium.

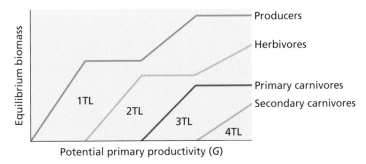

Figure 11.2 The predicted change in equilibrium biomass of different trophic levels in response to a change in potential primary productivity (*G*). These predicted responses are based on the model of Oksanen et al. (1981), which assumes that increases in potential productivity permit the addition of higher trophic levels (TLs). Complexities arising due to unstable consumer–resource interactions are ignored in this simplified treatment. After Mittelbach et al. (1988) and Leibold (1989).

Box 11.1

A linear version of the three-species consumer-resource model analysed by Oksanen et al. (1981) can be written as

$$\frac{dR}{dt} = S - a'NR - cR$$

$$\frac{dN}{dt} = N\left(a'b'R - a''P - c'\right) \quad \frac{dP}{dt} = P\left(a''b''N - c''\right)$$

where *R*, *N*, and *P* represent the abundances of the resource, consumer, and predator populations, respectively, *S* is the total supply of *R*'s resource, *a*'s represent attack rates, *b*'s convert food eaten into new consumers, and *c*'s are density-independent (non-consumptive) loss rates (Holt et al. 1994; Chase et al. 2000). Parameters without primes (*a*, *b*, *c*) are those of the resource population *R*, parameters marked with primes (*a'*, *b'*, *c'*) are those of the consumer *N*, and parameters with double primes (*a''*, *b''*, *c''*) are those of the predator *P*. As shown by Chase et al. (2000), the equilibrium numbers of resources, consumers, and predators in this system can be written as

$$R^* = \frac{Sa''b''}{a'c'' + ca''b''} \quad N^* = \frac{c''}{a''b''} \quad P^* = \frac{a'b'R - c'}{a''}$$

Note that the abundance of *R* increases with *S* and the abundance of *P* increases with *R*, but the abundance of *N* is a function of three constants and therefore does not vary with the environment (Chase et al. 2000). Thus, in this three-trophic-level system, the bottom and top trophic levels increase with productivity, whereas the middle trophic level remains constant (i.e., the pattern shown in region 3TL of Figure 11.2).

The above model assumes Type I (linear) functional responses, whereas the model of Oksanen et al. (1981) assumed a Type II (non-linear) functional response in the herbivore and carnivore populations. Such non-linear functional responses can lead to unstable dynamics and patterns of trophic level responses different from those shown in Figure 11.2 (Abrams and Roth 1994). However, these unstable dynamics were not the focus of Oksanen et al.'s analysis, and their main conclusions can be reasonably represented by the linear consumer-resource model described here.

In Box 11.1, we can see how solving for the density of each trophic level at equilibrium yields the alternating pattern of trophic-level response shown in region 3TL of Figure 11.2. The equilibrium abundances of the top and bottom trophic levels (primary carnivores and producers, respectively) are functions of resources available in the environment, whereas the equilibrium abundance of the middle trophic level (herbivores) is not (Chase et al. 2000). The model described in Box 11.1 is a linear version of

the consumer–resource model analyzed by Oksanen et al. (1981). Non-linear consumer–resource models with saturating functional responses can lead to unstable dynamics and patterns different from those shown in Figure 11.2 (Abrams and Roth 1994; Oksanen et al. 1981), as can other modifications to the model (Mittelbach et al. 1988). For now, however, we will ignore these complications and ask how well the predictions in Figure 11.2 correspond to HSS's intuition about the factors controlling trophic levels.

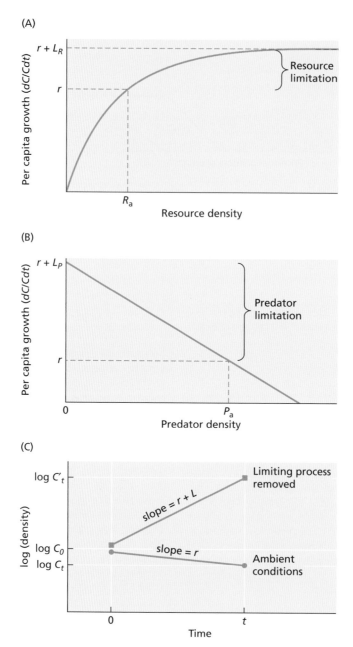

Figure 11.3 Graphic representation of the concepts of (A) resource limitation, (B) predator limitation, and (C) the assessment of limitation. (A) Per capita growth of the consumer population increases with resource density and is equal to r under ambient conditions (i.e., when $R = R_a$) and to $r + L_R$ in the absence of resource limitation (L_R). The difference between per capita consumer growth rates r and $r + L_R$ is a measure of resource limitation. (B) Per capita growth of the consumer population declines as predator density increases. The difference between per capita consumer growth rate in the absence of predators ($r + L_p$) and per capita consumer growth rate at ambient predator density (P_a) is a measure of predator limitation. (C) Limitation can be estimated with short-term field experiments as the difference between per capita growth under natural conditions (r) and under the conditions in which the focal (limiting) process has been removed ($r + L$)—that is, the difference between the slopes of the two lines in the bottom panel. After Osenberg and Mittelbach (1996).

A simple thought experiment illustrates the duality of top-down and bottom-up control

In a three-trophic-level system, herbivore biomass does not change with increasing G because all of the increased production of herbivores goes into carnivores, so herbivore abundance appears to be controlled by predation (Figure 11.2). Increasing G increases the equilibrium biomass of carnivores and producers; thus, the abundances of producers and carnivores appear to be controlled by competition (i.e., both producer and carnivore equilibrium biomasses increase with an increase in the productivity of their resources). Therefore, Figure 11.2 seems to support HSS's argument that the different trophic levels are controlled by different factors. However, in a more complete sense, the equilibrium biomass at each level below the top trophic level in Figure 11.2 represents the balance between the effects of predation and competition. To see this, consider the following "thought experiment." What would happen to the herbivore population in a three-trophic-level system if we were to remove the potentially limiting effects of predation and competition—say, by adding unlimited food or removing all predators? How would we expect the herbivore population to respond in the 3TL region of Figure 11.2? In the *short term*, herbivores should respond to unlimited food through an increase in per capita birth rates and an increase in herbivore biomass. Likewise, the removal of predators should lead to a *short-term* reduction in herbivore death rates and an increase in herbivore biomass. Experimentally, we could compare these short-term responses to food addition and predator removal with responses in control treatments in which the potential sources of limitation (resources, predators) are kept at natural levels (Figure 11.3).

As this simple thought experiment shows, the herbivore population in region 3TL is simultaneously limited in abundance by *both* competition (resource availability) and predation. It is not an either/or dichotomy. However, the degree to which resources and predators limit trophic-level biomass may not be equal, and we can estimate the relative degree of predator limitation and resource limitation by looking at the magnitude of population response to the removal of these limiting factors in the short term (see Figure 11.3). We will look more closely at the degree of resource (bottom-up) limitation and predator (top-down) limitation in real food webs in the next section of this chapter.

Some conclusions

What can we conclude from these models about the factors limiting trophic-level biomass? First, linear consumer–resource models predict that trophic levels should alternate in their responses to increases in potential productivity (resulting in the stepped pattern in Figure 11.2). Secondly, although this stepped response pattern appears to support HSS's hypothesis that competition and predation alternate in importance in controlling trophic levels, a more accurate assessment is that each trophic level (below the top level) is simultaneously limited by both competition and predation. Thirdly, the relative strengths of predator limitation and resource limitation can vary with trophic level and ecosystem productivity, and the strengths of those limiting factors can be estimated experimentally and compared with predictions.

Testing the predictions

This section explores tests of three important predictions of consumer–resource theory as applied to trophic levels:

1. In a food chain of a given length, an increase in potential primary productivity will increase the abundances of populations at the top trophic level and at alternating trophic levels below the top level; however, the abundances of populations at intervening trophic levels should not increase with potential productivity.
2. A reduction in the abundances of populations at the top trophic level will lead to an alternating increase and decrease in the abundances of populations at sequentially lower trophic levels. This prediction focuses on the transmission of effects down the food chain, which is often referred to as a "trophic cascade" (Paine 1980; Terborgh and Estes 2010).
3. An increase in potential productivity should lead to an increase in the number of trophic levels that can be supported in an ecosystem (i.e., an increase in food chain length).

Effects of productivity on tropic-level abundances

Kaunzinger and Morin (1998) tested the prediction of a stair-stepped response (see Figure 11.2) by varying nutrient inputs (potential productivity) in simple laboratory microcosms containing microbial communities with one, two, or three trophic levels. The results for the one- and two-trophic-level communities are shown in Figure 11.4. In a one-trophic-level community, the abundance of the bacterium *Serratia marcescens* increased directly with an increase in nutrients (Figure 11.4A). However, when a predator (the ciliate *Colpidium striatum*) was added to create a two-trophic-level community, *Serratia* abundance remained almost constant across the productivity gradient (lower line in Figure 11.4A), while the abundance of the predator increased (Figure 11.4B). Thus, the results of Kaunzinger and Morin's experiment match the predictions of simple consumer–resource theory applied to food chains (regions 1TL and 2TL in Figure 11.2). As ecologists we would like to know whether similar results hold true outside the laboratory. Wootton and Power (1993) provided such a field test in an innovative experiment conducted in the Eel River of northern California.

Wootton and Power established a primary productivity gradient by covering experimental enclosures in the river with different amounts of shade cloth, thereby manipulating light levels and the productivity of algae (periphyton) growing on the stream bottom (Figure 11.5). Trophic-level biomass in the enclosures responded as predicted. The biomass of the top trophic level (predators, including small fish and predatory invertebrates such as dragonfly and damselfly larvae) increased with increasing light, as did the biomass of the basal trophic level (i.e., attached algae; Figure 11.6). However, the biomass of the middle trophic level (herbivores, including mayfly nymphs, caddisfly larvae, midge larvae, and snails) showed no significant response to the productivity manipulation after 30 and 55 days. These experiments show a clear pattern of alternating trophic level response to increased productivity, supporting perhaps the strongest prediction of consumer–resource models applied to food chains. (See Aunapuu et al. 2008 for another example showing that Arctic food webs may also respond as predicted by the food chain model.)

Observations of trophic-level biomasses in many natural systems, however, are at odds with this pattern. Ginzburg and Akçakaya (1992) found that the abundances of all trophic levels (producers, herbivores, and carnivores) increased with potential productivity in North American lakes (Figure 11.7A). Chase et al. (2000) observed a similar pattern in grasslands, where the abundance of plants

(A)

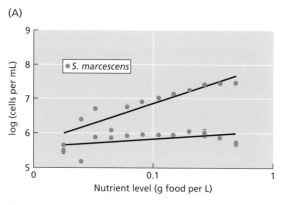

(B)

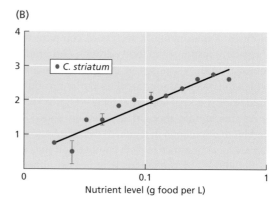

Figure 11.4 One- and two-trophic-level food chains respond differently to an increase in resources in a laboratory microbial community. (A) The abundance of the bacterium *Serratia marcescens* (blue symbols and upper black line) increases directly with an increase in the amount of resources when predators are absent (one trophic-level system). However, in a two trophic-level system containing the bacterium's predator, the ciliate protozoan *Colpidium striatum*, the abundance of *S. marcescens* is lower and remains constant (blue symbols and lower black line). (B) In the two trophic-level system, the abundance of the predator *C. striatum* (red symbols) increases directly with an increase in resources. After Kaunzinger and Morin (1998).

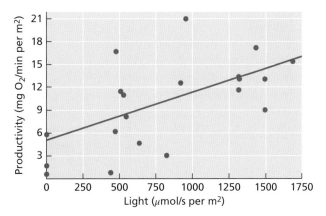

Figure 11.5 Shading of experimental enclosures in northern California's Eel River limited the productivity of algae on the stream bottom. Primary productivity in these systems was directly related to the amount of light transmitted to the stream. After Wootton and Power (1993).

and herbivores both increased with increasing rainfall (which is correlated with productivity; Figure 11.7B).

In a massive study of predator and prey abundances from >2000 communities worldwide, Hatton et al. (2015) found that the biomass of both predator and prey trophic levels increased in tandem across a gradient of increasing productivity. Most remarkable was their finding that the ratio of predator biomass to prey biomass followed a very consistent relationship (a power law with a 3/4 exponent) across nearly all of the communities studied, from terrestrial to marine to freshwater ecosystems, and for predators as different as lions and zooplankton. Figure 11.8 illustrates the predator–prey biomass relationship for large African carnivores and their prey and shows how the 3/4 scaling exponent means that the ratio of predator biomass to prey biomass becomes progressively smaller in communities with greater prey abundance (i.e., more productive communities). Hatton et al. (2015, p. 2) paint the picture vividly: "From the dry Kalahari desert to the teeming Ngorongoro Crater, there are threefold fewer predators per pound of prey, which leads to the question: where prey are abundant, why are there not more lions?". As will be seen later, a possible answer to this question is that in more productive environments the prey community is dominated by better-defended and less-edible prey. However, we are getting ahead of ourselves.

If natural systems show positive responses by adjacent trophic levels to increases in potential productivity, how do we reconcile these results with the predictions of consumer–resource theory and with the experimental results of Kaunzinger and Morin (1998) and Wootton and Power (1993)? One potential clue comes from looking closely at the types of species responding to increased productivity in natural systems. McCauley et al. (1988) and Watson et al. (1992) showed that increases in algal biomass with lake productivity were strongly correlated with an increase in the abundance of grazer-resistant algal species and that edible algae showed no significant increase in more phosphorus-rich lakes (Figure 11.9). These findings suggest that shifts in species composition within trophic levels (e.g., from relatively edible to relatively inedible species) may affect the pattern of trophic-level responses.

We can see how such shifts might work in a simple food web model. Consider a food web in which the middle trophic level is composed of two prey types (e.g., two herbivore species) that share a common resource (e.g., plants) and a common predator (Figure 11.10). The properties of this "diamond-shaped" food web have been well studied by theoretical ecologists (e.g., Holt et al. 1994; Leibold 1996). Figure 11.10A illustrates a case of **keystone predation**, where the better resource competitor (N_1) is also more vulnerable to predation (Paine 1966; Leibold 1996). Under these conditions, and assuming

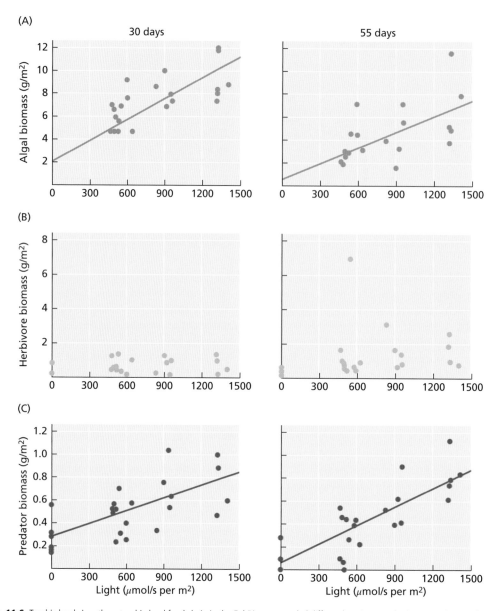

Figure 11.6 Trophic levels in a three-trophic-level food chain in the Eel River responded differently to increased primary productivity brought about by changes in resource (light) availability. The biomass of the bottom and top trophic levels (algae and predators, respectively; A and C) increased with increasing light availability, but the abundance of the middle trophic level (herbivores; B) did not change. This differential trophic level response is consistent with predictions of Oksanen et al.'s 1981 model (see Figure 11.2). The stability of the patterns after 30 and 55 days indicates that the system is in steady-state, although there is a suggestion of an increase in herbivore abundance with increasing resources after 55 days. Regression lines show significant relationships. After Wootton and Power (1993).

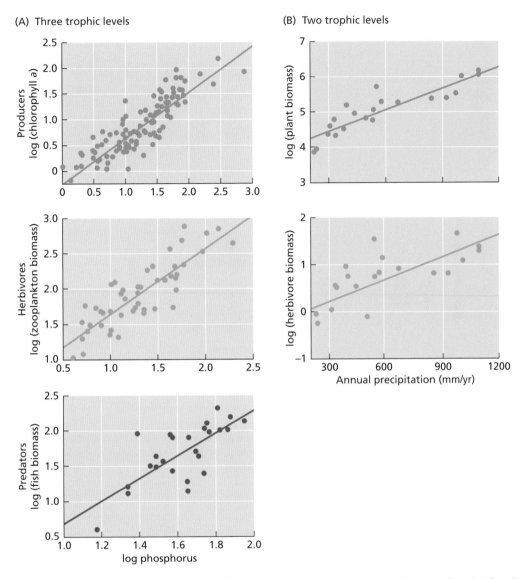

Figure 11.7 Correlations between trophic-level biomass and measures of resource availability in two natural systems indicate that all trophic levels show a pattern of positive response to increased resources, in contrast to the predictions of simple consumer–resource theory. (A) Data for three trophic levels from North American lakes in which phosphorus limits primary productivity. Each data point represents a single lake. (B) Data for plants and herbivores on North American grasslands where annual precipitation limits primary productivity. Each data point represents a grassland. (A) After Ginzburg and Akçakaya (1992); (B) after Chase et al. (2000).

there is no interference competition between prey types N_1 and N_2, an increase in the potential productivity of resource R (due to increased nutrient input) leads to a shift in dominance from the better resource competitor, N_1, to dominance by the more predator-resistant species, N_2. A region of intermediate productivity may allow for the coexistence of N_1 and N_2. Furthermore, if we look across the entire range of resource productivities modeled in Figure 11.10B, we see that the abundance of all three

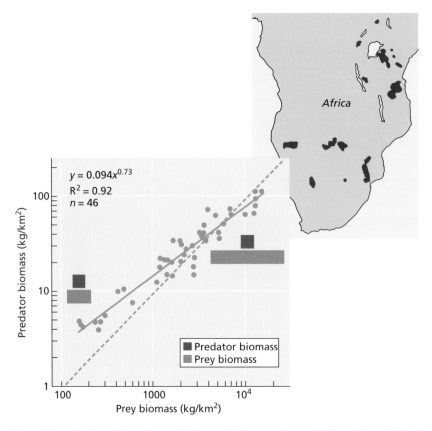

$$y = 0.094x^{0.73}$$
$$R^2 = 0.92$$
$$n = 46$$

Figure 11.8 Predators and prey abundances (biomass) are positively correlated in ecosystems worldwide and show a 3/4 scaling exponent on a log–log scale, meaning that the ratio of predator to prey biomass becomes progressively smaller in more productive ecosystems. Example shown is the relationship for large African carnivores and their prey. After Hatton et al. (2015).

trophic levels increases with nutrient input. Thus, this simple food web model shows how heterogeneity in species composition within a trophic level, coupled with a trade-off between competitive ability and vulnerability to predation, can lead to species replacements along productivity gradients, from good resource competitors to good predator resistors; and increases in the abundances of all trophic levels with an increase in the potential productivity of the bottom trophic level.

Such species turnover along productivity gradients has been observed in ponds, lakes, and streams (Rosemond et al. 1993; Osenberg and Mittelbach 1996; Leibold et al. 1997; Darcy-Hall 2006) and in grasslands (Milchunas and Lauenroth 1993; Chase et al. 2000), providing evidence that more predator-resistant species predominate in more productive systems. However, laboratory experiments provide the clearest examples of how an increase in potential productivity can favor more predator-resistant taxa and lead to an increase in biomass at all trophic levels.

Bohannan and Lenski (1997, 1999, 2000) and Steiner (2001) tested whether food chains (in which, recall, each trophic level is homogeneous) and food webs responded differently to increases in productivity by constructing simple microbial communities in the laboratory. Steiner's (2001) experiment contained two trophic levels: an herbivore trophic level (the grazing zooplankton *Daphnia*) and a producer trophic level. The producer level was composed of either a single species of edible algae (*Monoraphidium*) or a mixture of algal species, some of which were

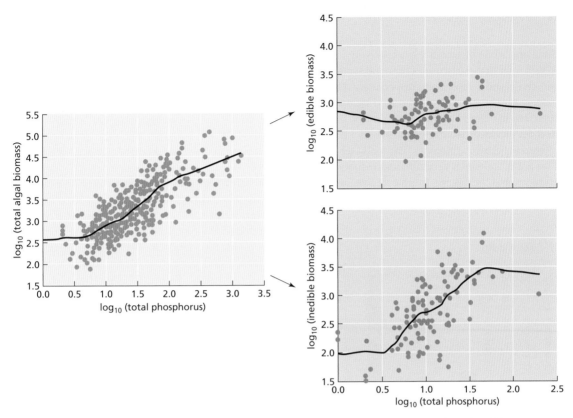

Figure 11.9 The overall positive relationship between phosphorus availability and total algal biomass in lakes (left-hand graph) has two separate components. Algal species small enough to be consumed by herbivores ("edible biomass"; above right) show no increase in abundance with increasing phosphorus input, whereas large, grazer-resistant algae ("inedible biomass"; below right) increase in abundance as phosphorus levels increase. All units are µg/L; the curve is fitted by LOWESS ("locally weighted smoothing") regression algorithms. After Watson et al. (1992).

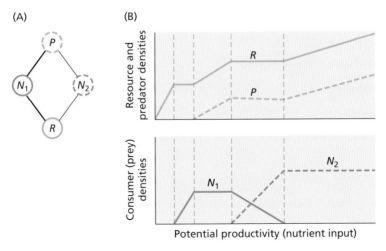

Figure 11.10 (A) An example of keystone predation, in which two prey types (N_1 and N_2) share a resource (R) and a predator (P), and there is a trade-off in the prey species' competitive ability and vulnerability to predation. The thick lines linking N_1 to R and P indicate that N_1 is a better competitor for resources and is more vulnerable to predation, whereas the thin lines linking N_2 to R and P indicate that N_2 is a poorer competitor, but is less vulnerable to predation. (B) The predicted effect of an increase in potential productivity on the abundances of resources (R), prey (N_1, N_2), and predators (P), based on a theoretical model of the diamond-shaped food web. An increase in productivity (increased input of resource R) leads to a shift in dominance from the better competitor, N_1, to the more predator-resistant species, N_2. A region of intermediate productivity may allow for the coexistence of N_1 and N_2. Looking across the entire range of resource productivities, we see that the abundance of all three trophic levels increases with nutrient input. (A) After Bohannan and Lenski (2000); (B) after Leibold (1996).

edible and some of which were relatively inedible. We will refer to these two experimental systems as a food chain and a food web, respectively. The food chain system responded as predicted by the model illustrated in Figure 11.2: nutrient addition had little effect on producer biomass, but caused a dramatic increase in herbivore biomass. The food web, on the other hand, showed a very different response: both producer and herbivore trophic levels increased significantly with nutrient input (Figure 11.11).

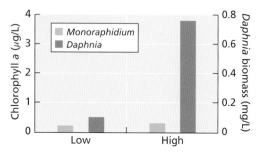

(A) Food chain

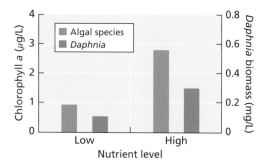

(B) Food web

Figure 11.11 Trophic-level biomass responds differently to nutrient inputs in food chains and in food webs. In Steiner's laboratory microcosms, food chains included only a single species of algae (*Monoraphidium*) and a single species of herbivorous zooplankton (*Daphnia*), whereas a food web comprised *Daphnia* and a mixture of multiple algal species, some edible and some relatively inedible. (A) The food chain system responded as predicted by Oksanen et al.'s (1981) model: nutrient addition had little effect on average algal biomass at the end of the experiment, but resulted in a dramatic increase in average herbivore biomass. (B) In the food web, both the algae and the herbivore increased in average biomass with increased nutrient input. Note that the size of *Daphnia*'s response to nutrient addition was less in the food web than in the food chain, which is expected if the composition of the producer trophic level is shifted towards less edible species at high nutrient concentrations. After Steiner (2001).

Note that the size of the *Daphnia* response to nutrient addition was less in the food web than in the food chain, which would be expected if the composition of the producer trophic level is shifted toward less edible species at high nutrient concentrations (Leibold 1996).

Bohannan and Lenski (1997, 1999, 2000) found a very similar result, but with an interesting twist. They grew bacteria (*E. coli*) in continuous-culture chemostats with glucose as the resource and the bacteriophage T4 as the predator. The system responded to an increase in glucose as predicted by a two-link food chain model: predator abundance increased and prey (*E. coli*) abundance remained (nearly) constant (Figure 11.12A). However, under these experimental conditions, phage-resistant mutants of *E. coli* arise spontaneously (and predictably) after a small number of bacterial generations. When predator-resistant *E. coli* entered the community, they changed the dynamics of the system from that of a food chain to that of a food web; *E. coli* abundance increased and phage abundance decreased (Figure 11.12B). Over time, phage-resistant *E. coli* came to dominate, and both trophic levels responded positively to increased glucose (Figure 11.12C).

These results support the predictions of the food web model described above, in which species occupying an intermediate trophic level (between a resource and a predator) experience a trade-off between competitive ability and resistance to predation. If species at an intermediate trophic level are completely invulnerable to predation, this will prevent any bottom-up flow of biomass, as all increases in potential productivity will go into increasing the biomass of inedible species (Phillips 1974; Abrams 1993; Leibold et al. 1997). Davis et al. (2010) recently observed a similar response during a 5-year enrichment experiment in a forest stream. Nutrient enrichment led to a positive biomass response by predators and prey in the first 2 years, but predator biomass eventually declined to pre-treatment levels as the prey population became dominated by large-bodied, predator-resistant species Adaptive foraging behavior, in which foragers respond to increased resource availability by reducing their foraging time in favor of increased safety, and thus make resources less available to the next

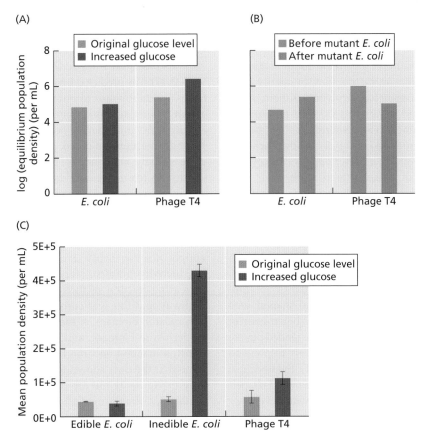

Figure 11.12 Bohannan et al. performed experiments with a simple microbial food chain maintained in continuous-culture chemostats, with glucose as the resource for the bacterium *E. coli* (the prey) and the bacteriophage T4 as the predator. (A) The experimental system responded to an increase in glucose as predicted by a two-link food chain model; that is, predator abundance increased, while prey abundance remained nearly constant. (B) When phage-resistant mutants of *E. coli* arose spontaneously, bacterial abundance increased and phage abundance decreased. (C) Over time, phage-resistant *E. coli* came to dominate the bacterial population, at which point *both* trophic levels responded positively to increased glucose. The invasion of the phage-resistant bacteria thus changed the dynamics of the system from that of a food chain to that of a food web. After Bohannan and Lenski (1997, 1999).

trophic level up, can also limit the bottom-up response of trophic levels to increased productivity (Abrams 1993), as can inducible defenses in species occupying intermediate trophic levels (Vos et al. 2004).

We have seen that the responses of natural systems to nutrient additions and increases in potential productivity are often very different from those predicted by simple consumer–resource models applied to food chains (Figures 11.2 vs Figure 11.7). Shifting species composition within a trophic level can prevent predator control and maintain the importance of resource limitation, even in high-nutrient systems. However, other factors may also lead to positive responses by adjacent trophic levels to increasing productivity, including direct density dependence (interference competition) within a trophic level (Wollkind 1976), complex life histories and invulnerable life stages (Mittelbach et al. 1988; de Roos et al. 2003; Darcy-Hall and Hall 2008), ratio-dependent predation (Arditi and Ginzburg 1989, 2012), and adaptive foraging and omnivory (Abrams 1993). Thus, we cannot infer a particular mechanism or model of trophic interaction based on patterns observed in nature. What we *can* say is that resource limitation and predator limitation should often act in concert to limit the abundances of organisms in ecosystems.

Food chains with parallel pathways of energy flow

A linear food chain is an extreme simplification of nature. In the real world the flow of energy from primary producers to top carnivores is more realistically described as a complex web of linkages between consumers and resources, as we saw in Chapter 10 (Figures 10.1 and 10.2). While much progress has been made in describing and analyzing the properties of complex food webs, their very complexity makes them ineffective for understanding the drivers of trophic-level abundances (i.e., bottom-up and top-down control). The "diamond-shaped" food web shown in Figure 11.9 adds a bit more realism to a food chain model and takes a small step toward the complexity of a full food web. As we have seen, even the slight increase in complexity introduced by the diamond-shaped web allows us to better understand how heterogeneity within a trophic level (e.g., edible vs defended plants) can affect how different trophic levels respond to changes in enrichment. More recently, theoreticians have embraced an expansion of the diamond-shaped food web to include two parallel pathways of energy flow through an ecosystem (e.g., Figure 11.13). These parallel energy pathways (or channels) are linked at the bottom by a shared resource pool and are often reconnected at upper trophic levels by generalist predators (creating what are termed "looped webs"). Importantly, this approach of two parallel energy pathways allows us to consider energy inputs to the ecosystem that are detritus-based as well as living producer based (the so-called "brown" and "green" food webs). The benefit of the parallel pathways approach is that it is more tractable for studying top-down and bottom-up effects than a full-scale food web model, but more realistic than a simple linear food chain.

Figure 11.13A shows the predicted response to nutrient enrichment in a looped food web containing a detritus-based (brown) pathway and a plant-based (green) pathway. The two pathways are linked at the bottom because detritus is produced from decaying plant biomass and linked at the top by a generalist predator that eats both detritivores and herbivores. For the detritus-based pathway the equilibrium biomass of all trophic levels is predicted to increase with increased productivity of resources, whereas the equilibrium biomass of herbivores in the grazing pathway is predicted to decrease (predictions based on Attayde and Ripa 2008; Wollrab et al. 2012). The predicted inverse response of detritivores and herbivores to nutrient enrichment is the result of predator preference for herbivores over detritivores and shared predation (energy flow through the detritus-based pathway results in an increase in the abundance of the top predator, which in turn decreases the abundance of herbivores; i.e., apparent competition). Ward et al. (2015) tested the predictions in Figure 11.13A with data gathered from 23 marine food webs situated across a natural gradient in primary production and found general support for the theory (Figure 11.13B). Across the productivity gradient, top predators derived an increasing fraction of their diet from the detritus pathway and this was correlated with an increase in predator biomass, a (non-significant) decrease in herbivore biomass, and a positive response in detritus biomass and the biomass of detritivores. Note that the data shown for just the grazer pathway in Figure 11.13B are consistent with the predictions of simple food chain theory as well (i.e., positive responses by alternating trophic levels to resource enrichment; Figure 11.2), which the authors acknowledge. They suggest, however, that the overall biomass response seen in both the detritus-based and grazer-based pathways shows that the response to enrichment is likely the result of multichannel control.

At the end of Chapter 10, we discussed how the presence of "fast" and "slow" energy channels in food webs can promote ecosystem stability. In the marine food webs analyzed by Ward et al. (2015), the flow of energy through the grazer pathway is a "fast" channel as the primary producers (phytoplankton) are small, fast-growing, and readily consumed by herbivores. The flow of energy originating from detritus, however, moves through a "slow" channel due to the slow breakdown of plant material and the relatively poor food quality of detritus. Therefore, if detritus pathways are able to sustain higher trophic levels during temporal fluctuations in adjacent grazing channels this will help stabilize the entire food web (Moore et al. 2004; Rooney et al. 2006; Rooney and McCann 2012).

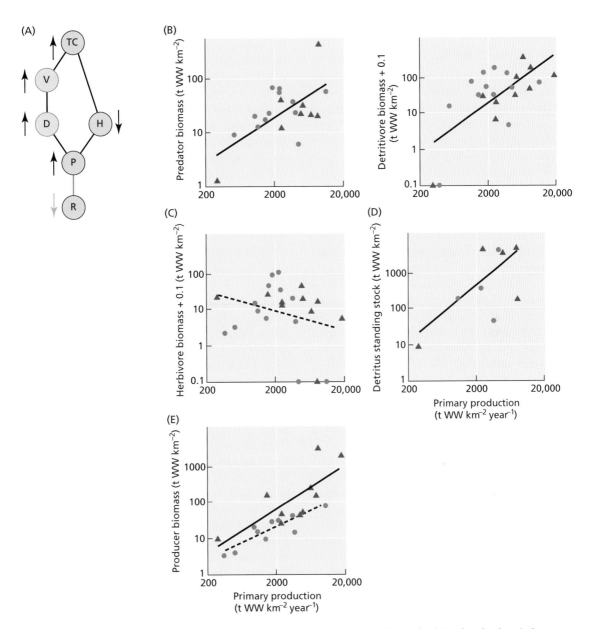

Figure 11.13 (A) Conceptual diagram of a "looped" food web containing two energy flow pathways. The detritus-based pathway is shown in brown and the plant-based pathway in green. The two pathways are linked by top consumers (TC) and plants (P). Additional symbols: (D) = detritus, (V) = detritivores, (H) = herbivores. Arrows show predicted change in abundance by each trophic level in response to an increase in resource (R) productivity. (B) Observed abundance of each trophic level in 23 marine food webs situated across a natural gradient in resource productivity. Circles indicate mixed benthic/pelagic ecosystems and triangles indicate pelagic ecosystems. 0.1 t WW km^{-2} (representing a detection limit) was added to estimates of herbivore and detritivore biomass. After Ward et al. (2015).

Trophic cascades and the relative importance of predator and resource limitation

Initial attempts to characterize ecosystem regulation as top-down vs bottom-up presented a false dichotomy. However, the relative strengths of resource and predator limitation at any trophic level may differ depending on the nature of interactions between consumers and resources or on ecosystem productivity. Recall that we employed a "simple thought experiment" earlier in this chapter to illustrate how short-term trophic-level responses to resource additions or predator removals might estimate the relative strengths of limitation by these two factors. In fact, ecologists have conducted many such experiments in the natural world, and in a variety of ecosystems. Recent meta-analyses of these experiments reveal some very interesting features of predator limitation and resource limitation in different ecosystems (these analyses are often placed within the context of top-down and bottom-up control). The results from these meta-analyses will be examined shortly. First, let's look at some examples of strong top-down control in nature, focusing on **trophic cascades** and the impact of top predators on the abundance and distribution of species at lower trophic levels.

Hairston et al. (1960) sparked the study of trophic cascades by proposing that the consumption of herbivores by predators keeps the terrestrial world green. The first experimental demonstrations of trophic cascades came not from terrestrial ecosystems, however, but from studies of marine and freshwater communities. Paine's (1966) pioneering experiment showing the consequences of removing a predatory starfish (*Pisaster*) for the structure of an intertidal community is often cited (Lafferty and Suchanek 2016) as the first experimental demonstration of a trophic cascade (a term Paine himself coined in 1980). Other examples from freshwater systems are equally dramatic and well known; impacts of piscivorous fish, for example, cascade down to the producer trophic level, affecting algal abundance and water quality (e.g., Carpenter and Kitchell 1993; see Shurin et al. 2010, for a review, and Figure 10.8 for an example).

One of the most remarkable of all trophic cascades, both in terms of the geographic scope of the inter-

action and the charismatic nature of the players, is the effect of sea otters on marine kelp forests in the Pacific Northwest. Sea otters (*Enhydra lutris*) ranged widely across the North Pacific Rim until the late 1700s, when their discovery by European explorers sparked a fur trade that led to their being hunted to near extinction. Pockets of sea otter populations survived the decimation in Russia, southwestern Alaska, and central California. Since an international ban on their hunting was imposed in 1911, otter populations have become successfully re-established in many areas, and sea otters are again found from Russia to Alaska to California. Jim Estes and his colleagues (e.g., Estes and Palmisano 1974; Estes and Duggins 1995) compared islands in the Aleutian archipelago where sea otter populations had recovered to islands where they had not. They found that islands with abundant otter populations had low densities of herbivorous sea urchins (the otter's favored prey; Figure 11.14) and extensive stands of kelp (*Macrocystis*). However, in the absence of otters, urchin populations boomed, and their grazing resulted in the creation of "urchin barrens" that were devoid of kelp and other attached algae. Thus, the very existence of the kelp forest along these islands, as well as the diverse fauna of fish and invertebrates found therein, was shown to be linked to the presence of sea otters—a dramatic example of a cascading effect (Figure 11.15, left). Studies in other parts of the otter's range, however, found somewhat different results. For example, in the kelp forests off Southern California, a diverse array of predators (including lobsters and fish) appears able to control urchin populations even in the absence of sea otters (Estes et al. 1989; Tegner and Dayton 1991; Steneck et al. 2002). Recent studies have shown that increasing predator diversity may strengthen trophic cascades in these kelp forests by modifying herbivore behavior (Byrnes et al. 2006). We discussed non-lethal (non-consumptive) effects of predators in Chapter 6 and, as we will see later, these trait-mediated effects can play a large role in generating trophic cascades and controlling trophic-level biomass in a variety of ecosystems.

There is one more element to the fascinating interaction between sea otters, sea urchins, and kelp. After nearly a century of recovery from overhunting, sea otter populations in the Aleutian Islands began

Figure 11.14 A sea otter consuming a sea urchin, one of its favorite prey. Photograph © Tom Mangelsen/Naturepl.com.

suffering dramatic declines in the 1990s, with otter densities dropping by an order of magnitude in some areas (Estes et al. 1998). The cause of these declines is controversial, but the evidence once again points to a cascading effect. Orcas (killer whales, *Orcinus orca*) in the Aleutians appear to have shifted their foraging behavior to include sea otters in their diet. Calculations by Williams et al. (2004) suggest that such a change in feeding behavior by even a very small number of orcas could have caused the observed decline in otter numbers, resulting in a four-level instead of a three-level trophic cascade (right side of Figure 11.15). Industrial whaling and the collapse of the great whale populations in the North Pacific—which eliminated an important food source for the orcas—has been proposed as a reason why orcas have expanded their diets to include sea otters, seals, and sea lions (Springer et al. 2003; but see DeMaster et al. 2006; Wade et al. 2009 for counter arguments).

Strong trophic cascades have been suggested to be less common in terrestrial than in aquatic ecosystems. For example, Strong (1992) argued that trophic cascades should be weak in terrestrial ecosystems because the primary producers in most terrestrial ecosystems (vascular plants) are larger, longer-lived, and better defended against herbivory than are most of the primary producers in aquatic ecosystems

(e.g., phytoplankton and macroalgae). It has also been suggested that terrestrial food webs contain more generalist consumers than aquatic food webs, and therefore that the pathways of interaction between trophic levels are more diverse and reticulate (branching) in terrestrial than aquatic ecosystems, again potentially weakening the propagation of top-down effects (Polis and Strong 1996; Hillebrand and Shurin 2005; Frank et al. 2006). Differences between terrestrial and aquatic systems in the relative sizes of producers and consumers (Cohen et al. 2003), the degree of omnivory in food webs (Strong 1992), and the nutritional quality of plants (stoichiometry; Elser et al. 2000; Hall et al. 2007) have also been hypothesized to lead to stronger trophic cascades in aquatic than in terrestrial ecosystems (see Shurin et al. 2010 for a discussion of these hypotheses). Are top-down effects and strong trophic cascades largely a property of aquatic systems—or, as Strong (1992) quipped, "Are trophic cascades all wet"? A number of meta-analyses have tested this idea by comparing the strengths of trophic cascades in experiments that manipulated consumer densities in freshwater, marine, and terrestrial ecosystems. The results of these meta-analyses have been synthesized and discussed in an excellent paper by Shurin et al. (2010), whose findings are summarized later.

1972–1990

1991–1997

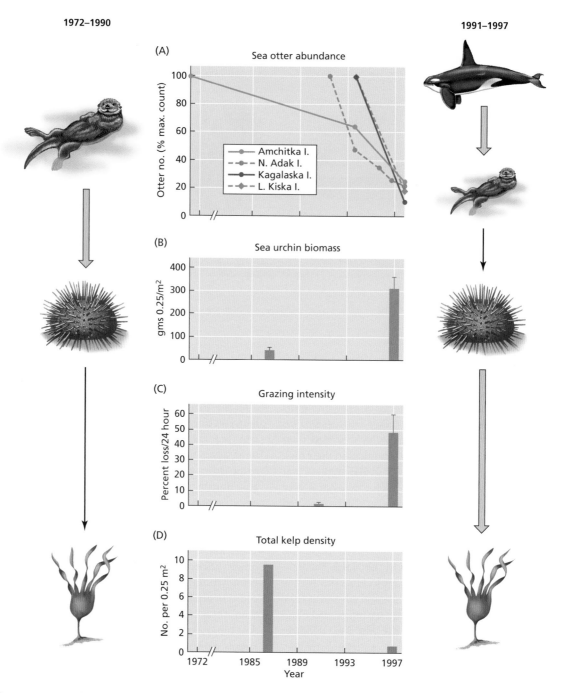

Figure 11.15 Examples of strong trophic cascades in a nearshore marine ecosystem. In the 1970s, abundant sea otter populations along the Aleutian Islands limited the numbers of sea urchins, reducing urchin grazing intensity and promoting dense stands of kelp (left). However, what appears to be a shift in the foraging behavior of a small number of orcas (killer whales) in the 1990s led to a dramatic decline in sea otter numbers, an increase in urchin biomass and urchin grazing, and a decline in kelp (right). After Estes et al. (1998).

How common are trophic cascades in different ecosystems?

Consistent cascading effects of predators on plants were reported by Schmitz et al. (2000) in their analysis of more than 40 terrestrial predator removal studies. They concluded that the magnitude of these cascading effects in terrestrial ecosystems was similar to that observed by others in aquatic ecosystems. Furthermore, they concluded that the indirect effects of predators on plant damage were much greater than their indirect effects on plant biomass and reproduction (Schmitz et al. 2000). Shurin et al. (2002) focused solely on the effects of predator removal on plant biomass, and compared effect sizes across terrestrial, marine, and freshwater ecosystems. They found that predator removal had a greater impact on plant community biomass in aquatic than in terrestrial ecosystems; on freshwater than on marine plankton; and in marine benthic (bottom) habitats than in marine open-water habitats. Gruner et al. (2008), in their meta-analysis of 191 factorial (cross-classified) manipulations of herbivores and nutrients, also found evidence that aquatic systems showed stronger trophic cascades. Herbivore removal generally increased producer biomass in both freshwater and marine systems, but its effects were inconsistent on land. Recent work by Jia et al. (2018), however, found strong and consistent negative effects of herbivores on terrestrial plant abundance, biomass, survival and reproduction in a meta-analysis of 123 native animal exclusions in terrestrial ecosystems worldwide.

Although the weight of the evidence suggests that top-down effects are more pronounced in aquatic than in terrestrial ecosystems, it is clear that trophic cascades are commonly found in both habitats. For example, a recent meta-analysis showed that birds had significant positive effects on plant survival and plant leaf damage (Mäntylä et al. 2011). Moreover, Shurin et al. (2010) noted that some of the best examples of terrestrial cascades (e.g., Schoener and Spiller 1999; Terborgh et al. 2001; Ripple and Beschta 2007) were not included in the meta-analyses discussed above because they did not report plant community biomass. Top-down effects on ecosystem properties other than plant biomass and plant leaf damage have not been well quantified. Schmitz

(2006) showed that spiders had little effect on plant biomass in an old-field community, but significantly affected the relative abundances of plant species. It is unknown whether species turnover in response to changing predator densities (Leibold 1996) is greater in some ecosystems than in others.

Although trophic cascades provide strong evidence for the importance of top-down processes (Schoener 1989; Terborgh and Estes 2010) and, although they have been viewed as supporting the predictions of HSS (e.g., Power 1990; Spiller and Schoener 1990), the existence of a trophic cascade says little about the *relative* importance of limitation imposed by predators versus resource limitation. As noted earlier in this chapter, the relative importance of predator and resource limitation can be measured experimentally through short-term removals of these limiting factors (see Figure 11.3). In one of the most comprehensive meta-analyses to date, Gruner et al. (2008) compared 191 factorial (cross-classified) manipulations of herbivores and nutrients from freshwater ($n = 116$), marine ($n = 60$), and terrestrial ($n = 15$) ecosystems in a quantitative assessment of the relative effects of fertilization and herbivory on plant biomass (Figure 11.16). Because the experimental treatments were cross-classified,

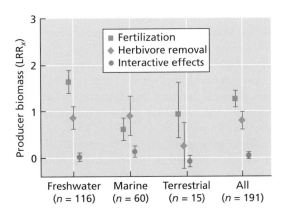

Figure 11.16 Results of a meta-analysis of 191 factorial manipulations (cross-classified for all possible combinations of fertilization and herbivore removal) show that herbivores and nutrients control plant community biomass to similar degrees across freshwater, marine, and terrestrial systems, and that these two processes act together in an additive fashion (i.e., interactive effects were minimal). LRR (natural log response ratio) measures effect size of fertilization or herbivore removal or their interaction on plant biomass. After Gruner et al. (2008).

Gruner et al. could also assess whether the effects of herbivores and nutrients acted independently (in an additive fashion) or if there was a positive or negative synergy (a significant interaction between treatments). They concluded that herbivores and nutrient supply control plant community biomass to similar degrees across freshwater, marine, and terrestrial systems, and that these two processes act together in an additive fashion (i.e., that interactive effects are minimal). They then concluded (Gruner et al. 2008, p. 748) that "the near equivalence and marked independence of fertilization and herbivore effect sizes…provides additional justification to retire the antediluvian notion that *either* top-down or bottom-up forces predominantly control plant biomass within major ecosystem types." Vidal and Murphy (2017) reached a similar conclusion, based on their meta-analysis comparing the strength of top-down and bottom-up forces on the fitness of herbivorous insects, the world's most diverse group of primary consumers. They found clear evidence of both top-down and bottom-up control on herbivore abundance and survival, with top-down forces being generally stronger than bottom-up forces.

Trophic cascades and non-consumptive (trait-mediated) effects

As we learned in Chapter 6, predators have both lethal (consumptive) and non-lethal (non-consumptive) effects on their prey. Predator-induced changes in prey behavior, habitat use, and morphological and chemical defenses can dramatically alter the nature of the interaction between a prey population and the prey's own resources. Thus, although HSS and subsequent studies (Paine 1980; Oksanen et al. 1981) focused on top-down effects in food chains that were driven by changes in predator and prey densities (or, more generally, consumer and resource densities), it is now clear that trophic cascades often result from a combination of consumptive (also called density-mediated) and non-consumptive (also called trait-mediated) effects (Abrams 1995; Schmitz 2010).

An early example of a non-consumptive trophic cascade was shown by Turner and Mittelbach (1990) who conducted an experiment with bluegill sunfish in ponds with and without the bluegill's predator,

the largemouth bass. The bass had no significant effect on bluegill mortality, as the bluegill retreated to the protection of the nearshore vegetation when bass where present. However, this predator-induced habitat shift significantly reduced bluegill predation on herbivorous zooplankton found in the open water (e.g., *Daphnia*), leading to a trophic cascade in which the bass had positive indirect effects on the abundance of herbivorous zooplankton (Turner and Mittelbach 1990). Similarly, He and Kitchell (1990) documented a non-consumptive trophic cascade in a whole-lake experiment, in which adding piscivorous northern pike (*Esox lucius*) caused a rapid habitat shift in the prey fish population: 50–90% of the total change in prey fish numbers resulted from the emigration of prey fish through an outlet stream.

These and other early studies of behaviorally induced trophic cascades in fresh waters (e.g., Power et al. 1985) were followed by a wealth of empirical and theoretical studies examining the non-consumptive effects of predators in terrestrial and aquatic food webs (reviews in Werner and Peacor 2003; Abrams 2010; Schmitz 2010; Terborgh and Estes 2010). Attempts to measure the relative impacts of density-mediated and trait-mediated cascading effects across studies have concluded that trait-mediated indirect effects are as important as density-mediated indirect effects (e.g., Bolnick and Preisser 2005; Preisser et al. 2005), and that a simple foraging gain vs predation risk trade-off can explain much of the variation in the strength of trophic cascades in terrestrial ecosystems (Schmitz et al. 2004). However, as discussed in Chapter 6, estimating the relative importance of consumptive and non-consumptive effects of predators in natural food webs is difficult. Abrams (2010) has concluded that the experimental and analytical methods employed to date probably overestimate the relative importance of trait-mediated effects. Still, as Abrams (2010, p. 15) notes, "There is little doubt that adaptively flexible foraging and defense behaviors have the potential to alter the population, community, and evolutionary dynamics of species that comprise food webs."

Trait-mediated trophic cascades need not affect plant biomass to have a strong and significant impact on plant community structure and ecosystem functioning. Schmitz (2003, 2010) provides an excellent example of this in his experimental studies of the impacts of predatory spiders in old-field food webs. Schmitz's (2010) study system focuses on strong interactions between spiders (predators), grasshoppers (herbivores), and plants (grasses and forbs). The spider *Pisaurina mira* is a sit-and-wait predator whose presence causes grasshoppers (*Melanoplus femurrubrum*) to switch from feeding on their preferred plant species (a grass, *Poa pratensis*) to feeding on a less nutritious (but safer) plant species, the goldenrod *Solidago rugosa*. Thus, these spiders exert most of their top-down control by altering grasshopper foraging behavior, rather than grasshopper population density. Because *S. rugosa* is a competitively superior plant species in this system, the net result of the predator-induced shift in herbivore foraging behavior is a change in plant diversity and productivity. In the absence of *P. mira*, *S. rugosa* dominates the plant community. However, where this predatory spider is present, herbivores suppress the growth of *S. rugosa*, reducing its competitive effects on other plants species. Thus, the spider's presence increases plant diversity (measured as species evenness), but decreases the overall productivity of the plant community (Figure 11.17A).

Schmitz (2008) went on to show how the impact of this trait-mediated trophic cascade depended on the hunting mode of the predator. When an actively roaming spider (*Phidippus rimator*) was substituted for the sit-and-wait spider *P. mira*, grasshoppers did not switch to feeding on *Solidago*; instead, they fed on their preferred resource, the grass *P. pratensis*. As a result, there was a 14% reduction in plant species evenness, a 33% increase in nitrogen mineralization (a measure of nutrient recycling), and a 163% increase in above-ground productivity, all due to a simple difference in the foraging mode of the top predator and the behavioral response of its prey (Figure 11.17B). More work is needed to determine whether such dramatic cascading effects of predator foraging mode occur in more complex communities where a variety of species (with differing traits) are found at each trophic level.

There are many other examples of trait-mediated cascading effects of predators in terrestrial and aquatic ecosystems (Werner and Peacor 2003; Schmitz 2010). For example, in subtidal marine systems, an increase in predator species richness has been shown to decrease herbivore foraging activity or herbivore per capita feeding rate, leading to a positive indirect effect of predators on plants (e.g., Byrnes et al. 2006; Stachowicz et al. 2007). Adaptive foraging behaviors and other trait-mediated responses can also change the bottom-up effects of resource enrichment on trophic-level abundances. For example, if consumers at intermediate trophic levels face a foraging gain–predation risk trade-off, they may reduce their foraging effort in favor of increased safety when food becomes more abundant. Such a behavioral response will limit the bottom-up flow of resources to upper trophic levels (Abrams 1984, 2010). However, unlike top-down interactions, in which the trait-mediated effects of predators on their prey's resources are always expected to be positive, the bottom-up effect of resources on predator abundance may (in theory) be either positive or negative, depending on the behavioral response of the prey (Abrams 1984, 2010; Werner and Anholt 1993). To date, there has been little empirical work on this question, and we cannot say which outcome is more likely (Abrams 2010).

What determines food-chain length?

It is rare to find natural ecosystems with fewer than three trophic levels (producers, herbivores, and carnivores), and those ecosystems occur only at the very lowest productivity levels, such as near the limits of vegetation in deserts or the high Arctic (Aunapuu et al. 2008; Terborgh et al. 2010). However, food chains with four, five, six, or more trophic levels exist in a number of places (Moore and de Ruiter 2012). Ecologists have long questioned what limits the length of food chains in nature (e.g., Elton 1927). As we have seen in this chapter, the number of food chain links can have important effects on the pattern of trophic-level regulation (e.g., see Figure 11.4) and the nature of cascading effects. In addition, the extent of bioaccumulation of contaminants in the food chain depends on the number of trophic links from a basal species to a top predator (e.g., Cabana and Rasmussen 1994).

Four major hypotheses have been proposed to explain why food chain length differs among ecosystems (Post 2002):

(A)

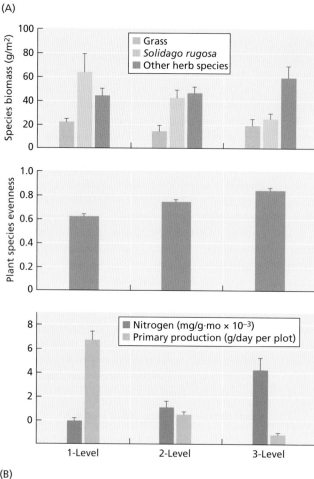

(B)

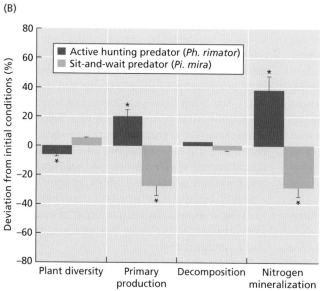

Figure 11.17 Behavioral responses to predators can lead to strong cascading effects. The spider *Pisaurina mira* is a sit-and-wait predator whose presence causes grasshoppers to switch from feeding on their preferred plant species (a grass, *Poa pratensis*) to feeding on goldenrod (*Solidago rugosa*). Because *S. rugosa* is competitively dominant, the net result of the predator-induced shift in herbivore foraging behavior is a change in plant diversity and productivity. (A) In the absence of *P. mira* (a two-trophic-level system), *S. rugosa* is abundant in the plant community. But in the presence of the spider (a three-level system), grasshoppers suppress the growth of *S. rugosa*, reducing its competitive effect on other plant species, and thus increasing plant diversity (measured as species evenness), but decreasing the overall productivity of the plant community. (B) This trait-mediated trophic cascade depends on the hunting mode of the predator. When an actively roaming spider (*Phidippus rimator*) was substituted for the sit-and-wait spider *P. mira*, grasshoppers did not switch to feeding on *Solidago*; instead, they fed on *P. pratensis*. As a result, there was a 14% reduction in plant diversity in the actively hunting predator treatments compared with the sit-and-wait predator treatments; a 163% increase in above-ground productivity in the active predator treatments compared with the sit-and-wait treatments; no difference in plant matter decomposition rates between treatments; and a 33% increase in nitrogen mineralization in the active as opposed to the sit-and-wait predator treatments. Values are mean ±1 SD. Asterisks indicate significant treatment effects ($P < 0.05$) by *t*-test. After Schmitz (2008, 2010).

1. The **energy limitation hypothesis (aka productivity hypothesis)** notes that energy is lost in the transfer between trophic levels (Lindeman 1942); therefore, food chain length should be limited by available energy, and more productive environments should support longer food chains (Elton 1927; Hutchinson 1959; Pimm 1992).
2. The **dynamic stability hypothesis (aka disturbance hypothesis)** takes as its basis a prediction from theoretical food chain models that longer food chains are less resilient to disturbance; therefore, disturbance frequency or intensity should limit food chain length (Pimm and Lawton 1977).
3. The **ecosystem size hypothesis** proposes that food chains will be longer in larger ecosystems because larger ecosystems (greater total area) will have greater total resources and will support more individuals and thus more species (Post et al. 2000).
4. A modification of the ecosystem size hypothesis, the **productive space hypothesis**, states that ecosystem size and productivity act together (productivity × ecosystem size) to determine food chain length (Schoener 1989).

Food chain length is a straightforward concept in theory, but it is complicated to measure in practice, given that multiple species may occupy each trophic level and that species may feed at multiple levels (omnivory). Still, a variety of methods have been used successfully to estimate average food chain length, including estimating the average trophic position of top predators in a food web from stable isotope analysis (Post 2002; Takimoto et al. 2008).

Takimoto and Post (2013) conducted a formal meta-analysis of 13 published studies from natural systems (12 freshwater and 1 terrestrial/island), and found strong support for the productivity and ecosystem size hypotheses, but not the disturbance hypothesis. Productivity and ecosystem size both had significant positive effects on food chain length that were similar in magnitude (effect size), whereas the effect of disturbance on food chain length was not significant. The strong support for the productivity hypothesis in the meta-analyses of Takimoto and Post (2013) contrasts with an earlier review by Post (2002) who concluded that there was little evidence that resource availability and food chain

length are positively correlated in nature. A separate analysis of food chain length in tropical terrestrial ecosystems also supports the productivity hypothesis, but not the ecosystem size or productive space hypotheses (Young et al. 2013). Overall, the dynamic stability hypothesis (disturbance) has received little support as a determinant of food chain length (Townsend et al. 1998; Takimoto et al. 2008; Walters and Post 2008; Takimoto and Post 2013; but see Parker and Huryn 2006; McHugh et al. 2010). Thus, while the question of what determines food chain length in nature is by no means settled, the evidence suggests that both productivity and ecosystem size, perhaps in combination, are the primary drivers. Readers interested in exploring more of the theory underlying the determinants of food chain length should consult Holt (2002, 2010), Kondoh and Ninomiya (2009), Takimoto et al. (2012), Ward and McCann (2017).

Conclusion

This chapter started by asking whether complex networks of species interactions might be simplified into chains of linked consumer–resource populations in order to address the question of what controls the abundances of plants, herbivores, and carnivores in ecosystems. Not surprisingly, Hairston et al.'s 1960 explanation for our "green world" is deficient in many ways. However, their fundamental insight—that we can apply our knowledge of density-dependent processes and species interactions to ecosystem-level questions—opened up a whole new way to look at food webs and trophic dynamics. We have seen how top-down and bottom-up forces combine to determine the abundances of organisms at different trophic levels, and how the strength of these forces may vary across gradients in ecosystem productivity. Trophic cascades are evidence of the importance of top-down control in many ecosystems, and they highlight the consequences of the loss of top predators from ecosystems worldwide (Estes et al. 2011). Moreover, simplifying complex food webs into model food chains has brought to light the important role that species diversity within a trophic level plays in determining a community's responses to resource enhancement or to the addition or removal of top predators. Particularly important are species' traits related to competitive ability and predator avoidance and whether species at intermediate trophic levels may respond behaviorally to changes in resource or predator abundance (trait-mediated effects).

One important aspect of food webs that we have not addressed in this chapter is the question of system stability in the face of environmental change. In Chapter 15, we will see how top-down and bottom-up forces may lead to dramatic changes in the structure and functioning of communities and ecosystems (e.g., regime shifts and alternate stable states), and we will look at the consequences of variable environments for species interactions, species diversity, and community stability.

Summary

1. The "world is green hypothesis" of Hairston, Smith, and Slobodkin ("HSS") proposed that in the absence of higher level predation, carnivores should be limited by competition for their food (herbivores) and, consequently, herbivore populations should be held below their carrying capacity and have little impact on plants; and finally, in the absence of control by herbivores, plants should be dense and limited by competition.

2. Oksanen et al. applied consumer–resource equations to interacting trophic levels. They modeled food chains as linked consumer–resource interactions. This model predicts that whether a trophic level responds to an increase in potential productivity depends on the number of trophic levels in the system, and that adjacent trophic levels will show an alternating pattern of response.

3. Although the stepped response pattern predicted by Oksanen et al. appears to support HSS's hypothesis that competition and predation alternate in importance in controlling trophic levels, a more accurate assessment is that each trophic level below the top one is simultaneously limited by competition and predation. Initial attempts to characterize regulation as top-down versus bottom-up presented a false dichotomy; the relative strengths of predator limitation and resource limitation vary with trophic level and ecosystem productivity.

4. Experiments show that organisms arrayed in food chains and food webs respond to changes in potential productivity as predicted by consumer–resource models.

5. Most natural systems show increases in the abundances of all trophic levels with an increase in potential productivity. A simple diamond-shaped food web model illustrates how heterogeneity in species composition within a trophic level, coupled with a trade-off between competitive ability and vulnerability to predation, can lead to this result, as well as to species replacements along productivity gradients, from good resource competitors to good predator resisters.

6. Shifting species composition within a trophic level can prevent predator control and maintain the importance of resource limitation, even in high-nutrient systems.

7. Although the weight of the evidence suggests that top-down effects are more pronounced in aquatic than in terrestrial ecosystems, trophic cascades are commonly found in both habitats. Trophic cascades provide strong evidence for the importance of top-down processes, but the existence of a trophic cascade says little about the relative importance of predator limitation versus resource limitation.

8. Strong cascading effects result from both the consumptive (density-mediated) and non-consumptive (trait-mediated) effects of predators.

9. Natural ecosystems with fewer than three trophic levels are rare and occur only in very unproductive environments (deserts or the high Arctic), but food chains with four, five, six, or more trophic levels can be found in a variety of places. The question of what determines food chain length in nature is by no means settled, however, the evidence suggests that both productivity and ecosystem size, perhaps in combination, are the primary drivers.

CHAPTER 12

Community assembly and species traits

[O]nly a fraction of the forms that could theoretically do so actually form a community at any one time.
Charles Elton 1950, p. 22

Community assembly is the study of the processes that shape the identity and abundance of species within ecological communities. **Nathan Kraft and David Ackerly 2014, p. 68**

There is perhaps no more fundamental question in ecology than what determines the number and kinds of species found in a community and their relative abundances. This idea dates back to seminal publications by Elton (1946, 1950), Williams (1947), and Patrick (1967). As noted in Chapter 1, Elton (1950) proposed that communities have limited membership, stating that in any prescribed area, "only a fraction of the forms that could theoretically do so actually form a community at any one time." Ecologists have approached the question of what determines the composition of a local community (i.e., its "limited membership") in two very different ways. The first is a reductionist approach using models of species interactions to predict the number and types of species that can coexist in a local community. Chapters 5, 7, and 8 exemplify this approach. The power of this method is the deep, mechanistic understanding it provides in specifying why community membership is limited to a certain number and type of species in a given environment. Its weakness, however, is that it is difficult (impossible?) to apply to most real-world communities, which may be composed of dozens, hundreds, or

even thousands of species. A second approach, which is less mechanistic, but more readily applied to the study of very diverse communities takes a more holistic view by proceeding from the regional scale of biodiversity to the scale of the local community in two steps. First, we start with the regional **species pool**, which includes all the species that could potentially colonize a local site or community over ecological time. We then seek to specify the **assembly rules** that explain why only a limited subset of the species found in the regional pool may occur in a given local community (Figure 12.1). We will explore examples of these assembly rules and the processes involved shortly. However, before we do we will briefly describe the species pool and the processes that limit membership.

Species pools

Community assembly starts with a regional species pool. The regional species pool has been variously defined as, "the set of species that could potentially contribute individuals to a local assemblage" (Zobel 1997; Carstensen et al. 2013), or "all species available

Community Ecology. Second Edition. Gary G. Mittelbach & Brian J. McGill, Oxford University Press (2019).
© Gary G. Mittelbach & Brian J. McGill (2019). DOI: 10.1093/oso/9780198835851.001.0001

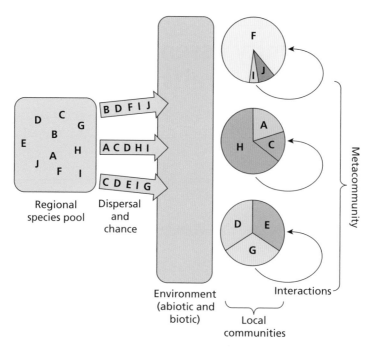

Figure 12.1 A conceptual model of community assembly, where local communities are composed of a subset of species from the regional species pool that have passed through environmental filters. After Mittelbach and Schemske (2015).

to colonize a focal site" (Cornell and Harrison 2014), or "the set of species that could potentially colonize a local site or community over ecological time" (Mittelbach and Schemske 2015). Such definitions are beguilingly simple, but hard to put into practice. How do we determine which species could potentially reach and colonize a local site, and over what time scales? This is a challenging question to answer empirically and early attempts to define species pools were often quite subjective (e.g., all species found within some (arbitrary) area, or the list of species found in a particular habitat type). More recently, ecologists and biogeographers have developed more standardized methods to characterize the pool of potential community members based on species occurrence data at geographical scales (e.g., the "biogeographic species pool"; Carstensen et al. 2013). These methods include distance-based clustering analysis, network modularity analysis, and assemblage dispersion fields. Each of these methods has its strengths and weaknesses and each may be more or less appropriate depending on the specific question of interest (see Vellend 2010; Lessard et al.

2012; Cornell and Harrison 2014). You may wonder why we should care about the particulars of how the species pool is defined? The most important reason is because we want to be able to quantitatively compare the species composition of the local community (i.e., the successful colonists of the community) to that of the regional species pool (i.e., the potential members of the community). Systematic differences between the number and types of species found in the local community and those in the regional pool should shed light on the processes governing community assembly.

Three assembly processes limit membership in the local community

Ecologists have identified three broad categories of processes that determine which potential colonists can arrive, survive and persist in a community (Keddy 1992). These are:

- **Biotic**—this incorporates all aspects of species interactions including competition (the presumed process behind many of the assembly rules proposed historically), predation, parasitism and disease, and mutualism.

- **Abiotic**—this incorporates the fit of the species to the environment including adaptations to heat, cold, drought, soggy soils, vegetation structure, etc.
- **Dispersal**—this incorporates the ability of a species to make it to the target local community and the rate at which its propagules arrive. A species might not have dispersed to a community (or disperse at a lower rate) due to physical limits (dispersal barriers), or simply because the appropriate dispersal event has not yet occurred (dispersal limitation).

Note that these three assembly processes are quite phenomenological in nature in comparison to many of the more detailed, mechanistic models studied previously (e.g. the consumer-resource model of competition is just one of many conceptualizations of a biotic process). Invoking high-level processes like these have both advantages and disadvantages (Vellend 2010). Many authors refer to the three assembly processes listed above as filters (e.g. Keddy 1992). Others refer only to the biotic and abiotic processes as filters (HilleRisLambers et al. 2012; Mittelbach and Schemske 2015). The filter metaphor is vivid, but it is not perfect. For example, most physical filters act passively and sequentially in doing their work (such as the progressively finer-grained filters used to sieve soils). By analogy, it is sometimes suggested that there is a temporal ordering to the three assembly processes listed above: first one must see if a species can get to a community (dispersal), then if it could survive the climate (abiotic) and only then if it can handle the species interactions (biotic). In reality, all three processes interact with each other simultaneously (Cadotte and Tucker 2017). Which species competitively excludes the other may depend on both the climate in which the contest occurs but also whether one species is receiving propagules from outside (rescue or mass effect). It is also unlikely that the three assembly processes will act equally on all species in a community. For example, common species may be more heavily filtered by biotic and abiotic processes, while rare species may be more heavily filtered by

dispersal (Maguran and Henderson 2003; Supp et al. 2015; Umaña et al. 2017).

Although the notion of three general assembly processes controlling which subset of the regional pool appears in a local community is an emerging paradigm, very few empirical studies have been set up to fully assess this framework. For example, it is not hard to imagine a study that experimentally increases or decreases the relative strengths of biotic, abiotic, and dispersal processes to see how community structure and composition change. Experiments looking at one or two of these factors at a time are common, but we know of none that have systematically done this across all three assembly processes. Instead of experimentally manipulating the three assembly processes, to date the study of community assembly has focused mainly on finding patterns of which subset of species are admitted to the local community and then trying to deduce from that pattern which processes were involved. This is admittedly a bit backwards and has serious limitations, but it has been a major thread of ecology for several decades.

Pattern-based assembly rules

Early attempts at specifying community assembly rules were based on describing allowable combinations of species in a local community, given the regional species pool. In perhaps the first proposed assembly rule, Williams (1947) suggested that because closely related species are ecologically similar and are thus probably strong competitors, there should be fewer species per genus than expected by chance (i.e. species in a local community would be spread out across many genera). Instead, he found that co-occurring species were more likely than just by chance to come from the same genus. In Hutchinson's landmark 1959 paper entitled "Homage to Santa Rosalia, or why are there so many kinds of animals?", he notes a number of observed patterns where pairs of species don't co-occur (i.e. only one or the other is found in a given community). As discussed in Chapter 1, he then goes on to speculate that competition limits how similar two species can be and still coexist, and suggests that co-occurring species must differ from each other by a factor of two in body mass (or the better known equivalent of approximately 1.3:1 in linear measurements) (Figure 1.3). In support of this body-size ratio rule, he presents a table of ten diverse taxa where this 1.3:1 ratio in body length appears to hold. Hutchinson's body-size ratio is typical in that while specific processes and mechanisms are invoked, it is ultimately a pattern-based assembly rule. Diamond (1973) followed up on the spatial aspects of species being variously present or absent that motivated Hutchinson by proposing the idea of a "checker-board" distribution. The checker-board distribution occurs when several similar species could live in a given habitat, but in any given local community only one of the species is found to live there. Diamond gives the example of eight species of *Lonchura* finches in New Guinea that have colonized a novel human-created habitat at mid-elevation levels and notes that "Each midmontane area supports only one species...but the areas inhabited by a given species are often scattered hundreds of kilometers apart." (Diamond 1973, p. 762).

Building on these ideas, Fox (Fox 1987; Fox & Brown 1993) presented an especially strongly predictive assembly rule. Fox proposed that "There is a much higher probability that each species entering a community will be drawn from a different genus (or other taxonomically related group of species with similar diets) until each group is represented, before the rule repeats". For example, a community might obtain one underground insectivore, one above-ground insectivore, one underground root herbivore, and one above ground (leaf) herbivore before it will obtain a second in any of these guilds/ genera (in the small mammals Fox studied genus and diet guild are highly correlated). Cavendar-Bares and colleagues (2004) similarly found that while the species changed along an environmental gradient, at any point along the gradient one species each from the red-oak subgenus and the white-oak subgenus were more likely to co-occur than two species from one sub-genus, linking co-occurrence to evolutionary relatedness. We return to this theme later and in Chapter 16 on Evolutionary Community Ecology as it helped launch the modern field of community phylogenetics.

Precisely predictive statements are rare in ecology and, therefore, appealing. However, the above pattern-based approaches to community assembly

rules have been strongly criticized (e.g., Strong et al. 1979; Simberloff and Conner 1981) who, worried by the proclivity of the human mind to find patterns everywhere, noted that these seeming patterns could have been produced by chance. The proposed solution was to test the pattern of interest (body-size ratios, checker-boardness, constancy of species-genus ratios, etc.) against null models. These null models took several basic properties of communities from the empirical data (e.g. the total number of species in the regional pool, the number of species in each local community, and the occupancy rate of each species) as constraints, and then performed random shuffling of species locations (i.e. which local community a species is found in) subject to these constraints. The null-model approach would identify a pattern (e.g., checkerboardness) as ecologically interesting if the pattern was found to be stronger than found in 95% of the random reshufflings (i.e., $p<0.05$). While this approach is statistically rigorous, the proper constraints for a "null-model" are not obvious and the odds of Type II errors (erroneously rejecting ecologically interesting patterns) were ignored in early applications. A rigorous and acrimonious debate ensued. In the end, many of the initially identified community patterns were found to be no different than expected by chance, whereas some of the patterns were statistically non-random (e.g., Figure 1.3, see Gotelli and Graves 1996 for an extended discussion). Another example of a "real" pattern is a study of 36 grassland bird species in the Great Plains region of the United States. Veech (2006) developed a C-score as a measure of checkerboardness and performed randomizations which showed that 40–60% of the possible species pairs were not significantly different from random associations, but the rest were non-random. Somewhat surprisingly, 37–54% of the species showed a positive relationship (tending to co-occur with each other more often than chance— the opposite of Diamond's checkerboard pattern) and <5% showed a negative relationship (tend to avoid each other). Other meta-analyses, while also finding many pairs of species that are independent, find more negative associations (checkerboard) than positive and that the total number of negative associations is greater than expected by chance (Gotelli & McCabe 2002; Gotelli & Ulrich 2012).

These differences in results may be due to the different spatial scales of the studies, with more negative relationships being found at small (local) spatial scales and more positive relationships being found at larger spatial scales.

Comparisons between local and regional species richness

A completely different approach to community assembly rules is to look not at which subset of species from the regional community are found in a local community, but rather to look at how many species occur in a local community compared to how many species are in the regional species pool. The reasoning is as follows. If local communities have limited membership (Elton 1950) and species interactions such as competition and predation restrict which species are able to coexist in a community, than we would expect species richness in local communities to **saturate** as regional richness increases (the type II curve in Figure 12.2). In other words, there may be a limited number of niches in any community, so that no matter how high the regional species richness (and therefore, no matter how large the potential pool of colonists), niches fill up and the local community can support only so many species. This approach makes the same assumptions as the pattern-based rules in the last section, but studies it by comparing relative species richness instead of pairs of co-occurring species. Alternatively, if interspecific competition or other species interactions are unimportant in determining community composition, then local species richness should increase linearly with the richness of species in the region, resulting in the type I curve in Figure 12.2.

Cornell and Lawton (1992) conducted an initial analysis of published local–regional richness relationships. They tentatively concluded that there was little evidence for saturation and that the principal direction of control for species richness was from regional to local. Later analyses tended to support Cornell and Lawton's initial conclusion that linear local–regional richness relationships are more common than nonlinear ones (e.g., Lawton 1999; Hillebrand and Blenckner 2002; Shurin and Srivastava 2005; Cornell et al. 2008; Harrison and Cornell 2008).

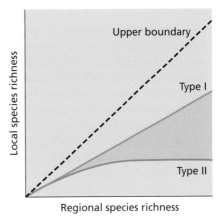

Figure 12.2 Theoretical curves for the relationship between local and regional species richness. In the Type 1 curve, local species richness increases linearly (proportionally) with regional species richness. In the Type II curve, the relationship is nonlinear and saturating because of biotic interactions and niche-filling in the local community. Real communities are likely to fall somewhere in the continuum between these two extremes (shaded area). The "boundary" represents the upper limit of the relationship, as local species richness cannot exceed regional species richness. After Cornell and Lawton (1992).

However, Szava-Kovats et al. (2013) showed that there was an inherent bias in earlier regression analyses of local-regional richness relationships, which they show can be removed by applying a log-ratio transformation to the species richness data. When the authors re-examined the literature using the unbiased log-ratio analysis, they found roughly equal numbers of type I and type II patterns. In other words, saturating local-regional richness relationships are as common as unsaturating relationships.

Although the log-ratio transformation proposed by Szava-Kovats et al. (2013) eliminates an important source of bias in earlier analyses of local-regional species richness, many issues remain to muddle our interpretation of the patterns. For example, a contrast between linear and saturating local–regional richness relationships was originally framed in terms of noninteractive versus interactive (niche-filling) communities. However, it turns out that a combination of processes can result in different patterns. At a methodological level, "pseudosaturation" and the scale at which local and regional richness are measured have been shown to affect the predicted and observed local–regional species richness relationship (Hillebrand 2005; Shurin and Srivastava 2005). At a more fundamental level, theoretical analyses have shown that there may be a continuum of local–regional richness relationships (from saturating to linear), depending on the relative differences between rates of local species extinction and regional dispersal (e.g., He et al. 2005; Shurin and Srivastava 2005). If local extinction is the dominant process, the relationship will be non-linear, but a linear pattern will result if dispersal dominates. Likewise, if species are maintained in local communities via source–sink dynamics (e.g., dispersal from the outside maintains species in the local community that would otherwise go extinct due to biotic or other processes), then we would expect to see saturation of local diversity at low dispersal rates, but more linear local–regional patterns at high dispersal rates (Shurin and Srivastava 2005). Moreover, if higher regional species richness results in higher rates of species dispersal to local communities, then local richness should scale positively with regional richness, even when there is strong species sorting at the local scale (Ptacnik et al. 2010). Thus, it is risky to infer the processes regulating local species diversity only from plots of local–regional species richness, and comparisons of patterns should be accompanied by more detailed studies focused on the mechanisms underlying community assembly (see Chapter 14; also Shurin 2000; Shurin et al. 2000).

Trait-based community assembly

As we saw in the discussions of pattern-based rules, assembly rules can be described either taxonomically or phenotypically. Thus, Diamond's checker-board pattern focused on pairs of species (a taxonomic rule), while Hutchinson's body-size ratio focused on what combinations of phenotypes could co-occur. Over the last decade, a coherent approach to phenotype-based ecology, including phenotype-based community assembly, has developed under the name "trait ecology" or, more specifically, "functional trait ecology".

A trait is "a well-defined, measurable property of organisms, usually measured at the individual level and used comparatively across species. A **functional trait** is one that strongly influences organismal performance" (McGill et al. 2006). Examples of functional traits measured on individuals include body mass, beak size, seed size, specific leaf area, wood density, and relative growth rate. Selection is based on population level fitness, so it is common in traits to focus on performance. Performance consists of the key aspects of a species success that can be measured at the individual level such as growth rate, survival, or fecundity (Arnold 1983). To be most useful in predicting community composition, functional traits should vary more between than within species (McGill et al. 2006), but in some communities like tropical trees there is more variation within than between species and functional-trait based assembly is more predictive about the phenotypes of individuals irrespective of species.

A focus on species traits

A key reason to focus on functional traits rather than species identity is that it can lead to more general rules of community assembly. Hutchinson's trait-oriented claim that body lengths must differ by a factor of about 1.3 is a more universal and strong claim than the specific example that motivated him—a taxonomically-oriented claim about whether two species of water boatmen (*Corixa dentipes* and *C. affinis*) co-occur. In particular, if we can develop trait (or phenotype) based assembly rules, we may also be able to apply them in conservation contexts where species are poorly studied. As a result, recent years have seen an explosion of interest in compiling species traits (Kattge et al. 2011; Parr et al. 2017; Froese & Pauly 2000) and relating these traits to the process of community assembly (McGill et al. 2006; Litchman and Klausmeier 2008; Lebrija-Trejos et al. 2010; Kraft and Ackerly 2014).

Traits and limited community membership

If membership in a community is limited by traits, then one should see the same distribution of traits across nearby local communities with similar environments. A clear example of this is given by Messier and colleagues (2010). They examined traits of tropical trees in 4–8 20 m × 20 m plots at each of three different sites along a strong precipitation gradient on the Isthmus of Panama. They found that the community means did not vary between plots within a site—i.e. 0% of the variance was across plots within the same site based on a formal variance partitioning across six different scales (leaf to geographic). Moreover, not just the mean, but the entire distribution of a trait (mean, variance, skew and higher moments) was strongly conserved between plots at a site (Figure 12.3). Because this is in a tropical forest, the overlap in species composition of different plots within a site is very low (0–33%), meaning that the distribution of traits is constant even when the contributing species change. Unlike some other pattern based studies of assembly, trait-based assembly has made strong efforts to identify the processes involved. We will now look at the link between traits and first abiotic and then biotic processes.

Traits and abiotic processes

It has long been known that categorical traits of organisms (e.g., plants with broad- vs needle-like leaves) vary across environmental gradients; indeed, such categorical traits help define ecological "biomes" (Walter & Breckle 2002). More recently, ecologists have studied how the quantitative traits of organisms found in a community also vary in relation to their environment. A common approach is to calculate the mean trait for a community (i.e. averaged across the species present). Ideally, the average is weighted by the abundance of each species (or equivalently averaged across the individuals in the community). This is referred to as the community-weighted mean (CWM). For a large number of plant traits and environmental gradients, the CWM varies in systematic ways with environmental gradients within regions (Cornwell and Ackerly 2009; Fortunel et al. 2014; ter Steege et al. 2006; also see Figure 12.3), or across regions (Niinemets 2001; Wright et al. 2004; Šímová et al. 2015; Barton et al. 2013; Figure 12.4). Similar correlations of community weighted mean traits with environment can also be found for animal traits. Two well-known examples are the variation in morphology and coloration of guppies in response to a predation gradient in streams in Trinidad (Endler 1980; Magurran 2005) and changes in wing morphology of hummingbirds in relation to varying air density along an elevational gradient (Altshuler

2005). Such environment-trait correlations are suggestive of the importance of abiotic processes, but they give minimal information about how these processes work.

Laughlin et al. (2012) developed a TRAITSPACE model that combines three empirical pieces of data:

- how the CWM traits changes along an environmental gradient;
- the distribution of traits observed for each species regardless of location;
- the regional pool of species and their relative abundances.

These three pieces of data produce a model that predicts well the changing abundance of species along an environmental gradient, which in turn is strong evidence for the role of abiotic processes in driving the changing abundance of species along environmental gradients.

Shipley et al. (2006) provided a model of abiotic processes interacting with traits based on maximum entropy, a principle widely used in physics. In ecology, maximum entropy identifies the configuration of species abundances that maximizes Shannon diversity (see Box 2.1) subject to the constraints that the CWM trait values match those expected based on the position of a plot along an environmental gradient. This maximization approach successfully predicted the relative abundance of species in a community and also how these changed along an environmental gradient of field successional age.

Both of these models successfully predict the abundance of species within a community—a strong prediction by the standards of most community ecology models. Both work by constraining the local community to be a sample of the regional pool such that the mean trait values of several traits vary along environmental gradients in the expected way. In a recent review of trait-based abiotic processes, Laughlin and Messier (2015) emphasize two important points. The first is that abiotic processes affect the assembly of communities when the performance or fitness conferred by a trait depends on and varies with the environment. The second is that we need to move beyond looking at single traits (unidimensional) and move to a multidimensional view of traits (also see Kraft et al 2015).

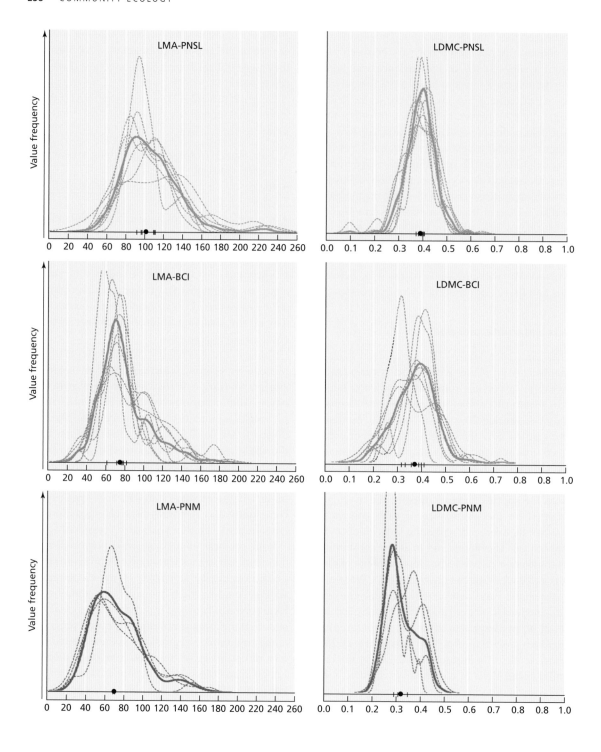

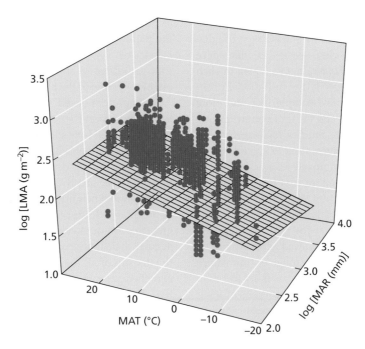

Figure 12.4 The trait of leaf mass per unit area (LMA), basically leaf thickness, varies systematically with both mean annual temperature (MAT) and mean annual rainfall (MAR) as evidenced by the orientation of the plane relative to these two axes. However, there remains a great deal of variation within a single community (the vertical lines of dots). After Wright et al. (2004), fig. 3.

Traits and biotic processes

If solely abiotic processes were operating, one would expect to find the traits in a community tightly gathered around the optimum trait value for that environment. In fact, the range of trait values within a community is very large (Figures 12.3 and 12.4). In their formal variance partitioning exercise, Messier and coworkers (2010) found that only 16% and 30% of the variation in two traits was found along the major precipitation gradient, while 84% and 70% of the variation in each trait was found within a single 20 m × 20 m plot. If abiotic processes play a strong role in driving community

assembly, why would only 15–30% of the variation in traits occur along strong environmental gradients? The classic explanation for this result invokes the notion that biotic processes are also important. Specifically, competition limits the number of species in a community that are overly similar in trait values, causing the coexisting species to "spread out" and be more distinct from each other in trait values (Figure 12.5). The simplest measure of spreading is whether the range of traits occupy a larger volume in trait space (Ricklefs and Travis 1980; Cornwell et al. 2006; Figure 12.5). However, spreading-out due to biotic processes like competition can also affect how the species are packed

Figure 12.3 Each panel shows the distribution of one trait (leaf mass per unit area or leaf dry matter content) at three locations along a steep precipitation gradient (PNSL, BCI, PNM). Within a panel, thin lines show the distribution of traits of individual plants within a 20 × 20 m plot with 4–8 plots sampled per site. The thick line shows the average trend. Note that the mean trait (denoted by the dots on the *x*-axis) does vary along the climatic gradient. Also note that among plots within a site, not only is there very little variation in the mean trait value for a plot, but the overall distribution of traits (e.g. variance and skew) is also strongly conserved between plots within a site. After Messier et al. (2010), fig. 1. ←

within a given volume of trait space. Packing in trait space is often studied by statistics on the distance between neighboring species in trait space (Ricklefs and Travis 1980; Kraft et al. 2008). In such studies it is critical to specify the null model properly because the larger the sample (e.g. the more species richness), the greater the trait volume purely due to sampling effects. Studies looking for evidence of biotic processes resulting in spreading out in trait space have found mixed results. In a study of six important functional traits of tropical trees in the Amazon on a 25-ha plot, Kraft and colleagues (2008) found evidence of both abiotic and biotic processes. They found that for many but not all traits, the mean trait values varied across quadrats (20 m × 20 m) and the absolute range of trait values within a quadrat (lowest to highest) was smaller than you would expect by chance (expectations of what patterns could occur by chance was determined by a null model of randomly sampling trees from the entire 25-ha plot). They interpreted these two findings as evidence for abiotic processes, specifically soil properties due to topography, limiting membership in local communities. However, they also found that many, but not all of the traits were more platykurtic and evenly spread-out (constant distance from neighbors) than expected by chance. This was interpreted

as due to competition and limits to similarity. Thus, the pattern of traits observed was interpreted as a balance between concentration around an optimum trait value due to abiotic processes and spreading out of traits due to the biotic process of competition.

Such pattern-based inference of processes has limits. Two recent studies with forest trees looked more directly at how species traits and neighborhood competition influenced individual growth rates (and survival in one study). The null model assumed that forest tree survival and growth depends only on the number of neighbors, independent of what kind of neighbors. To test the null model they regressed growth rate (or survival rate) of a focal individual against a distance- and size-weighted average of the density of all the other individual trees nearby and then repeated this for every tree as the focal tree. They also tested different hypotheses that the kind of neighbor could further depress growth and survival if:

- neighboring trees were the same species as the target;
- neighboring trees were similar in traits to the target;
- neighboring trees were phylogenetically closely related to the target.

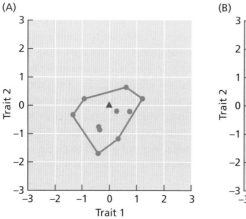
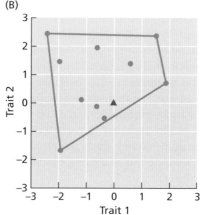

Figure 12.5 Calculating functional diversity by looking at volume in trait space. (A) Hypothetical example when abiotic processes dominate. All species (blue dots) are tightly clustered around the optimum for the environment (red triangle). (B) Hypothetical example when competition is also important. The species are centered round the same optimum (red triangle) but are more spread out. "Spread out" is reflected in the larger volume of area covered (the blue line is the smallest convex polygon containing all points and is called the convex hull) and in the points being more evenly distributed within the space. One common measure of "evenly spread out" is variation in the average distance to the nearest neighbor in trait space for each species.

These hypotheses were incorporated by adding additional explanatory variables to the regressions. Again, results were mixed. In Fortunel et al. (2016), the null model (all individuals equal in impact once size and distance are controlled for) was the best predictor of growth rates over 70% of the time, but various models with impact related to trait-similarity models were best most of the rest of the time. In Uriarte et al. (2010), null models were the best predictor of growth or survival 1/3 of the time, and trait-similarity models were best 2/3 of the time. The amount of variance explained by the models was low for predicting growth (typically 5–20%), but higher for survival (typically 60–99%). In both studies, phylogenetic distance showed up only rarely as most important. Thus, having neighboring trees that share similar traits often, but not always directly reduces growth and survival of the target tree.

Results across the pattern-based and more direct tests for the importance of traits in influencing competitive outcomes are similar. There is clear evidence that traits can and do influence competitive outcomes, but also suggestions that often they do not. There are at least two big reasons why the methods above might not observe impacts of traits on competitive outcomes, even when they are there. First, the preceding analyses examined one trait at a time. However, there are often strong and complex correlations between traits, so no one trait is filtered in isolation. For example, filtering might favor a higher leaf thickness due to herbivory, but filtering might also favor shorter-lived leaves due to the seasonal environment. Since leaf thickness and leaf life span are positively correlated, what type of filtering on leaf thickness would be expected? Many such pulls in opposite directions due to the correlations among traits can make a signal harder to find. Secondly, a deeper, co-existence based understanding of competition (Chapter 7) shows that competition can actually produce opposite patterns of trait spread. Mayfield and Levine (2010) give the example of root traits adapting to different soil grain sizes (and thus moisture regimes) vs the trait of maximum height. Differences in root traits related to soil moisture represent adaptations to different niches, enabling coexistence. Thus, the presence of competition would tend to cause spread along an axis of a root trait related to soil moisture. But differences in maximum height represents differences in competitive ability (for light), where competition will tend to favor only the tallest plants, and thus competition leads to a filter that represents a clustered subset of all possible trait values (e.g., tall plants). Thus, competition can cause either spread or clustering in traits depending on how those traits link to co-existence processes of competition.

Assessment and prospectus of trait-based assembly research

For all of the above reasons, it is increasingly important to turn to even more mechanistic models of trait-based biotic limitations to membership in order to know what patterns of traits to expect. Marks and Lechowicz (2006) conducted a computer simulation of plant seedling growth and survival based on 34 physiological traits. They found strong evidence of both abiotic and biotic filtering based on traits. They also found that there were multiple combinations of traits that were equally successful, but that these combinations were far from each other in trait space. In a similar study, Kleidon and Mooney (2000) found that the number of distinct combinations of traits that could survive a given climate closely paralleled the variation in species richness with climate along the latitudinal gradient. Litchman and Klausmeier (2008) developed detailed models of phytoplankton growth and competition based on traits. Kraft and colleagues (2015) studied grassland annual plants. By using a specific population dynamic model they were able to directly quantify niche differences (stabilizing effects) and competitive ability differences of each of 18 different species. They found stabilizing niche differences were poorly correlated with any one trait, but well-predicted by combinations of multiple traits. In contrast, competitive ability differences were well-predicted by several individual traits related to phenology and plant size. Because Kraft et al. could calculate competitive ability differences and niche differences from just the traits of the species, they were able to apply coexistence theory (Chapter 8) to predict which pairs of species would coexist based just on their traits. The predictions were largely accurate, and many of the predicted non-co-occurrences that did occur could be attributed to other forms of coexistence such as spatial and temporal heterogeneity.

Overall, the trait-based assembly research program has had several successes. It appears membership in communities can be limited based on the traits of species, and that both abiotic and biotic processes can be linked to the traits that are selected for in-community assembly. Initial efforts to link both abiotic and biotic assembly processes to traits were based largely on studies of patterns (such as patterns of covariation of CWM traits with the environment or the spread of species in trait space). However, modern studies have used both theory and experiment to explore more mechanistically the processes of abiotic filtering and trait-based coexistence (i.e. biotic filtering). Although we did not cover studies linking traits to dispersal processes, this is also a rapidly growing field.

Functional diversity

Once one begins to think about traits, it becomes clear that another approach to studying biodiversity is possible. Instead of thinking about the number of species that are present, as done with traditional species richness (Chapter 2), one can explore the range and variety of traits present. This notion is called functional diversity. One common measure of functional diversity looks at the volume of the smallest convex polygon in trait space containing all the species present (known as a convex hull volume; Figure 12.5). Numerous other measures are also used (Weiher 2011). A key point when discussing functional diversity is whether functional diversity is adjusted for the number of species present. The more species present, the greater the likelihood that a species outside the current range of trait space will occur, expanding the functional diversity. Without controlling for the number of species, functional diversity and species richness are strongly positively correlated (and thus largely redundant) (e.g., Swenson and Weiser 2014). Thus, it is common to use a form of rarefaction where functional diversity is compared across communities by randomly sampling the functional diversity for the same number of species in each community (Petchey et al. 2007).

Compare this with using species-individual curves to standardize communities with different numbers of individuals before comparing for species richness (Chapter 2).

A nice illustration of these points is given by Petchey and colleagues (2007, Figure 12.6). Petchey examined real-world communities of birds in Britain. They looked at a number of bird traits like diet preference, foraging mode, body size, and foraging time-of-day. They used an alternative method for volume in trait space (see paper for details) to calculate functional diversity (FD). Then they compared the FD of these real-world communities to the FD of null model communities (communities with the same taxonomic richness as the real-world communities, but with species randomly sampled from the regional pool of birds). What they found was that real-world communities had lower functional diversity than randomly-assembled communities. This suggests that some combination of the three assembly process (abiotic, biotic, dispersal) favor communities with species that are functionally similar to each other, with abiotic filtering being a likely guess.

Swenson and colleagues (2012) found that functional diversity of local communities of trees largely paralleled the latitudinal gradient in species richness, with more functional diversity in the tropics than the temperate zone. After controlling for species richness, they found the tropics were even more functionally diverse than would be expected based on the number of species, while the temperate zones had less functional diversity than expected. Lamanna et al. (2014) found a similar result for local communities. However, when looking at regional functional diversity rather than local, they found that the temperate zone actually has greater functional diversity than the tropics. Thus, while one community in the tropics is more functionally diverse than one community in the temperate zone, the regional pool of plants in the tropics show less range of functional diversity than the regional pool of plants in the temperate zone (Figure 12.7). Comparing trees in North America and Europe, Swenson et al. (2016) showed that the two continents had very similar functional diversity (and a similar saturation curve of functional diversity plotted against species richness) even though North America is well-known to

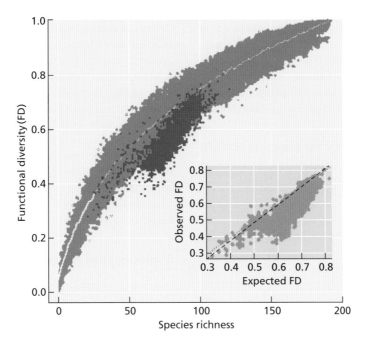

Figure 12.6 Comparison of functional diversity (FD) expected by chance (random sampling of species in the regional pool) shown in blue compared to actual bird assemblages (shown in red). Real communities of British birds showed less functional diversity than expected by chance, suggesting that the assembly processes were selecting for species that were functionally similar to each other. After Petchey et al. (2007).

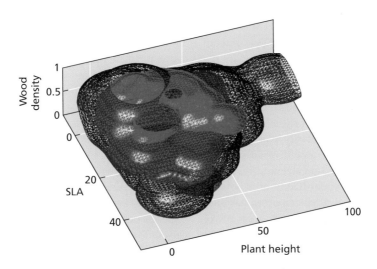

Figure 12.7 The trait volume (functional) diversity for the regional pool of plots in the tropics (red) and temperate zone (blue). Plotted in a three dimensional trait space (plant height vs specific leaf area (SLA) vs. wood density). Each species is represented by a single point in this 3-D space. The volumes shown are skins wrapped around all of the species found in a region (tropics or temperate zone). Note that neither the tropics nor temperate zone functional space is wholly nested inside the other and that the temperate zone functional space is larger (higher regional functional diversity). Data from Lamanna et al. (2014).

have much higher species richness. The higher species richness in North America appears to be due to genera unique to North America but which appear to be functional redundant to the genera shared between the two continents. In other words, the extinctions of trees during glaciation in Europe mostly eliminated functionally redundant genera.

Phylogeny as a proxy for functional traits

Darwin (1859) was the first to frame community assembly rules explicitly in an evolutionary context, stating that "As species of the same genus have...some similarity...the struggle will generally be more severe between species of the same genus, when they come into competition with each other, than between species of distinct genera." Aside from inspiring Williams (1947) to study species/genus ratios as an assembly rule, this insight led to the idea that phylogenetic distance can serve as proxy or shortcut for measuring trait similarity; i.e., all of the hypotheses given above about trait similarity can be turned into hypotheses about phylogenetic similarity. Thus, community assembly can be explored as a matter of taxonomic similarity (e.g. the work by Diamond), a matter of phenotypic/trait similarity (the preceding section) or a matter of phylogenetic similarity. When community assembly is studied from the point of view of phylogeny it is called community phylogenetics (additional aspects of using phylogenies in the study of community ecology are discussed in Chapter 16). Several tests based on phylogenetic similarity were already mentioned in this chapter. Both Uriarte et al. (2010) and Fortunel et al. (2016) found that phylogenetic similarity was usually a much poorer predictor of competitive effects than traits (or null models based on size and distance). On the other hand, Cavendar-Barres et al.'s work (2004) on oaks found that phylogeny (red oak subgenus vs white oak subgenus) was a good proxy for trait similarity and for predicting the outcome of biotic processes. Because phylogenetic distance is intended to serve as a proxy for trait similarity, community phylogenetics suffers from all the limitations and challenges already mentioned for traits [especially the fact that opposite patterns can emerge

from the same process per Mayfield and Levine (2010)]. However, phylogenetic assembly rules also face the additional challenge that phylogenetic proximity is a signal of trait-similarity only if the trait is evolutionarily conserved along the phylogeny (Losos 2008a; Mayfield and Levine 2010; Chapter 16). Many traits (e.g. flower color or many aspects of bird song) are evolutionarily labile and so species that are phylogenetically far apart can be more similar than phylogenetically close species. This can further blur the signal of functional assembly rules. An exciting open question is to identify scenarios where taxonomic-based, trait-based, and phylogeny-based assembly rules will work best.

Regional pool processes

Although most of the work on community assembly to date has focused on the processes leading to limited local community membership, there is a second, equally important implication of community assembly. Namely, community assembly implies that the structure of the regional pool has a strong impact on the local community. This means we need to pay much more attention to the processes structuring the regional pool (Mittelbach and Schemske 2015). Although no one can precisely describe the scale of the regional pool, it is orders of magnitude larger than a local community. Thus, there is no reason to assume that the processes structuring a local community are the best descriptions of the processes structuring regional pools. For example, dispersal probably still matters for structuring regional pools, but it now becomes the rare inter-regional or inter-continental dispersal events that matter. And speciation and phenotypic evolution emerge as important processes in structuring the regional pool. These regional evolutionary processes play out over large heterogeneous areas and over very long time scales. Thus, to understand how species pools are generated requires an understanding of the processes of speciation, extinction and dispersal (Mittelbach and Schemske 2015). Models of evolution in metacommunities provide one approach to modeling the evolution of regional species pools (Leibold and Chase 2018). Thompson's geographical mosaic idea (Thompson 1999) is also relevant, as are discussions of the factors limiting species range expansion and secondary sympatry (see Chapter 16). It is important to note here the notion that species interactions are intermittent at the regional scale. Species A will sometimes interact with species B, but in another site, species A may instead be interacting with species C. These interactions will take place across a spatially and temporally heterogenous environment. Thus, while species interactions at broad spatial scales clearly influence selection, phenotypic evolution and even speciation, they are likely more diffuse and intermittent than typically envisioned (MacArthur 1972, p. 29; McGill et al. 2006).

Community assembly as an organizing framework for ecological theory

It is useful to note that the regional pool/three assembly processes view of community assembly encompasses part or all of many models in community ecology (Mittelbach and Schemske 2015) For example, island biogeography and neutral theory emphasize the dispersal process, while co-existence theory emphasizes the biotic process. Metacommunity theory (Chapter 14) is also converging on this paradigm, starting with an emphasis on the dispersal process, but with recent theories also incorporating the abiotic and sometimes even the biotic filter (Leibold and Chase 2018). In a computer simulation, Chave and colleagues (2002) explicitly incorporate models of dispersal, abiotic, and biotic processes. They then tuned the model to incorporate different strengths of these processes and they found that emergent patterns, such as species abundance distributions and species–area relationships looked very similar across the different mixes of filters. Thus, they recommend against using patterns to infer processes (as has often been stressed in this chapter). It is interesting to note that niche theory has gradually evolved to parallel the three process paradigm. Under Hutchinson (1957), the realized niche was a combination of the biotic and abiotic processes, while the fundamental niche was only the abiotic filter (Chapter 1). Pulliam (2000) has more recently pulled in the dispersal niche as an aspect of defining the niche.

Conclusions

Community assembly asks what limits the membership in a local community, or in other words why do we get a particular subset of regionally available species in a local community. This question has been central throughout the history of community ecology (Chapter 1). Although approaches to this question have varied through time, many efforts have been pattern based, seeking to infer processes from observed patterns. However, decades of experience suggest that pattern-based approaches to studying community assembly are weak in their ability to identify the processes involved. The approach described in this chapter, focusing on three processes that can limit membership from the regional pool, is proving a useful framework for studying community assembly and is helping to unify a variety of more specific theories. Incorporating trait-based and phylogenetic approaches provide additional dimensions to this framework.

Summary

1. Understanding what limits the membership in a local community has been a central organizing question throughout the history of community ecology. Community assembly starts with the set of species potentially able to populate a local community (the regional pool). Then, broadly, three processes can limit membership in a community: abiotic factors (environment), biotic factors (competition, predation, etc.), and dispersal (a failure to get there). These three processes are often envisioned as filters, especially the first two, but there are limits to this analogy. Nevertheless, the three-process view is a powerful way to think about community assembly. Many theories in community ecology can be classified according to the relative emphasis they place on these three different processes.
2. Many pattern-based limits on community assembly have been proposed including ones based on body size ratios, species–genus ratios, checkerboard patterns in space, and guild limits. Recent efforts have tended to move from pattern-based assessments to more process-based analyses.

3. Comparing local richness to regional richness has been used to try to identify the processes involved in community assembly. A saturating relationship has been thought to suggest niche-based processes, while a linear relationship has been thought to involve more random or dispersal-limited processes, but this view is overly simplified. Both kinds of patterns are found in nature.

4. Traits have become an increasingly important part of the study of community assembly. Traits are at once more general than taxonomic-based assembly rules and also more mechanistic. It is easy to demonstrate that traits vary along gradients and that abiotic processes limit membership based on traits. Although the results are more mixed, clear examples of biotic processes (primarily competition to date) limiting membership based on traits also exist. The study of traits needs to continue to grow to be less pattern-based and more mechanistic and to deal with multiple traits simultaneously.

5. Functional diversity is a trait-based view of biological diversity. Functional diversity is largely redundant with species richness, unless richness is first controlled for. At local scales functional diversity seems to vary along latitudinal gradients in the same way as species richness, but at larger scales, they often tell different stories.

6. Phylogenetic proximity can be thought of as a proxy for trait similarity, allowing the study of the phylogenetic structure of a community as an outcome of assembly processes.

7. A community assembly view implicitly highlights the importance of the regional pool from which communities are assembled. We need to pay much more attention to what structures regional pools. These are likely to be different from what structures local communities.

PART IV

Spatial Ecology: Metapopulations and Metacommunities

CHAPTER 13

Patchy environments, metapopulations, and fugitive species

Species interactions take place in a spatial context. **Martha Hoopes et al., 2005, p. 35**

It is conceivable that species…may actually be able to exist primarily by having good dispersal mechanisms, even though they inevitably succumb to competition with any other species capable of entering the same niche. Such species will be termed fugitive species.

G. Evelyn Hutchinson, 1951, p. 575

Up to this point, we have focused on populations and species without considering their spatial distribution in the environment. By assuming that individuals mix freely in a homogeneous environment, we have been able to use simple models to understand the fundamental machinery of predation, competition, and facilitation, and we have been able to combine these species interactions into modules to explore the functioning of food webs and other ecological networks. However, it is time for us to relax the assumption of environmental homogeneity, as there are important ecological phenomena that depend critically on the fact that populations and species may be distributed heterogeneously across the landscape. We begin our exploration of these phenomena by considering the case of a single species in a patchy environment.

Metapopulations

H. G. Andrewartha and Charles Birch, in their landmark 1954 text *The Distribution and Abundance of Animals*, were the first to advocate thinking about populations in a spatial context. Their experience

with studying insects suggested that local populations experience large fluctuations in abundance, potentially disappearing for years only to be reestablished at a later time by colonists from other populations. Their premise—that populations occur in **patches** of suitable habitat surrounded by areas of unsuitable habitat, that such local populations are connected by the migration of individuals, and that local populations may be re-established by colonists from other populations after going extinct—describes what we now call a **metapopulation,** or "population of populations." Unfortunately, Andrewartha and Birch's early insights into metapopulations appear to have been lost in an ongoing debate over density dependence and density independence (Hanski and Gilpin 1991; see Chapter 1), and 15 years passed before Richard Levins (1969, 1970) formally introduced the concept of the metapopulation and developed a mathematical model to describe its dynamics.

The classic Levins metapopulation model

The **Levins metapopulation model** is simple in form (Levins 1969, 1970) and serves as a foundation for

Community Ecology. Second Edition. Gary G. Mittelbach & Brian J. McGill, Oxford University Press (2019).
© Gary G. Mittelbach & Brian J. McGill (2019). DOI: 10.1093/oso/9780198835851.001.0001

all other metapopulation models (Hanski 1997). Like the logistic equation, it can be viewed as a paradigm of population growth, but in this case, population growth in a patchy environment (Hanski and Gilpin 1991). The basic assumptions of the Levins model are as follows:

1. The environment is composed of a large number of discrete patches, all identical and all connected to each other via migration (i.e., dispersal is global).
2. Patches are either occupied or not (actual sizes of populations within patches are ignored and it is assumed that each colonized patch quickly reaches its carrying capacity). Species abundance is described by the fraction of total patches occupied (P).
3. Populations within patches have a constant (per patch) rate of extinction (m).
4. The rate of patch colonization is proportional to the per patch colonization rate (c) times the fraction of currently occupied patches (P) times the fraction of currently empty patches ($1-P$), which are the targets of colonization.

Given the above assumptions, the rate of population growth can be described as

$$dP/dt = cP(1-P) - mP \quad \text{[Eqn 13.1]}$$

Rearranging this equation yields

$$dP/dt = (c-m)P\left(1 - \frac{P}{1-m/c}\right) \quad \text{[Eqn 13.2]}$$

Equation 13.2 shows that the Levins model is directly analogous to the well-known logistic equation (see Chapter 4). Like the logistic model, the Levins model has a globally stable equilibrium point, and the population size at equilibrium (i.e., the fraction of patches occupied when $dP/dt = 0$) is equal to

$$\hat{P} = 1 - m/c \quad \text{[Eqn 13.3]}$$

The major conclusions (Nee and May 1992) of the Levins metapopulation model are:

1. At equilibrium (steady state), there will always be some unoccupied patches in the environment. The fraction of occupied patches at equilibrium is $\hat{P} = 1 - m/c$, and the fraction of unoccupied patches is $1 - \hat{P} = m/c$.

2. Decreasing the number of patches in the environment (for example, if habitat destruction permanently removes a fraction D of the patches from the system) will decrease the rate of population growth as

$$dP/dt = cP(1-D-P) - mP \quad \text{[Eqn 13.4]}$$

3. The population reaches an "extinction threshold" (where the equilibrium fraction of occupied patches is zero) when the fraction of patches permanently removed from the system is

$$D = 1 - m/c.$$

The Levins model is simple in form, but it captures the essence of metapopulation dynamics and is very general in its application. Like other general models we have studied (e.g., the logistic growth model, the Rosenzweig–MacArthur predator–prey model), the Levins model greatly simplifies the real world. It is useful, however, in that it offers important insights into the roles of dispersal and local extinction in determining the dynamics and persistence of patchily distributed populations and species. As we will see when we discuss metapopulations and metacommunities (Chapter 14), **dispersal limitation** occurs when a species is unable to occupy all suitable patches in the environment, and the degree to which species are dispersal-limited has important consequences for population dynamics, species coexistence, and community structure.

Implications of the metapopulation model for conservation biology

Metapopulation theory has largely replaced the theory of island biogeography (MacArthur and Wilson 1967; see Chapter 2) as a conceptual underpinning for the field of conservation biology (Merriam 1991; Hanski and Simberloff 1997). In particular, the Levins model helps us understand how habitat loss may affect species persistence. Its first important insight is that we should expect to find unoccupied patches of suitable habitat in all metapopulations. In the Levins model, the fraction of unoccupied patches at equilibrium is $1 - (1-m/c) = m/c$. Thus, a significant fraction of unoccupied patches should be found in metapopulations of species that have low rates of colonization (c) relative to their rate of extinction within patches

(m). Moreover, the model shows that the presence of unoccupied patches in the landscape should not be viewed as an indication that there are "spare" or "extra" habitats into which a species might expand, or that some of these unoccupied habitats could be eliminated with no consequence to the species. Instead, decreasing the number of patches decreases the rate of population growth (Equation 13.4). Furthermore, if the number of patches is decreased sufficiently (that is, to the point where $D = 1 - m/c$) the species will disappear from the landscape.

The second important insight from the Levins model is that the disappearance of a metapopulation due to habitat loss does not happen immediately, but may occur generations after the critical loss of habitat. Tilman et al. (1994) referred to this delay as the **extinction debt**; Hanski (1997) conjures up an even more chilling description by describing the members of metapopulations facing slow extinction as the "living dead." Therefore, as Hanski (1997, p. 88) notes, "Conservationists should dismiss the false belief that protecting the landscape in which a species now occurs is necessarily sufficient for long-term survival of the species."

In other words, preserving existing habitat may not be enough to save a species whose dynamics are those of a metapopulation. For example, Hanski and Kuussaari (1995) suggest that about 10% of the resident butterfly species in Finland may be facing extinction debt. On the bright side, however, is the fact that the decline toward extinction can be slow, giving conservationists time to intervene and potentially save a species by increasing the number of available habitat patches, or by increasing the rate of dispersal among patches (e.g., by establishing habitat corridors or assisting the migration of individuals between patches). As Equation 13.3 shows, species abundance in a metapopulation increases as the rate of dispersal between patches increases. If dispersal between habitat patches is restricted by a surrounding landscape of unfavorable or impenetrable habitat (e.g., roads, agricultural fields, urbanized areas), then the establishment of "movement corridors" that allow individuals to more readily disperse between patches should increase the overall population size. We discuss an example of this from a lizard meta-population later in this chapter.

Even though ecologists and conservation biologists were quick to recognize the potential for corridors of favorable habitat to help maintain species in fragmented landscapes, scientific evidence supporting this idea was slow to materialize (Simberloff et al. 1992; Van Der Windt and Swart 2008). Simberloff et al. (1992) were particularly critical of the early, unbridled acceptance of movement corridors as a conservation tool, noting that corridors also have potential negative consequences, such as increasing the spread of disease or of competitor and predator species, or enabling the invasion of exotic species (Simberloff and Cox 1987; Resasco et al. 2014).

In the decades since Simberloff et al.'s 1992 paper, there have been a number of experimental tests of the efficacy of corridors in promoting species dispersal and persistence. For example, Gillies and St. Clair (2008) found that corridors facilitated the movement of tropical bird species between isolated forest patches. They began their study by capturing territorial individuals of barred antshrikes (*Thamnophilus doliatus*) and rufous-naped wrens (*Campylorhynchus rufinucha*) in a highly fragmented tropical dry forest of Costa Rica. They translocated the birds 0.7–1.9 km from their "home" territories, placing them in several different habitat types. Then, using radiotelemetry and GPS, they follow the movements of the individual translocated birds. Barred antshrikes, which are forest specialists (normally found in only the most intact forest in this region), returned to their territories faster, and with greater success, in treatments containing corridors of forest trees. However, return rates of rufous-naped wrens, which are forest generalists (found in both intact and degraded forests), were about the same whether forest corridors were available or not (Figure 13.1). A meta-analysis (Gilbert-Norton et al. 2010) of 78 corridor experiments from 35 studies concluded that corridors increase movement between habitat patches by approximately 50% compared with patches that are not connected by corridors. Gilbert-Norton et al. also found that natural corridors (i.e., those existing in landscapes prior to the studies) facilitated movement more than corridors created and maintained for the studies.

Movement corridors also may serve to increase species richness. In a large-scale experiment, Damschen et al. (2006) examined the long-term

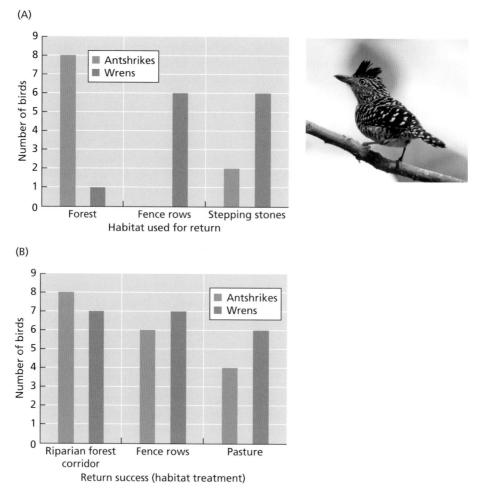

Figure 13.1 The dispersal of tropical forest-dwelling birds is enhanced by the existence of habitat corridors. Individual barred antshrikes (*Thamnophilus doliatus*; photo) and rufous-naped wrens (*Campylorhynchus rufinucha*) were removed from their territories in a tropical dry forest in Costa Rica and transported 0.7–1.9 km distant into three habitat treatments: (1) riparian forest corridors, (2) fence rows, and (3) pastures. Ten birds of each species were released in each treatment; the birds were followed using radiotelemetry and GPS. (A) Antshrikes, which are forest specialists, primarily used riparian forest corridors to return to their territories. Wrens, which are forest generalists, used a combination of habitat types (including fencerows and small patch "stepping stones") to return to their territories. (B) The return success of antshrikes was significantly greater when riparian forest corridors were available than in the other two treatments. Wrens showed no difference in return success by treatment type. After Gillies and St. Clair (2008); photograph courtesy of Brian Gratwicke.

effect of corridors on plant species richness by experimentally creating patches of longleaf pine habitat at the Savannah River Ecological Laboratory in South Carolina (Figure 13.2A). They found that habitat patches connected by corridors retained more native plant species than did isolated patches, and that this difference increased over time (Figure 13.2B). Thus, corridors can be a significant tool in the conservation of populations in fragmented landscapes

(Gilbert-Norton et al. 2010). This is not to say that there can be no negative effects of corridors. In the same experimental system studied by Damschen et al., researchers found that corridors increased invasion by the polygyne form of the fire ant (*Solenopsis invicta*), resulting in a 23% decrease in native ant species richness. However, while corridors can have negative effects on biodiversity by enhancing invasions of exotic species or increasing the spread

(A)

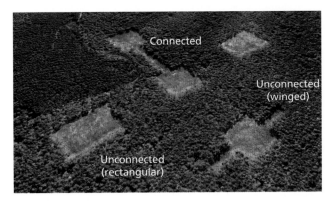

(B)

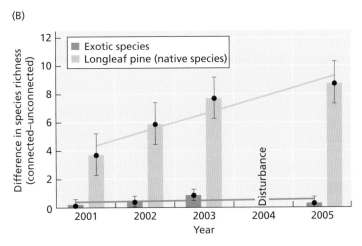

Figure 13.2 The presence of movement corridors can affect species richness. (A) One of six experimental study landscapes where Damschen et al. constructed large (100 m × 100 m) patches of longleaf pine habitat within a surrounding matrix of pine plantation. The patches mimic native longleaf pine habitat, which is characterized by an open and species-rich understory maintained by frequent low-intensity fires. Some patches were connected by corridors to other patches; others were not. (B) Patches connected by corridors were found to be richer in plant species native to longleaf pine habitat than unconnected patches, and that this difference in native species richness increased over time. The number of exotic plant species was unaffected by the presence of corridors. Data were unavailable for 2004, when patches were burned by the U.S. Forest Service as part of restoration management. (A) courtesy of USDA Forest Service; (B) after Damschen et al. (2006).

of predators or disease (Simberloff and Cox 1987), more studies have shown beneficial effects of corridors on species dispersal and biodiversity, rather than negative effects (Haddad et al. 2014).

The assisted dispersal of species, which has been proposed as a way of ameliorating the effects of climate change, is a much more controversial measure. Global climate change is occurring at a rapid rate, and there is clear evidence that species are shifting their ranges in response to changing environments. The ranges of a variety of organisms have shifted

poleward at average rates of 6–17 km per decade over the past 50 years (e.g., Parmesan and Yohe 2003; Chen et al. 2011). For some species, these rates of range shift have kept pace with changing environments. However, there is concern that less mobile species may not be able to keep up with climate change. Studies of past extinctions (such as those of the Pleistocene) show that species with small ranges are particularly vulnerable to rapid climate change (Sandel et al. 2011). Moreover, past rates of climate change mostly have been slower than those taking place today. For sure, the landscapes species had to move through to track past climate changes were far less fragmented than they are today, which can greatly slow down the rate of movement (Lazarus and McGill 2014; Miller and McGill 2018). Therefore, some ecologists and conservation biologists have advocated the active translocation of at-risk species to areas where future climates are projected to be more suitable (see discussions in Hunter 2007; McLachlan et al. 2007; Hoegh-Guldberg et al. 2008). Such **assisted dispersal** (also called assisted colonization or assisted migration) carries with it many risks and uncertainties, however, as evidenced by the well documented negative impacts of unintended species introductions on native communities. Thus, Ricciardi and Simberloff (2009, p. 252) suggest that "assisted colonization is tantamount to ecology roulette." Assessing the pros and cons of assisted dispersal and developing sound conservation strategies to reduce the impacts of global climate change on species with limited dispersal ability is a major challenge facing ecologists and conservation biologists in the very near future.

Parallels between metapopulation models and epidemiology

There are many parallels between the spread of infectious organisms (viruses, bacteria, metazoan parasites) and the dynamics of a metapopulation. For example, the host can be seen as analogous to a patch, infection as akin to colonization, and host recovery, death, or immunization as akin to patch extinction (Lawton et al. 1994). The threshold theory of epidemiology has long recognized that a minimum number of susceptible host individuals is required to sustain a disease outbreak (Kermack and McKendrick 1927; Anderson and May 1991). This number is directly analogous to the minimum number of patches required to maintain a metapopulation in the Levins model (Lawton et al. 1994). Furthermore, such metapopulation thinking demonstrates that eradicating a disease does not require immunizing all the hosts (or eliminating all disease-transmitting vectors, such as mosquitoes for malaria). Instead, one only needs to reduce the number of susceptible hosts (or the density of vectors) to a level at which the disease is no longer able to persist

(i.e., that specified by $D = 1 - m/c$ in the Levins model). Thus, it has been possible to eradicate smallpox without vaccinating every person, and to eliminate malaria from some regions of the world without eliminating every mosquito (Nee et al. 1997). The extinction threshold in the Levins metapopulation model is analogous to the eradication threshold in epidemiology models that assume weak homogeneous mixing (i.e., that new infections are proportional to the number of susceptible hosts; Anderson and May 1991; Nee 1994). In both cases, the threshold for extinction/eradication occurs when the fraction of patches destroyed (or the fraction of hosts vaccinated) equals $1 - m/c$ (Lande 1988a; Nee 1994; Bascompte and Rodríguez-Trelles 1998). Note that if the system is at equilibrium, the quantity $1 - m/c$ is equal to the fraction of patches occupied (Equation 12.3).

Empirical examples of metapopulation dynamics

Are there real-world populations that behave as predicted by the simple Levins metapopulation model? The answer is yes, although their number seems limited. Studies of frogs (Sjögren 1988; Sjögren Gulve 1991), butterflies (Hanski et al. 1994; Hanski and Kuussaari 1995), birds (Verboom et al. 1991), zooplankton (Pajunen and Pajunen 2003), and plants (Menges 1990) all show the signatures of a classic metapopulation: populations exist in a network of occupied and unoccupied patches, and local populations undergo colonization and extinction events in each generation. One of best examples of a classic metapopulation comes from the work of Templeton and colleagues who studied the distribution and dynamics of Eastern collared lizards (*Crotaphytus collaris collaris*) living in glades—dry, open, rocky habitats embedded in a woodland matrix—in the Ozark highland region of Missouri (USA). These glades support a diverse community of dry-adapted organisms that originally colonized the Ozarks from the American southwest some 8000 years ago (during the Xerothermic maximum; Cole 1971). As the climate cooled and became wetter, eastern collared lizards in the Ozarks were restricted to dry, rocky glades, which functioned as habitat patches within a woodland matrix. In the past, frequent fires maintained an open woodland habitat that allowed lizard dispersal between glades (Hutchison and Templeton 1999). However, a policy of fire suppression in the 1990's lead to the growth of dense forest between glades, eliminating lizard dispersal and disrupting gene flow between lizard populations occupying the glades. By 1980, an estimated 75% of the collared lizard populations in the eastern Ozarks had become extinct (Templeton et al. 1990).

In 1994, a policy of managed woodland fires was initiated with the goal of restoring historic dispersal dynamics to allow the recovery of the eastern collared lizard in particular and enhance regional biodiversity in general. Almost immediately, collared lizard populations in the burned areas exhibited a strong, positive response to the controlled burns (Figure 13.3). Dispersal rates increased, leading to

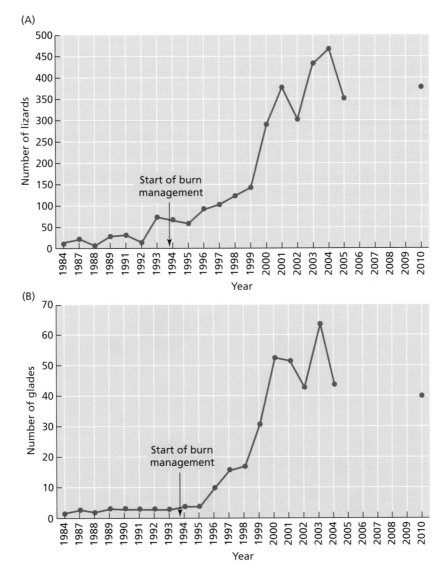

Figure 13.3 The population size (A) and the number of habitat patches (glades; B) occupied by eastern collared lizards on Stegall Mountain (Missouri, USA) increased dramatically following the implementation of habitat management (prescribed woodland fires) that allowed lizards to disperse more readily between habitat patches. After Neuwald and Templeton (2013).

the colonization of previously unoccupied glades, an increase in total lizard population size, and an increase in genetic diversity (an example of "genetic rescue", see Chapter 16). By 2000, the collared lizard metapopulation had returned to stability in terms of total population size, the number of glades occupied, and the balance of glade extinction versus recolonization (Templeton et al. 2011; Neuwald and Templeton 2013; Sites 2013).

The eastern collared lizard study provides an excellent example of the classic metapopulation structure envisioned by Levins, Hanski, and others. However, as we might expect, nature is more varied than our simple models. As ecologists have delved further into the study of metapopulations, they have uncovered a range of different spatial structures beyond those assumed by the classic Levins model. Harrison (1991) categorized these spatial structures into four main types (Figure 13.4):

1. **Classic metapopulations**, as characterized by the Levins model, in which all local populations are similar in size and type. All populations may go extinct, but have the potential to be reestablished by colonization.
2. **Mainland–island metapopulations**, in which a large population or patch (the "mainland") exists without significant risk of extinction and smaller surrounding populations are supported by immigrants from that mainland population.
3. **Patchy populations**, in which individuals within a single interbreeding population are clumped in space, but the clumps do not exist as separate populations (i.e., the degree of movement and gene flow within the population is high).
4. **Non-equilibrium populations**, in which extinction is not balanced by recolonization. If local populations are effectively isolated from one another, then the extinction of local populations eventually leads to regional species extinction (as in the case of boreal mammals isolated on desert mountaintops following post-Pleistocene climate change; Brown 1971).

In reality, these different classifications grade into one another, and we can think of them as points along a

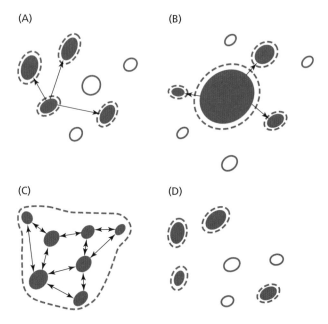

Figure 13.4 Stylistic representation of four different kinds of metapopulations. Circles represent habitat patches: filled circles represent occupied patches, open circles represent vacant patches. Arrows indicate migration, and dashed lines indicate the boundaries of local populations. (A) The classic Levins metapopulation. (B) Mainland–island (or source–sink) metapopulation. (C) Patchy population. (D) Non-equilibrium population; this differs from (A) in that there is no recolonization. After Harrison (1991).

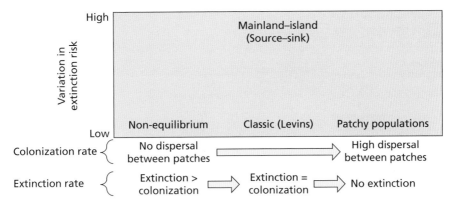

Figure 13.5 The four metapopulation types shown in Figure 12.3 can be viewed as points along a continuum characterized by varying degrees of extinction risk due to variation in patch size and/or quality; degree of dispersal between patches; and extinction rate relative to colonization rate. After Driscoll (2007).

continuum of spatially structured populations that are connected by dispersal (Driscoll 2007; Figure 13.5). The Levins model assumes that all patches are equal and that all patches have a finite probability of population extinction. The mainland–island model assumes that habitat patches differ in size and that populations on the mainland have essentially zero probability of extinction. It is conceptually similar to models of **source–sink dynamics** (Pulliam 1988), which assume that habitat patches differ in quality. High-quality habitats (**source habitats**) support populations where birth rates exceed death rates, leading to an export of individuals to lower-quality habitats (**sink habitats**), where birth rates do not keep up with death rates. Populations in source habitats have a very low probability of going extinct, whereas populations in sink habitats are maintained only by the flow of individuals dispersing from productive source populations [the "rescue effect" of Brown and Kodric-Brown (1977) or the "mass effect" of Shmida and Ellner (1984)].

The challenge facing ecologists and conservation biologists today is to determine where a particular population falls on this continuum of metapopulation types (Figure 13.6). That categorization can guide us in applying the knowledge gleaned from simple metapopulation theory, as well as from more sophisticated metapopulation models (e.g., Hanski 1994; Keymer et al. 2000; Hastings 2003), to predict the dynamics and persistence of a species [e.g., the classic work of Lande (1987, 1988a,b) on the dynamics of the Northern spotted owl]. Even though relatively few populations in nature exactly match the assumptions of the classic Levins model or the mainland–island model, the concept of the metapopulation is of great importance because it focuses our attention on the flow of individuals between populations (Baguette 2004). This exchange of individuals between habitat patches has consequences not only for the preservation of a metapopulation, but also for the maintenance of species diversity within a landscape.

Fugitive species: competition and coexistence in a patchy environment

G. Evelyn Hutchinson, more than anyone else, directed our attention to the problem of how species diversity might be maintained in the face of interspecific competition (Hutchinson 1959). His suggestion that niche partitioning could ameliorate the effects of competition and allow species to side-step the competitive exclusion principle laid the foundation for a theory of species diversity that dominated ecological thought for decades (see Chapters 1 and 8). Hutchinson also realized that the dispersal of species between habitat patches could provide another mechanism for the maintenance of species diversity.

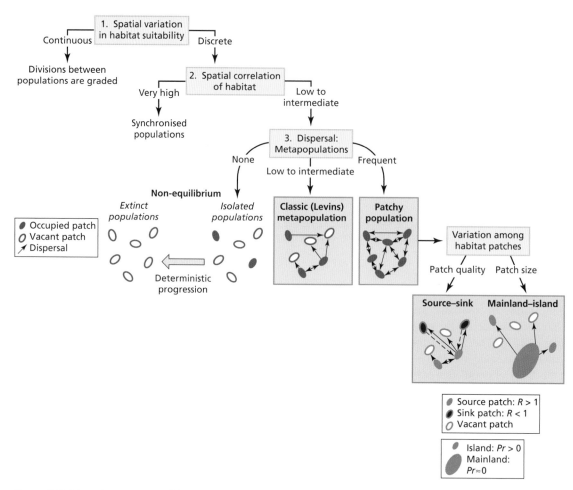

Figure 13.6 Schematic representation of the conditions in which metapopulation theory (metapopulation types as defined in Figure 13.4) is most useful. Variation in patch size or quality is not a condition for metapopulation dynamics, but does impact the properties of metapopulation systems. R is population growth rate in the absence of dispersal; Pr is probability of extinction. After Schtickzelle and Quinn (2007).

The competition/colonization trade-off

Hutchinson published his insight in a short note with the curious title "Copepodology for the Ornithologist." His writing is so clear and inspired that it is worth quoting at length (Hutchinson 1951, p. 575):

It is conceivable that species…may actually be able to exist primarily by having good dispersal mechanisms, even though they inevitably succumb to competition with any other species capable of entering the same niche. Such species will be termed *fugitive species*. They are forever on the move, always becoming extinct in one locality as they succumb to competition and always surviving by

reestablishing themselves in some other locality as a new niche opens. The temporary opening of a niche need not involve a full formal successional process. Very small, seemingly random changes in the physical environment might produce locally a new niche suitable for the fugitive, or some unfavorable circumstance might cause the temporary disappearance of a competitor from such a suitable niche over a small region. Any such change would give the fugitive species its chance, for a time, until some more slowly moving competitor caught up with it. Later it would inevitably become extinct, but some of its descendants could occupy transitorily a temporarily available niche in some other locality. A species of this sort will enjoy freedom from competition so long as small statistical fluctuations in the environment give it a refuge into which it can run from competitors.

Recall that some patches remain unoccupied when a metapopulation is at equilibrium. Thus, as Hutchinson noted here, it may be possible for an inferior competitor to coexist with a superior competitor if the inferior competitor possesses the superior ability to consistently move on to colonize open patches—that is, it is a **fugitive species**.

Simple models similar to the Levins (1969) metapopulation model show that two (or more) species may coexist in a patchy environment when there is a trade-off between competitive ability and dispersal ability (the **competition/colonization trade-off**; e.g., Skellam 1951; Levins and Culver 1971; Horn and MacArthur 1972; Armstrong 1976; Hastings 1980; Tilman 1994). For example, we can easily modify the Levins model to consider two competing species. Let us assume that the environment consists of a large number of habitat patches, all of the same type, and that the superior competitor (species 1 in this case) always displaces the inferior competitor (species 2) from a habitat patch. Thus, species 1's population dynamics is unaffected by species 2. The inferior competitor (species 2), however, can colonize only those patches where species 1 is absent. If the two species have equal mortality rates (m), then species 2 will persist in the system only if it is a sufficiently better colonist than species 1. The relationship describing the condition for two species' coexistence when the species have equal mortality rates is

$$c_2 > c_1^2 / m$$

where c_1 and c_2 are the colonization rates for species 1 and species 2, respectively (Hastings 1980). Thus, even though coexistence within a patch is impossible, species 1 and 2 can coexist regionally in a system of patches due to the fact that species 2 is the better colonizer. Tilman (1994) shows how this simple two-species model can be generalized to include multiple coexisting species that differ in their colonization and mortality rates.

The competition/colonization trade-off is a potentially strong stabilizing mechanism for maintaining biodiversity in a patchy environment (Chesson 2000a; Muller-Landau 2008). Theoretically, a large number of species could coexist via this mechanism if:

- species exhibit a competitive hierarchy;
- the better competitor always wins within a patch;
- there is a strict trade-off between the ability to compete within a patch and the ability to colonize a patch (Tilman and Pacala 1993; Tilman 1994).

Do natural systems fit these expectations?

A number of studies have proposed that species diversity in terrestrial plant communities may be maintained via a competition/colonization trade-off (e.g., Platt and Weiss 1977; Tilman 1994; Rees 1995; Rees et al. 1996). For example, Tilman and colleagues documented what appears to be a competition/colonization trade-off in five grass species in abandoned agricultural fields (Figure 13.7) as a result of differential allocation of biomass to roots and reproductive structures. A competition/colonization trade-off in plants may also result from differential allocation of resources to seed number and seed size. That is, for a given level of reproductive investment, plants may either make many small seeds or a few large seeds each time they reproduce. A **seed size/seed number trade-off** is well documented in species with similar adult sizes and life histories, as seen in Figure 13.8 (e.g., Shipley and Dion 1992; Greene and Johnson 1994; Rees 1995; Turnbull et al. 1999; Jakobsson and Eriksson 2003; Moles et al. 2004). Note, however, that the relationship between seed size and lifetime fecundity can be different if large-seeded plant species are bigger or longer lived. Small-seeded species are expected to have a greater probability of colonizing open patches, both because small seeds disperse further and because plants that produce small seeds have higher fecundity (Harper et al. 1970; Howe and Westley 1986; Greene and Johnson 1993). Large-seeded species, on the other hand, are expected to be better competitors because they have higher germination success and higher survival through the seedling stage (Gross and Werner 1982; Westoby et al. 1996; Eriksson 1997; Turnbull et al. 1999).

The previous arguments make a good case for the existence of competition/colonization trade-offs in plants and, thus, we might expect to find evidence for this mode of species coexistence in a variety of systems. Surprisingly, however, the evidence is mixed. Support for a competition/colonization

(A)

(B)

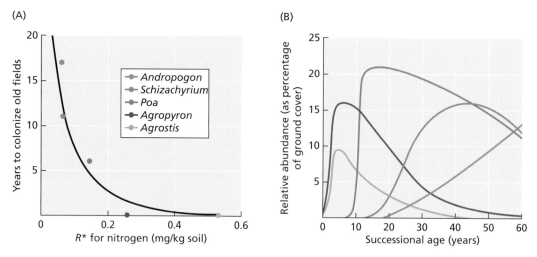

Figure 13.7 Grass species growing in abandoned agricultural fields ("old fields") in Minnesota exhibit an apparent trade-off between their ability to compete for soil nitrogen and their ability to colonize old fields (measured as the number of years required to colonize the fields). (A) Species competitive ability is expressed as the observed R^* for soil nitrogen (determined for each species grown in monoculture). Species that are better competitors (i.e., have lower R^* values) are slower to colonize old fields. (B) The approximate dynamics of succession for these five grass species at the Cedar Creek Ecosystem Science Reserve, based on a chronological sequence of old fields. After Tilman (1994).

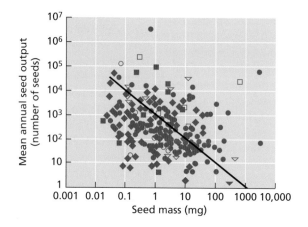

Figure 13.8 A trade-off between seed size and seed number exists for a wide variety of plant species that are similar in adult size and life history. Each data point represents a species; different symbols refer to species from different habitat types. After Moles et al. (2004).

trade-off in plant communities has come from studies by Platt and Weiss (1977) on a native prairie, Tilman and Wedin (1991b) on a grassland/savanna, Rees et al. (1996) on sand dune annuals, and Turnbull et al. (1999) on a limestone grassland. On the other hand, studies by Thompson et al. (2002), Jakobsson and Eriksson (2003), Leigh et al. (2004),

Muller-Landau (2008), and Pastore et al. (2014) have found little support for the hypothesis in other systems. For animal communities the results are similar; only a handful of studies document species coexistence via what appears to be a competition/colonization trade-off (e.g., Hanski and Ranta 1983; Hanski 1990; Marshall et al. 2000).

Probably the best empirical examples of species coexistence via a competition/colonization trade-off come from two laboratory studies conducted with microorganisms. Cadotte et al. (2006b) found a clear competition/colonization trade-off in a group of aquatic microbes commonly used in laboratory experiments (protozoans and rotifers; **Figure 13.9**). Livingston et al. (2012) directly manipulated the strength of the competition/colonization trade-off in two strains of the bacterium *Pseudomonas* and tested whether the strength of the competition/colonization trade-off correctly predicted the outcome of competition. They found that the competitive outcome shifted from competitive exclusion of the better colonizer by the superior competitor under conditions of a weak trade-off to exclusion of the superior competitor by the better colonizer when the competition/colonization tradeoff was strong and dispersal rates where high (Figure 13.10A). Coexistence, however, was rare and occurred in only one of three experiments where it was predicted (Figure 13.10B).

Why do we have so few real-world examples of a species coexistence mechanism suggested by Hutchinson (1951) more than 60 years ago—a mechanism that has been rigorously examined in theoretical models, tested in simple laboratory environments and that continues to hold a prominent place in ecological thinking? This question is puzzling indeed. Hutchinson's basic insight about fugitive species is no doubt correct: some species are "forever on the move…becoming extinct in one locality as they succumb to competition and always surviving by reestablishing themselves in some other locality." However, it is also likely that the "patchiness" of real-world communities is much more varied than envisioned in models focused on a single patch type, where species coexistence occurs via a competition/colonization trade-off. It is also possible that competition/colonization trade-off occur more commonly in some community types than others (e.g., Bracewell et al. 2017). As we will see later, when the environment contains a variety of patch types (e.g., differences in soil type or nutrient availability for plants), additional mechanisms of species coexistence come into play.

Consequences of patch heterogeneity

If we allow the landscape to vary in the spatial arrangement of patches, or in the types of environment in the patches available, we open up many other avenues for species coexistence via colonization-related trade-offs and niche partitioning (Pacala and Tilman 1994; Levine and Rees 2002; Muller-Landau 2008). In a heterogeneous environment, species may coexist via combinations of different colonization and competitive abilities, such that each species does best in a different place or time (Chesson 2000a; Muller-Landau 2008). For example, if we modify an environment composed of identical patches and add spatial heterogeneity, such that some patches are close together and some are far apart, then species may coexist via a **dispersal/fecundity trade-off**. The more fecund species will be more successful in areas of high patch density, and the better dispersers will be more successful in areas of low patch density. Yu et al. (2001, 2004) provide a nice example of species coexistence via this mechanism in a community of Neotropical ants that live symbiotically on a single plant species, *Cordia nodosa*. The plants produce structures called domatia ("domiciles") that house and protect the ants (Figure 13.11A,B), while the ants are thought to help protect the plant from herbivory. Ants of the

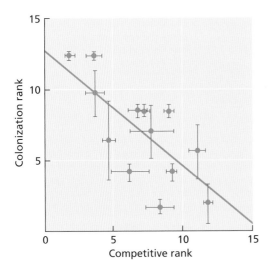

Figure 13.9 Microbial species (protozoans and rotifers) coexisting in laboratory microcosms show a competition/colonization trade-off. Data points are means ±95% confidence intervals. After Cadotte et al. (2006b).

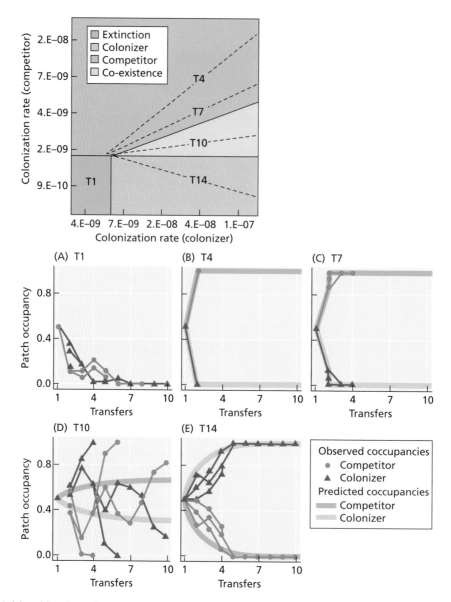

Figure 13.10 (A) Model predictions for two species of bacteria in a laboratory environment showing how an increase in the colonization rate of the superior competitor (treatments T14 to T4) changes the predicted outcome of competition from dominance by the better colonizer (pink region) to coexistence (yellow region) to dominance by the better competitor (blue region). At very low colonization rates (T1) both species go extinct. (B) Observed outcome of competition (population trajectories over time) in three replicates of the five treatments shown in (A). Observations matched model predictions well, except for the prediction of two-species coexistence (T10). After Livingston et al. (2012).

genus *Azteca* are stronger fliers, and their foundresses are able to disperse over longer distances, whereas the ant species *Allomerus* cf. *demerarae* has a higher per capita fecundity and is more successful at colonizing nearby patches. Looking across a gradient in host plant (i.e., patch) density, Yu et al. (2004) found that the relative abundance of *Azteca* increases as host plant density decreases (Figure 13.11C), as would be predicted via a dispersal/fecundity trade-off.

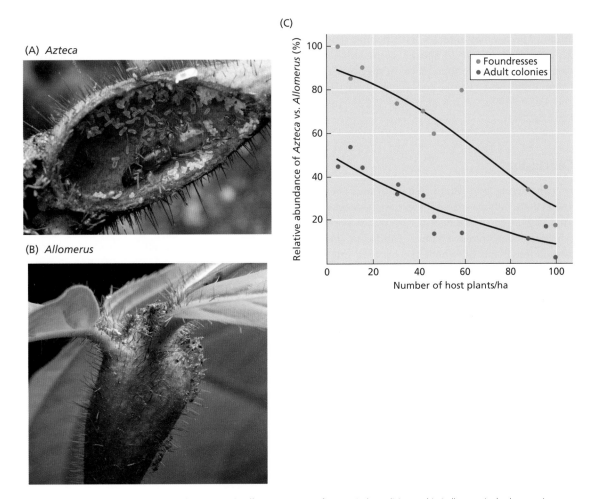

Figure 13.11 An example of a dispersal/fecundity trade-off in a community of neotropical ants living symbiotically on a single plant species (*Cordia nodosa*). *Azteca* ants are able to disperse over longer distances, whereas *Allomerus* is more likely to colonize nearby patches successfully. (A) An *Azteca* queen with a brood inside a plant domatium (cutaway view). (B) *Allomerus* workers enter a domatium. (C) The relative abundance of *Azteca* foundresses and adult colonies increases as host–plant density decreases, as would be predicted when there is a dispersal/fecundity trade-off. Photographs © Alexander Wild; © after Yu et al. (2004).

If we expand our thinking to consider an environment containing different patch types, then we open up the opportunity for species to differ in their tolerance of or performance in a patch type (i.e., for species to exhibit niche differences), as well as in their probability of reaching a patch (dispersal ability). This situation describes a more complex competition/colonization trade-off, which Muller-Landau (2008) refers to as the **tolerance/fecundity trade-off**. In this case, species coexistence is promoted via niche differences (by means of habitat

partitioning); however, differences in colonizing ability (fecundity) can modify the conditions for coexistence and strongly affect the relative abundances of species within the community (Levine and Rees 2002). As Muller-Landau (2008) notes, there is an extensive literature (empirical and theoretical) on species coexistence via habitat partitioning (see Chapter 8) and a parallel literature focusing on species coexistence via competition/colonization traits in a homogeneous environment. What we need next is a combination of theory and empirical work exploring how differences in colonization ability among species may interact with habitat heterogeneity to contribute to species coexistence and determine the relative abundances of species within communities (e.g., Pacala and Tilman 1994; Hurtt and Pacala 1995; Levine and Rees 2002; Muller-Landau 2008).

Conclusion

By considering population dynamics and species interactions within a spatial context, we have opened up a rich array of mechanisms by which populations may be maintained across a landscape (metapopulations), and by which species may interact and coexist (patch dynamics and fugitive species). Metapopulation theory provides the conceptual foundation for much of conservation biology and, in addition, is fundamentally linked to models of epidemiology. Metapopulation theory continues to develop at a rapid pace. As we will see in the Chapter 14, it is natural to extend our thinking about two species interacting as metapopulations in a patchy landscape to thinking about local communities linked by the dispersal of species into metacommunities. By incorporating the effects of species dispersal into community dynamics, we begin to come full circle, back to a question introduced at the beginning of this book: how do local and regional processes interact to produce and maintain the diversity of life?

Summary

1. A metapopulation is a set of local populations, each of which occupies a patch of suitable habitat surrounded by unsuitable habitat, that are connected by the migration of individuals. Local populations that go extinct may be re-established by colonists from other populations.
2. The Levins metapopulation model is a simple model that offers insights into the roles of dispersal and local extinction in determining the dynamics and persistence of patchily distributed populations and species. In this model, decreasing the number of patches decreases the rate of population growth and may lead to the extinction

of the species. Thus, the presence of unoccupied patches is to be expected and should not be viewed as an indication that these habitats could be eliminated with no consequence to a species.
3. The Levins model shows that the extinction of a species with metapopulation dynamics does not happen immediately, but may occur generations after a critical habitat patch is lost. For this reason, the preservation of existing habitat may not be enough to save the species. Such a delay in extinction is referred to as extinction debt.
4. The Levins model also shows that abundance in a metapopulation increases as the rate of dispersal between patches increases. Recent evidence supports the effectiveness of movement corridors between patches in promoting dispersal, and therefore species persistence, in fragmented landscapes.
5. The extinction threshold in the Levins metapopulation model is analogous to the eradication threshold in epidemiology models.
6. Evidence for classic metapopulation dynamics is found in nature, with detailed studies of the eastern collared lizard in the Ozark highlands (USA) providing a notable example. However, natural populations also show a range of different spatial structures beyond those assumed by the classic Levins model.
7. Metapopulations can be characterized into four main types:
   • classic metapopulations, as characterized by the Levins model;
   • mainland–island metapopulations, in which a large population that has a low risk of extinction supports smaller surrounding populations by emigration;
   • patchy populations, within which movement and gene flow are high;
   • non-equilibrium populations, in which extinction is not balanced by recolonization.
   In reality, these four types grade into one another.
8. It may be possible for an inferior competitor to coexist with a superior competitor if the inferior competitor is better at colonizing open patches—a "fugitive species". This competition/colonization trade-off is a potentially strong stabilizing mechanism for maintaining biodiversity in a patchy environment.

9. Evidence for the competition/colonization trade-off as a mechanism of species coexistence in natural communities is mixed, probably because models of this trade-off focus on a single patch type. When the environment contains a variety of patch types, additional mechanisms of species coexistence come into play.

10. Where some habitat patches are close together and some are far apart, species may coexist in those patches via a dispersal/fecundity trade-off—the more fecund species will be more successful in areas of high patch density, and the better dispersers will be more successful in areas of low patch density.

11. If species differ in their tolerance of or performance in a patch type (i.e., in their niches) as well as in their probability of reaching a patch (i.e., in dispersal ability), they may coexist via a tolerance/fecundity trade-off. In this case, species coexistence is promoted via habitat partitioning, but the difference in colonizing ability can modify the conditions for coexistence and strongly affect the relative abundances of species within the community.

CHAPTER 14

Metacommunities

Metacommunity ecology…involves an expansion and enrichment of traditional community ecology, not a replacement for it.

Marcel Holyoak et al., 2005, p. 466

The concept of the metacommunity follows closely from the study of metapopulations. Just as the dispersal of individuals may link the dynamics of populations separated in space, the dispersal of species among communities may link local communities into a metacommunity. A **metacommunity** thus may be defined as *a set of local communities linked by the dispersal of one or more of their constituent species* (Leibold et al. 2004). The local community, in this framework, includes all species that potentially interact in a single locality. (This chapter will use "locality" and "patch" interchangeably to describe the space occupied by a local community.)

The study of metacommunities is a relatively young, but rapidly expanding field, extending many of the ideas developed from the study of metapopulations to the study of multispecies communities (Hanski and Gilpin 1991). This chapter will examine ways in which the dispersal of species between communities may have significant impacts on local and regional diversity, as well as affecting the ability of communities to track environmental change. Four different perspectives characterize how dispersal rates and environmental heterogeneity interact to influence species diversity in metacommunities (Leibold et al. 2004; Holyoak et al. 2005; Leibold and Chase 2018):

1. The **patch dynamics perspective** is an extension of metapopulation models to more than two

species. Species interact in a patchy environment (patches are commonly assumed to represent similar or identical environments).

2. The **species sorting perspective** emphasizes differences in species abilities to utilize different patch types in a heterogeneous environment (patches differ in their environment) and dispersal rates are low.

3. The **mass effects perspective** also emphasizes differences in species abilities to utilize different patch types in a heterogeneous environment, and incorporates source–sink dynamics and rescue effects (dispersal rates are high).

4. The **neutral perspective** assumes that species are functionally identical and that niche differences are unimportant (because species have identical ecologies, environmental heterogeneity doesn't matter).

A more comprehensive and synthetic theory may someday unite these different perspectives into a single mathematical framework describing metacommunity dynamics. However, this goal is a long ways off and may or may not be achievable (Leibold and Chase 2018). In the meantime, examining these perspectives separately helps us understand how dispersal, in combination with the traits of species and spatial heterogeneity in the environment, act to determine patterns of species distribution and diversity in metacommunities. Let's start by considering

Community Ecology. Second Edition. Gary G. Mittelbach & Brian J. McGill, Oxford University Press (2019).
© Gary G. Mittelbach & Brian J. McGill (2019). DOI: 10.1093/oso/9780198835851.001.0001

(A)

Species attributes	Spatial environment	
	Homogenous	Heterogeneous
Different niches	1. Patch dynamics (see Chapter 13)	2. Mass effects/source–sink dynamics 3. Species sorting
Demographically equivalent	4. Neutral theory	

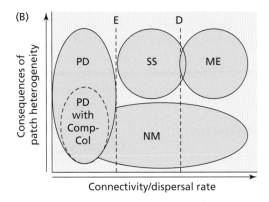

Figure 14.1 (A) Matrix classification of the four perspectives for approaching the study of metacommunities described by Leibold et al. (2004) and Holyoak et al. (2005). These perspectives apply to different types of environments (homogeneous or heterogeneous localities) and to different species attributes (niche differences or demographic equivalence). The fourth quadrant of the matrix is empty because species that are demographically equivalent are assumed to experience all environments equally. (B) The four perspectives of metacommunity ecology in relation to the amount of interpatch habitat heterogeneity (the degree of environmental difference between habitat types) and the connectivity of patches (e.g., species' dispersal rates). After Leibold and Chase (2018).

PD = patch dynamics, SS=species sorting, ME=mass effects, NM=neutral model. Most patch dynamics (PD) models assume little to no patch heterogeneity (the dashed-dotted oval).

how each perspective fits into a matrix that partitions environment and species along two major axes (Figure 14.1A). The first axis distinguishes the spatial environment of a metacommunity as being either homogeneous or heterogeneous (Amarasekare 2003). In a homogeneous environment, a species' performance (e.g., its competitive success) is the same in all patches; in a heterogeneous environment a species' performance differs between patch types (see Chapter 13). How well a species performs in a given environment is determined by its traits (McGill et al. 2006) and how it interacts with other species in the

environment. The second axis characterizes the species occupying a metacommunity as either having niche differences (traits) that affect their tolerance of or performance in different patch types (or their ability to disperse between patches), or as being ecologically or demographically equivalent (Hubbell 2006).

In their recent book on *Metacommunity Ecology*, Leibold and Chase (2018) distinguish the four metacommunity perspectives somewhat differently using the axes of connectivity/dispersal rate and environmental heterogeneity (Figure 14.1B). Dispersal rate is a continuous variable describing how frequently species disperse between patches or local communities. To the left of the line E, dispersal is limited and the extinction of local populations is high relative to their colonization. To the right of line D there is dispersal excess and immigration is high relative to local extinction. The y-axis refers to the consequences of environmental (patch) heterogeneity in the four different metacommunity perspectives. Patch heterogeneity is, again, a continuous variable, but most metacommunity models describe the environment as either homogeneous or heterogeneous. Patch heterogeneity plays a major role in the models epitomized by SS=species sorting and ME=mass effects, but not the NM=neutral model. Patch dynamic (PD) models focusing on competition/colonization trade-offs in the absence of patch heterogeneity are illustrated with a dashed oval to highlight the fact that most PD models are of this type (Leibold and Chase 2018; see also Chapter 13).

Metacommunities in homogeneous environments

The patch dynamics perspective

We have already encountered the patch dynamics perspective; it was introduced in the latter half of Chapter 13 when we extended metapopulation models to include multiple competing species. There we saw that multiple species can coexist in an environment composed of homogeneous habitat patches when there is a trade-off between dispersal ability and competitive ability (the competition/colonization trade-off). We saw that species can also coexist in a homogeneous environment if there is spatial variation in the density of patches and there is a trade-off between dispersal ability and fecundity (Yu et al. 2004). Thus, as Hutchinson (1951) hypothesized, and as Skellam (1951), Levins and Culver (1971), and others showed theoretically the addition of spatial patchiness to an otherwise homogeneous environment could act to increase biodiversity. However, as also seen in Chapter 13, there are few documented cases of species coexisting via a strict competition/colonization trade-off or a dispersal/fecundity trade-off.

The notion of trade-offs—that is, that no species can be the best at everything (no species can be a

"superspecies," *sensu* Tilman 1982)—is fundamental to all branches of ecology, from life history evolution to functional morphology to nutrient cycling and ecosystem processes. When considered in a community context, trade-offs correspond to niche differences among species (Chase and Leibold 2003a; Kneitel and Chase 2004). For example, a root/shoot trade-off implies that a plant species that is good at competing for soil nutrients may be poor at competing for light (see Chapter 8). Likewise, an activity level/mortality risk trade-off may make a species that is good at avoiding predators poor at finding resources or mates (see Chapter 6). Trade-offs play major roles in species coexistence; we have explored some of those roles in non-spatial models of competition and predation (see Chapters 7 and 8). The mass effects and species-sorting perspectives also recognize the importance of niche differences and trade-offs in promoting species coexistence in spatially heterogeneous environments.

The neutral perspective

The neutral perspective stands in stark contrast to the other three metacommunity perspectives, as it assumes that all individuals, regardless of species and regardless of environmental context, have equal fitness (i.e., equal probabilities of giving birth, dying, migrating, or giving rise to new species). Thus, under the neutral perspective, environmental heterogeneity is irrelevant to the question of species coexistence, since all species experience the environment equally (see Figure 14.1). The idea that species within a community are ecologically equivalent may seem far-fetched, but as we will see later in this chapter, the neutral perspective yields surprisingly good fits to some real-world patterns of biodiversity (Hubbell 2001) and provides a "null model" against which to contrast the effects of niche differences between species.

Metacommunities in heterogeneous environments

Environmental heterogeneity is common in nature, and most species experience a world that is both patchy *and* heterogeneous. The species sorting and mass effects perspectives on metacommunities, both of which incorporate environmental heterogeneity, open up a variety of mechanisms for species coexistence via niche differences and ecological trade-offs.

The species sorting perspective: "Traditional" community ecology in a metacommunity framework

The species sorting perspective shares much in common with what might be termed "traditional" community ecology in that it focuses on the outcome of species interactions occurring within local communities (habitat patches) that differ in their abiotic and biotic properties. Under this perspective, species are assumed to differ in their performance/fitness in different habitat types. Dispersal occurs at a rate that is low enough that the movement of individuals between local communities does not have a direct effect on population abundance within a patch or the outcome of species interactions within local communities (i.e., there are no mass effects), but at a rate high enough so that species are able to reach every locality where they are capable of persisting (Chase et al. 2005). Thus, the species sorting perspective adopts the classic niche-based framework of community ecology, but expands it to consider species diversity at the local scale and in the whole metacommunity. In addition, by allowing for a low rate of species input to local communities, the species sorting perspective permits local communities to track environmental change and potentially respond to changing environments as "complex adaptive systems" (*sensu* Levin 1998; Leibold and Norberg 2004).

We can illustrate how species input to the local community can facilitate "environmental tracking" with an example we encountered back in Chapter 11 when we addressed the question of how trophic-level biomass responds to a change in productivity at the base of the food chain or food web (Leibold 1996 Leibold et al. 2005). Recall that in simple food chains, in which each trophic level is composed of a single species or of a group of species that all respond similarly to food and predators, an environmental change that increases the productivity of the basal resource (e.g., nitrogen deposition in

grasslands or nutrient inputs into lakes) is predicted to increase the biomass of the top trophic level and alternating trophic levels below it (see Figure 11.2). Thus, in a three-trophic-level system, the abundances of predators and plants should increase with an increase in productivity, but the abundance of herbivores should remain constant. Simple experimental food chains show this alternating trophic level response, as do short-term manipulations of productivity in natural systems. However, most natural communities show an increase in the biomass of all trophic levels (producers, herbivores, and predators) when potential productivity is increased. As Leibold (1996) and others have shown, the input of species with differing competitive abilities and differing mortality risks can change the prediction of alternating trophic level responses, leading instead to a positive biomass response by all trophic levels as basal resources are increased (see Figure 11.10B). Thus, the observed positive biomass response of trophic levels in natural communities (see Figure 11.7) may be brought about by changing species composition within a community due to the dispersal of new species from the surrounding metacommunity. Such compositional shifts have been observed in laboratory experiments (Bohannan and Lenski 1997, 1999) and natural systems (Watson et al. 1992; Davis et al. 2010).

How much of the variation in species composition between communities (β diversity) can be attributed to species sorting (e.g., niche differences) and how much is due to spatial process (e.g., dispersal)? One way to address this is by partitioning the variation in species composition into its environmental and spatial components (and their interactions). The species sorting perspective assumes that the composition of local communities is determined primarily by environmental differences between communities and the fit of different species' niches to these environments. The fraction of β diversity attributable to the environment should, therefore, reflect the influence of species sorting, whereas the fraction of β diversity attributable to spatial differences between communities should reflect the influence of dispersal (e.g., the patch dynamics and neutral perspectives on metacommunities). Cottenie (2005) and Soininen (2014) used this variation partitioning approach to assess the influence of species sorting on metacommunity composition and found that 22% and 26% of the variation in species composition, respectively, was explained by the environment when the influence of space was controlled. These and other studies of variation partitioning make a number of important assumptions that can affect how we interpret their results (see discussion in Leibold and Chase 2018). Therefore, it would be presumptive to say at this point that species sorting accounts for a specific fraction (e.g., 20–30%) of the variation in species composition among communities. What we can more safely conclude is that "traditional" (niche-based) community ecology plays a major role in determining the variation in species composition observed between communities in nature (we return to this point in our discussion of the neutral perspective later in this chapter).

The mass effects perspective: diversity patterns in source–sink metacommunities

Like the species sorting perspective, the mass effects perspective includes environmental heterogeneity and differences in species niches (Figure 14.1), but it also allows for the dispersal of individuals from high-quality patches (source habitats, where populations have positive growth rates) to low-quality patches (sink habitats, where population growth rates are negative; see Chapter 12) to have significant effects on population dynamics and species diversity in local communities ("mass effects," *sensu* Smida and Elner 1984). A mathematical model developed by Mouquet and Loreau (2003) nicely shows how the amount of dispersal between local communities may affect the three components of species diversity: alpha (within-community) diversity, beta (between-community) diversity, and gamma (regional) diversity (see also Loreau 2010a). In their model, Mouquet and Loreau make the following assumptions:

1. Each local community contains up to S species found in the region, which compete for a limited proportion of vacant patches.
2. The metacommunity consists of N local communities that differ in their environmental conditions. Species are differentially adapted to these local conditions (i.e., due to niche differences), and

resulting in higher reproductive output in those local communities to which they are best adapted.
3. At the regional scale, a constant proportion (a) of each local population disperses between communities. This proportion represents the fraction of the local reproductive output that emigrates. In the model, the dispersing fraction a is assumed to be equal for all species.

Using this model, Mouquet and Loreau (2003) ran numerical simulations for a system of 20 species competing in 20 local communities until the system reached equilibrium. They then measured alpha, beta, and gamma diversity. Figure 14.2 demonstrates how these three diversity components vary as a function of the fraction of individuals dispersing between communities (a). At zero dispersal, each local community is dominated by the best local competitor; thus, alpha diversity is at a minimum (one species per local community), while gamma and beta diversities are at their maximum.

As the proportion of dispersing individuals increases, alpha diversity increases rapidly, the result of species being rescued from competitive exclusion in local communities by immigration from other local communities where they were dominant (a mass effect). Beta diversity decreases with increasing dispersal because local communities become more similar in their composition, whereas gamma diversity remains constant because all species remain in the metacommunity. As the fraction of individuals dispersing increases above a_{max}, local diversity decreases because the best competitor at the scale of the region comes to dominate each community and other species are progressively excluded. Finally, when dispersal is very high, the metacommunity functions as a single large community in which the regionally best competitor excludes all other species and local and regional diversities are at their minimums (Mouquet and Loreau 2003). Note, too, how variation in the proportion of individuals dispersing from the local community affects the theoretical relationship between regional and local species richness (Figure 14.3).

Empirical studies provide general support for Mouquet and Loreau's predictions. For example, the diversity of aquatic invertebrates living in rock

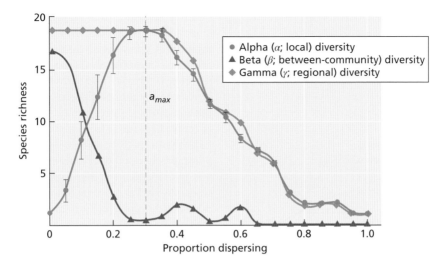

Figure 14.2 Simulation results from Mouquet and Loreau's model of source-sink metacommunities shows how species richness at different spatial scales (alpha, beta, and gamma diversity) should vary as a function of the proportion the local reproductive output that disperses between communities. Local species diversity is maximized at a_{max}. After Mouquet and Loreau (2003).

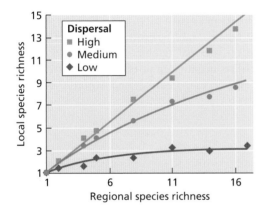

Figure 14.3 The expected relationship between local and regional species richness at different degrees of dispersal between communities in Mouquet and Loreau's metacommunity model. When dispersal between communities is relatively high, local richness is proportional to regional richness (a linear function). As dispersal decreases, the local-to-regional richness relationship becomes curvilinear and saturating due to competitive exclusion within communities. After Mouquet and Loreau (2003).

NB: the "high" dispersal case in this figure is $\alpha = 0.1$ which is still well below the value of α_{max} in Figure 14.2.

pools in South Africa varied with pool isolation (the average distance separating pools; Vanschoenwinkel et al. 2007). The within-pool (alpha) richnesses of all invertebrate taxa and of taxa that disperse passively were highest at intermediate levels of pool isolation, corresponding to intermediate levels of dispersal between pools. However, actively dispersing species, with their potentially high rates of dispersal between pools, showed no relationship between species richness and isolation distance (Figure 14.4).

Experiments with invertebrate communities living inside the leaves of pitcher plants also support the hypothesis that local species richness is maximized at intermediate levels of dispersal. Kneitel and Miller (2003) manipulated the dispersal rates of protozoans inhabiting pitcher plant leaves by withdrawing an aliquot of the water from each of five pitchers, pooling the five aliquots, and then adding back a portion of the pooled mixture to each pitcher. They performed this procedure weekly (low-dispersal treatment) and biweekly (high-dispersal treatment); a no-dispersal treatment was included as a control. The results show protozoan species richness within pitcher plant leaves to be highest at the low dispersal rate—that is, the intermediate condition, which is consistent with theoretical predictions (see Figure 14.2). They found, however, that dispersal rate had no effect on species richness when predaceous mosquito larvae were included in the community (Figure 14.5). Similar results were observed by Howeth and Leibold (2010), who studied zooplankton species diversity in experimental pond metacommunities with and without predation by bluegill sunfish.

A laboratory study that manipulated dispersal rates of benthic microalgae also found that α diversity peaked at intermediate dispersal frequencies

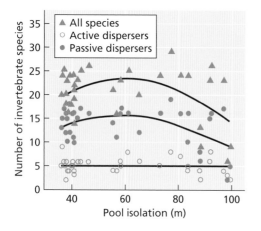

Figure 14.4 Studies of invertebrate species richness in South African rock pools provide qualitative support for Mouquet and Loreau's model, showing that local species richness peaks at intermediate levels of pool isolation (measured as the average distance separating pools) for passively dispersing species and for all species combined. After Vanschoenwinkel et al. (2007).

(Matthiessen and Hillebrand 2006), whereas two laboratory experiments with protozoan metacommunities (Cadotte et al. 2006a; Davies et al. 2009) and a field mesocosm experiment with pond invertebrates (Pedruski and Arnott 2011) found that increasing the rate of dispersal between communities increased alpha diversity. Although a positive effect of dispersal rate on alpha diversity is inconsistent with the prediction of a hump-shaped relationship between those parameters (see Figure 14.2), it is possible that the dispersal rates used in these studies revealed only the left-hand portion of the predicted hump. The authors of the experimental studies attributed the increase in alpha diversity with increasing dispersal rate to mass effects and the maintenance of species in sink habitats via continued dispersal, although more complicated effects of dispersal on species dynamics are possible (Fox 2007). Experiments also show that beta diversity tends to decrease with increased dispersal rate (e.g., Warren 1996; Matthiessen and Hillebrand 2006; Pedruski and Arnott 2011), as predicted by the model (see Figure 14.2), whereas the effects of dispersal rate on regional (gamma) diversity are variable (positive, negative, or no effect have been observed; see references in Pedruski and Arnott 2011).

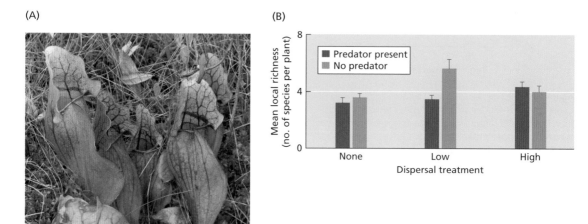

Figure 14.5 (A) The highly modified leaves of *Sarracenia purpurea* contain a characteristic community of invertebrates (e.g., protozoans, rotifers, insect larvae), many of which are obligate pitcher plant dwellers. (B) In the absence of predatory mosquito larvae in the community, mean species richness within a given pitcher plant ("mean local richness") peaked when dispersal was intermediate (the "low dispersal" treatment). When predators were present, however, the frequency of dispersal had no effect on local species richness. Error bars are ±1 SE. (A) Courtesy of David McIntyre; (B) after Kneitel and Miller (2003).

Measuring dispersal in metacommunities

As we have seen, dispersal rates play fundamental roles in metacommunity dynamics: they determine patterns of local and regional diversity (e.g., Figure 14.2; Mouquet and Loreau 2003), and they distinguish the relative impacts of species-sorting vs mass effects on community dynamics and composition. Dispersal, however, is a notoriously difficult process to study (Jacobson and Peres-Neto 2010). Consequently, we have relatively little information on actual rates of movement of individuals and species between communities. For some taxa, dispersal rates can be estimated by directly observing their rates of colonization of new habitats. For example, Hanly and Mittelbach (2017) measured the rate of dispersal of 179 plankton taxa by placing water-filled containers (50 L) near each of 10 freshwater ponds in Michigan and compared the measured dispersal rates to the abundance of the plankton taxa in the nearby pond. For most taxa, dispersal rates into the experimental containers were positively related to taxon abundance in the pond. Also, the assembly of plankton communities in experimental containers that were initially devoid of species was strongly related to dispersal from the local pond. However, dispersal had no measurable impact on the species composition of pre-assembled mesocosm communities. Studies of dispersal at larger landscape scales using newly created freshwater ponds similarly show surprisingly high rates of movement of passively dispersed organisms such as zooplankton and considerable variation in dispersal ability among species (Jenkins and Buikema 1998; Louette and De Meester 2007; Louette et al. 2008).

An alternative to directly observing dispersal is to measure some proxy that correlates with dispersal. For example, population geneticists have long used estimates of genetic dissimilarity between spatially separated populations ("isolation by distance") to estimate rates of gene flow (Slatkin 1993). The idea here is that the greater the rate of gene flow between populations, the more similar they should be in genetic composition. Likewise, ecologists have begun to measure the degree of dissimilarity in species composition between local communities to estimate rates of species dispersal (Shurin et al.

2009). Species that are relatively poor dispersers are expected to show more clumped spatial distributions or greater spatial autocorrelation in community composition than good dispersers do. Studies employing measures of community dissimilarity with distance (i.e., the **rate of distance decay of similarity**; Nekola and White 1999; Soininen et al. 2007) have documented dispersal limitation in a variety of systems, including tropical and temperate forests (Tuomisto et al. 2003; Gilbert and Lechowicz 2004) and freshwater lakes and ponds (Beisner et al. 2006; Shurin et al. 2009). In a few cases, they have compared these measures for different taxa in the same ecosystem (e.g., Beisner et al. 2006; Astorga et al. 2012). In addition, by regressing community similarity against environmental and spatial distance, one can estimate the relative effects of local environmental variation (and species-sorting) vs dispersal limitation in determining the change in community species composition with distance (e.g., Tuomisto et al. 2003; Green et al. 2004). Ecologists are beginning to elucidate differences in dispersal tendencies related to species traits (e.g., body size; Beisner et al. 2006; Shurin et al. 2009). However, we are still a long way from knowing the dispersal rates of a significant fraction of the constituent species for any given metacommunity.

An insightful approach for empirically estimating the impact of species dispersal on community structure and metacommunity dynamics was employed by Luc De Meester, Karl Cottenie, and their colleagues in a study of a series of 34 interconnected ponds in Belgium. The authors estimated that an average of 3600 zooplankton individuals per hour dispersed though the overflows and rivulets that connect each pair of ponds in this system, a rate corresponding to a population turnover time of 13 days (Michels et al. 2001). Despite this relatively high rate of species dispersal, the ponds exhibited distinct plankton communities that were predictably related to the local environment, particularly to the presence or absence of fish and macrophytes (Cottenie et al. 2003; Cottenie and De Meester 2003; Vanormelingen et al. 2008). Moreover, these strong and predictable influences of the biotic environment on zooplankton community structure were confirmed via a transplant experiment (Cottenie and De Meester 2004). Thus, this metacommunity showed the clear impact of species sorting, despite relatively high rates of species dispersal. The effects of dispersal, however, were not insignificant. Dispersal increased the species richness of cladoceran zooplankton by an average of three species (a mass effect), and dispersal made the response of the community to the local environment more predictable (i.e., the local communities were acting as adaptive systems).

The neutral perspective

Ecologists have long focused on the importance of niche differences in maintaining biodiversity. Niche differences not only allow species to coexist (thus maintaining biodiversity), but their proliferation, through the processes of character displacement and adaptive radiation, is hypothesized to be the principal generator of species diversity over evolutionary time (see Chapter 16). As Schluter (2000, p. 2) explains, "Adaptive radiation is the evolution of ecological diversity within a rapidly multiplying lineage. It is the differentiation of a single ancestor into an array of species that inhabit a variety of environments and that differ in traits used to exploit those environments."

Thus, the idea that species differ in the ways in which they use the environment (i.e., in their niches) seems fundamental to community ecology and to the evolution and maintenance of species diversity. That is why, when Steve Hubbell's *Unified Neutral Theory of Biodiversity and Biogeography* was published in 2001, it created quite a stir. In Hubbell's vision, niches are absent and all species are functionally equivalent. Stated more formally, species are assumed to be demographically identical on a per capita basis in terms of their vital rates of birth, death, and dispersal regardless of environmental context (Hubbell 2006).

Before concluding that Hubbell is some harebrained theoretician, who has never set foot in the field or seen the variety of life firsthand, we need to realize that he has a long history of working in one of the most diverse ecosystems in the world—the tropical rainforest. Hubbell was among the first to recognize the value of mapping all the trees in a tropical forest, and he did this (along with Robin Foster and other colleagues) in a 50-ha forest plot on Barro Colorado Island in Panama (Figure 14.6). This plot contained about 300 species of trees and about 250,000 individuals with a stem diameter of at least 1 cm at breast height (Hubbell and Foster 1983; Condit 1998). Since that time (1980), dozens of other 50-ha forest plots from around the world have been mapped, yielding a wealth of new information (see the Center for Tropical Forest Science website, www.ctfs.si.edu).

The evolution and maintenance of high species diversity in tropical wet forests is an enduring puzzle to ecologists (see Chapter 2). It seems implausible that the 300 or so tree species on Barro Colorado Island could occupy 300 different niches. How can there be that many ways for trees to differ in habitat, resource use, etc.? Rather than searching for subtle niche differences between tropical tree species, Hubbell took exactly the opposite approach. He wondered instead whether tree diversity might be maintained if the species were functionally equivalent; that is, if their interactions were neutral to any differences between species. This shift is very similar to what happened to population genetics in the 1960s, when the application of electrophoresis revealed an unexpectedly high degree of allele polymorphism. Many population geneticists concluded that observed levels of allele polymorphism were too high to be maintained by natural selection. Therefore, Kimura (1968) and others reached the conclusion that most alleles are selectively neutral—mutations arise that are neutral in their effects, and the frequencies of such mutations in a population fluctuate at random. In the ensuing years, a large body of theory developed around the idea of neutral molecular evolution, and the idea of a "molecular clock," which proposes that the rate of molecular evolution is constant, was one of the outcomes (Zuckerkandl and Pauling 1965). The neutral theory of biodiversity (Bell 2001; Hubbell 2001) is analogous to the neutral theory of population genetics; Leigh (1999, 2007) and Nee (2005) provide insightful discussions of the parallels between the two theories.

Assumptions of the neutral theory

In their initial development of the neutral theory of biodiversity, Hubbell and Foster (1986) proposed a model for tropical forests based on disturbance and

(A)

(B)

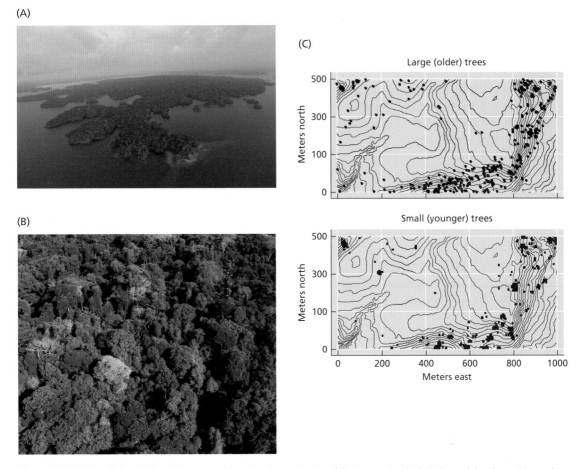

(C)

Figure 14.6 (A) Barro Colorado Island, Panama, was formed by the construction of the Panama Canal, which dammed the Chagres River and created Gatun Lake. (B) Scientists have mapped and repeatedly censused all woody plants larger than 1 cm in diameter at breast height in the Forest Dynamics Plot, a 50-ha experimental plot of tropical rainforest on Barro Colorado Island. (C) Mapped distributions of small and large trees of the species *Ocotea whitei* in the Forest Dynamics Plot. The distribution of *O. whitei* is significantly associated with slope at both life stages. Photographs © Christian Ziegler/Minden Pictures; (C) from Comita et al. 2007.)

subsequent recolonization by species with similar competitive abilities. Their model makes several assumptions:

1. The number of individuals in the community is constant. Space is a limiting resource, and all space is occupied. Therefore, gains in abundance by one species must be balanced by losses in the abundances of other species. Biodiversity is a "zero-sum game."

2. All individuals and species have an equal probability of colonizing open space. The colonizer of a site vacated by death (disturbance) is a random draw from the individuals present, making the probability of a given species being selected equal to a species' relative abundance.

3. Death occurs at a constant and fixed rate.

Under these assumptions, the long-term (equilibrium) outcome of competition in a local community is a random walk to dominance by a single species. A population geneticist would call this fixation to a single species via drift. However, if the community contains a large number of individuals that die at a relatively slow rate, this random walk to single-species dominance can take a very long time (in fact, maybe too long a time, a point we will return to later). Hubbell and Foster argued that, although their model did not result in the equilibrium coexistence of species, the time over which species are maintained in the local community may be as long as the time it takes for new species to arise via speciation.

In his book, Hubbell (2001) extended the neutral model presented in Hubbell and Foster (1986) to include local communities that are connected by the immigration of species to a much larger metacommunity. This was a significant advance. In Hubbell's words (2001, p. 5), the metacommunity contains "all trophically similar individuals and species in a regional collection of local communities." Each local community, as in the original Hubbell and Foster model, consists of J total individuals whose offspring compete for sites that become available when an individual dies. Every individual (regardless of species) in the local community has an equal probability of colonizing an open site (the "neutrality" assumption). In addition, all open sites are assumed to be filled, and community size is constant (the "zero-sum" assumption). Furthermore, offspring from the local community compete for sites not only with other individuals from the local community, but also with immigrants from the metacommunity. The parameter m defines the probability that an open site will be colonized by an immigrant from the metacommunity. Dispersal limitation occurs when $m < 1$. The net effect of immigration from the metacommunity is the same as that of modelling multiple patches in a metapopulation. Species that are absent from the local community can colonize and establish (or re-establish) there. This constant introduction of new species counterbalances the random walk to fixation to a single species that occurs if we model only the local community without a metacommunity. Thus, local species richness is the outcome of a balance between extinctions due to random drift (fixation at zero) and species introductions.

Meanwhile, in the metacommunity, a random walk to extinction is avoided when new species arises via speciation. Thus, the metacommunity provides a pool of species at a regional level; the dispersal of these species into the local community may counteract the steady loss of species through random drift. The total number of individuals in the metacommunity (J_M) is assumed to be large and

constant, and the number of species in the metacommunity is in stochastic balance between the rate of extinction and the rate of speciation.

Hubbell (2001) considers two speciation mechanisms, "point mutation" and "random fission." In the "point mutation" model (to which Hubbell devotes most of his analysis), ν (Greek nu) is the per capita speciation rate (the probability of a speciation event per individual in the metacommunity). Hubbell then goes on to define θ, the rate at which new species appear in the metacommunity, as

$$\theta = 2\rho A_M \nu$$

where ρ is the mean number of individuals per unit area in the metacommunity and A_M is the area of the metacommunity. Thus θ, which Hubbell refers to as the "fundamental biodiversity number," is a compound parameter composed of metacommunity size (the total number of individuals in the metacommunity, $N = \rho A_M$) times ν (the per capita speciation rate). For species with discrete generations, $\theta = 2N\nu$; for species with overlapping generations, $\theta = N\nu$ (Leigh 1999).

Testing the predictions of the neutral theory

Had Hubbell proposed his neutral theory without linking it to real-world data, it probably would have had little impact. Indeed, neutral models for communities were not new (e.g., see Caswell 1976). However, one of the strengths of Hubbell's theory is that it makes specific predictions about patterns of biodiversity that can be compared with data from natural systems. Most notably, his model predicts the expected distribution of species abundances in both the local community and the metacommunity. As noted in Chapter 2, species abundance distributions (SADs) were one of the first biodiversity patterns examined by ecologists (Fisher et al. 1943; Preston 1948). However, until the publication of Hubbell's book, niche-based theories of species diversity had been strangely silent about the predicted forms of SADs (except for a brief flurry of activity in the 1960s; see Chapter 2). Hubbell's theory, assuming speciation by point mutation, predicts that the SAD for a metacommunity will correspond to Fisher's log series distribution. In fact, θ is equivalent to the parameter α in Fisher's log series. For a local community that receives species from the metacommunity at a migration rate of m, the SAD is similar to a lognormal distribution (see Magurran 2005a for an excellent discussion of SADs and the neutral theory).

Hubbell (2001; see also Volkov et al. 2003) examined SADs from a number of local communities, particularly those from tropical trees on Barro Colorado Island in Panama (Figure 14.7) and from birds in Great Britain. He found that the neutral model predicted those SADs very well. However, other researchers soon responded, showing that other models (including those that focus specifically on niche differentiation) also successfully predict observed SADs (i.e., Chave et al. 2002; McGill 2003). After much analysis, the bottom line is that a variety of models fit the data on SADs almost equally well (Chave 2004; McGill et al. 2006). Further, McGill et al. (2006, p. 1415) argues that "the (neutral theory) is unusually flexible in its ability to fit SADs...because each parameter is independent. J sets the scale, θ controls the shape to the right of the mode and m

controls the shape to the left of the mode." Therefore, although Hubbell's initial focus on predicting SADs is laudable (SADs were too long forgotten by community ecologists), it is now clear that looking at goodness of fits to empirical data will not allow us to distinguish among competing theories of biodiversity.

One of the strengths of the neutral theory is that it makes quantitative and testable predictions about species abundances in space and time. If the abundance of each species within a community drifts randomly through time, then we can specify the expected difference in species composition between communities separated in space (beta diversity), and we can specify how long it should take species, on average, to change in abundance within a community (including the time to extinction and the time to become common after invasion). Ecologists have examined both of these predictions in detail. For example, several studies have compared the species compositions of communities separated in space in tropical forests (Terborgh et al. 1996; Condit

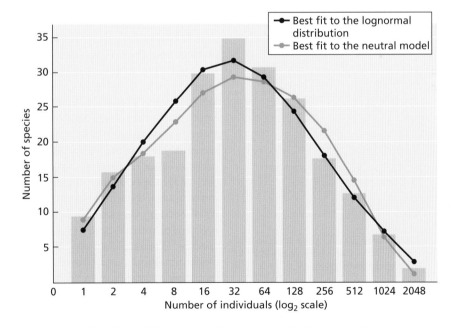

Figure 14.7 Hubbell's neutral model and Preston's lognormal model both provide excellent fits to the empirical data on the species abundance distribution of tropical trees in the Forest Dynamics Plot on Barro Colorado Island (see Figure 14.6). The bars are observed numbers of tree species, binned in $\log_2$ abundance classes (meaning that species abundance doubles in each class). After Volkov et al. (2003). See http://communityecologybook.org/neutral.html for an online dynamic model of the neutral theory.

et al. 2002) and in temperate forests (Gilbert and Lechowicz 2004). In each case, the authors found that species turnover over short distances was greater than that predicted by neutral theory, as was the number of widespread species shared among communities. Working at smaller spatial scales, Wootton (2005) devised an experimental test of the predictions of the neutral theory and applied it to the community of sessile organisms found in the marine rocky intertidal zone on Tatoosh Island, off the coast of Washington State. He found, as have others, that the neutral theory provided a good fit to the general shape of the species abundance distribution in the community. However, when Wootton tried to use the theory to explain the dynamics of individual species over the 7–11 years' duration of the experiment, it did a poor job of predicting the observed community response (Figure 14.8). In another test of the neutral theory's predictions at small spatial scales, Adler (2004) compared patterns of species abundances among plots in a Kansas grassland, looking at the joint effects of space and time, and found that the neutral theory failed to explain the observed patterns.

Ricklefs (2006) argues that the most unambiguous tests of Hubbell's models address time itself.

Population geneticists long ago worked out the time required for a neutral mutation to increase or decrease to a certain level within a population of size N (Fisher 1930; Kimura and Ohta 1969). Drift is a slow process, and the larger the population, the longer it takes. Ecologists and evolutionary biologists have applied the theories of population genetics to estimate the expected life spans of the most common species in a community, the expected time to extinction, and the time to community equilibrium under the assumptions of the neutral model. In all cases, the predicted time scales are inconsistent with those in real communities—species come to dominance, and species disappear, much more quickly in nature than predicted based on a theory of neutral dynamics (Leigh 1999, 2007; Lande et al. 2003; Nee 2005; Ricklefs 2006; McGill et al 2005). For example, estimates based on neutral dynamics would place the ages of common tropical tree species back before the origin of the angiosperms (Nee 2005), and would make the time to extinction for abundant species such as oceanic plankton or tropical insects or trees greater than the age of Earth (Lande et al. 2003).

Proponents of the neutral theory recognize these problems and have proposed a modification to the original neutral theory whereby speciation in the metacommunity occurs via a protracted process that produces a recognizable new species only after a transition period of some number of generations (Rosindell et al. 2010). By introducing a lag time between the origination of incipient species from a single ancestor and the actual formation of recognizable species, Rosindell et al. were able to produce more realistic speciation rates and numbers of rare species than the original formulation of the neutral theory. This modification to the neutral theory, however, doesn't completely solve the problem of unrealistically great ages for common species (Rosindell et al. 2010). In addition, Desjardins-Proulx and Gravel (2012) showed that modeling speciation as a more explicit allopatric or parapatric process within the neutral theory severely limits the number of species that can be supported. Thus, the neutral theory's incompatibility with realistic modes and rates of speciation presents a fundamental challenge to its ability to explain biodiversity.

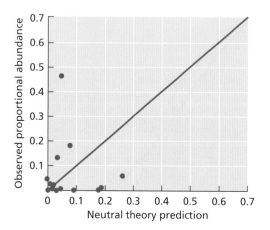

Figure 14.8 A comparison between species-specific predictions of the neutral model parameterized for sessile marine species in the rocky intertidal on Tatoosh Island, Washington State, and the actual proportional abundances of species as observed in experimental plots. The diagonal line is the expected relationship for a perfect fit of experimental data to model predictions. After Wootton (2005).

The value of the neutral theory

You may wonder why we have devoted so much of this chapter to Hubbell's neutral theory of biodiversity, given that the weight of empirical evidence is against it. There are at least four reasons why the neutral theory is worth studying in detail:

1. The theory has been enormously influential; it has caused ecologists to re-examine areas of ecology that were almost forgotten (e.g., SADs). It will continue to play a major role in community ecology for years to come, as a null model if nothing else (Gotelli and McGill 2006; Leigh 2007).
2. The neutral theory is mathematically elegant and tractable. Because of this, it can be applied to many problems in ecology outside the reach of current niche theory. Thus, for these problems, we can see what the solution would look like if species were "ecologically neutral."
3. A number of ecologists have suggested that real communities represent a continuum of niche and neutral interactions; that is, that communities are composed of a mix of species, some with niche-based dynamics and some with neutral dynamics (e.g., Gravel et al. 2006; Leibold and McPeek 2006; Adler et al. 2007; Chase 2007; McPeek 2008; Ellwood et al. 2009).
4. Because it brushes aside species differences, neutral theory focuses our direct attention on how dispersal and regional species pools may determine local community structure.

Interestingly, work by Zillio and Condit (2007) and Chisholm and Pacala (2010) suggests that the assumption of "neutrality" itself may not be the dominant force behind the success of the neutral theory in explaining SADs. McGill and Nekola (2010) reach a similar conclusion, suggesting that of the three most important aspects of the neutral theory of biodiversity, neutrality appears dispensable, while dispersal limitation and the input of species from the metacommunity (regional replacement) appear to be critical in determining SADs for local communities.

While the neutral theory is fundamentally flawed in assuming that all species are ecologically equivalent, it has pointed us in new directions, it has highlighted critical processes for the maintenance of biodiversity, and it has invigorated our science. As Hubbell (2006, p. 1388) noted, "Perhaps the [unified neutral theory's] most significant and lasting contribution will be its explicit incorporation of linkages between the ecological processes of community assembly on local scales and evolutionary and biogeographic process on large scales, such as speciation and phylogeography." Or, as Ricklefs (2006, p. 1430) put it, "Except for its abandonment of ecology, Hubbell's view of the world incorporates much of the regional and historical perspective that I would advocate. The fact that many ecologists have been attracted to this idea gives hope that the discipline is ready to embrace a more comprehensive concept of the species composition of ecological systems."

Niche-based and neutral processes in communities

To what extent does species identity matter in the dynamics of communities? Neutral theory says that it does not matter. Niche-based theory says it does. Ecologists are just beginning to address this question, which has close linkages to the notion of coexistence mechanisms discussed in Chapter 8. Specifically, in a neutral, niche-free world coexistence is enhanced, made possible for extended periods of time, but not made permanent by equalizing effects, where the fitness of species are very similar (or exactly identical in neutral theory). In contrast, niche theory invokes stabilizing effects whereby a species has stronger density-dependent regulation from individuals of its own species than other species. Although both niche and neutral can lead to coexistence for extended periods of time, only niches can lead to permanent (equilibrial) coexistence.

A simple experimental design to discriminate between niche-based and neutral processes is to manipulate both the total and relative abundances of co-occurring species occupying the same trophic level (i.e., potential competitors). If neutral mechanisms govern the dynamics of co-occurring species, then we would expect each species' performance to be equal and to depend on total abundance across all

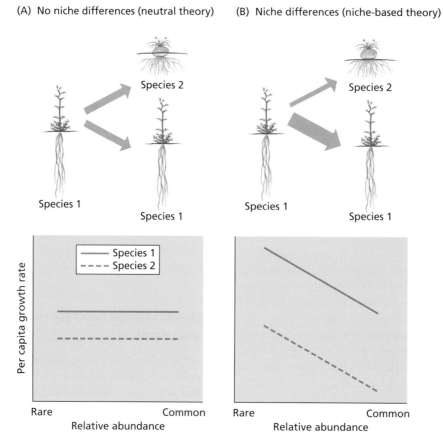

Figure 14.9 Niche differences (such as differences in rooting depths in plants) cause species to limit their own growth more than the growth of other species (size of arrows in upper drawings indicate strength of negative competitive effects on growth). If there are no niche differences, species limit themselves and other species equally. (A) When niche differences are absent, a species' relative abundance has no effect on its per capita growth rate. (B) When niche differences are present, a species' per capita growth rate declines with an increase in its relative abundance in the community. After Levine and HilleRisLambers (2009).

species, but not on its own specific abundance. That is, all species should show similar density-dependent responses to manipulations of total abundance, but no response to manipulations of relative abundances when total abundance is held constant (Figure 14.9A). On the other hand, if niche-based mechanisms govern the dynamics of co-occurring species, we would expect each species to respond strongly to changes in its relative abundance (Figure 14.9B). Species responses to density manipulations are most appropriately measured as changes in per capita population growth rates, although other measures correlated with per capita population growth rate are also informative. This simple approach

allows us to assess whether niche-based or neutral processes are operating in a community, although it does not distinguish their relative importance (Adler et al. 2007; Levine and HilleRisLambers 2009; Siepielski et al. 2010).

Mark McPeek and colleagues used this experimental approach to explore niche-based and neutral mechanisms of species coexistence in damselflies, focusing on interactions at the larval stage. Larval damselflies of the genus *Enallagma* are generalist predators that live in the shallow waters of ponds and lakes. In North America, the damselfly genus *Enallagma* (38 species) differs ecologically from its sister genus *Ischnura* (14 species). Larval *Enallagma* are better at avoiding predation than are larval *Ischnura*, but *Ischnura* larvae are better at converting prey into their own biomass than are *Enallagma* (McPeek 1998). Thus, there are pronounced niche differences between the genera that promote coexistence at this level. Within the genus, species are very similar: all of them spend 10–11 months of the year as aquatic larvae living in the vegetation of lakes. From 5 to 12 *Enallagma* species may co-occur in lakes throughout eastern North America (Johnson and Crowley 1980; McPeek 1998). Negative density-dependent feedbacks are well documented at the larval stage—*Enallagma* larvae experience density-dependent growth rates due to competition and density-dependent predation by fish and invertebrate predators (McPeek 1998). Do these density-dependent feedbacks act to regulate individual species in a stabilizing fashion?

McPeek and Siepielski (Siepielski et al. 2010) addressed this question by experimentally manipulating the absolute and relative abundances of two *Enallagma* species in the field. They found that the species responded strongly to manipulations of total abundance, but showed no responses to changes in their relative abundances when total abundance was held constant (the responses measured were larval per capita mortality and growth rates over a 3-month period). Furthermore, in a survey of 20 lakes located close to one another, they found that per capita mortality and growth rates of multiple *Enallagma* species were not correlated with the species' relative abundances in a lake, and that the pattern of co-occurrence of species among lakes did not differ from what would be expected by chance. Thus, Siepielski et al. (2010) concluded that within the genus *Enallagma*, the species appear to be ecologically equivalent.

A problem with explaining diversity via equalizing mechanisms alone, however, is the question of how ecologically similar species arise and become established if they have no competitive advantage when rare. The evidence suggests that *Enallagma* diversification may not have involved the invasion of intact communities by rare species. Much of the diversification in the genus *Enallagma* occurred during periods of repeated glaciation during the Quaternary, when the landscape was fragmented into isolated lake communities and species may have

arisen and expanded their ranges into relatively unoccupied lakes (Turgeon et al. 2005). Moreover, it appears that speciation in this damselfly group was driven primarily by sexual selection for differentiation in reproductive structures, with little ecological differentiation among species (McPeek et al. 2008). Thus, average fitness equality and very long times to competitive exclusion (helped by large population sizes and low levels of dispersal between lakes) may explain much of the diversity in *Enallagma* in the absence of strong niche differences and little evidence for stabilizing mechanisms (McPeek 2008; see also see Chapter 8 for discussion of how fitness equalization can delay extinction times, but not create a permanent stable equilibrium).

In contrast to the damselflies, two studies of plant communities showed that the relative abundances of species with differences in average fitness were maintained by stabilizing mechanisms. Levine and HilleRisLambers (2009) constructed an experimental community of annual plant species that occur naturally on serpentine soils in California. They varied the relative abundances of ten focal species in field plots while keeping the total abundance of all plants in the plots constant. If niche properties stabilize coexistence, then we would expect each species' population growth rate to decrease with an increase in its relative abundance. However, if species identity does not matter and the system dynamics are neutral, then a species' growth rate should not change with a change in its relative abundance as long as total abundance remains constant. The most abundant species in the annual plant community showed strong declines in population growth with increasing relative abundance, demonstrating strong stabilizing mechanisms at work. Other species showed weaker declines in growth rates with increased relative abundance, whereas the growth rates of the three rarest species actually increased with an increase in relative abundance. The authors concluded that niche differences stabilized species diversity, especially for the dominant species in the community.

Levine and HilleRisLambers (2009) and Adler et al. (2010) also tested for niche-based and neutral mechanisms in natural plant communities by fitting demographic models to species' observed population dynamics and then using these models to look at the predicted response of the community when stabilizing mechanisms were removed. Both studies found that removing the effects of stabilizing mechanisms caused diversity within the communities to decline rapidly. In the serpentine plant community studied by Levine and HilleRisLambers (2009), species diversity (measured as a function of species richness and species evenness, or Shannon's index of diversity) decreased after only two simulated generations when niche-based stabilizing mechanisms were removed (Figure 14.10). Stabilizing mechanisms were also prevalent in the sagebrush steppe community studied by Adler et al. (2010). In this study, the authors analyzed spatially explicit models fit to 22 years of demographic data to:

- quantify the stabilizing effect of niche differences;
- compare the relative strengths of stabilizing mechanisms and fitness differences;

- partition the stabilizing mechanisms into fluctuation-dependent and fluctuation-independent processes.

They found that average fitness differences between species were small and that stabilizing mechanisms were much larger than required to overcome the fitness differences. They also showed that fluctuation-dependent mechanisms were less important than fluctuation-independent mechanisms in this community. Thus, Adler et al. (2010) concluded that niche differences are crucial to maintaining species diversity in this community.

The demographic modeling approach used by Levine and HilleRis-Lambers (2009) and Adler et al. (2010) is a powerful tool for quantifying the relative importance of niche-based and neutral mechanisms in maintaining biodiversity in communities. However, it does not identify the niche differences

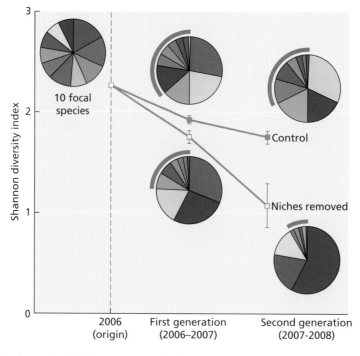

Figure 14.10 Two simulated generations (2006–2007, 2007–2008) of change in the diversity and composition of a serpentine plant community stabilized by niche differences ("control") vs one in which the demographic influence of niche differences was removed. Pie charts show the average proportion of total community seed mass for each of 10 focal species in each treatment and generation. Arcs at the edges mark the collective abundances of the seven rarest species. Species diversity (data points) was measured as a function of species richness and evenness, or Shannon's diversity index (mean ± 1SE). Diversity decreased after only two simulated generations when niche-based stabilizing mechanisms were removed. After Levine and HilleRisLambers (2009).

between species that promote coexistence (e.g., resource partitioning, frequency-dependent predation, the storage effect). Instead, it evaluates the collective importance of multiple coexistence mechanisms acting in concert and points to whether further investigation into those mechanisms would be worthwhile. Clark (2010), reasoning along somewhat different lines, suggests that niche differences acting along many axes ("high niche dimensionality") are, in fact, crucial for the coexistence of diverse communities such as forest trees (see also Clark et al. 2007). Using 6–18 years of data on tree dynamics in 11 forests in the southeastern United States, Clark modeled the responses of individual trees and tree species to environmental variation. He concluded that individuals responded in different ways to environmental fluctuations, depending on fine-scale variation in resources (e.g., moisture, nutrients, and light) and their genotypic differences. However, because individuals of the same species responded more similarly to one another than to individuals of different species, intraspecific competition was stronger than interspecific competition—the hallmark of stabilizing mechanisms of coexistence (see further discussion in Chesson and Rees 2007; Clark and Agarwal 2007).

Conclusion

By viewing communities as interconnected entities within a landscape, metacommunity ecology focuses our attention on the importance of dispersal, regional species pools, and demographic stochasticity (in the case of neutral theory) in determining species composition and species abundances within communities. These exciting developments will move us toward a better understanding of how biodiversity and community structure are determined at both local and regional scales. However, while the theory of spatial ecology marches forward at a heady pace, empirical studies lag behind; Amarasekare (2003, p. 1109) notes that, while "theory is in advance of data in most areas of ecology, nowhere is it more apparent than in spatial ecology." This is not surprising; there are enormous challenges in quantifying the important ecological processes that occur across spatial scales (e.g., dispersal rates) and in linking theoretical concepts such as "patch" and "local

community" to real-world entities that we can define and study. Despite these challenges, great progress has been made in a very short time.

Summary

1. A metacommunity is a set of local communities linked by dispersal of one or more of their constituent species.
2. Ecologists have studied metacommunities from several perspectives. Differences between these perspectives hinge on the degree of environmental heterogeneity and species niche differences found in the metacommunity.
3. The patch dynamics perspective extends metapopulation models (Chapter 12) of species dynamics in a system of homogeneous patch types to include more than two species.
4. The mass effects perspective applies a multispecies version of source–sink dynamics and rescue effects in a system of heterogeneous patch types. A model of source–sink metacommunities devised by Mouquet and Loreau shows that the rate of dispersal between local communities may affect diversity at multiple spatial scales. At a dispersal rate of zero, each local community is dominated by the best local competitor, and gamma and beta diversities are at their maximum. As dispersal rate increases, alpha diversity increases rapidly as species are rescued from competitive exclusion by immigration. Beta diversity decreases, because local communities become more similar in their composition, whereas gamma diversity remains constant. When the dispersal rate is very high, the metacommunity functions as a single large community in which the regionally best competitor excludes all other species. Empirical studies provide general support for the predictions of Mouquet and Loreau's model.
5. The species-sorting perspective emphasizes differences in species' abilities to utilize different patch types in a heterogeneous environment. Like the mass effects perspective, species-sorting assumes a heterogeneous environment and interspecific niche differences. It also assumes dispersal occurs at a low rate, such that dispersal has no direct effect on population abundances or

the outcome of species interactions within local communities (i.e., there are no mass effects), but dispersal rate is high enough that all species are able to reach every locality where they are capable of persisting.

6. A low, but non-zero rate of dispersal potentially allows communities to track environmental change and to respond to changing environments as "complex adaptive systems." The metacommunity provides a source of renewal (and new species) to local communities, and its importance for sustaining ecosystem function in changing environments may be large.

7. The rate of dispersal between local communities is a key parameter in all metacommunity models, but is difficult to measure in nature. For some taxa, dispersal rates can be estimated by observing their rates of colonization of new habitats. An alternative to directly observing dispersal is to measure some proxy that correlates with dispersal, such as the degree of dissimilarity in species composition between local communities.

8. The neutral perspective posits that niche differences between species are unimportant. Species are assumed to be demographically identical in terms of their per capita birth, death, and dispersal rates.

9. In Hubbell's neutral theory of biodiversity, every individual in the local community (regardless of species) has an equal probability of colonizing an open site (the "neutrality" assumption). All open sites are assumed to be filled, and community size is constant (the "zero-sum" assumption). Individuals compete for sites with other individuals from their local community, and also with immigrants from the metacommunity. New species arise in the metacommunity via speciation. Dispersal of these species into the local community may counteract the loss of species through random drift; thus, the metacommunity provides a pool of species at a regional level.

10. The neutral theory successfully predicts real-world species abundance distributions, but other models, including those based on niche differentiation, predict them equally well. Furthermore, species come to dominance, and species disappear, much more quickly in nature than predicted by neutral dynamics. Despite these issues, neutral theory focuses our attention directly on how dispersal rates and regional species pools may determine local community structure, and thus embodies the essence of the metacommunity perspective.

11. The niche vs neutral contrast found in differing models of metacommunities has close ties to the distinction between equalizing vs. stabilizing mechanisms of coexistence theory (Chapter 8). A simple experimental design that compares responses to total community abundance vs abundance of one's own species can distinguish these. One can find examples of both equalizing mechanisms (*Enellegma*) and stabilizing mechanisms (multiple plant studies) being predominant.

PART V

Species in Changing Environments: Ecology and Evolution

CHAPTER 15

Species in variable environments

The only constant in life is change.   **Heraclitus, c. 505** BCE

Scientists often make simplifying assumptions to bring certain process into focus. For example, your physics professor may have said, "Consider two balls traveling on a frictionless plane", or your biology professor might have said, "Imagine a spherical cow". We know, of course, that friction is real and spherical cows are not, but sometimes it helps us understand a process better if we ignore certain aspects of the real world. The same is true for the study of ecological communities. Until now, we have generally considered the environment in which species interact and in which communities assemble to be relatively unchanging. This has allowed us, for example, to look at the outcome of species interactions at steady state or equilibrium. In reality, however, species and communities exist in a dynamic state of change in response to changing environments. This chapter will explore some of the consequences of a varying environment to species interactions and community structure. In particular, how disturbance can result in the succession of ecological communities, how disturbance may promote (or hinder) species coexistence, how a varying environment can promote species coexistence through a mechanism called the storage effect, and how communities may shift between alternative states in response to environmental change will all be examined. In exploring these topics, we will reconnect with some of the ideas introduced in Chapter 8 (Species Coexistence and Niche Theory) and in Chapter 14 (Metacommunities).

The empirical existence of temporally variable populations and communities is extremely well established and the variance tends to be large. Temporal variability comes in two broad forms, either random variation around an apparent set point; or long term trends (Figure 15.1). For example, a study of North American birds found the median coefficient of variation in the abundance of a population is 0.86 meaning the standard deviation of abundance is greater than 86% of the mean population size for over half the bird species. The same study found that over a 30-year time period 72.8% of the populations showed a statistically significant trend up or down (McGill 2013).

Today, essentially all species exist in changing environments due to humankind's influence on the global climate. How climate change will affect the distribution and abundance of species, their interactions, and biodiversity and ecosystem function is perhaps **the** major question facing ecologists now and in the future. However, the question of how species and communities will respond to global climate change is vast—much too vast for a single book chapter. Our goal here is more modest. We focus instead on ways in which environmental variation (including disturbance) may affect the outcome of species interactions and the process of community assembly at local and regional scales. A solid grounding in these topics, however, will help you better understand the myriad ways that global climate change may affect ecological communities.

Community Ecology. Second Edition. Gary G. Mittelbach & Brian J. McGill, Oxford University Press (2019).

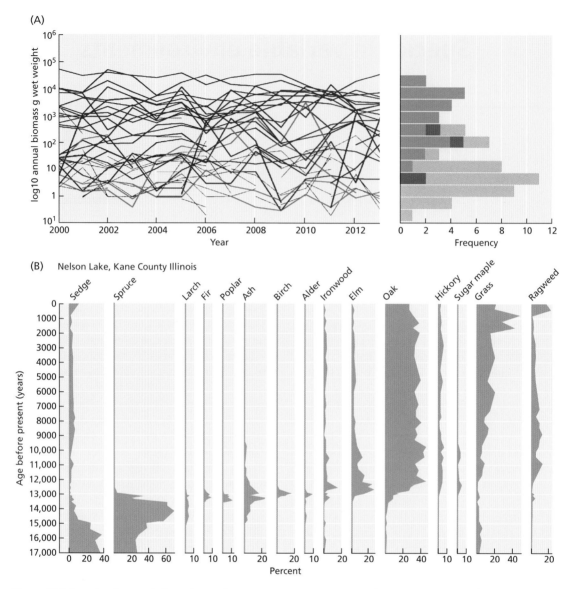

Figure 15.1 Examples of constantly fluctuating populations in communities in response to fluctuating environments. (A) Fish population abundance in an estuarine in the UK varies over time (note the log scale for abundance which makes the fluctuations look smaller than they are). All fish species show marked variability and many of the rare species actually appear, disappear and sometimes reappear in the community (after Henderson and Magurran 2014). The right panel shows the frequency distribution of the average abundance. Blue species show density-dependence, red species do not, gray species are transient. (B) The relative abundance of different species of plants for the last 17,000 years as measured by pollen grains. The long-term trends were driven by climate change (the retreat of the glaciers), causing several fairly abrupt shifts in the dominant species. But note that there are also fluctuations on the scale of hundreds of years. Redrawn from Curry et al. (1999).

Ecological succession

Among the first community ecologists were botanists who studied patterns in plant species associations along environmental gradients (e.g., Cowles 1899; Clements 1916; Gleason 1917, 1927; Cooper 1923; Tansley 1939). These early plant ecologists were particularly interested in the process of **succession**, defined as a pattern of community change following a disturbance that removes organisms/biomass from the landscape. Among the classic studies of ecological succession are those of Cowles (1899) and Olson (1958), who described regular patterns in the succession of plant species on sand dunes along the shores of Lake Michigan, USA (Figure 15.2). By studying dunes of different ages, Cowles, Olson, and others deduced that the initial colonists of a newly created sand dune were species that could tolerate nutrient poor conditions (particularly low nitrogen) and shifting sand. As the sand stabilized and nutrients (especially nitrogen) accumulated, these initial colonists were replaced by forbs, shrubs and eventually trees (e.g., white pine, oak, sugar maple). Frederic Clements (1936) famously claimed that, for a given climate and species pool, succession would always end with the same set of species, known as the **climax community**. While there is a more or less predictable pattern of dune succession, Olson (1958) and Lichter (2000), and others noted that there was no single endpoint or final climax community; rather, the species composition attained at the end of succession depended on the slope, aspect, and original parent material of the dune. In addition, succession exhibits a great deal of stochasticity from site to site because colonization and establishment are necessarily probabilistic events.

The succession of plant species on Lake Michigan sand dunes is an example of **primary succession**, where disturbance exposes bare mineral substrate that is essentially devoid of plant roots, seeds, and other organic material. Volcanic eruptions and glacial retreats are other examples of disturbances that lead to primary succession. **Secondary succession** occurs when a disturbance removes an existing plant community (e.g., a forest clear-cut or an abandoned agricultural field), allowing species to colonize bare substrate. In contrast to primary succession, this bare substrate is likely to consist of organic soil that contains vestiges of previous plant communities (e.g., seeds, roots). Thus, the first colonists in secondary succession are often species whose seeds have remained dormant and buried in soil for decades. It is important to recognize that the distinction between primary and secondary succession is also a matter of degree (i.e., how much does disturbance remove the existing community and the underlying organic soil; completely or partially)?

Succession occurs at all spatial scales. An extremely important form of succession in most forests occurs at the scale of an individual mature tree in the canopy (roughly 10 m × 10m) and is known as **gap dynamics**. Although the death of a canopy tree does not necessarily cause a disturbance that kills the individuals under it, the suppression of other species caused by shade is removed. This newly found source of light in the open canopy sets off a very succession-like series of plant species replacing each other until ultimately another canopy tree closes off the light. Many forest species can be classified as specialists in early, middle, or late stages of this gap-dynamic process. The end result is that a forest is a spatial mosaic of patches at different stages

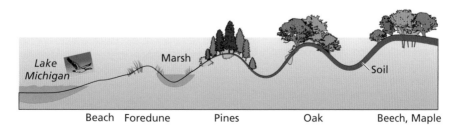

Figure 15.2 An illustration of succession on sand dunes.

of gap dynamics succession. This is an important mechanism to increase the heterogeneity and diversity of forests (Brokaw 1985; Whitmore 1989).

Studies of succession pioneered many current research areas focused on species and communities in variable environments (although succession itself is less of a research focus today because the processes governing primary and secondary succession are for the most part understood; see Box 15.1).

Current research areas that trace their origins to the study of succession include; studies of species trade-offs (e.g., competition and colonization, seed size and seed number), models of species coexistence in variable environments (e.g., intermediate disturbance hypothesis, Chesson and Warner lottery model, Hubbell and Foster neutral model), studies of regime shifts and alternative stable states.

Box 15.1 Mechanisms of succession

What causes succession to occur? Connell and Slatyer (1977) provide an essential summary of successional mechanisms and Chapin et al. (1994) carefully documented many of these mechanisms in plant primary succession at Glacier Bay, Alaska following the retreat of a glacier (Chapin et al. 1994). Recall that succession is a more or less orderly one-directional sequence of sets of species replacing previous sets of species. Thus, we need to explain two main questions:

- Why doesn't succession move backwards and forwards?
- Why doesn't succession just jump to the end?

As discussed at the beginning of this chapter, it is common for communities to show variability through time. Succession is different in being a trajectory rather than a cloud of points around an equilibrium. Why doesn't succession return to past states? The answer lies primarily in competitive exclusion (Chapter 7). Succession involves increasingly more competitively dominant plants, excluding the plants previously there. Much of this competitive dominance involves light competition. Taller plants shade out the shorter plants. The ability to tolerate shade during the juvenile stages is also an important attribute of late successional species.

The second question is if succession is going to end up with one set of tall, competitively dominant plants, why does it go through several stages of smaller plants instead of immediately jumping to the final stage? There are at least four different mechanisms that lead to this sequencing:

- **Facilitation**—early successional species modify the physical environment to make it more conducive to the growth of later successional species. Common examples are improving the structural or nutrient qualities of the soils, or providing a canopy that retains moisture and prevents excessive sun for seedlings. This was originally thought to be the primary mechanism causing succession but we now know it is only one of many.
- **Inhibition**—if the early arrivers can slow down the establishment of later species, but cannot outright prevent the establishment of late successional species, this can magnify small initial differences in arrival and extend the time frame over which succession occurs. This commonly occurs when early successional species compete for water, nutrients and light with the germination and seedling stages of later successional species.
- **Faster dispersal**—early successional species may simply arrive to a disturbed patch faster than late successional species. For example, weedy early successional species typically produce many small seeds, while late successional species often produce few large seeds to give them the resources they need to survive as a seedling in a crowded environment. Acorns and beech nuts are examples of the latter but they are poor dispersers compared to the small windborne seeds of many early successional plants. This is a key part of the competition-colonization trade-off discussed in Chapter 13. Conversely, disturbance and successional dynamics are an important part of some temporal coexistence mechanisms.
- **Relative growth rates**—early and late successional species may arrive at a disturbed site at the same time, but the early successional species may grow to maturity faster than the late successional species.

Note that many of these mechanisms are plant-specific and the study of succession is often focused on plants. Interesting questions exist as to what degree animal communities also show succession upon appearance of a newly available patch and whether animals have internal successional dynamics with mechanisms similar to plants or if animals merely track the changes in the plant community (Gittings and Giller 1998).

The intermediate disturbance hypothesis

The **intermediate disturbance hypothesis** (**IDH**; Grime 1973a,b; Connell 1978) is one of the oldest and best-known explanations for the maintenance of species diversity in a fluctuating environment. In brief, the IDH proposes that intermediate levels of environmental disturbance will allow more species to coexist than low or high rates of disturbance. Initial statements of the IDH were quite general (Grime 1973a,b; Connell 1978), with the basic idea being that disturbance interrupts the process of succession, such that good colonizers are favored at high rates of disturbance, good competitors are favored at low rates of disturbance, and intermediate rates of disturbance promotes the coexistence of both types of species. "Disturbance" may be any factor that causes the destruction of biomass (Grime et al. 1987) and makes space and resources available for use by new individuals (Roxburgh et al. 2004). "Intermediate" most often refers to the frequency of disturbance (its time scale), but the term has also been applied to the intensity, extent, and duration of disturbance as well. Only the frequency of disturbance over time will be considered here.

Ecologists have most often applied the IDH to species coexistence in patchy environments, where disturbance resets succession within a patch by opening up space and freeing up resources, thereby allowing a number of species to coexist in a system of patches at different successional states (e.g., Sousa 1979a,b). This application of the IDH to a patchy environment, which has been termed the "successional mosaic hypothesis" (Chesson and Huntly 1997), generally interprets species coexistence as the result of a trade-off between competitive ability and dispersal ability (see the discussion of this competition/colonization trade-off in Chapters 13 and 14). The IDH has also been applied to environments without explicit spatial structure (e.g., Bartha et al. 1997; Collins and Glenn 1997). In a spatially homogeneous context, disturbance affects all organisms simultaneously, regardless of their spatial location (Roxburgh et al. 2004). Because there is no spatial structure, the competition/colonization trade-off does not apply in this case, although other niche differences between species may allow for differential responses to disturbance.

Roxburgh et al. (2004; see also Shea et al. 2004) provide an excellent discussion of the mechanisms underlying the effects of intermediate disturbance frequencies on species coexistence in spatially heterogeneous or homogeneous environments. They show that coexistence in a patchy environment depends on the competition/colonization trade-off; without this trade-off, the superior competitor wins at all disturbance frequencies at which species can maintain positive population growth. With this trade-off, coexistence is promoted at intermediate disturbance levels. Intermediate disturbance frequencies can also promote coexistence in spatially homogeneous environments, but only if niche differences exist between species that differentiate their responses to disturbance (e.g., life history differences that trade off with competitive ability). When these niche differences are present, disturbance may promote coexistence via two fundamental mechanisms: relative-nonlinearity (Chapter 7) and the storage effect (Chesson 2000a; Roxburgh et al. 2004; discussed in the following section, "Fluctuation-dependent mechanisms of species coexistence"). In the absence of such niche differences, disturbance can only slow the rate of competitive exclusion (as in the models of Huston 1979; Hubbell and Foster 1986); it cannot lead to long-term coexistence.

There are some excellent empirical studies demonstrating that species richness may be increased by intermediate levels of disturbance. One of the best of these studies is the classic work by Sousa (1979a,b), who showed that the number of algal species living on rocks in a intertidal boulder field near Santa Barbara, California, peaked at intermediate boulder sizes. Sousa convincingly demonstrated that boulders of intermediate size were disturbed (rolled over) by storms at intermediate frequencies, and that this intermediate level of disturbance prevented them from being dominated by the best algal competitors. In this system, there appears to be a trade-off between competitive ability and dispersal ability. However, despite the intuitive appeal of the IDH and its long tenure in ecological thought, there is surprisingly little evidence supporting its general importance in nature. In a literature review, Shea et al. (2004) found that only 18% of 250 studies supported the IDH. Similarly, in their meta-analysis of 116 species richness–disturbance relationships, Mackay and Currie (2001) found that only 16% of those relationships showed a peaked response. Bongers et al. (2009) recently examined the IDH in tropical forests, the ecosystem for which the IDH was first proposed (Connell 1978). Using a large data set of tree inventories from 2504 1-ha plots, Bongers et al. (2009, p. 798) found that "while diversity indeed peaks at intermediate disturbance levels little variation is explained outside dry forests, and that disturbance is less important for species richness patterns in wet tropical rain forests than previously thought."

As discussed in Chapter 8, stable coexistence requires mechanisms that favor the recovery of species when they become rare (Chesson and Huntly 1997). Environmental fluctuations that affect species equally are not stabilizing (Chesson and Huntly 1997). This distinction, however, is not always recognized. Instead, it is common to hear ecologists express the idea that environmental disturbance or harshness reduces the intensity of competition and thereby promotes species coexistence. The intermediate disturbance hypothesis is sometimes portrayed in this light, as are the ideas expressed by Huston (1979) and Menge and Sutherland (1987). However, as Chesson (2000a, p. 354) emphatically notes, "no credence can be given to the idea that disturbance promotes stable coexistence by simply reducing population densities to levels where competition is weak" (see also Holt 1985; Chesson and Huntly 1997; Chase et al. 2002; Barabás et al. 2018). Mortality by itself cannot promote species coexistence, because its positive effect of lowering competition is countered by its negative effect on population growth. For environmental fluctuations (including disturbance) to stabilize species coexistence, they must affect species differentially. This differential effect on species need not be immediate, however, but may act in a chain of events that differentially affect species' recruitment and/or mortality (as will be seen later).

Fluctuation-dependent mechanisms of species coexistence

Recall from the studies of competition (Chapters 7 and 8) that the strength of intraspecific competition relative to interspecific competition is key to promoting stability among interacting species. Stabilizing mechanisms such as resource partitioning and frequency-dependent predation operate whether the environment fluctuates or not; these stabilizing mechanisms are **fluctuation-independent mechanisms**. Other mechanisms that can stabilize the coexistence of species depend critically on fluctuations in population densities and environmental factors (Chesson 1985, 2000a). Such stabilizing factors are **fluctuation-dependent mechanisms**.

Chesson (2000a) recognized two classes of fluctuation-dependent stabilizing mechanisms: those involving *the relative non-linearity of competition* and those involving *the storage effect*. You are already familiar with the first of these classes. In Chapter 7, it was seen that two (or more) consumer species can coexist on a single limiting resource if there are non-linearities in the consumers' functional responses (e.g., Figure 7.5) and if there are fluctuations in resource abundance. This stabilizing mechanism operates because some species are favored when resource abundances are high and other species are favored when resource abundances are low. Variation in the physical environment (i.e., seasonal changes) can cause resource abundances to fluctuate over time. In the case of biotic resources (living species), the interaction between consumers and resources can also generate endogenous temporal variations in resource abundance—i.e., stable limit cycles. When consumer and resource densities fluctuate in a stable limit cycle, two (or more) consumer species can coexist indefinitely on a single resource (Armstrong and McGehee 1980). An example of such a trade-off was discussed in Chapter 7 when two types of consumers were considered—gleaners and opportunists (Frederickson and Stephanopoulos 1981; Grover 1990). Species interactions due to shared predation (apparent competition; see Chapter 7) may also be stabilized by fluctuations in predator densities when at least one of the consumer species has a non-linear mortality response to predator numbers (Kuang and Chesson 2008).

The second type of fluctuation-dependent stabilizing mechanism, **the storage effect**, was discovered by Chesson in the early 1980s, when he set out to rigorously explore a hypothesis, proposed by Peter Sale, that temporal variation in recruitment could allow the coexistence of many species of fish competing for territories on a coral reef (Sale 1977). Chesson and others have since shown that the storage effect can operate in a wide variety of taxa and communities, and not just in situations where species compete for space (e.g., Chesson 2000b; Kuang and Chesson 2010). However, competition for space among reef fishes will be used to introduce the storage effect, as it provides a clear and intuitive example of this stabilizing mechanism. This will then be expanded to show the generality of the mechanism, although it will steer away from the mathematical details. Those wishing a rigorous mathematical treatment of the storage effect in variable environments should consult Chesson (1994, 2000a,b).

The storage effect

The life histories of many coral reef fishes include an adult phase that is sedentary and territorial, and thus tightly tied to a single reef, as well as a widely dispersed larval phase, which can colonize multiple reefs. Competition for space on a reef is intense, but an individual that succeeds in obtaining a territory on the reef is likely to retain that territory until it dies. Sale (1977) hypothesized that this system could allow for the coexistence of a large number of fish species, even if those species were very generalized in their diets and habitat use, so long as there was random temporal variation in the availability of territories (caused by the deaths of adults) and in the settlement of larval fish of different species on those territories. Sale (1977, p. 354) envisioned that "reef fishes are adapted to this unpredictable supply of space in ways which make interspecific competition for space a lottery in which no species can consistently win. Thus, the high diversity of reef fish communities may be maintained because the unpredictable environment prevents development of an equilibrium community."

Chesson and Warner (1981) formalized Sale's verbal argument in a more rigorous mathematical model that they initially termed the "lottery model," based on the way competition for space was determined. Since that time, the theory behind the lottery model has been expanded to include a much broader domain of study and is now more generally referred to as models of the storage effect, in reference to the mechanism of coexistence.

Figure 15.3 illustrates a simplified version of the lottery model for two fish species. Species 1 initially holds 90% of the 100 territories on the reef, and species 2 holds the remaining 10% of the territories. Each species suffers 10% adult mortality between recruitment events. The key features of Chesson and Warner's model as applied to the competition for space between reef fish are:

1. Adult fish hold territories on the reef, and an individual holds its territory until it dies. Adults are iteroparous (reproduce more than once), and each species has overlapping generations.
2. To survive and reproduce, a larval fish must obtain a territory. Space is limiting, and there are always more larvae attempting to establish territories than space allows. Allocation of territories is on a first-come first-served basis; the first individual to arrive at a suitable site can establish a territory there.
3. Larvae are produced in abundance and dispersed; therefore, the number of larvae seeking territories at a site is not directly related to the number of adults already present at that site.
4. Larval survival depends on environmental conditions. Environmental conditions vary over time, and species differ in their responses to those conditions, such that species have relatively "good" and "bad" years for larval survival and recruitment.

How would each species respond to a "good" recruitment year, in which one species obtains all the available territories and the other species obtains none? Differences in good and bad recruitment years among species are expected, for example, if variation in the physical environmental affects the number of larvae surviving to potentially recruit to the reef. The hypothetical data in Figure 15.1 illustrate that a good recruitment year has a disproportionate positive effect on the species that is at lower abundance. This temporal variation in recruitment, coupled with a relatively long-lived adult stage that can "store" the population contributions of good years (or, similarly, buffer against population loss in bad years), allows temporal fluctuations to act as a stabilizing mechanism—a concatenation of qualities that has come to be known as the **storage effect**.

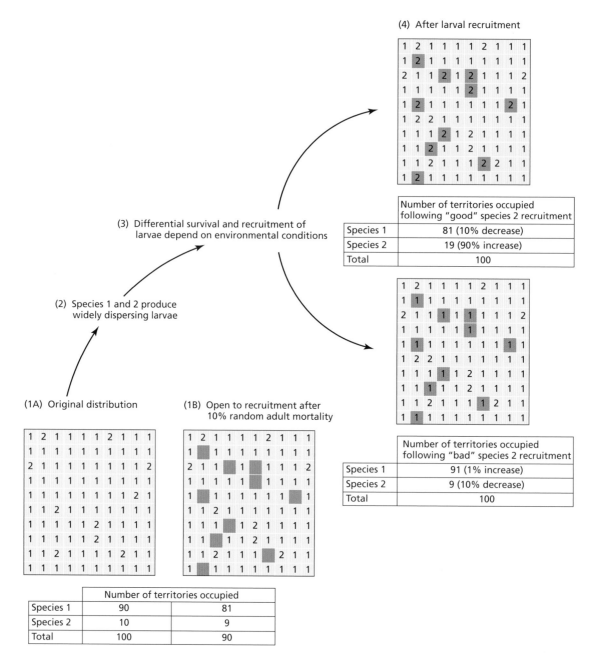

Figure 15.3 A simple characterization of Chesson and Warner's lottery competition model and the storage effect. In this hypothetical case, individuals of two fish species occupy territories on a coral reef, represented by 1s and 2s in the "territorial squares" of the matrices. Colored squares are unoccupied by either species following random adult mortality. The two matrices at the far right show possible results of larval recruitment to the unoccupied territories, with the environment favoring recruitment by either species 2 (above) or species 1 (below). The storage effect of the lottery model is illustrated in the tabular data below each matrix, which show that rare species can make big proportional gains in "good" years and may persist through "bad" years.

The storage effect model is potentially relevant to any group of organisms that have over-lapping generations and a relatively long-lived life stage. Annual plants with long-lived seeds (Pake and Venable 1995; Sears and Chesson 2007; Angert et al. 2009), zooplankton with diapausing (resting) eggs (Cáceres 1997) and trees with long-lived adults (Usinowicz et al. 2012) have been used to test its predictions. Chesson and colleagues have further shown that the storage effect is quite general and that it can act to promote species coexistence in a wide variety of situations where there is temporal or spatial variation in the environment (Chesson 2000b; Chesson and Kuang 2010).

Environmental variation (fluctuation) may come from many factors, including resource availability, rainfall, temperature, and predation levels. For the storage effect to function, this variation must change the birth, survival, or recruitment rates of species from year to year (or place to place). Critically important is that species differ in their responses to this environmental variation—that is, they must differ in their niches (Chesson 1991). Environmental variation that affects all species equally, such as density-independent mortality caused by disturbance, harsh environmental conditions, or nonselective predation, is not a stabilizing mechanism (Chesson and Huntly 1997; Chase et al. 2002). In order for environmental variation to stabilize species coexistence, species must respond differently to the environment, such that some species are favored (e.g., have high larval survivorship) at particular times or places and other species are not. This is the first component of the storage effect—*species-specific environmental responses*.

The second component of the storage effect is a long-lived life stage (e.g., diapausing eggs, dormant seeds, long-lived adults). A long-lived life stage allows species to store the positive effects of "good" years and thus buffer the negative effects of "bad" years. Otherwise, a species would go extinct after a short run of "bad" years. Chesson (2000a) calls this second component *buffered population growth*.

The third component of the storage effect is *covariance between environment and competition*. That is, the effects of competition must covary with the environment such that negative environmental effects decrease competition and positive environmental effects increase competition. It is easy to see this covariance operating in the simple reef fish example presented in Figure 15.3. Because the two species respond differently to the environment (component 1), we expect that intraspecific competitive effects will be strengthened when the environment favors a species' population growth and will be reduced when the environment is unfavorable for a species and it becomes rare.

Adler et al. (2006a) tested for the three components of the storage effect in a long-term (30-year) data set that mapped individual prairie plants in permanent plots in Kansas, focusing their analysis on the three grass species that made up more than 95 percent of the vegetative cover. All three species had long life spans and thus met the requirement for buffered population growth (Figure 15.4A). In addition, there were differential species-specific responses to environmental variation (e.g., variation in precipitation and mean annual temperature) such that pairwise correlations in species' population growth rates were absent or weak (Figure 15.4B). Finally, Adler and colleagues found evidence for the third component of the storage effect: the effects of competition were more severe in more favorable years (i.e., covariation between environment and competition). Using a population growth model parameterized for each species, they showed that neighboring plants limited a species' population growth in favorable years, but had weak or even facilitative effects in unfavorable years (Figure 15.4C). Thus, all three requirements of the storage effect were met. Adler et al. (2006a) also showed via simulations that each species' low-density population growth rate was higher in a variable environment than in a constant environment (Figure 15.5), further evidence that the storage effect promoted coexistence in these prairie grasses. Adler et al. (2006b) applied a similar analysis to a sagebrush steppe community. However, in this case, they found little covariation between environmental conditions and competition, leading them to conclude that climate variability had only a weak effect on species coexistence in this ecosystem.

The storage effect has been shown to contribute to species coexistence in a variety of systems; e.g., desert annual plants (Pake and Venable 1995; Angert et al. 2009), freshwater zooplankton (Cáceres 1997),

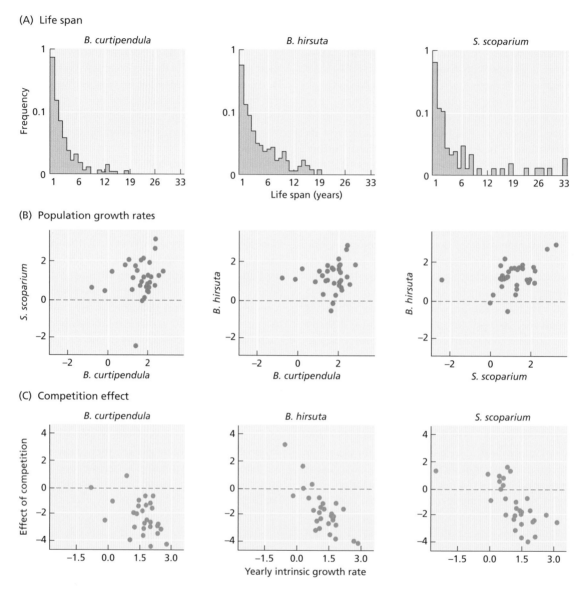

Figure 15.4 Evidence that the three components of the storage effect are operating in a community of coexisting prairie grasses dominated by *Bouteloua curtipendula*, *B. hirsuta*, and *Schizachyrium scoparium*. (A) Each of the three species has the potential for long life spans, which buffers population growth. (B) Weak correlation of population growth rates between species pairs demonstrates that the species respond differently to environmental variation. (C) For each species, competition has stronger negative effects on population growth in more favorable years (years with higher intrinsic growth rates). After Adler et al. (2006a).

coral reef fishes (Chesson and Warner 1981), and tropical trees (Runkle 1989; Usinowicz et al. 2012, 2017). These studies demonstrate that the storage effect is an important force in nature, but we would also like to know much it contributes to species coexistence relative to other mechanisms. This

can be a challenge. The few studies that have quantified the storage effect's contribution to coexistence relative to other coexistence mechanisms (e.g., Angert et al. 2009; Usinowicz et al. 2012) had to derive complex mathematical models to conduct their analysis. Recently, Ellner et al. (2016)

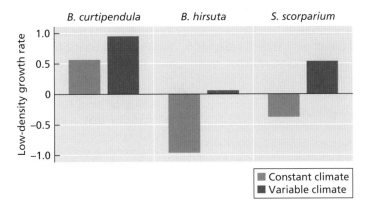

Figure 15.5 Estimated long-term low-density growth rates in constant and variable climate conditions based on model simulations for the three species of prairie grass in Figure 15.3. All three performed better in variable than in constant environments, indicating that the storage effect may be acting to stabilize species coexistence. After Adler et al. (2006a).

developed a simulation-based approach to overcome this limitation and they applied their methods to two studies of competition in fluctuating environments:

- algal species in a chemostat with variable temperatures (Descamps-Julien and Gonzalez 2005);
- three grass species and a shrub in a sagebrush steppe community (Chu and Adler 2015).

In each case, they found that environmental variability was critically important for species-coexistence, but that the storage effect's contribution to coexistence was relatively small. Letten et al. (2018) also found that the storage effect and relative non-linearity contributed similarly to species coexistence in a series of laboratory experiments with nectar yeasts. Future studies will hopefully expand on this work to better illuminate how important the storage effect and other fluctuation-dependent mechanisms are to the coexistence of species in nature. A very interesting study by Usinowicz et al. (2017) showed that the storage effect makes a greater contribution to species coexistence in tropical forests than in temperate forests, thus linking this fundamental mechanism of species coexistence to the latitudinal diversity gradient in trees.

Regime shifts and alternative stable states

We have seen how environmental variation may promote species coexistence through mechanisms

that act via niche differences (broadly defined) to strengthen the effects of intraspecific density dependence relative to interspecific density dependence. Environmental variation can have other important consequences for communities as well. For example, over geologic time scales, climatic fluctuations drive glacial cycles, shift species ranges, cause species extinctions, and influence patterns of biodiversity at global scales (Figure 15.1B). We touched on some of these long-term effects of climate variation in Chapter 2 when we examined broad-scale patterns in species diversity, including the latitudinal diversity gradient. We live today in an era of rapid, human-driven climate change, where shifts in species ranges, species extinctions, and changing patterns of biodiversity are observable even in human lifetimes. The scope of global climate change and its impacts on ecological systems is beyond what we can cover in a book focused on patterns and processes in community ecology. However, the study of species interactions and the mechanisms that structure communities leads naturally to the questions of whether communities might occur in alternative states, depending on the outcomes of species interactions, and whether environmental variation can cause communities to shift from one state to another.

The idea that communities or ecosystems might occur in alternative stable states was first proposed theoretically by Lewontin (1969), explored mathematically by May (1977b), and linked empirically to ecology by Holling (1973) and Sutherland (1974). In theory, even small environmental changes can lead to dramatic shifts in community state. Ecologists are rightfully concerned about these shifts, as they have important implications for the preservation of biodiversity, the management of biotic resources, and the restoration of ecosystems.

Before beginning our discussion of this important topic, we need to define some terms. **Alternative stable states** describes a situation where communities may exist in different configurations (e.g., of species richness, species composition, food web structure, size structure) under the same environmental conditions and these different community configurations persist unless a major perturbation causes a transition to another (May 1977b; Beisner et al. 2003; Petraitis 2013). Alternative stable states may be defined mathematically as different basins of attraction (May 1977b; Scheffer et al. 2001). The concept of alternative stable states is often represented graphically as a landscape of hills and valleys (Figure 15.6). The community is represented by a "ball" on the landscape, and different valleys (or basins) represent different potential community states. A community tends to remain in a stable state (i.e., in a valley) until environmental disturbance or change causes it to shift to a new state (Figure 15.6).

A change from one community state to another is termed a **regime shift** or **phase shift**, and the critical threshold at which a system undergoes a regime shift is referred to as a **tipping point**. Pimm (1984) defined the strength of perturbation needed to shift a community from one state to another as **resistance**

and the speed at which a community recovers from a perturbation as **resilience**. The concept of resilience, however, has been used in multiple ways since it was first employed by Holling (1973). It is now common to see resilience defined either as:

- **ecological resilience**—the magnitude of a perturbation that a system can withstand before undergoing a regime shift: or as
- **engineering resilience**—the time it takes for a system to recover from a perturbation (see reviews in Gunderson 2000; Nyström et al. 2008).

There are many examples of communities or ecosystems showing dramatic regime shifts in response to environmental change. These include shifts from coral-dominated reefs to reefs dominated by fleshy macroalgae (McCook 1999; Nyström et al. 2008; Hughes et al. 2010; Mumby et al. 2013), shifts from clear to turbid shallow lakes (Scheffer 1998; Scheffer and van Nes 2007), shifts from grass-dominated to tree-dominated savannas (Wilson and Agnew 1992; Bond 2010), shifts in arid ecosystems from vegetated communities to deserts (Rietkerk et al. 2004; Kéfi et al. 2007a), abrupt shifts in fish communities and sudden collapses in harvested fish stocks (Hutchings and Reynolds 2004; Daskalov et al. 2007; Persson et al. 2007), and even shifts in the microbial community of the human gut following surgery or treatment with antibiotics (Hartman et al. 2009; Dethlefsen and Relman 2011). Many of these examples of regime shifts are the result of anthropogenic environmental changes that cause ecosystems to shift to an undesirable state. For example, increased grazing may reduce vegetation cover in arid ecosystems, reducing the facilitation of local water availability by plants and shifting the system toward desertification (Schlesinger et al. 1990; Rietkerk et al. 2004). Thus, an important question in both basic and applied ecology is how predictable regime shifts are and whether they are reversible (Suding et al. 2004; Scheffer 2009). Simple graphic models can help us understand why regime shifts occur and why they may be difficult to reverse. The presentation that follows borrows from the work of Scheffer and colleagues (e.g., Scheffer et al. 2001; Scheffer 2009) and Kéfi (2008).

We expect that a slow and steady change in environmental conditions will result in a smooth

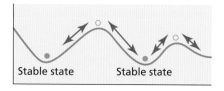

Figure 15.6 "Hill and valley" illustration of alternative stable states. A community or ecosystem (represented by the ball) will remain in the same state (valley or basin) unless it is perturbed by some environmental disturbance to shift to another state, or the landscape changes shape due to changes in the external environment. The steepness of a valley corresponds to the size of the perturbation required to shift the community into a new state and to the speed at which a community recovers from a perturbation.

transition from one community state to the next—for example, the loss or gain of species in response to a change in temperature or nutrients. Such a community response to environmental change might be very gradual and linear, as shown in Figure 15.7A. Or, a community might be relatively insensitive to environmental changes across a range of conditions, but respond strongly around some threshold condition (Figure 15.7B). In this case, we might observe a change in community state that is sufficiently large and abrupt to be considered a regime shift. A more extreme kind of change at a threshold occurs when a system has alternative stable states (Figure 15.7C). In this case, the curve that describes the community's response to environmental change folds back on itself. When this occurs, the system has two alternative stable states, separated by an unstable equilibrium that marks the border between basins of attraction.

A variety of mechanisms can theoretically lead to alternative stable states in communities, including positive feedbacks, nonlinear species interactions, size-based refuges from predation, and priority effects (e.g., Lewontin 1969; Chase 2003; van Nes and Scheffer 2004; Kéfi et al. 2007b). We will not go into the theory underlying the formation of alternative stable states, but it is important that we understand how environmental variation can affect communities that have alternative stable states. There is a crucial distinction between the abrupt response exhibited by the system in Figure 15.7B and the alternative stable states shown in Figure 15.7C. We can best appreciate this distinction by following the response of a community as environmental conditions change.

In the situation depicted by Figure 15.7B, a change in environmental conditions (along the x axis) will cause the equilibrium state of the community to track smoothly along the path shown by the arrows in Figure 15.7D. There may be regions of more or less rapid community change, but a change in environmental conditions in one direction, followed by a reversal of that change, will cause the state of the community to transition smoothly along a single path. One the other hand, if a community exists in alternative stable states, there is a tipping point (a point of critical transition) between community states (Figure 15.7C). When the community is in the state represented by the upper branch of

the folded curve in Figure 15.7C, it cannot pass to the lower branch smoothly. Instead, a change in environmental conditions past the tipping point c_2 will cause the community state to shift abruptly to the lower branch. In the theory of dynamic systems, this instantaneous shift from one system state to another is known as a *bifurcation point*. If the environmental conditions return to their previous state (i.e., they move back along the x axis in Figure 15.7C), there is a lag in the community response. Only after the environmental conditions reach the tipping point c_1 does the community shift back to its previous state. This delayed response is illustrated in Figure 15.7E, where we see that the state of the community tracks along two different paths in response to an environmental change in one direction followed by a reversal of that change. This delayed response to a forward and backward change in environmental conditions is termed **hysteresis**, and it is a distinguishing feature of systems that exist in alternative stable states.

Another way to visualize the transitions between alternative stable states is to use the familiar hill and valley diagram. Figure 15.8 uses this approach to show how a vegetated community in an arid environment might shift to a desert state due to overgrazing. The community is represented by a ball in a landscape of environmental conditions (hills and valleys). Starting with environmental conditions near c_1, we see that the landscape contains only one stable state, which has a deep basin of attraction (panel A at the bottom of Figure 15.8). At this point, the community is stable and will respond to a perturbation by quickly returning to its original state. However, as environmental conditions change (moving from c_1 to c_2), the shape of the landscape changes. At the point along the x axis where the environmental condition is $c = 2.3$, there are two alternative stable states (panel C). A sufficiently strong perturbation will push the community from one alternative state to the other (represented by the "dotted" ball in panel C). However, in the face of a small perturbation, the community will remain in its current state. As environmental conditions move farther along the x axis toward condition $c = 2.5$ (approaching the tipping point c_2 in the top diagram), the basin of attraction in which the community rests becomes very small and shallow (panel D). At

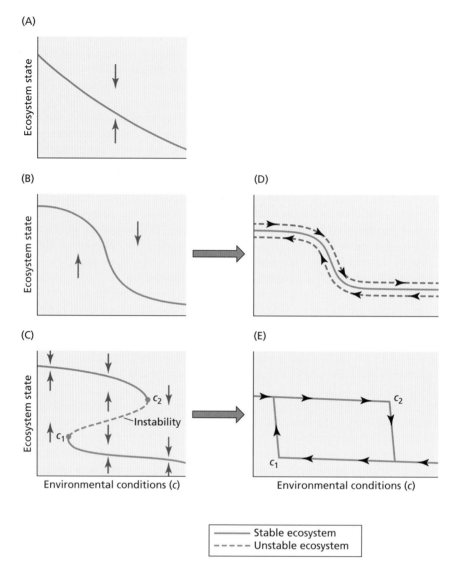

Figure 15.7 Communities and ecosystems may respond smoothly or abruptly to changing environmental conditions. Abrupt responses may lead to regime shifts and alternative stable states. (A–C) Responses in the equilibrium state of a community (blue curve) to a change in environmental conditions can be classified into three types: smooth (nearly linear) responses (A); abrupt (highly non-linear) responses (B); and catastrophic responses (C), resulting in alternative stable states. Red arrows indicate the system's movement toward its equilibrium state. In (A) and (B), only one equilibrium community state exists for each environmental condition. In (C), however, the equilibrium curve is folded back on itself, so that multiple equilibria exist for each environmental condition. Equilibria on the dashed line are unstable. If the system is on the upper branch of the equilibrium curve near point c_2, a slight increase in environmental conditions (a move to the right on the x axis) will cause the system state to jump catastrophically to the lower branch (an alternative stable state). (D–E) A change in environmental conditions over time can reveal alternative stable states. (D) As environmental conditions decrease and then increase (black arrowheads), the state of the system responds smoothly along the solid line (although there may be regions of rapid change). (E) A gradual increase in environmental conditions causes the community to shift state at a tipping point (c_2). If the change in the environmental condition is subsequently reversed, the system shows hysteresis, and a shift back occurs only when environmental conditions return to the tipping point (c_1). The existence of such hysteresis loops is strong evidence for the existence of alternative stable states. After Scheffer et al. (2001, 2009).

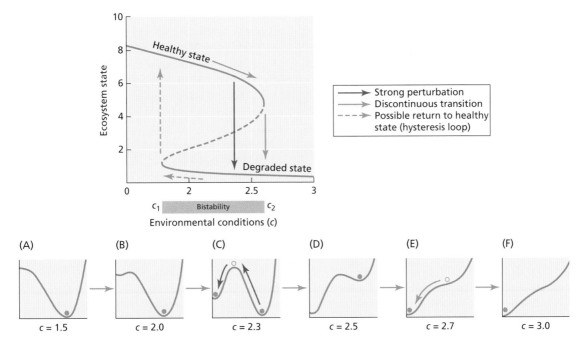

Figure 15.8 Two ways in which a community may shift from one stable state to another. The upper diagram is a bifurcation plot similar to that in Figure 15.7C. The solid portions of the equilibrium curve correspond to stable states and the dashed portion to instability. The x-axis represents environmental conditions, c. Tipping points c_1 and c_2 bound the parameter space where two stable states co-occur (bistability). (A–F) How the stability landscape changes with changing environmental conditions; the ball represents the community state. For small values of c, the landscape has only one valley. With increasing c, a second valley appears (at c_1), and the community can exist in either of two alternative states. At tipping point c_2, the first valley disappears and the landscape is again reduced to a single valley. Once the system has made the shift, recovery to the initial state may occur if environmental conditions return to a point below c_1 (hysteresis). After Kéfi (2008).

this point, the system will be slow to recover from a perturbation, and even a small perturbation may send it into a new state. At tipping point c_2 (c = 2.7, panel E), the community shifts abruptly to the alternative stable state (the lower arm of the equilibrium curve in the top diagram). If we were to now reverse the environmental conditions and move from c_2 toward c_1, the community would remain in its new stable state until environmental conditions again reached c_1; in other words, the system displays hysteresis.

You might wonder why we should be so concerned about whether communities respond to environmental change as predicted by the theory of alternative stable states. After all, the difference between a non-linear community response (Figure 15.7B) and a shift between alternative stable states (Figure 15.7C) might seem to be only a matter of degree. In a sense, that is true, making it difficult in practice to distinguish between a strong nonlinear response to

environmental change and a true shift to an alternative stable state (Scheffer and Carpenter 2003). For this and other reasons, some ecologists have long questioned the evidence for alternative stable states in nature (Connell and Sousa 1983). However, there are good reasons why ecologists and conservation biologists should be concerned about the existence of alternative stable states. If communities or ecosystems have tipping points at which they shift abruptly to different states in response to environmental change, this has obvious consequences for predicting the impacts of climate change and environmental degradation. More importantly, if communities or ecosystems occur in alternative stable states, reversing the impacts of environmental change becomes very difficult; that is, returning the environment to its previous condition will not cause the community or ecosystem to return to its previous state if there is hysteresis. In Hawaii, for example, the invasion of woodlands by non-native grasses promotes fire and alters nitrogen cycling, which positively affects the invasive grasses at the expense of native shrubs, creating an internally reinforced state that is very difficult to reverse (Mack and D'Antonio 1998; Mack et al. 2001). Suding et al. (2004), Kéfi (2008), Martin and Kirkman (2009), and Firn et al. (2010) provide additional examples of degraded or changed ecosystems that are resilient to traditional restoration efforts in ways that suggest alternative stable states, and necessitate new management practices to disrupt feedbacks and address constraints on ecosystem recovery.

How prevalent are alternative stable states in nature? As mentioned previously, showing definitively that alternative stable states exist in any real community or ecosystem is a challenge (Connell and Sousa 1983; Mumby et al. 2013; Capon et al. 2015). Documenting hysteresis in response to a slow forward and backward change in environmental conditions is perhaps the strongest test (Scheffer and Carpenter 2003). This approach has been used in a few experimental studies (e.g., Handa et al. 2002; Schmitz et al. 2004). However, Schröder et al. (2005), in a literature review, found no field experiments that had manipulated a natural system over sufficient time to demonstrate full hysteresis. Experimental manipulations are difficult to apply at large spatial scales and over long periods. Instead, ecologists have followed long-term community responses to environmental changes induced by human activities, "natural experiments," or a combination of events. For example, Scheffer and Carpenter (2003) noted that vegetation cover in a shallow lake in the Netherlands demonstrated hysteresis in its response to an increase and subsequent decrease in nutrient concentrations (Figure 15.9A). Daskalov et al. (2007) found hysteresis loops in the responses of some marine organisms to long-term changes in fishing pressure in the Black Sea (Figure 15.9B,C), although some organisms they studied did not respond in this way.

In our own work, Mittelbach et al. (2006) examined changes in the zooplankton (cladoceran) community of a Michigan lake in response to a long-term (16-year) decrease and then increase in planktivorous fish density. Planktivore densities in Wintergreen Lake, Michigan, have varied over two orders of magnitude following the elimination of two fish species (largemouth bass and bluegill) due to a low-oxygen winterkill in 1978 (a natural event) and the sequential reintroduction of those species in the following years (bass in 1986 and bluegill in 1997). The loss of bass (the top predator in the lake) led to a dramatic increase in the abundance of planktivorous golden shiners, resulting in a "high-planktivory" state. When bass were reintroduced, they consumed all the golden shiners (see Figure 10.8), resulting in a "low-planktivory" state. The reintroduction of bluegill returned the lake to a "high-planktivory" state.

The change in planktivorous fish densities caused dramatic shifts in the species composition of the zooplankton community. The "low-planktivory" zooplankton community was dominated by two species of large-bodied *Daphnia*, whereas the "high-planktivory" zooplankton community was composed of a suite of small-bodied species (Figure 15.10). Moreover, these two community states (high and low planktivory) shared no zooplankton species in common (i.e., there was complete species turnover). Plotting the dissimilarity of the zooplankton community from the initial community state in 1989 for each year as a function of planktivore density allows us to see the pattern of community disassembly and reassembly over time and to look for hysteresis in the community response (Figure 15.11). This long-term data set shows that the pronounced regime shift in community composition occurred smoothly and predictably; the replacement of species followed a very similar trajectory through both a decrease and an increase in planktivore abundance, with little suggestion of hysteresis.

Despite great interest in the topic of regime shifts, tipping points, and multiple stable states, convincing evidence for alternative stable states in nature is limited. Recent literature reviews of freshwater ecosystems (Capon et al. 2015), coral reefs (Dudgeon et al. 2010; Mumby et al. 2013), and a variety of laboratory and field studies (Schröder et al. 2005) all agree that few studies satisfy the stringent criteria laid down to demonstrate the existence of alternate stable states in nature. Rather, it seems more likely that alternative or multiple stable states represent the end of a continuum of responses to environmental

(A)

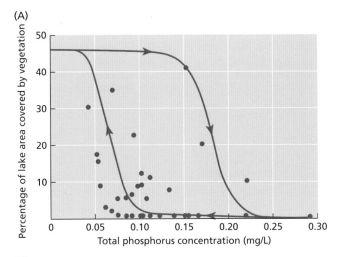

(B)

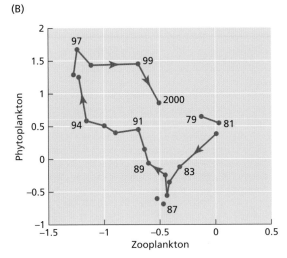

(C)

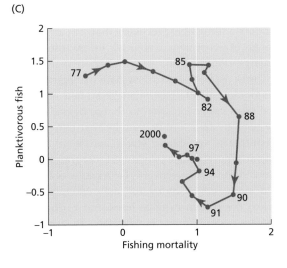

Figure 15.9 Examples of ecosystems showing hysteresis loops and potential alternative stable states. (A) Vegetation cover in a shallow lake changed with nutrient (phosphorus) concentration; arrows show the direction of change over time. Note how the abrupt response in vegetation cover resembles that predicted for systems displaying alternative stable states (see Figure 15.7E). (B,C) Time series of changes in resource biomass in response to changes in consumer abundance in the Black Sea. (B) Change in phytoplankton biomass as a function of zooplankton biomass. (C) Change in planktivorous fish biomass as a function of fishing mortality. Numbers on the plots are years. In each case, the trajectory of resource biomass over time in response to changing consumer abundance suggests a hysteresis loop. (A) after Scheffer and Carpenter 2003; (B,C) after Daskalov et al. (2007).

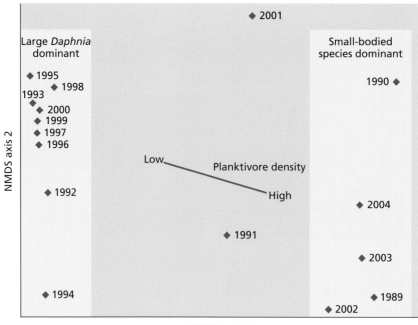

Figure 15.10 The zooplankton (cladoceran) community of Wintergreen Lake, Michigan, shifts between two states, depending on the density of planktivorous fish. Changes in zooplankton species abundances over the years 1989–2004 were visualized using non-metric multidimensional scaling (NMDS), an ordination technique which reveals that the zooplankton community (represented by the red diamonds) occurs in two distinct states (gold shading). In years when the lake contained few planktivorous fish, the zooplankton community was dominated by large-bodied *Daphnia* (left column). In years of abundant planktivores, the zooplankton community was made up of small-bodied zooplankton (e.g., *Ceriodaphnia* and *Bosmina*), and large *Daphnia* were completely absent (right column). The years 1991 and 2001 had intermediate planktivore density and were transitional between the two zooplankton community states. The angle and length of the joint plot line of planktivore density (center) indicates the direction and strength of the relationship of planktivore density with the NMDS scores (Pearson's $r = 0.74$), and shows that the separation in zooplankton community types was strongly correlated with the density of planktivorous fish. After Mittelbach et al. (2006).

change (e.g., Fung et al. 2011); for example, we refer back to Figure 15.7. If the "fold" in the equilibrium curve in Figure 15.7C is relaxed and made smaller the response of the system becomes smoother and more like the response shown in Figure 15.7B. The shift between community states in Figure 15.7B is still large and rapid around the tipping point—there is a regime shift—but there are no alternative stable states. Thus, whether a community responds to a changing environment with a steep threshold response or with a bifurcation is important theoretically, but this distinction may be difficult to demonstrate empirically, and in the end it may make little practical difference when applying these concepts to ecosystem management or restoration. Managers rarely have the time or resources to determine whether their system exhibits threshold or hysteresis dynamics

(Kinzig et al. 2006; Suding and Hobbs 2009). Yet understanding the processes that can drive the range of ecosystem responses to environmental change (Figure 15.6) and their potential consequences is critical to ecosystem management or restoration. In the past, managers often relied on the process of succession to restore degraded ecosystems to their former state by first mitigating a disturbance or abiotic stress and then "letting nature take its course". Today, we recognize that the smooth and steady recovery of a degraded ecosystem via successional dynamics may be impossible when the system exhibits non-linear or threshold dynamics (Suding and Hobbs 2009). In such cases, restoration will require active management of the recovering ecosystem, including perhaps a larger and more forceful alteration of environmental

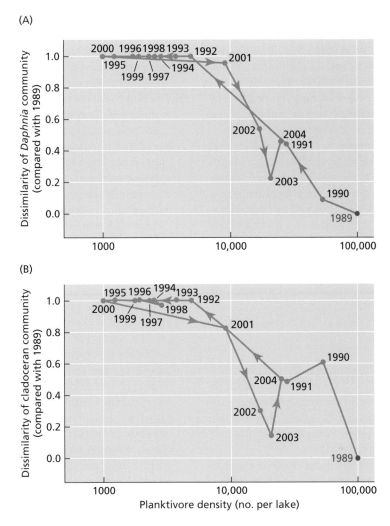

Figure 15.11 Zooplankton community composition changes smoothly with changes in planktivore density in Wintergreen Lake, Michigan, 1989–2004. Change in community state refers to the degree of dissimilarity in species composition relative to the initial community state in 1989 (red), measured as Bray–Curtis distances, where 0 indicates identical species composition and 1.0 indicates no species in common (maximum dissimilarity). Time trajectories are indicated by arrows and years placed next to data points. (A) Change over time for the *Daphnia* community. (B) Change over time for the entire cladoceran assemblage (eight species). After Mittelbach et al. (2006).

conditions to move the system past a tipping point or by introducing species that are no longer present in the species pool (Suding et al. 2004; Suding and Hobbs 2009).

Conclusion

Environments change over time and these changes profoundly affect a species' abundance and population dynamics, its interactions with other species,

and structure of communities. Successional dynamics are driven in large part by environmental change that favors species that are good colonizers early in succession and good competitors late in succession. Over time, community composition changes due to competitive exclusion. Hutchinson's (1961) proposed that competitive exclusion may be prevented in an environment that shifts periodically to favor different species at different times. This early hypothesis set the stage for more rigorous mathematical models, showing how variable environments may promote species coexistence via relative non-linearity of competition and the storage effect. In each case, niche differences among species play important stabilizing roles in promoting species diversity.

We live today in an era of rapid environmental change, and understanding how communities and ecosystems respond to changing environments is an area of intense research. Theory predicts that communities may respond to gradual environmental change via smooth transitions, or via regime shifts—abrupt and startling shifts between states. Ecologists are exploring the causes and consequences of regime shifts, looking for early warning signals of critical transitions (Scheffer et al. 2009; Carpenter et al. 2011; Dakos et al. 2011), and examining the potential impacts of regime shifts and tipping points on restoration and conservation (Suding and Hobbs 2009).

Summary

1. Populations and communities show a great deal of natural variability. This variability may include random fluctuations in multiple directions or may involve directional trends, which may involve smooth or abrupt transitions, the latter of which may or may not be reversible.

2. Succession was a foundational topic in community ecology, and remains related to many key ideas like the effects of disturbance, competitive exclusion, forest gap dynamics, and competition-colonization trade-offs.

3. The intermediate disturbance hypothesis proposes that disturbance interrupts the process of succession, so that good colonizers are favored at high rates of disturbance, good competitors are favored at low rates of disturbance, and intermediate rates of disturbance promote the coexistence of both types of species. There is surprisingly little evidence supporting the general importance of this hypothesis in nature.

4. Species within a community may co-occur for long periods as a result of slow rates of competitive exclusion, but if there are no mechanisms promoting the recovery of species that become rare, species coexistence will be unstable, species will eventually disappear, and community richness will decline. Stable coexistence requires that species tend to recover from low densities and that species densities do not show long-term trends.

5. The three components of the storage effect are:
   • differential, species-specific responses to environmental variation;
   • buffered population growth, in which a long-lived life stage allows species to "store" the positive effects of "good" recruitment years and buffer the negative effects of "bad" years;
   • covariance between environment and competition such that intraspecific competitive effects are strengthened when the environment favors a species' population growth and reduced when the environment is unfavorable and the species becomes rare.

6. A community's response to slow and steady environmental change might be gradual and linear or a community might be relatively insensitive to environmental changes across a range of conditions, but respond strongly around some tipping point. Such a large, abrupt change is referred to as a regime shift. The strength of perturbation needed to shift a community from one state to another is called resistance, and the speed at which a community recovers from a perturbation may be referred to as resilience, although resilience has other definitions as well.

7. Communities may exist in alternative stable states—different configurations that represent different equilibrium states. Alternative stable states may be defined mathematically as different basins of attraction and represented graphically as a landscape of hills and valleys. A community tends to remain in a stable state (in a valley or basin) until a perturbation shifts it to another state, or until the landscape changes shape due to a change in the external environment. The community's delayed response to a forward and backward change in environmental conditions is termed hysteresis. There are a few good examples of alternative stable states in nature, although the number of well-documented cases remains limited.

CHAPTER 16

Evolutionary community ecology

Nothing in biology makes sense except in the light of evolution.

Theodosius Dobzhansky, 1964, p. 449

While it has long been recognized that adaptive evolution occurs in an ecological context, it is now clear that evolutionary change feeds back directly and indirectly to demographic and community processes. **Scott Carroll et al., 2007, p. 390**

Despite the potential for a directional influence of evolution on ecology...we still don't know if the evolution–ecology pathway is frequent and strong enough in nature to be broadly important.

Thomas W. Schoener, 2011, p. 426

Ecology and evolution go hand in hand, as noted by Hutchinson (1965) in his famous essay entitled *The Ecological Theater and the Evolutionary Play*. Hutchinson's metaphor describes a traditional way of viewing the interplay between ecological and evolutionary processes: ecology provides the selective environment in which evolution acts. Because evolution occurs over relatively long time scales, ecologists had thought it unlikely that evolutionary events could affect the dynamics of populations or the outcome of species interactions or the structure of local communities. This view is changing, however. There are now at least three active areas of research in which ecologists are recognizing that evolutionary processes directly feedback on ecology:

1. In some cases **evolutionary processes are rapid enough** to occur on the same time-scales as ecological processes like population dynamics. One common scenario is when predation creates strong selection pressure, and prey evolve in just a few generations, thereby bringing ecological and

evolutionary time-scales into alignment. Another is when rapid environmental change occurs, requiring a population to evolve quickly or go extinct, sometimes leading to what is known as evolutionary rescue.

2. From the community assembly (Chapter 12) and metacommunity (Chapter 14) viewpoints, local communities are a subset of all possible species in the regional pool. Evolutionary history in the form of a phylogeny could be informative about how these filtering processes work. In particular, closely related species might have more niche overlap and compete more strongly with each other, a notion that goes back to Darwin. Using phylogenies to study the structuring of local communities is known as **community phylogenetics**.

3. The community assembly and metacommunity perspectives also highlight the importance of studying and understanding the large-space, long-term **evolutionary processes that create the regional pools** that local communities are drawn from (as called for in Chapter 12).

Community Ecology. Second Edition. Gary G. Mittelbach & Brian J. McGill, Oxford University Press (2019).
© Gary G. Mittelbach & Brian J. McGill (2019). DOI: 10.1093/oso/9780198835851.001.0001

We will now explore the three viewpoints highlighted above where the Hutchinsonian metaphor is broken and where the feedback between evolution and ecology is more direct.

Rapid evolution and eco-evolutionary dynamics

Biologists have long appreciated how the interplay between ecology and evolution can lead to the generation of biodiversity over relatively long time scales (e.g., Darwin 1859; Lewontin 2000). Adaptive radiation is a classic example of how ecological interactions can drive the process of speciation (Schluter 2000; Grant and Grant 2008; Losos 2010), resulting in an increase in the diversity of form and function within a community, which in turn has important consequences for community and ecosystem processes. Only recently, however, have ecologists come to recognize that evolutionary changes in species traits driven by selection over a relatively few generations can result in eco-evolutionary feedbacks in contemporary time. Post and Palkovacs (2009, p. 1629) define **eco-evolutionary feedbacks** as "the cyclical interaction between ecology and evolution such that changes in ecological interactions drive evolutionary change in organismal traits that, in turn, alter the form of ecological interactions." Post and Palkovacs (2009) focus on the ecological consequences of eco-evolutionary feedbacks in species that have especially strong effects on communities and ecosystems (e.g., keystone species, foundation species, and ecosystem engineers). Other studies of eco-evolutionary feedbacks have focused on how rapid evolutionary changes in species traits affect population dynamics and the outcome of species interactions over ecological time (e.g., Hairston et al. 2005; Strauss et al. 2008; terHorst et al. 2010; Carlson et al. 2011; Hiltunen et al. 2014; Hendry 2017).

Rapid evolution, and its consequences for population dynamics and species interactions

Theory suggests that evolutionary changes in prey vulnerability can markedly affect the stability of predator–prey interactions, particularly the tendency for populations to cycle (Ives and Dobson 1987; Abrams and Matsuda 1997; see reviews in Abrams 2000; Fussmann et al. 2007). For example, if traits that confer reduced vulnerability to predators come at the cost of lower population growth in the prey (e.g., there is a trade-off between being well defended and being a good competitor), and if predators have a saturating (type II) functional response, then the evolution of reduced vulnerability will generally increase the propensity for prey and predator populations to cycle (Abrams 2000). Might such effects of eco-evolutionary feedbacks on predator–prey dynamics occur in the real world? Hairston, Ellner, and colleagues (Yoshida et al. 2003; Becks et al. 2010) developed an elegant experimental system using two species of freshwater plankton—the green alga *Chlamydomonas reinhardtii* and the grazing rotifer *Brachionus calyciflorus*—to answer this question.

Phytoplankton (algae) possess a number of traits that can reduce their vulnerability to grazing by zooplankton (such as rotifers), including spines, toxic chemicals, gelatinous coatings, and especially increased size. One common way in which algae can increase in size is to form colonies. *Chlamydomonas reinhardtii* normally grow as individual biflagellate cells. When grazing zooplankton are present at high densities, however, *Chlamydomonas* form palmelloid colonies (clumps of dozens of unflagellated cells) that are too large for rotifers to eat. There is a cost to this defense, however, as the clumped form of *Chlamydomonas* is a poorer competitor for resources and has a lower population growth rate than the single-celled form (see Yoshida et al. 2003; Becks et al. 2010 and references therein). Becks et al. (2010) showed that the formation of colonies in *Chlamydomonas* is a heritable trait subject to natural selection. Thus, this rotifer–algal system provides an excellent model system in which to test for eco-evolutionary dynamics.

Jones and Ellner (2007) and Jones et al. (2009) showed theoretically that when prey exhibit a trade-off between defensive ability and competitive ability, qualitatively different population dynamics will result depending on whether the initial prey population contains a wide or a narrow range of genotypes along the trade-off curve. If the range of variation among the prey genotypes present is large, then oscillations in predator and prey densities will create temporal changes in the direction and rate of evolution that result in the maintenance of variation in prey defensive traits. However, if the initial range of prey genotypes present is small (alternative types are close together on the trade-off curve), then the population will evolve toward a single genotype, with the better-defended prey trait going to fixation (Becks et al. 2010). Thus, Becks et al. were able to

make and test clear predictions about the role of eco-evolutionary dynamics in maintaining adaptive variation in a prey population and in generating alternative predator–prey dynamics (e.g., cycling versus stable coexistence).

Becks et al. (2010) initiated some of their laboratory microcosms with a population of *Chlamydomonas* that had not been exposed to grazing by rotifers and

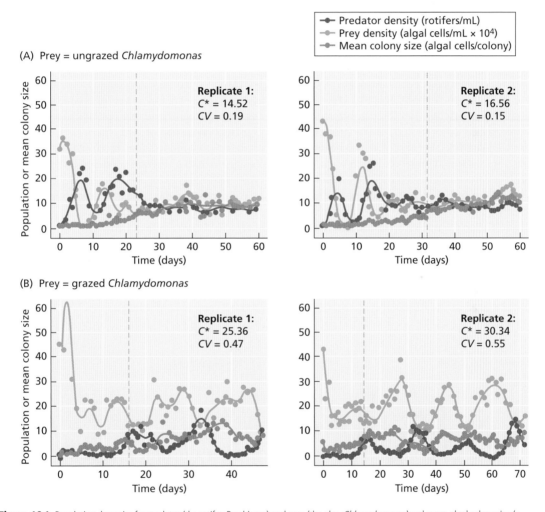

Figure 16.1 Population dynamics for predator (the rotifer *Brachionus*) and prey (the alga *Chlamydomonas*) and mean algal colony size (a measure of algal defensive ability) in two replicate chemostat runs with "ungrazed" and "grazed" *Chlamydomonas* as prey. (A) "Ungrazed" *Chlamydomonas* were obtained from laboratory cultures grown for thousands of generations in the absence of any predators. (B) "Grazed" *Chlamydomonas* came from laboratory cultures continuously exposed to predation by *Brachionus* for 6 months. C^* is a measure of algal defensive ability; higher values indicate greater resistance to predation. CV is the average coefficient of variation for the predator and total prey abundances (higher values indicate stronger predator–prey cycling). C^* and CV were computed using the dates after the vertical dashed line in each panel. After Becks et al. (2010).

thus had relatively little heritable variation in their defensive ability (measured here by colony size). They found that after one full predator–prey oscillation, in which prey were driven to very low densities, the prey population evolved toward a single, moderately well-defended genotype, and predator and prey populations settled into an equilibrium (Figure 16.1A). This is exactly what theory predicted in the presence of eco-evolutionary feedbacks. On the other hand, when microcosms were started with a *Chlamydomonas* population that had been exposed to grazing—and thus had a large amount of genotypic variation in defensive ability—the predator and prey populations cycled, with the dominant algal genotype alternating over time between well-defended large colonies and vulnerable small colonies or single cells (Figure 16.1B).

Again, this is exactly what the theory predicted; high predator densities should lead to strong selection for well-defended genotypes, which results in a decrease in predator abundance, which leads to the replacement of well-defended prey genotypes with less well-defended, but better competitor genotypes, which leads to an increase in predator densities and the initiation of another cycle. Moreover, Becks et al. showed that average prey defensive ability (mean colony size) was greater in the cycling populations than in the populations at equilibrium (Figure 16.1), as predicted by theory. In a recent review of the literature, Hiltunen et al. (2014) found

evidence for significant eco-evolutionary dynamics in 13 of 21 published studies of multi-generational predator–prey experiments conducted in the laboratory. Interestingly, these studies included some of the "classic" predator-prey experiments in ecology (e.g., Gause 1934), conducted well before the concepts of rapid evolution and eco-evolutionary dynamics were developed.

Eco-evolutionary feedbacks and predator–prey cycles

As we saw in Chapter 5, the interaction between a predator and its prey (or more generally, a consumer and its biological resource) can produce population dynamics that range from stable coexistence to coupled oscillations of predator and prey densities through time (e.g., predator–prey cycles) to extinction of one or both species. Conventional predator–prey models (i.e., those without adaptive change in either the prey or predator) produce cycles that have a characteristic form - the population response of the predator lags approximately one quarter period behind that of its prey (as in Figure 16.2B; see also Fig. 5.6A). On the other hand, predator-prey models that include adaptive change in the prey population produce predator-prey cycles that are "antiphase" with a half period lag (Figure 16.2A); i.e., the maximal abundance of the predator population coincides with the minimum

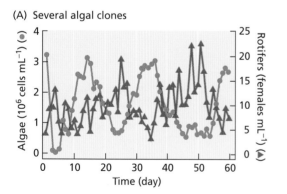 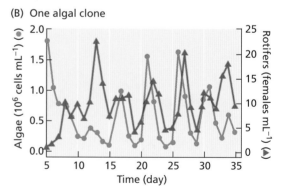

Figure 16.2 Examples of "classical" and "evolutionary" predator-prey cycles. (B) Classical predator–prey cycles with a quarter-period lag in a rotifer–algal laboratory culture, with a single prey type. Solid red circles are the rotifer predator and open green circles are the algal prey (Yoshida et al. 2003). (A) Evolutionary (anti-phase) cycles in laboratory cultures with a rotifer predator and multiple algal clones as prey (Yoshida et al. 2003). After Ellner (2013).

abundance of the prey (Abrams and Matsuda 1997; Jones and Ellner 2007).

Rapid evolution of prey defense and competitive ability can produce antiphase cycles. When predator numbers are high, there is strong selection for effective predator defense, leading to an increase in the abundance of well-defended prey genotypes. This causes predator numbers to decline and remain low, even when (well-defended) prey are abundant. However, because prey defense comes at the cost of reduced competitive ability, there is selection for more competitive (but less-defended) prey genotypes when predator numbers are low. This, in turn, leads to an increase in the abundance of less-defended prey, followed by an increase in predator density, starting the antiphase cycle over again. An important feature of antiphase evolutionary dynamics is that defended and competitive prey genotypes oscillate in frequency along with changes in predator abundance.

In the algal-rotifer interaction studied by Hairston, Ellner, and colleagues, the defended and competitive prey genotypes are different clones of the algae, *Chlorella vulgaris*. Indeed, most of the species shown by Hiltunen et al. (2014) to exhibit antiphase cycling in the laboratory reproduce asexually by binary fission. Thus, "evolution" in these systems is synonymous with clonal selection. The origin of the different genotypes (clones) varied among the study systems reviewed by Hiltunen et al. (2014). In one study, multiple genotypes (clones) were purposefully introduced at the start of the experiment (e.g., Yoshida et al. 2003) and in some studies they were known to arise by de novo mutation (Bohannan and Lenski 1997; see discussion of this experiment in Chapter 11), but in most studies the origin (and, indeed, existence) of different genotypes (clones) was unknown. If defended genotypes arise in predator-prey interactions by mutation or selection of initially rare genotypes, then the system should quickly transition from 'classic' predator–prey oscillations to antiphase dynamics. Hiltunen et al. (2014) found evidence for such transitions in a number of published studies.

Eco-evolutionary feedbacks in nature

The experimental work of Hairston, Ellner, and colleagues described in this chapter is a striking example of how eco-evolutionary feedbacks may affect population dynamics, the stability of species interactions, and the maintenance of genetic variability within populations. Few studies in any area of community ecology exhibit such close and elegant ties between theoretical predictions and experimental tests in a model system. In addition, Hiltunen et al. (2014) found evidence of eco-evolutionary type cycles in about half of published laboratory studies using simple predator–prey systems (e.g., protozoans and bacteria, rotifers, and algae). However, as we discussed in Chapter 8, ecologists want to know whether species in nature (which are exposed to a variety of uncontrolled abiotic and biotic effects) also respond as theory and model systems predict. Hiltunen et al. (2014, p. 922) emphasize this point as well, stating that "… a pressing question in the study of eco-evolutionary dynamics is the extent to which they occur in natural populations, embedded in complex communities …". To date, relatively few studies have examined the ecological consequences of rapid evolution in the field, although the number is steadily increasing (Fussmann et al. 2007; Schoener 2011). We highlight a few eco-evolutionary field studies below. Andrew Hendry's (2017) book *Eco-evolutionary Dynamics* also presents a number of field studies showing different aspects of eco-evolutionary feedbacks and a recent symposium volume collected the results of many recent (and ongoing) eco-evolutionary field studies with fish, amphibians and lizards (Hendry and Green 2017).

1. Bassar et al. (2010) found that guppies (*Poecilia reticulata*) from Trinidadian streams with few or abundant predators evolved different morphological phenotypes and feeding behaviors. Guppies from high-predation streams exhibited greater food selectivity, consuming more invertebrates and less algae and detritus, than guppies from low-predation streams. These heritable differences in guppy phenotypic traits were correlated with ecological differences in guppy population density and population size structure between high- and low-predation streams. When Bassar et al. experimentally varied guppy phenotypes and abundances independently in field mesocosms (holding population size structure constant), they found that phenotypic differences between

populations had marked effects on several eco-system-level properties; for example, mesocosms containing guppies from high-predation streams had greater algal standing stocks, lower invertebrate biomasses, lower leaf decomposition rates, and higher NH_4 excretion rates than mesocosms containing guppies from low-predation streams. Thus, the evolutionary responses of guppies to high (HP) and low predator (LP) densities led to marked ecosystem changes in experimental mesocosms in the field. Importantly, Simon et al. (2017) showed that excluding HP and LP guppy phenotypes from areas of a natural stream in Trinidad had ecosystem effects similar to those observed by Bassar et al. (2010) in their experimental mesocosms.

2. Ecotypes ("benthics" and "limnetics") of the three-spined stickleback, *Gasterosteus aculeatus*, have evolved in a number of north-temperate lakes since the last ice-age (Schluter 1994). Harmon et al. (2009) showed that these two stickleback ecotypes can differentially affect primary production, dissolved organic matter, and prey species diversity when stocked into experimental mesocosms (although a similar mesocosm experiment using a lower fish density observed much smaller effects; Des Roches et al. 2013). Best et al. (2017) also showed how two different evolutionary lineages of sticklebacks from Switzerland had strong and contrasting impacts on ecosystem properties (e.g., prey community structure) in experimental mesocosms and that these ecosystem impacts affected the survival of juvenile stickleback. Finally, in a combination of field and mesocosm studies, Rudman and Schluter (2016) showed that the collapse of the two stickleback ecotypes in Enos Lake (British Columbia, CA) into a single hybrid population ("reverse speciation" or introgressive extinction) lead to marked changes in the lake invertebrate prey community consistent with the loss of the benthic and limnetic forms. Thus, the evolution of phenotypic differences among stickleback populations strongly affects ecosystem properties that may in turn affect selection in subsequent generations.

3. In New England, the construction of dams has resulted in the formation of migratory and land-locked forms of the alewife (*Alosa pseudoharengus*). These fish, which can act as keystone predators in lakes, have evolved morphological differences in response to being resident or migratory as the result of eco-evolutionary interactions with the lake zooplankton community (Post et al. 2008; Palkovacs and Post 2009). In lakes where alewife are present year-round, large-bodied zooplankton are rare (due to size-selective predation by alewife) and the resident alewife have evolved to become more efficient at feeding on small-bodied prey (e.g., smaller mouth gape and finer gill raker spacing). Interestingly, bluegill sunfish that occur in lakes with resident alewife also display morphological traits and feeding behaviors that make them better at feeding on small-bodied zooplankton compared to bluegill from lakes with migratory alewife (Huss et al. 2014). The extent to which among-lake differences in bluegill behavior and morphology are the result of genetic divergence or phenotypic plasticity is unknown.

These field experimental studies, as well as observational field studies analyzing long-term data (e.g., Grant and Grant 2008; Ezard et al. 2009), provide valuable insights into the potential consequences of eco-evolutionary feedbacks in nature (Hendry 2017). However, no experimental field study to date has demonstrated the multigenerational, ongoing process of eco-evolutionary feedback to the same convincing degree that laboratory experiments have (e.g., Fussmann et al. 2005; Becks et al. 2010; terHorst et al. 2010).

Quantifying the ecological consequences of rapid evolution

Ellner et al. (2011) developed analytical methods to quantify the relative effects of evolutionary trait change, non-heritable trait change (i.e., phenotypic plasticity), and the environment on the outcome of species interactions and the structure and functioning of communities and ecosystems. Their approach to studying the ecological consequences of rapid evolution is an advance over earlier analytical methods developed by these and other collaborators (i.e., Hairston et al. 2005), which assumed that all trait change was the result of evolution (i.e., no phenotypic plasticity). Ellner et al. (2011) applied their analytical methods to a number of empirical studies of fish, birds, and zooplankton. They found that the contribution of rapid evolution to the outcomes of ecological interactions can be substantial, but that this contribution can vary widely among study systems (see also Ellner 2013). Moreover, they found that rapid evolution of traits often acts to oppose or reduce the impact of environmental change, suggesting that rapid evolution may be most important when it is least evident. For example, in analyzing

the results of the guppy experiments by Bassar et al. (2010) described above, Ellner et al. (2011) showed that half of the ecosystem response variables measured were affected more by guppy evolutionary responses to predation than by predator-driven changes in guppy densities, and that in a number of cases the strong contribution of evolutionary change was opposite in direction from the ecological contribution (Figure 16.3). Govaert et al. (2016) provide a useful summary and comparison of the different metrics used to quantify how ecology and evolution contribute to trait change within a community.

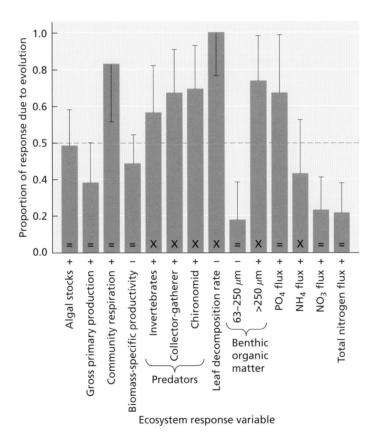

Figure 16.3 Ellner and colleagues analyzed the experiments of Bassar et al. (2010) to quantify the ecosystem effects of the evolutionary and ecological responses of guppy populations to high and low predator densities. Exposure to predators caused an evolutionary change in guppy phenotypes and an ecological change (roughly a two-fold decrease) in guppy population density. Histogram bar heights measure the absolute value of the "evolution" contribution to an ecosystem response divided by the sum of the absolute values of the "evolution" and "ecology" contributions (i.e., the proportion of the total response that is due to "evolution"). The range of possible values is 0–1; a value of 0.5 (shown by the horizontal dashed line) indicates an equal effect of guppy evolution and guppy density on the ecosystem response variable. Symbols below the bars indicate the "ecology" contribution to the ecosystem response variable was positive or negative; symbols within the bars indicate whether the "ecology" and "evolution" contributions are in the same (=) or opposite (X) directions. After Ellner et al. (2011).

Eco-evolutionary dynamics in diverse communities?

Many questions remain about the importance of rapid evolution to the outcome of ecological interactions, population dynamics, and the structure and functioning of ecological systems. Addressing these questions in natural systems won't be easy (Ellner et al. 2011; Schoener 2011; Strauss 2014). Nevertheless, the study of rapid evolution and eco-evolutionary dynamics is making impressive progress. One question that has not been addressed is how species diversity within a community may affect the potential for rapid evolution within species or its ecological consequences. For example, consider the rotifer–alga predator–prey system studied by Hairston, Ellner, and colleagues. In this system, we saw that when prey exhibit a trade-off between defensive ability and competitive ability, qualitatively different dynamics will result depending on whether the prey population contains a wide or a narrow range of genotypes along the trade-off curve. However, what if the prey community contains multiple species arrayed along the trade-off curve of defensive and competitive ability, rather than the single prey species studied by Becks et al. (2010)? We might expect the range of trait space occupied by multiple species to be far greater than the range of trait space occupied by multiple genotypes of a single species. How would this affect the potential for rapid evolution and its impact on population dynamics and species interactions? Would we see eco-evolutionary dynamics between predator and prey, or would we see species sorting and dominance by different species under different environmental conditions, or some combination of the above? These and other questions in eco-evolutionary dynamics are just beginning to be explored (de Mazancourt et al. 2008).

Evolutionary rescue

Faced with a changing environment, species must migrate, adapt, or go extinct. Evolutionary rescue describes the process by which adaptive evolutionary change may prevent extinction and restore positive growth to a population declining as the result of a changed environment (Carlson et al. 2014; Bell 2017). Gomulkiewicz and Holt (1995) developed the mathematical framework for evolutionary rescue in a short paper entitled "When does evolution by natural selection prevent extinction?". Here they show that evolutionary rescue in the face of environmental change should often result in a U-shaped pattern of population density through time (Figure 16.4) and that the key element of successful evolutionary rescue is whether adaptive change is fast enough to restore positive population growth before a population declines to the point of impending extinction. Gomulkiewicz and Holt's

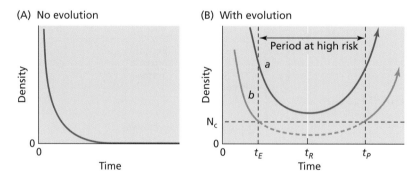

Figure 16.4 The process of evolutionary rescue, whereby evolution by natural selection may prevent extinction in the face of environmental change. (A) In the absence of evolution a maladapted population declines to extinction. (B) With evolution an initially maladapted population may eventually achieve positive population growth. Trajectory *a* shows a population that evolves quickly enough to remain above critically low densities (below N_c). Trajectory *b* shows a population that adapts more slowly and declines to a critically low density where it is vulnerable to extinction before it may achieve positive population growth. After Gomulkiewicz and Holt (1995).

(1995) analysis suggests that only relatively large populations that are not too severely maladapted to their new environment are likely to escape extinction via evolutionary rescue.

Evidence for evolutionary rescue has been demonstrated in laboratory experiments with yeast and bacteria. For example, in a series of elegant experiments by Bell and Gonzalez (2009, 2011) and Samani and Bell (2010), yeast populations were subject to a novel, and maladaptive environment (high concentration of salt). These experiments confirmed theoretical predictions that evolutionary rescue is more likely when populations are large, when environmental change is gradual, and when populations have some prior evolutionary history with the environmental stressor. Evidence of evolutionary rescue in nature, however, is still relatively rare [but see Bell (2017) for some examples, particularly in microbes]. Carlson et al. (2014) present two putative examples of evolutionary rescue in macroorganisms:

- the rapid evolution of resistance to a novel pathogen in rainbow trout (*Oncorhynchus mykiss*; Miller and Vincent 2008);
- the rapid evolution of "silent" male field crickets (*Teleogryllus oceanicus*) in Hawaii in response to an invasive parasitoid fly that selectively preys on male crickets calling to attract mates (Zuk et al. 2006; Tinghitella 2008).

In each case, a sharp population decline following the introduction of a new mortality agent was mitigated by the rapid evolution of an adaptive trait. However, much more work is needed to determine how often evolutionary rescue occurs in nature and whether it is likely to be a potent force in reducing species loss due to environmental change (Bell 2017).

Community phylogenetics

No ecological study can fail to benefit in some way from an understanding of the phylogenetic relationships of its taxa. C. O.   **Webb et al. 2002, p. 497**

A **phylogeny** describes the hypothesized pattern of evolutionary relationships among a set of organisms (generally represented as a branching tree with living taxa at the tips of the branches). A simple phylogenetic tree with the major features labeled is

presented in Box 16.1. Although evolutionary relationships and phylogenies have been studied for a very long time (Darwin produced one of the first evolutionary trees in his 1859 book *On The Origin of Species by Means of Natural Selection or the Preservation*

Box 16.1 A phylogenetic tree

A **phylogenetic tree**, also known as a phylogeny, illustrates the inferred evolutionary relationships among a group of organisms (taxa) that are descended from a common ancestor. The tips of the tree represent descendant taxa (A–F in the example below), these are often species, but could be higher taxonomic units such as genera or families. A **node** represents a taxonomic unit (terminal nodes are existing taxa and internal nodes represent shared ancestors). Speciation events are inferred to occur at the internal nodes. The branches of the tree define the relationships among the taxa; the pattern of branching is referred to as the tree's **topology**. Branch lengths may or may not have meaning. If a phylogeny is constructed based on molecular data, such as DNA nucleotide substitutions, then the branch lengths may represent the number of nucleotide substitutions. If the substitution rate is calibrated to time (a molecular clock), then the branch lengths may represent evolutionary time in years. In this example, a scale bar denotes the branch length corresponding to 0.1 units of time. Most phylogenies include an **outgroup**, or distantly related taxon, which is useful for rooting the tree (i.e., determining the group's common ancestor or **root**) and defining the relationships among the taxa of interest (species F is the outgroup in this example). A **monophyletic lineage** within a phylogeny—one that includes an ancestor and all its descendants—is called a **clade** (the dashed line encircles one of the clades in this example). By definition, a clade can be separated from the root of

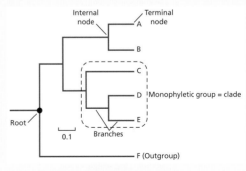

continued

Box 16.1 *Continued*

the tree by cutting a single branch. Two descendants that split from the same node are called **sister taxa**. In this example, species A and B are sister species (they are each other's closest relatives). Likewise, the clade including species A and B and the clade including species C, D, and E, are sister clades (the two clades share a node). For more information on phylogenies and phylogenetic methods, see Baum 2008 and the following online sources: University of California Museum of Paleontology, *Understanding Evolution*, http://www.evolution.berkeley.edu/; National Center for Biology Information, *A Science Primer*, http://www.ncbi.nlm.nih.gov/About/primer/phylo.html.

---

of Favoured Races in the Struggle for Life), the application of phylogenetic analysis to the study of comparative biology is surprisingly recent, beginning in earnest in the 1980s (e.g., Brooks 1985; Felsenstein 1985). Felsenstein's 1985 paper is a landmark in that it was among the first to show why studies of comparative biology often need to incorporate the evolutionary relationships among organisms (their phylogeny), as well as providing a method for doing so. Consider this simple example. Suppose we hypothesize that larger beaks in seed-eating birds evolved as an adaptation to allow them to eat bigger and harder seeds. We might test this hypothesis by looking at the relationship between beak size and the average seed size eaten by different bird species. Figure 16.5 shows such a relationship for five hypothetical bird species (species A–E). The tight, positive relationship between beak width and the average seed size eaten in these five species

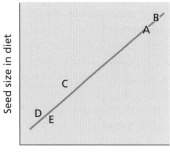

Figure 16.5 A hypothetical relationship between beak size and the size of seeds eaten by five bird species (species A–E).

would seem to support our hypothesis. However, what if the bird species in our hypothetical community are related as shown in the phylogeny in Box 16.1? In this example, species with similar beak sizes are also very closely related (e.g., the large-beaked species A and B, and the small-beaked species D and E). Therefore, using each bird species as a data point in our analysis (Figure 16.5) violates the statistical assumption that the data points are independent of each other. Evolutionarily, they are not independent. Rather, it is possible that some species have similar beak sizes because they share a recent common ancestor, not because they evolved a similar beak size in response to selection. However, it is possible that the species in Figure 16.5 are clustered for other reasons and beak size may still vary adaptively with seed size. For example, if the evolution of coloration or foot structure causes species A–B to occupy one habitat type (e.g., forest) and C–E another habitat type (e.g., grassland), this could cause species A–B to encounter (and eat) different sizes of seeds than species C–E. In this scenario, having larger beaks might still be an adaptation for eating larger seeds. Felsenstein (1985) provided a method (phylogenetic independent contrasts or PIC) to account for the evolutionary relationships among species when conducting statistical tests of evolutionary hypotheses using comparative data. Since then evolutionary biologists have developed a robust set of phylogenetic tools for testing evolutionary hypotheses (see O'Meara 2012; Pennell and Harmon 2013) and today the application of phylogenetic comparative methods (PCMs) is a near requirement in studies of comparative biology and ecology. Note that these phylogenetic methods give primacy to the evolutionary history hypothesis rather than direct selective hypotheses (for a quick introduction to the debate see: Westoby et al. 1995; Harvey et al. 1995; Ricklefs and Starck 1996; Martins 2000).

The application of phylogenetics to the study of ecological communities got its start in the early 2000's, despite the fact that over 100 years ago Darwin (1859, p. 76) suggested one of the first applications of evolutionary analysis to community ecology. Darwin noted that closely related species might compete strongly with one another because "species of the same genus have usually, though by no means invariably, some similarity in habits and constitution, and always in structure." In modern parlance, Darwin recognized that species' functional traits may be conserved across phylogenies. Such strong phenotypic similarities among closely related species are an example of what Derrickson and Ricklefs (1988) term **phylogenetic effects. Or, as is often stated in the ecological literature, phylogenetic niche conservatism** (Wiens and Graham 2005).

In Chapter 12 we saw that one of the first modern assembly rules was from Williams (1947), who built on Darwin's observation that evolutionary relatedness is indicative of ecological (niche) relatedness. Specifically, Williams suggested that a local community should have fewer species per genus than expected by chance. We saw that Williams actually found the opposite (species in a community are concentrated in fewer genera than expected by chance). Perhaps this was due to Williams simplification of evolution to the very binary categorization of in-same-genus/not-in-same-genus. Phylogenies give us a much more powerful tool to move from the binary to a continuous measure of relatedness; namely, the total branch length between two species and their most recent shared common ancestor. In a phylogeny constructed from molecular data, branch lengths provide a measure of time. Once molecular phylogenies became widely available, ecologists were fairly quick to apply them to the study of community assembly (Webb et al. 2002). The assembly view of both biotic and abiotic limits to species membership in a community also led to an improvement over Williams approach. Namely, Williams assumed that similar species would be less likely to co-occur due to competition (a biotic filter), but it could be equally likely that similar species would be more likely to co-occur due to environmental filtering (abiotic filter).

Community phylogenetics has two main assumptions, the second of which has a biotic filtering and an abiotic filtering version. These asssumptions are:

1. Species that are more closely related to each other (i.e. closer to each other in phylogenetic distance) will be more similar to each other in phenotype and thus ecological niche (species traits are evolutionarily conserved).
2. Similarity of ecological niche is predictive of community assembly. Species that are more similar to each other in ecological niche will either:
   A. be more likely to be found in the same community due to similar requirements for environmental conditions (be it climate, habitat,

   mutualistic partners, predator context, etc.) (habitat filtering or environmental filtering)
   B. be less likely to be found in the same community due to the increased likelihood of competition and thus competitive exclusion.

We will discuss the validity of Assumption #1 (phylogenetic niche conservatism) in more detail at the end of this section. Assumption #2 can be viewed in terms of niche theory - species' niche relationships are expected to drive community structure (Chapter 1, Chapter 8). However, like much of niche theory, this idea has not been strongly formalized. Assumption #2B is known as the "competitive-relatedness hypothesis" (Cahill et al. 2008) or the "phylogenetic limiting similarity hypothesis" (Violle et al. 2011). Assumption #2A in principal could be due to either abiotic (climate, habitat) or biotic (positive interactions, predator context) factors, but in practice is normally viewed as acting through abiotic factors and thus is often associated with the idea of environmental (habitat) filtering.

If assumptions #1 and #2B hold, then we would expect community composition to show **phylogenetic overdispersion**; that is, species should be more evenly dispersed in phylogenetic space than would occur by chance (i.e., in a community assembled by a random draw from the species pool) (Webb et al. 2002; Kraft et al. 2007; Cavender-Bares et al. 2009; **Figure 16.6A2**). The concept of phylogenetic overdispersion is similar to the concept of niche overdispersion discussed in Chapter 8 (Hutchinsonian ratios, shown in Figure 1.3, are an example of niche overdispersion). On the other hand, if assumptions #1 and #2A hold, then we would expect communities to contain more closely related species than would occur by chance, or **phylogenetic clustering** (Figure 16.6A1). Of course, if either assumption #1 or assumption #2 is false, then we wouldn't necessarily expect any phylogenetic pattern to the species found in a community (e.g., Figure 16.6B4). Thus, phylogenetic analysis (along with phenotypic analysis; see later) has the potential to help resolve this question and reveal the relative importance of competition vs habitat filtering in determining the species composition of communities.

The evidence for Assumption #2A (environmental filtering) has been discussed already in Chapter 12 as it relates to patterns of community assembly

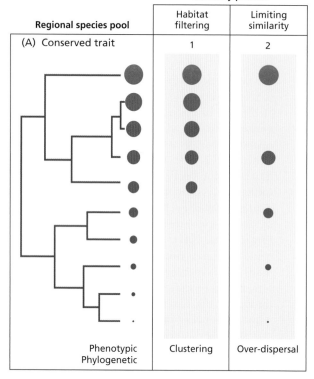

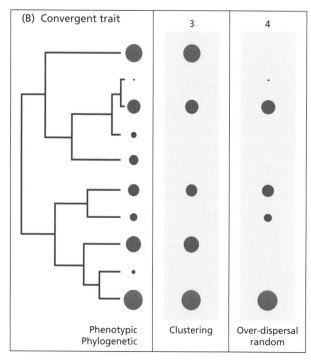

Figure 16.6 An illustration of how processes at the regional and local levels may interact to determine the phylogenetic structure of a community. The red dots represent a quantitative trait (e.g., seed size, specific leaf area, mouth size) and the size of the dot represents the value of that trait (e.g., species with similar-sized dots have similar trait values). In the regional species pool (left panels), the trait may be (A) conserved or (B) convergent across a phylogeny. The results of two different community assembly processes, habitat filtering, and limiting similarity (interspecific competition) are illustrated in the right-hand portion of the figure. The shaded boxes represent four hypothetical communities (numbered 1–4), each containing five species selected from the regional pool of 10 species. Habitat filtering favors species with similar trait values (communities 1 and 3), thus generating phenotypic clustering within communities; the phylogenetic structure within a community depends on whether traits are conserved or convergent. Limiting similarity prevents species with similar trait values from co-occurring, producing phenotypic overdispersion in communities 2 and 4. The phylogenetic structure is over-dispersed in community 2, but can be clustered, random, or over-dispersed when traits are convergent (community 4). After Pausas and Verdú (2010).

based on species traits. The experimental evidence for stronger competition amongst closely-related species (Assumption #2B) will be discussed later. We will then examine the community phylogenetic patterns that have been found and how to interpret these results. Finally, because it provides a nice transition to the third major way evolution influences ecology—evolutionary processes determining the regional species pool—we examine the evidence for phylogenetic conservatism (assumption #1) last.

Assumption #2A: are closely related species stronger competitors?

There have been relatively few tests of the assumption that closely related species are stronger competitors than distantly related species. In a comprehensive analysis, Cahill et al. (2008) found that the relationship between the intensity of interspecific competition and the evolutionary relatedness of the competitors was weak in 142 plant species surveyed, especially compared with the strong and consistent relationships observed between competitive ability and plant functional traits such as size. Cahill and colleagues note that one explanation for this result is that plant traits related to competitive ability may not be strongly conserved across phylogenies, at least not among the species they examined (i.e. phylogenetic conservatism or assumption #1 may be false, also see Mayfield and Levine 2010). On the other hand, Burns and Strauss (2011) found that more closely related plant species were more similar in their ecologies (e.g., rates of germination and early survival) and (under some conditions) were stronger competitors. Violle et al. (2011) conducted a multigenerational laboratory experiment with 10 freshwater protist species (ciliate bacterivores) that demonstrated a strong negative relationship between phylogenetic distance and the frequency of competitive exclusion. More closely related protist species were less likely to coexist in pairwise competition (Figure 16.7A). In addition, the time to competitive exclusion (Figure 16.7C) and the impact of competition on population density (in cases of species coexistence) were both correlated with phylogenetic distance. Thus, Violle et al.'s 2011 study provides strong support for the idea that more closely related species are stronger competitors, at least in the laboratory.

Mayfield and Levine (2010) stressed that studies examining the relationship between phylogenetic relatedness, interspecific competition, and species coexistence should use modern coexistence theory as a conceptual framework. Under this framework, we want to know how the evolutionary relatedness of species affects their stabilizing niche differences and their equalizing fitness differences (Chesson 2000a see Chapter 8). Greater niche differences should promote coexistence, whereas greater fitness differences should hinder coexistence. This means that competitive coexistence processes could create either phylogenetic clustering or phylogenetic overdispersion depending on whether the differences across a phylogeny are related to stabilizing or equalizing mechanisms, respectively. This becomes a twist on assumption #2B, whereby competing species that are more similar are actually *more* likely to coexist due to the equalizing component of coexistence theory. Two recent studies with annual plants have made these comparisons. Godoy et al. (2014) studied 18 species of California annual plants in the field and found that niche differences were unrelated to phylogenetic distance, whereas fitness differences were negatively associated with phylogenetic distance. That is, more closely related species were more similar in fitness in a given environment. Germain et al. (2016) also tested for stabilizing niche differences and equalizing fitness differences in 30 Mediterranean annual plant species in the greenhouse. They found that species from the same geographical region (sympatric species) showed a positive relationship between phylogenetic distance and stabilizing niche differences, which they interpreted as evidence for evolved niche partitioning that might promote coexistence. However, in the same set of species, more distantly related species also were more dissimilar in their fitness, which should hinder coexistence. Thus, phylogenetic distance is not a clear predictor of the outcome of competition in these species. Finally, Narwani et al. (2013) found no support for a positive relationship between phylogenetic distance and stabilizing niche differences in freshwater algae. Clearly, more studies are needed to determine if and how phylogenetic distance is related to the components of contemporary coexistence theory. Taken as a whole, however, the evidence to date provides at best weak support for the hypothesis

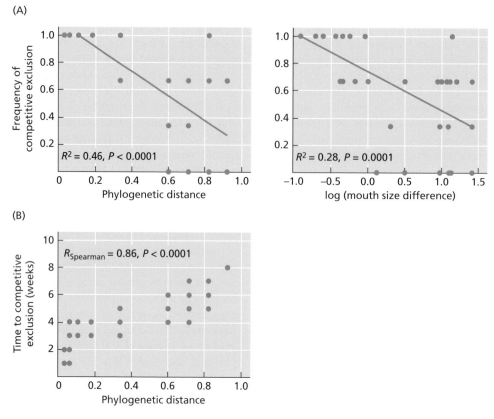

Figure 16.7 Competitive exclusion was measured in 8-week laboratory experiments conducted for all two-species combinations of 10 freshwater protist species. (A) The frequency of competitive exclusion declined as the phylogenetic distance (left) and niche difference (mouth size; right) between species increased. Lines and statistics are based on linear regressions. (B) The time to competitive exclusion increased as a function of phylogenetic distance between the competing species. Results are listed for Spearman's rank correlation. After Violle et al. (2011).

that more closely related species compete more strongly and are therefore less likely to coexist (Darwin's hypothesis).

Patterns of phylogenetic structure in communities

In contrast to the few studies that have tested the assumptions of the competitive-relatedness hypothesis, many studies have examined the community patterns predicted by this hypothesis (summarized in Figure 16.6A). For example, Graham et al. (2009) compared the phylogenetic structures of 189 hummingbird communities in tropical Ecuador. Hummingbird communities are excellent systems in which to test the competitive-relatedness hypothesis; these birds compete strongly for nectar, and there is abundant evidence linking hummingbird traits such

as bill length, bill curvature, and wing length/body size ratio to a bird's ability to hover and harvest nectar in different environments. There is also some evidence that these functional traits are evolutionarily conserved. In Ecuador, hummingbirds are found in a variety of habitats, ranging from moist tropical lowlands to seasonally arid lowlands to high mountains (up to 5000 m), and more than 25 species of hummingbirds may co-occur at a site. Graham et al. found that 37 of 189 hummingbird communities (28%) exhibited significant phylogenetic clustering, but only one community (2%) showed significant phylogenetic overdispersion.

The rarity of significant phylogenetic overdispersion is somewhat surprising, given the strong evidence for interspecific competition in this taxonomic group. However, when Graham et al. (2009) analyzed the spatial distribution of the communities that fell in

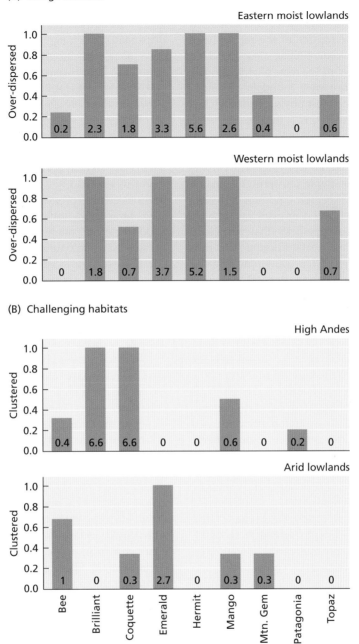

Figure 16.8 Hummingbird communities in Ecuador show the greatest phylogenetic dispersion in the environmentally benign moist lowlands; they show the greatest phylogenetic clustering in the challenging environments of the high Andes and the dry western lowlands. (A) Histograms show hummingbird clades represented in the top tenth percentile of phylogenetically over-dispersed communities that exist in the eastern and western moist lowlands. (B) Histograms show the clades in the top tenth percentile of phylogenetically clustered communities existing in the high Andes and the dry regions west of the Andes. Hummingbird clades are listed along the x axis. The y-axis shows the proportion of communities where a given clade is represented; the numbers in the bars are the mean numbers of species per community. Over-dispersed and clustered communities tend to be represented by different clades in a manner consistent with their ecologies. After Graham et al. (2009).

the top 5% of the most phylogenetically dispersed communities and the top 5% of the most phylogenetically clustered communities, they found an interesting spatial pattern. The most phylogenetically dispersed communities were found in the moist lowlands, whereas the most phylogenetically clustered communities were found in the harsh environments of the high Andes and in the seasonally dry lowlands west of the Andes (Figure 16.8). Other studies also have found more phylogenetic clustering (as well as phenotypic clustering) in harsh environments than in benign environments (e.g., Swenson and Enquist 2007; Cornwell and Ackerly 2009), supporting the hypothesis that we would expect stronger environmental filtering (and thus phylogenetic clustering) in harsh environments and stronger biotic filtering (and thus phylogenetic overdispersion) in benign environments (Weiher and Keddy 1995).

The application of phylogenetic analysis to understanding the relative contributions of habitat filtering and interspecific competition (or other biotic interactions such as facilitation) to the structuring of communities is an active area of research. Note that contrasting predictions can be made depending on whether traits are evolutionarily conserved or evolutionarily convergent (see Figure 16.6B). Such contrasting predictions highlight the importance of testing for a phylogenetic signal in the traits of interest in any phylogenetic analysis of community assembly (see next section).

Cavender-Bares et al. (2009), Pausas and Verdú (2010), Swenson (2011), and Gerhold et al. (2015) provide insightful summaries of the literature on the phylogenetic structure of communities and guides to the methods employed. General conclusions are still tentative, but the evidence suggests that studies conducted at broad taxonomic and spatial scales tend to find phylogenetic clustering because more closely related species have similar habitat affinities (e.g., Helmus et al. 2007). The filtering of regional phenotypic (trait) diversity into local communities follows a similar trend, although the outcome depends on the types of traits examined (Swenson 2011; Gerhold et al. 2015). Studies conducted at smaller spatial scales (within local communities), however, find mixed results. Sometimes phylogenetic diversity within communities is greater than expected by chance (phylogenetic overdispersion), which may be evidence for the importance of interspecific competition (e.g., Cavender-Bares et al. 2006; Swenson et al. 2006, 2007; Helmus et al. 2007; Vamosi et al. 2009; Kraft and Ackerly 2010). However, there are also examples showing a lack of phylogenetic overdispersion at a local scale (e.g., Webb 2000; Silvertown et al. 2006; Zhang et al. 2017). These mixed results (and untested assumptions) lead Gerhold et al. (2015) to conclude in their recent review that "… phylogenetic patterns may be little useful as proxy of community assembly".

To summarize the state of community phylogenetics, a wide variety of patterns have been observed. This is not surprising. Assumption #1 is only sometimes true (see next section). In addition to #2A (phylogenetic clustering due to environmental filtering) and #2B (phylogenetic overdispersion due to niche overlap leading to competitive exclusion), we have to add a 3rd option #2C that suggest more similar species that compete with each other will be phylogenetically clustered due to the equalizing component of coexistence. Then we have to recognize that #2A, #2B, and #2C are all going on at the same time and pulling in opposite directions. Thus, forces causing phylogenetic clustering and forces causing phylogenetic overdispersion may in combination result in a net outcome of no phylogenetic signal. Thus, the current consensus suggests that it is not surprising to find no phylogenetic signal (when phylogenetic conservatism is weak or when #2A, #2B, and #2C oppose each other). We would most likely find phylogenetic clustering when phylogenetic conservatism is strong and when either equalizing mechanisms are strong or when environmental filtering is strong such as along environmental gradients or in harsh environments. Phylogenetic overdispersion would be most likely when phylogenetic conservatism is strong and strong competition among species with similar fitness occurs emphasizing the stabilizing components of coexistence (e.g. in local communities among closely related species).

Phylogenetic niche conservatism (assumption #1)

As we transition from community phylogenetics to the evolutionary forces structuring regional species pools, we turn to examine the role of phylogenetic conservatism which is central to both concepts. The idea that closely related species are ecologically similar underlies another current topic in evolutionary ecology: the concept of phylogenetic niche conservatism, or simply niche conservatism (Wiens and Graham 2005; Losos 2008a; Warren et al. 2008; Wiens et al. 2010). **Phylogenetic niche conservatism** can be defined as "the tendency of species to retain ancestral ecological characteristics" (Wiens and Graham 2005, p. 519). A stricter definition, such as proposed by Losos (2008), requires that closely related species should resemble each other more than expected by neutral genetic drift (Münkemüller et al. 2015). Wiens and Graham (2005) listed four potential causes of niche conservatism:

- stabilizing selection for functional traits;
- limited genetic variation in the traits of interest;
- gene flow from the center of a population's range to the edge, which may limit opportunities to adapt to new environmental conditions;
- pleiotropy and trade-offs among traits.

There has been considerable controversy over the concept of niche conservatism—whether it is a pattern or a process, how best to test for its existence, and at what spatial and temporal scales it applies (Losos 2008a,b, 2011; Warren et al. 2008; Wiens 2008; Wiens et al. 2010). As Wiens and Graham (2005) pointed out, we expect functional traits to be conserved to some extent among related species, but rarely they will be identical. Therefore, proponents of niche conservatism argue that the most important question is not whether niches are more or less conserved than some particular process (e.g., the rate of divergence predicted by a Brownian motion model of evolution) would lead us to expect, but rather, what the consequences of niche conservatism are for ecological and evolutionary patterns and processes. For example, those who study niche conservatism are interested in how the conservation of traits affects the formation of regional biotas, the distribution of species across landscapes, the generation of diversity gradients (e.g., the latitudinal diversity gradient), and the maintenance of biodiversity (Wiens et al. 2010). Interpreting the role of niche conservatism in generating biodiversity patterns is challenging, however, because the

consequences of niche conservatism depend on the spatial and temporal scale of the questions explored. The difficulty is best illustrated by an example.

Figure 16.9A shows a hypothetical phylogeny for a group of eight species. In this case, six of the eight extant species are found in tropical habitats and two species are found in temperate habitats. Evolutionary relationships among the species (inferred, say, from some combination of molecular data, fossils, and species traits) suggest that the earliest ancestor of the group was found in tropical habitats and that the two species currently found in temperate habitats are the result of a recent niche shift from tropical to temperate habitats. If the six tropical species are more similar in their functional traits than are the two temperate species, we might conclude the following. First, the conservation of ancestral traits (niche conservatism) is responsible for the present-day pattern of species distribution across habitats—six of eight extant species are found in the habitat of origin (the tropics), and they share similar traits. Secondly, niche conservatism, combined with increased time for speciation, has resulted in the greater current species richness in tropical than in temperate habitats.

Including a more distantly related group of species in the phylogeny, however, reveals a potentially different story (Figure 16.9B). Now we see that the earliest common ancestor of the clade occurred in temperate habitats and that the diverse tropical component of the clade was the result of a very early niche shift, followed by a higher rate of diversification in the tropics than in temperate habitats. If the phylogeny depicted in Figure 16.8B traces the true evolutionary history of the group, then it appears that a shift from temperate to tropical habitat very early in the clade's history was crucial in producing the current pattern of species diversity across habitats. Moreover, the current high tropical diversity is the result of early colonization of the tropics, a higher rate of tropical diversification, and the tendency of the tropical clade to remain in this habitat after the initial habitat shift.

The example in Figure 16.9 is hypothetical. However, Losos (2008b) and Wiens (2008) discuss a real-world example of lizard species distributions in the Caribbean and in the continental United States that illustrates some of the same challenges

(A)

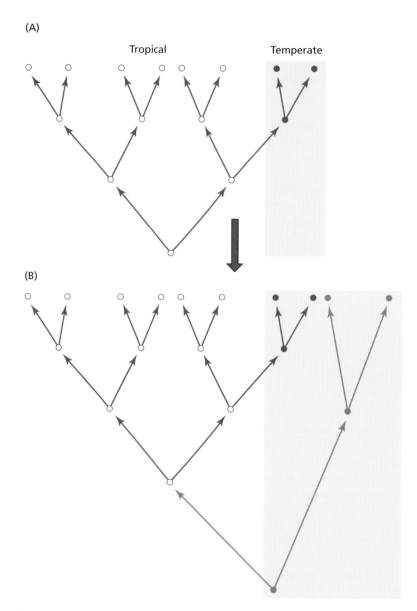

Figure 16.9 A hypothetical phylogeny illustrating the difficulties in interpreting patterns of niche conservatism. (A) Six of the eight extant species in this clade are found in tropical habitats and two species are found in temperate habitats. The evolutionary relationships among the species suggest that the earliest ancestor of the clade was found in tropical habitats, and that the two species currently found in temperate habitats are the result of a recent niche shift from tropical to temperate habitats. Assuming that the six tropical species are more similar in their functional traits than are the two temperate species, we might conclude that phylogenetic niche conservation contributes to the present-day pattern of species distribution across habitats. (B) An expanded phylogeny for the same clade that includes two distantly related species with a temperate-zone distribution. This expanded phylogeny suggests that the earliest common ancestor of the clade occurred in temperate habitats and that the diverse tropical component resulted from a very early niche shift from the temperate zone to the tropics, followed by a high rate of diversification in the tropics compared with the temperate zone and a subsequent recent niche shift back to the temperate zone by two species. These two phylogenies illustrate the difficulties in interpreting the underlying cause of the difference in tropical and temperate species richness in this clade: niche conservatism and time for speciation, as suggested by (A), or differences in rates of diversification between habitats, as suggested by (B).

in interpreting the role of niche conservatism in structuring species assemblages at different spatial and temporal scales. Condamine et al. (2012) provide an impressive example of how fossil-calibrated molecular phylogenetic data, combined with ancestral state reconstruction (based on host–plant associations) and ancestral area reconstruction (based on paleogeographic data and species ranges) can yield a synthetic picture of the underlying evolutionary and ecological drivers of broad-scale diversity patterns in swallowtail butterflies (Papilionidae).

The extent to which species traits are conserved evolutionarily and the degree to which niche conservatism determines contemporary distributions of species and clades are active areas of research that have yielded significant insights into patterns of biodiversity, including the latitudinal diversity gradient (e.g., Wiens et al. 2009) and the composition and richness of local communities (e.g., Harrison and Grace 2007). How species and communities will respond to global climate change is also intimately tied to the question of niche conservatism and the degree to which species can adapt to changing environments (Pearman et al. 2008). Rapid evolution of adaptive traits may ameliorate, in part, the negative impacts of global climate change by allowing species to respond to changing environments (Thomas et al. 2001; Jump and Peñuelas 2005; Lavergne et al. 2010). On the other hand, evolutionary conservation of climatic tolerances could require species to shift their ranges or go extinct (Thuiller et al. 2005; Parmesan 2006). As Sexton et al. (2009, p. 430) note, "Rapid climate change will teach us volumes about the stability or mobility of range limits." Global climate change is also likely to "teach us volumes" about the ability of species to adapt to new environments via rapid evolution or to be constrained by niche conservatism to environments rooted in their evolutionary past.

Evolutionary processes that structure regional species pools

The third type of evolutionary impact on ecological processes that we will examine involves looking at the evolution of the regional pool of species, and specifically the process of evolutionary radiation that sets the diversity of the regional species pool. Adaptive radiation—the diversification by adaptation of a single phylogenetic lineage into a variety of forms occupying different ecological niches (Simpson 1953; Schluter 2000)—is thought to play a prominent role in the evolution of diverse communities. Classic examples of adaptive radiations include Darwin's finches on the Galápagos archipelago (Grant and Grant 2008); cichlid fishes in the African Great Lakes (Fryer and Iles 1972; Seehausen 2006); *Anolis* lizards on islands in the Caribbean (Losos 2009); and silverswords (plants; Figure 16.10), honeycreepers (birds), and *Drosophila* (fruit flies) in the Hawaiian Islands (Carlquist et al. 2003; Pratt 2005; Losos and Ricklefs 2009). It is no accident that these classic examples of adaptive radiation come from isolated islands and lakes. Adaptive radiations are strongly associated with ecological opportunity and often occur when an ancestral species colonizes a relatively unoccupied environment where resources are underutilized (Losos 2010). Other ways in which a lineage may "stumble upon an array of unexploited niches" (Schluter 2000, p. 163) are to survive an extinction event or to acquire a novel evolutionary character—a "key innovation"—that allows the lineage to exploit resources in a new way (Simpson 1953; Roelants et al. 2011). In the rest of this section we will look first at radiation in island-like systems that have been a favorite study subject of both ecologists and evolutionists. Then, we will attempt to outline some basic theory in the much more poorly studied question of how evolution sets the diversity of mainland regional species pools.

Speciation and community assembly in island-like systems

Studies of adaptive radiation in island-like systems provide important insights into the ecological and evolutionary mechanisms that may underlie the assembly of communities over evolutionary time (Losos and Ricklefs 2009; Warren et al. 2014). For example, speciation on small islands is rare, perhaps because below a minimum island size there is insufficient opportunity for geographical isolation and that most speciation is allopatric (Kisel and

(A) *Madia sativa* (ancestral)

(B) *Argyroxiphium sandwicense*

(C) *Wilkesia gymnoxiphium*

(D) *Dubautia arborea*

Figure 16.10 The Hawaiian silversword alliance is a spectacular example of an adaptive radiation of plants. All silverswords are descended from a single ancestral tarweed (probably similar to the extant tarweed *Madia sativa*, shown in panel A) that colonized the Hawaiian Islands from western North America some 5 million years ago (Baldwin and Sanderson 1998). The three genera and 30-odd silversword species endemic to Hawaii show an extravagant diversity of life forms, including monocarpic rosettes (B,C), and woody vines and trees (D). They occur in habitats from rainforests to deserts. (A) © Steffen Hauser/botanikfoto/Alamy; (B) © Science Photo Library/Alamy; (C) © Dani Carlo/AGE Fotostock; (D) courtesy of Forest and Kim Starr.

Barraclough 2010). Secondly, niche filling on isolated islands is achieved by in situ speciation, but niche filling on islands near the mainland is generally the result of colonization (Emerson and Gillespie 2008). Thirdly, morphological diversification in some island taxa can be extensive (e.g., Hawaiian silverswords; Figure 16.10), whereas other taxa colonizing the same islands show little or no in situ diversification (Losos and Ricklefs 2009). Finally, evolution may result in replicated radiations in which morphologically similar species are produced independently, suggesting that niche filling has a deterministic component (Losos and Ricklefs 2009).

In one of the best-documented examples of replicated adaptive radiations, *Anolis* lizards on the four islands of the Greater Antilles (Cuba, Hispaniola,

Jamaica, and Puerto Rico) have diversified independently to produce the same set of habitat specialists (ecomorphs) on each island (Figure 16.11; Losos 2009). Similarly, land snails in the genus *Mandarina* have diversified into the same four niches (arboreal, semi-arboreal, sheltered ground, and exposed ground) on different islands in the Bonin Islands near Japan (Chiba 2004; reviewed in Losos 2009). In post-glacial lakes in the Northern Hemisphere (lakes can be viewed as "islands" in a terrestrial

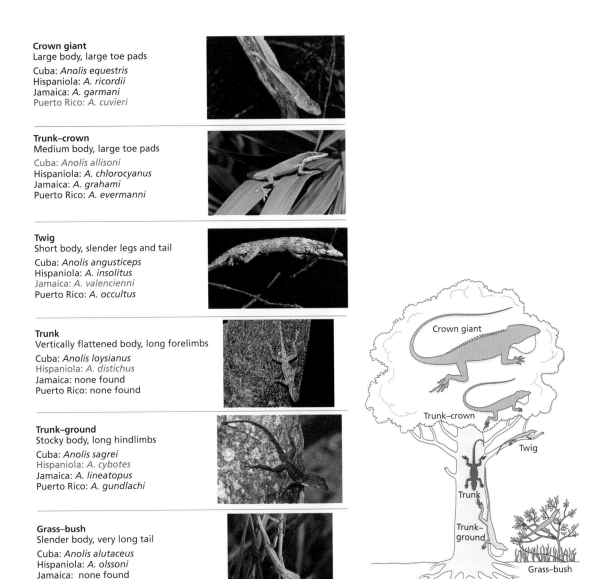

Crown giant
Large body, large toe pads

Cuba: *Anolis equestris*
Hispaniola: *A. ricordii*
Jamaica: *A. garmani*
Puerto Rico: *A. cuvieri*

Trunk–crown
Medium body, large toe pads

Cuba: *Anolis allisoni*
Hispaniola: *A. chlorocyanus*
Jamaica: *A. grahami*
Puerto Rico: *A. evermanni*

Twig
Short body, slender legs and tail

Cuba: *Anolis angusticeps*
Hispaniola: *A. insolitus*
Jamaica: *A. valencienni*
Puerto Rico: *A. occultus*

Trunk
Vertically flattened body, long forelimbs

Cuba: *Anolis loysianus*
Hispaniola: *A. distichus*
Jamaica: none found
Puerto Rico: none found

Trunk–ground
Stocky body, long hindlimbs

Cuba: *Anolis sagrei*
Hispaniola: *A. cybotes*
Jamaica: *A. lineatopus*
Puerto Rico: *A. gundlachi*

Grass–bush
Slender body, very long tail

Cuba: *Anolis alutaceus*
Hispaniola: *A. olssoni*
Jamaica: none found
Puerto Rico: *A. pulchellus*

Figure 16.11 Adaptive radiation among *Anolis* lizards in the islands of the Greater Antilles. Interspecific competition has led to a variety of niche specialists (inset drawing; lizard body outlines are drawn to approximate scale relative to one another). Lizards adapted to corresponding niches on the different islands look substantially similar, although radiation has been independent on each island and thus the species listed are not closely related. Photos illustrate one species (named in color type) exemplifying each niche. From Losos (2009, 2010); photographs courtesy of Jonathan Losos.

landscape), fish species have repeatedly diversified into pelagic (open water) and benthic (bottom habitat) specialists (Schluter 2000). Although these case studies suggest similarities in community development and niche filling across independent adaptive radiations, Losos (2010) argues from a review of the literature that replicated adaptive radiations, in the sense of strict species-for-species matching (Schluter 1990), are rare in nature and that those that occur involve closely related species. As Losos and Ricklefs (2009) note, islands provide excellent examples of both determinism and contingency in evolution, as the chance colonization of islands by different taxa has resulted in the evolution of markedly different communities. Some general trends in niche filling are evident, however; for example, woodpecker "equivalents" have repeatedly evolved on islands in the absence of true woodpeckers, which are poor dispersers.

Building a mainland regional species pool

Community assembly in ecological time is generally viewed as a process involving the dispersal of species from a regional species pool, followed by environmental filtering to establish the local community (Chapter 12). This conceptual framework treats the species pool as an independent and static entity in the assembly of communities. Thus, it ignores how regional species pools are formed, why they differ in diversity and composition, and perhaps most importantly, how ecological and evolutionary processes interact to generate species pools. Already in this chapter we have seen how eco-evolutionary dynamics operating at relatively small spatial scales and over a few hundred generations can have dramatic effects on consumer-resource dynamics, species interactions, and the functioning of ecosystems. Eco-evolutionary dynamics also occur at much larger spatial and temporal scales and it is at these regional/continental scales that species pools are formed. Thus, a more complete picture of community assembly needs to address the ecological and evolutionary feedbacks that generate species pools (Figure 16.12; Mittelbach and Schemske 2015).

Studies of adaptive radiation in island-like systems provide valuable insights into the mechanisms underlying the evolutionary assembly of communities (Gillespie 2015; Rominger et al. 2016); however, there are important ways in which the formation of continental species pools may differ from adaptive radiations observed on islands. For example, most island radiations involve diversification of one or a few clades in relative isolation. However, lineage diversification on a continental scale often occurs in multiple clades, where members of a diversifying lineage may interact with a host of unrelated species across a broad landscape. Moreover, while ecological opportunity, adaptation, and niche differentiation can promote the diversification of lineages via adaptive radiation in relatively species-poor environments, ecological interactions may also act to limit diversification in species-rich biotas. To see how this works, we need to consider more closely the geography of speciation.

Allopatric speciation occurs when a geographical barrier prevents gene flow between populations and reproductive isolation evolves as a byproduct of divergent natural selection. Traditionally, evolutionary biologists studying speciation have focused on the evolution of reproductive isolation, giving scant attention to the relationship between the origin of species and the origin of community diversity. However, if allopatric speciation is the dominant form of speciation in nature (as the evidence strongly suggests; Coyne and Orr 2004), this presents a conundrum for the development of regional species pools and, ultimately, for community development. Consider the process as conceptualized in Figure 16.13. Geographically isolated populations through time are expected to become adapted to their local environments, to diverge in their niches, and as a byproduct may become reproductively isolated (become separate species). This is a standard model for evolutionary diversification. However, for these newly-evolved sister species to become part of the same regional species pool and thereby increase species pool richness, they must again become sympatric. If sister species remain allopatric, it seems unlikely that they would meet the criteria of being in the same species pool (i.e., members of a set of species that can potentially colonize a local site over ecological time). Figure 16.13 illustrates three possibilities for achieving secondary sympatry.

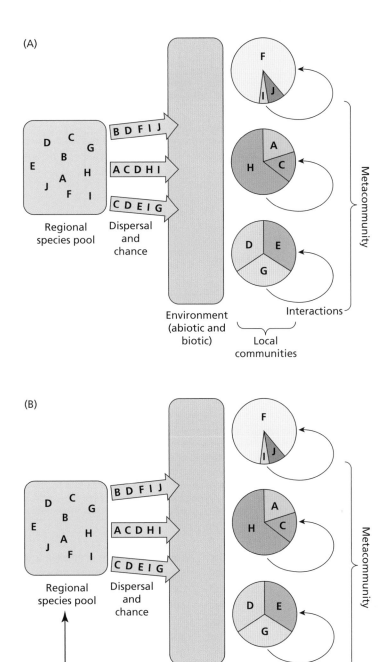

Figure 16.12 Two models of community assembly. (A) Local communities are comprised of a subset of species from the regional species pool that have passed through environmental filters. There is no feedback from the metacommunity (collection of local communities) to the regional species pool. (B) Local communities are assembled as in (A), but speciation adds new species to the pool, extinction removes others, and dispersal allows the persistence of species that might otherwise go extinct. After Mittelbach and Schemske (2015).

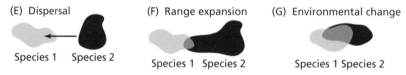

	Reproductive isolation	Niche divergence	Local adaptation
(A)	None	None	None
(B)	Weak	Weak	Weak
(C)	Strong	Strong	Strong
(D)	Complete	Very strong	Very strong

Time

Species 1   Species 2

Secondary sympatry and coexistence

(E) Dispersal

Species 1   Species 2

(F) Range expansion

Species 1   Species 2

(G) Environmental change

Species 1 Species 2

Figure 16.13 Consequences of allopatric speciation for community assembly. Panels (A–D) illustrate the process whereby populations becoming geographically isolated (A) and evolve reproductive isolation (become separate species) as a byproduct of niche divergence and adaptation to their local environments (B–D). Three mechanisms of achieving secondary sympatry and the possible consequences for coexistence are depicted in panels (E–G); (E) dispersal of species 2 into the range of species 1 does not result in coexistence because of strong local adaptation and/or competitive exclusion. (F) adaptation allows species 2 to expand its range and to coexist with species 1 because of diminished local adaptation and strong niche divergence, and (G) large-scale environmental change causes range shifts which erase local adaptation to ancestral habitats and coexistence is achieved because of prior niche divergence. Scenarios E-G focus on mechanisms of stable coexistence between species and do not consider the possibility that species 1 and 2 may overlap in distribution, but are undergoing slow competitive exclusion (i.e., neutral dynamics). After Mittelbach and Schemske (2015).

Achieving secondary sympatry is an important, but largely neglected aspect in the evolutionary development of communities (Price et al. 2014) and it can play a significant role in determining effective speciation rates (Price et al. 2010). For example, Rabosky and Matute (2013) found that speciation rates in birds and *Drosophila* were not closely associated with the time to achieve reproductive isolation, suggesting that other facts (e.g., range expansion and secondary sympatry) may limit the effective rate of speciation. Additional data suggests that the time to achieve secondary sympatry in birds and other taxa can be much longer than the typical time to achieve reproductive isolation (Weir and Price 2011; Price et al. 2014; Laiolo et al. 2017). Much more work is needed to determine how continental species pools evolve and how the geography of speciation affects this evolution. However, it is clear that strong feedbacks

can occur between species interactions (e.g., interspecific competition) in local communities and the development of regional species pools (Mittelbach and Schemske 2015).

Evolutionary radiations often show a burst of speciation and morphological diversification early in their history and a slowing of diversification thereafter (Gavrilets and Losos 2009; Mahler et al. 2010). For example, many fossil groups show a peak in morphological diversity soon after their appearance in the fossil record (Foote 1997) or differentiate widely and rapidly following a mass extinction (Erwin 2001). Phylogenetic analyses from a wide variety of taxa also show a general pattern of diversification "slowdowns" over time in the evolution of a lineage (e.g., Phillimore and Price 2008; Rabosky and Lovette 2008), although there are exceptions to this pattern (e.g., Condamine et al. 2012). Bursts of taxonomic and morphological diversification early in a clade's history, followed by slowdowns suggest strong ecological controls on speciation, with diversification rates declining as niches are filled (Phillimore and Price 2008; Rabosky and Hurlbert 2015). For example, in an extensive analysis of 245 species-level molecular phylogenies for both animals and plants, McPeek (2008) found that most clades showed a pattern of diversification slowdown. McPeek attributed this slowing in diversification not to a reduction in speciation rate as niches filled, but rather to increased rates of competitive exclusion and species extinction. By combining a model of competitive interactions and species coexistence with a model of clade diversification, McPeek showed that clades that diversify by strong niche differentiation exhibit decelerating lineage accumulation over time, whereas clades that diversify primarily by modes of speciation that generate little or no ecological diversification (e.g., sexual selection) show accelerating diversification. The predictions of the model were broadly consistent with the patterns of diversification observed in molecular phylogenies (McPeek 2008) and in the fossil record (McPeek 2007). However, whether and how often the process of evolutionary diversification is diversity-dependent remains controversial and processes other than niche filling can readily account for observed slowdowns in diversification rates inferred from phylogenies (Morlon 2014; Harmon and Harrison 2015). In addition, counter examples showing that high regional diversity may promote diversification (e.g., Machac and Graham 2017) support the idea that sometimes "diversity begets diversity" and that species can provide niches for the evolution of more species (Erwin 2008; Benton 2009; Schemske et al. 2009).

Conclusions

The fields of eco-evolutionary dynamics and community phylogenetics have exploded in recent years. The number of studies demonstrating contemporary evolution and its ecological consequences is expanding rapidly, although experimental field studies are still rare. Likewise, the range of application of phylogenetic analysis to ecological questions seems unbounded. Work to-date shows that the phylogenetic structure of many communities is non-random and that habitat filtering appears to play the more important role in structuring communities over large spatial scales where environmental gradients are strong, and that competitive exclusion/limiting similarity is more important at local scales where competition is strong. Similar results may apply to the filtering of regional trait diversity into local communities, although the evidence here is more limited and the results depend on the types of traits measured (Swenson 2011). Other recent applications of phylogenetics to community ecology have included:

- combining phylogenetic information with species abundance data to obtain better estimates of the elusive parameters of dispersal rates and speciation rates needed to test Hubbell's neutral theory (Jabot and Chave 2009);
- applying phylogenetic information to the study of food webs and other ecological networks (e.g., Cattin et al. 2004; Rezende et al. 2007);
- using phylogenetic analyses to understand patterns of species invasions (e.g., Strauss et al. 2006a,b; Gerhold et al. 2011);
- applying phylogenetics to metacommunities (e.g., Leibold et al. 2010; Leibold and Chase 2018);
- understanding the importance of phylogenetic cascades (co-diversification across multiple trophic levels) in the development of high tropical diversity (Forister and Feldman 2011).

Each of these areas illustrates the potential for using the rapidly expanding database of phylogenetic relationships and rapidly improving analytical techniques to better understand the role of evolution in determining the structure and functioning of communities.

It is important to remember, however, that phylogenies are hypothesized evolutionary relationships; they are not literal descriptions of the evolution of any taxon. In an elegant and thought-provoking essay, Losos (2011) discusses the limitations of the application of phylogenies to comparative biology (including ecology). He concludes that, while phylogenetic analysis is a powerful tool for understanding ecology in an evolutionary context, it is not omnipotent. Therefore, phylogenetics should be combined with other investigative tools (fossils, trait analysis, studies of ongoing evolutionary processes, ecological experiments) to best understand the evolution of communities. Losos (2011, p. 724) concludes with the admonition that "the study of evolutionary phenomena [is like] a detective story in which there is no smoking gun and in which disparate clues have to be drawn together to provide the most plausible hypothesis of what happened." Similarly, it is important to think about the broader context of current studies of rapid evolution and its ecological consequences.

We live in an era of rapid environmental change, presenting evolutionary biologists with unique opportunities to study the processes of selection and adaptation in nature. This chapter has examined examples of rapid evolution and eco-evolutionary feedbacks, documenting their effects on population dynamics in the laboratory, and illustrating their potential consequences for community structure and ecosystem functioning in nature. The study of eco-evolutionary dynamics is fertile ground, both for increasing our understanding of the processes that structure communities and for predicting the consequences of climate change for species distributions and biodiversity. Likewise, evolutionary rescue may play a role in preventing species loss in the face of environmental change, depending on the relative rates of environmental alteration and species adaptation.

An additional long-standing question in evolutionary biology is how the microevolutionary processes of selection and adaptation are related to the macroevolutionary processes of speciation, adaptive radiation, and the filling of ecological niches (Reznick and Ricklefs 2009). These macroevolutionary processes are of particular interest to ecologists because they focus directly on the production of biodiversity. Adaptive radiations highlight how species interactions, particularly interspecific competition, may lead to adaptive niche divergence and how ecological opportunity may drive the rate of diversification within lineages. However, much remains to be done to link Darwin's "struggle for existence" and adaptive divergence to a truly mechanistic understanding of how the microevolutionary processes of selection and adaptation translate into the macroevolutionary dynamics of speciation and the diversification of forms (Carroll et al. 2007; Jablonski 2008; Reznick and Ricklefs 2009). Adaptive

radiations provide excellent opportunities to study these links, and advances in molecular biology and the ability to study the genetic basis of adaptation promise to make the next few decades "a golden age in the study of adaptive radiation" (Losos 2011, p. 635). It is important to note, however, that while adaptive radiations in young environments (e.g., islands) have been well studied, much less is known about how ecology and evolution interact in older environments (e.g., continental radiations) to generate regional diversity. Advances in this area require specific consideration of the factors that influence the "cycle of speciation" (Grant and Grant 2008; Price 2008), whereby populations diverge in allopatry, evolve reproductive isolation, and then expand their ranges to become secondarily sympatric (Figure 16.13). Only recently have ecologists and evolutionary biologists focused their attention to the last step in this process. Interestingly, it is here that ecological interactions may turn out to be the rate-limiting step in the evolution of regional species pools (Price et al. 2014; Mittelbach and Schemske 2015; Cooney et al. 2017).

This chapter has only **begun** to explore the rich palette of interactions by which ecology and evolution color their joint dynamics to produce the patterns we see in ecological communities. Interested readers should consult recent book-length treatments of *Eco-evolutionary Dynamics* by Hendry (2017) and *Evolutionary Community Ecology* by McPeek (2017).

Summary

1. In some communities, the time scales of selection, adaptation, and ecological interactions are very similar, resulting in eco-evolutionary feedbacks that may affect population dynamics, species diversity, community structure, and ecosystem functioning.
2. Theory suggests that evolutionary changes in prey vulnerability can markedly affect the stability of predator–prey interactions, particularly the tendency for populations to cycle and the periodicity of these cycles, and that eco-evolutionary dynamics can maintain adaptive variation in prey populations. Laboratory experiments have borne out these predictions, but no experimental field study has yet demonstrated eco-evolutionary dynamics to the same degree.

3. New analytical methods applied to empirical studies have shown that the contribution of rapid evolution to the outcomes of ecological interactions can be substantial, but that this contribution can vary widely among study systems. Moreover, rapid evolution of traits may act to oppose or reduce the impact of environmental change, suggesting that rapid evolution may be most important when it is least evident.

4. Rapid adaptive evolution in response to a deteriorating environment can prevent the extinction of a population, a process known as evolutionary rescue.

5. Community phylogenetics uses the pattern of relatedness among species (a phylogeny) to better understand the processes the determine community structure (e.g., the abiotic and biotic filtering of species from the regional species pool into local communities). Predicted patterns of community structure depend on multiple assumptions, including whether species that are more closely related to each other will be more similar in phenotype and stronger competitors.

6. If species' functional traits are conserved across phylogenies, and if species with similar functional traits use resources and habitats in similar ways, then close relatives may compete more strongly with one another than with more distant relatives. The "competitive-relatedness hypothesis" predicts that if that is the case, then interspecific competition should lead to communities that show phylogenetic overdispersion. On the other hand, if closely related species share similar environmental requirements, then communities should show phylogenetic clustering due to habitat filtering.

7. There have been few direct tests of the assumption that closely related species are stronger competitors than distantly related species, and the evidence on this point is mixed.

8. Phylogenetic and phenotypic diversity in communities is often a nonrandom sample of regional diversity. The pattern observed across spatial scales seems to be phylogenetic clustering at regional scales (driven by habitat filtering) and phylogenetic overdispersion at local scales (driven by competitive interactions), although more work is needed to determine the generality of this pattern.

9. Habitat filtering and biotic interactions (including competition) may act together in the assembly of communities. Phylogenetic approaches that partition the relative effects of these different factors may prove useful in the future.

10. Macroevolution processes are of particular interest to community ecologists because they focus directly on the production of biodiversity.

11. **Adaptive radiation**, the diversification by adaptation of a single phylogenetic lineage into a variety of forms occupying different niches, is thought to play a prominent role in the evolution of biodiversity.

12. Adaptive radiations are well-documented on islands and in other island-like systems, and are strongly associated with ecological opportunity. Adaptive radiations often occur where resources are underutilized, as in a newly colonized, relatively unoccupied environment, or when a lineage acquires a "key innovation" that allows it to exploit resources in a new way.

13. The processes leading to the formation of regional (mainland) species pools differ in many ways from the adaptive radiations observed on islands.

14. Advances in understanding the evolution of regional species pools requires specific consideration of the factors that influence the "cycle of speciation", whereby populations diverge in allopatry, evolve reproductive isolation, and then expand their ranges to become secondarily sympatric.

CHAPTER 17

Some concluding remarks and a look ahead

Illuminating our understanding of nature, that's what it's about.

Peter A. Lawrence, 2011, p. 31

I am not embarrassed to be a puzzle solver.

Michael L. Rosenzweig, 1995, p. 373

We can imagine a student reaching the end of this book and saying,

That's all very interesting, really. I see how species diversity varies across space and time—fascinating stuff. And I understand how biodiversity can affect the functioning of communities and ecosystems and why the loss of biodiversity is such a critical concern. And I have learned a great deal about the nitty-gritty of species interactions and how these interactions impact species diversity. And I understand that species diversity in a community is a function of local interactions plus the movement of species between communities. But what I *don't* get is how everything fits together. That's what we are supposed understand, isn't it, if we want to use our ecological knowledge to help preserve the diversity of life in our world? So, just tell me—which of all the things that we studied is most important in determining the number and kinds of species I see around me?

There are days when we share this imaginary student's frustration. In the short history of our discipline, community ecologists have identified many (perhaps most) of the processes that regulate and maintain species diversity at local and regional scales. We understand (in principle) how these processes act together to form areas of high and low biodiversity, and we can predict (in a very general

way) how biodiversity will change with a changing environment. However, if you ask why the small Michigan lake outside one author's backyard is home to about 20 species of fish, rather than 2 species or 200 species, we cannot tell you, even though Mittelbach has spent most of his adult life studying fish. The same is true for how many species of trees are in the forest across the street from McGill's office, but maybe that's asking too much of our science. A colleague once said that we ecologists are always lamenting our inability to predict species diversity, when in fact we should celebrate our successes. For example, ecologists understand in a deep and fundamental way the factors that regulate populations, and we use this knowledge to manage our natural resources effectively (or at least we try to—politics often gets in the way). We understand the processes that govern species interactions and that determine the outcomes of these interactions under different environmental conditions. We understand the mechanisms that promote species coexistence, regulate biodiversity, and affect biogeochemical processes, even if we cannot yet predict precisely the number of species we will find in any one community. These are our successes.

Community Ecology. Second Edition. Gary G. Mittelbach & Brian J. McGill, Oxford University Press (2019).
© Gary G. Mittelbach & Brian J. McGill (2019). DOI: 10.1093/oso/9780198835851.001.0001

A few years back, a well-known ecologist, John Lawton (1999, p. 178), pronounced that "community ecology is a mess, with so much contingency that useful generalizations are hard to find." Contingency indeed abounds in ecological interactions; however, we hope that this book has shown how far community ecologists have come in lifting us out of "the mess" that Lawton perceived in the 1990s. Still, it may well be true that predicting the number and types of species found in any given community is an unreasonable goal. Indeed, Lawton highlighted that understanding the drivers of species richness is a problem with an awkward level of complexity. Physicists would call this the "intermediate numbers problem" (it is easier to make predictions about 2 or 1,000,000 objects in a system than 10 or 100—exactly the scope of the question of how many species are in a community.

We have worked hard to broaden the scope of the second edition of this book, expanding on many topics given short shrift in the first edition and highlighting some exciting new areas of study, but we know it is still incomplete. Some topics (e.g., infective agents and disease, herbivory, stoichiometry, invasive species, metabolic theory, species distribution modeling) still receive only a passing glance. It would be easy to say there just was not room to include them all, but that would not be honest. In reality, as Robert May's quote from the Preface notes, deciding what to emphasize in a book about community ecology is "necessarily quirky." This book represents our own quirkiness. Moreover, community ecology has always felt the tension between two different approaches. Approach #1 is process focused and reductionist, and defines community ecology as fundamentally about the study of species interactions and how the outcome of these interactions determine community level properties like species richness. Approach #2 is more pattern focused and emergent and is fundamentally about understanding and predicting the emergent properties of communities like diversity, composition and function. Even we, as authors of this text, wake up some days believing more in Approach #1 and other days in Approach #2. Obviously, a powerful tool is to use these two approaches in combination—a one–two punch if you will. This book has tried to do so, with mixed success, but that is the state of the field. It

may be that 50 years from now you or your students will look back and feel that ecologists have succeeded in merging the two approaches. However, it is also a distinct possibility (John Lawton thought this) that the two approaches are compatible and complementary, but not truly unifiable and 50 years from now students will still be confused by this dichotomy.

This highlights one area of community ecology that still needs to make progress. However, there are other areas where community ecology has made great advances in recent decades. In particular, it is now recognized that, except in rare cases, communities do not have distinct boundaries, nor are they closed to the immigration or emigration of species. This means that, although we might study a group of species found in a particular forest, or field, or lake of our choosing, we need to appreciate how that community is connected to the broader landscape in both space and time. Recognizing the connections between local and regional processes is what sets the current discipline of community ecology apart from its earlier days, when the focus was almost entirely on local processes and local diversity. Ecologists may wrestle at the moment with our concepts of "metacommunities," how to define and delimit them, how to measure the rate of species dispersal between communities, and how regional diversity arises and affects local diversity. However, we will never go back to studying communities in isolation.

How, then, might we go forward? That's a big question. It is tempting to stop now without sticking our necks out to discuss "future directions," but we would be remiss to end this book without at least giving it a try. Listed are some areas that we think will occupy the attention of community ecologists for the next decade or so and that may yield big dividends in terms of understanding the processes that structure communities and govern their functioning.

Looking ahead: issues to ponder

Metacommunities and the integration of local and regional processes

The study of metacommunity dynamics, broadly defined, will continue to play a large role in the development of community ecology. Hanski and colleagues once estimated that more than 50% of Finnish butterfly species may function as metapopulations (Hanski and Kuussaari 1995). Can we make such an estimate of the percentage of communities of different types that function as metacommunities? More importantly, can we determine how much of the structure and functioning of local communities is governed by the dispersal of species between communities and the size of regional species pools as compared with the impact of species interactions and species sorting within communities? At some point, the answer to these questions comes down to measuring the rates of fundamental processes. As depicted in Figure 1.5, four processes (selection, drift, speciation, and dispersal) feed into "the black box of community ecology" to determine the patterning of species diversity across space and time (Vellend 2010). Selection (species sorting in this context) and drift (random external effects on population abundances) are local processes, speciation occurs at a regional scale, and dispersal is the mechanism that ties local and regional dynamics together. Conceptually, the relative rates and strengths of dispersal vs selection (species sorting) seems to be something of master dial on types of communities (Chapter 14; Leibold and Chase 2018) That's probably as good as we can do (for now) in developing an overarching framework for community ecology. However, as remarked upon in Chapter 1, what we really want to do is look *inside* the black box and shed some light on the workings within. To do this, we need to know more about the actual rates of species movement between communities, of species selection (sorting) and drift within communities, and of species accumulation (or loss) within regions.

Drivers of regional biodiversity

We believe that the latitudinal diversity gradient, as well as other broad-scale gradients in biodiversity, will continue to serve as useful frameworks for understanding the processes that generate regional differences in biodiversity. Ecologists are able to map patterns of biodiversity (e.g., species richness and species traits) onto climatic variables, landforms, and historical and evolutionary processes

(e.g., paleoclimates, phylogenetic patterns) with a level of precision unimaginable 20 years ago. This capability has revitalized the study of biodiversity at broad spatial scales and has helped fuel the integration of biogeography, ecology, and evolution (Ricklefs and Jenkins 2011). The coming years may reveal answers to a question that has hounded ecologists for 200 years: "Why are the tropics so species-rich?" However, determining causal relationships between regional biodiversity and a host of potential drivers is difficult, especially at spatial scales where manipulative experiments are difficult or impossible and many possible explanatory variables are highly correlated with each other. Further insights into the evolutionary and ecological factors driving broad-scale biodiversity patterns will come from creative efforts to apply a combination of tools (e.g., phylogenetics, historical biogeography and paleontology, distributional mapping, trait analyses; e.g., Belmaker et al. 2012; Condamine et al. 2012: Pigot et al. 2016) to understand why biodiversity is distributed heterogeneously across Earth.

Although experimental studies over broad spatial scales are challenging, they can reveal much about the drivers of biodiversity. We have highlighted a few examples of replicated experiments conducted at multiple sites in previous chapters (e.g., see Figure 3.4, Hector et al. 2002; Figure 9.9, Callaway et al. 2002). More recently, large-scale experiments have tested the hypothesis that biotic interactions are more intense in the tropics than in the temperate zone. For example, by manipulating predator presence and absence in tropical and temperate intertidal communities (Freestone and Osman 2011; Freestone et al. 2011), by comparing predation rates on tethered natural prey in eelgrass communities across 37° of latitude (Reynolds et al. 2018), and by measuring predation rates on plasticine "model" caterpillars across >100 degrees of latitude (Roslin et al. 2017). Additional tests of the strength of biotic-interactions across environmental gradients have used latitudinal comparisons of herbivory rates in terrestrial and marine systems (e.g., Baskett and Schemske 2018; Longo et al. 2018) and measured the intensity of intraspecific density-dependence (LaManna et al. 2017) and the storage effect (Usinowicz et al. 2017) in temperate and tropical trees. These and other experimental manipulations conducted at multiple sites across the globe (e.g., Adler et al. 2011; Fraser et al. 2015) add an exciting new dimension to the integrated study of species interactions at local and regional scales. We hope to see many more of these collaborative, question-driven, cross-site experiments in the next decade.

Community assembly and functional traits

Two of the most active areas of research in community ecology today are the overlapping areas of community assembly and functional traits (Chapter 12). On the one hand, community assembly has largely been a background notion that is just now emerging to the fore. Will a focus on high-level processes of dispersal and abiotic and biotic selection develop into a robust and quantitative theory? Can we predict in which types of communities the different processes will predominate? On the other hand, functional traits have seen a gigantic explosion of research effort with thousands of papers. Yet the field is lacking a core set of general principals and questions, and is beginning to show a sense of drift. Will unifying theory and questions emerge? Will some of the missing features we highlighted in Chapter 12 like investigating traits in their full multivariate phenotype context, and the trait by environment interaction appear and advance the field? Will we begin to pick traits to study by the roles they fulfil in theory rather than by what is easy to measure?

Pathogens, parasites, and natural enemies

We have been asked several times if there would be a chapter on "disease ecology" in this book, and so far the answer has been no. However, we do believe we are seeing see a boom in research on the effects of pathogens and natural enemies in communities. In particular, there is a lot of current excitement over the idea that species-specific pathogens may promote species coexistence and diversity within communities (e.g., Ricklefs 2011), even though the foundation for this idea was set long ago (the Janzen–Connell hypothesis discussed in Chapter 8).

Evidence for Janzen–Connell effects is accumulating in a variety of ecosystems, including tropical forests (e.g., Bagchi et al. 2010; Comita et al. 2010; Mangan et al. 2010; Liu et al. 2012) and temperate grasslands (e.g., Bever et al. 2010; de Kroon et al. 2012), and the diversity-promoting effects of species-specific natural enemies have been suggested as a factor in animal communities as well (e.g., Ricklefs 2011). Recent studies also suggest an important role for soil pathogens in the relationship between plant diversity and ecosystem functioning (see Chapter 3): reductions in primary productivity at low plant species richness may be driven in part by the negative effects of species-specific soil pathogens on plants grown in monoculture or at low species richness (e.g., Maron et al. 2011; Schnitzer et al. 2011; deKroon et al. 2012). However, as Ricklefs (2011, p. 2045) notes, pathogens are difficult to study in nature, and "the impacts of pathogens on the coexistence, distribution and abundance of competing species in regional communities are poorly known." Chesson and co-workers have increasingly brought into focus the fact that all of the mechanisms of coexistence that exist within competitive systems also existent within predator–prey (or host–disease) systems (Chesson and Kuang 2008; Chesson 2011).

Biodiversity and ecosystem functioning

The study of how biodiversity affects ecosystem functioning has been one of the most active fields in community ecology for almost 20 years. It is hard to know whether this momentum and rate of discovery will be maintained for another decade or longer. Yet there continue to be important advances. As noted above, soil pathogens have been identified as playing an important role in determining the effects of diversity on plant productivity (Maron et al. 2011; Schnitzer et al. 2011; deKroon et al. 2012). Recently, Meyer et al. (2016) suggested that a decline in the functioning of low-diversity communities due to the build-up of species-specific pathogens is one important factor behind the observed strengthening of biodiversity effects on ecosystem functioning over time in the long-running Jena grassland experiment. They suggest that the other critical component leading to an increase in ecosystem functioning over time is selection for greater niche differentiation in mixed-species communities. There is increasing evidence that natural selection and rapid evolution are important in driving the differential response of species in mixtures and monocultures (e.g.,

Zuppinger-Dingley et al. 2014; van Moorsel et al. 2018). Here again, we see an increased awareness of the role of evolution and adaptation in determining population and community dynamics on ecological time scales (eco-evolutionary dynamics). Efforts are also underway to better integrate community assembly and biodiversity to understand ecosystem functioning (Bannar-Martin et al. 2017). Finally, as noted at the end of Chapter 3, many unanswered questions remain about the effects of biodiversity on ecosystem functioning. We will not repeat them here. Suffice it to say that the field shows little evidence of slowing and we suspect that a number of important connections between biodiversity, ecosystem functioning, and ecosystem services remain to be discovered.

Changing technology will change how we collect data

Most ecologists collect data in the same way it was collected 100 years ago; e.g., one or a few people go outdoors and count living organisms by trapping, tagging, visual survey, etc. The core data produced includes counts and species identification, and it is likely that ecologists will continue to need such data for years to come. However, technology is revolutionizing *how* ecologists collect these data. Eco-informatics is allowing for the assembly of data from hundreds of scientists all over the world into giant databases with easy-to-analyze formats. Citizen science allows leveraging the observational powers of hundreds or thousands of people to populate these databases. Sometimes the citizen scientists go into the field themselves. Other times they go online; one good example of online citizen science is the identification of mammals captured by camera traps. Automated image processing techniques are also getting better at doing this with no human involved. Remote sensing is becoming increasingly powerful. Especially with the advent of hyperspectral technology that breaks visible light into over 100 separate bands and with LIDAR, which can detect height and structure of the landscape. We are getting increasingly close to being able to automatically detect and identify species in remotely sensed images. DNA barcoding and

eDNA are also changing our ability to sense organisms that may be small, hard to identify to species, or even those that have just recently passed through an area. It is hard to predict exactly where this new technology will lead and how it might change the field of community ecology, but it is a good bet that the proportion of data collected in these new ways will rapidly increase.

Eco-evolutionary feedbacks and regional pool processes

Few ecologists a decade ago would have appreciated the potential for rapid evolutionary change to influence population dynamics and the outcome of species interactions in ecological time. Yet, as Chapter 16 showed, eco-evolutionary dynamics are common in many systems, even if we do not yet know whether "the evolution–ecology pathway is frequent and strong enough in nature to be broadly important" (Schoener 2011, p. 426). Schoener's question about the importance of eco-evolutionary feedbacks in nature is crucial, and we suspect that efforts to answer it will occupy the attention of many evolutionary community ecologists in the coming years. One important question that is not often raised with regard to rapid evolution and eco-evolutionary dynamics is how much of the evolutionary change observed over short (i.e., ecological) time scales is meaningful in the long run. Evolutionary biologists have found evidence for significant, ongoing directional selection in many (perhaps most) populations in nature (e.g., Kingsolver et al. 2001; Kingsolver and Diamond 2011), yet stasis in species traits is the general rule—that is, species tend to look the same year after year. Futuyma (2010, p. 1875) and others argue that such "sluggish evolution" and organism-wide stasis are the result of processes that limit adaptation, and that "without reproductive isolation (speciation), interbreeding will cause the adaptations of local populations to be ephemeral." We raise this issue as a cautionary flag for the study of rapid evolution and its consequences for species interactions and the structuring of communities. If the direction of selection on species traits varies over time and/or across environments, or if there is significant gene flow between populations, then rapid evolution may have minimal impact on the outcome of species interactions in the long run. Again, it is a question of rates.

Scaling up and understanding how the evolutionary dynamics at the regional pool scale and the ecological processes at the local community scale interact to produce the regional pool will be an increasingly important question going forward.

Climate change, and its effects on species distributions and species interactions

How climate change (and other types of global change like land use change) will affect the distribution and abundance of species, their interactions, and biodiversity and ecosystem functioning is *the* question facing ecologists in the next decade and beyond. Here again, most of the important questions and their answers focus on rates of change. How rapidly will species adapt to changing environmental conditions? Will the movements of species (and perhaps communities) in response to climate change be rapid enough to prevent extinctions? Are movement corridors and assisted dispersal useful tools for the conservation of species faced with fragmented landscapes and changing climates? Or are we playing Russian roulette with species introductions? How much will climate change impact upon the functioning of communities by changing relationships among species? The orthodoxy that climate is the only control of range boundaries of species is unravelling and there is an increasing recognition of the role of species interactions in setting range limits (e.g., Hargreaves et al 2014; Rich and Currie 2018). Thus, a community ecology perspective is growing in importance in the study of climate change impacts. Understanding anthropogenic impacts may necessitate ecologists to partially shift ecological goals from understanding to prediction.

The role of time

It is fair to say that the last 30 years of community ecology have been about embracing the importance of space, and the dispersal of individuals and species across space. The emergence of the metapopulation and metacommunity perspectives are a clear example. It might seem funny to say that time is not already incorporated in ecology. After all, the central theory like Lotka–Volterra equations are differential equations expressing change over time. However, those equations all result in nice stable equilibrial endpoints. The focus has always been on the endpoints, not the trajectories, but several of the future directions we have listed share in common a theme of time and change. Biological systems have always been subject to external forcing from the climate and this is accelerating in the Anthropocene. We will have to reconfigure our perspective to include a temporal view. Recent studies (e.g. Dornelas et al. 2014) have suggested that some community properties like richness stay fairly constant, while others such as the species composition of a community change rapidly. What variables do we expect to change or show stasis, and why? How much change is background and how much is attributable to humans? Do ecological systems mostly respond to external forcing in a gradual fashion, or do they show tipping points and abrupt change? It is likely that historical ecology and paleoecology will become increasingly important allied fields of community ecology. This can all be summarized by how much do communities change and how do they change?

In closing, we would like to say . . .

It is an exciting time to be a community ecologist. We have the power to illuminate our understanding of the natural world as never before, to help conserve Earth's biodiversity, and to help provide the knowledge needed to develop a more sustainable world. We look forward to seeing what ecologists can accomplish in the coming decade, and we hope this book might contribute in some small way toward facilitating those accomplishments, perhaps by inspiring more young scientists to take up the challenge.

Literature cited

Aarssen, L.W. (19970. High productivity in grassland ecosystems: affected by species diversity or productive species? *Oikos*, **80**, 183–4.

Abrahams, M.V. and L.M. Dill. (1989). A determination of the energetic equivalence of the risk of predation. **Ecology**, **70**, 999–1007.

Abrams, P.A. (1975). Limiting similarity and the form of the competition coefficient. *Theoretical Population Biology*, **8**, 356–75.

Abrams, P.A. (1982). Functional responses of optimal foragers. *American Naturalist*, **120**, 382–90.

Abrams, P.A. (1983a). Arguments in favor of higher order interactions. *American Naturalist*, **121**, 887–91.

Abrams, P.A. (1983b). The theory of limiting similarity. *Annual Review of Ecology and Systematics*, **14**, 359–76.

Abrams, P.A. (1984). Variability in resource consumption rates and the coexistence of competing species. *Theoretical Population Biology*, **25**, 106–24.

Abrams, P.A. (1987). Indirect interactions between species that share a predator: varieties of indirect effects. In Kerfoot, W.C. and Sih, A. eds. *Predation: Direct and Indirect Impacts on Aquatic Communities*, pp. 38–54. University Press of New England, Hanover.

Abrams, P.A. (1988). Resource productivity–consumer species diversity: simple models of competition in spatially heterogeneous environments. *Ecology*, **69**, 1418–33.

Abrams, P.A. (1989). Decreasing functional responses as a result of adaptive consumer behavior. *Evolutionary Ecology*, **3**, 95–114.

Abrams, P.A. (1990). The effects of adaptive behavior on the type-2 functional response. *Ecology*, **71**, 877–85.

Abrams, P.A. (1992). Why don't predators have positive effects on prey populations? *Evolutionary Ecology*, **6**, 449–57.

Abrams, P.A. (1993). Effects of increased productivity on abundances of trophic levels. *American Naturalist*, **141**, 351–71.

Abrams, P.A. (1995). Implications of dynamically variable traits for identifying, classifying, and measuring direct and indirect effects in ecological communities. *American Naturalist*, **146**, 112–34.

Abrams, P.A. (2000). The evolution of predator–prey interactions: theory and evidence. *Annual Review of Ecology and Systematics*, **31**, 79–105.

Abrams, P.A. (2001). The effect of density-independent mortality on the coexistence of exploitative competitors for renewing resources. *American Naturalist*, **158**, 459–70.

Abrams, P.A. (2004a). Trait-initiated indirect effects due to changes in consumption rates in simple food webs. *Ecology*, **85**, 1029–38.

Abrams, P.A. (2004b). When does periodic variation in resource growth allow robust coexistence of competing consumer species? *Ecology*, **85**, 372–82.

Abrams, P.A. (2007). Defining and measuring the impact of dynamic traits on interspecific interactions. *Ecology*, **88**, 2555–62.

Abrams, P.A. (2008). Measuring the impact of dynamic antipredator traits on predator–prey–resource interactions. *Ecology*, **89**, 1640–9.

Abrams, P.A. (2009a). Determining the functional form of density dependence: Deductive approaches for consumer-resource systems having a single resource. *American Naturalist*, **174**, 321–30.

Abrams, P.A. (2009b). The implications of using multiple resources for consumer density dependence. *Evolutionary Ecology Research*, **11**, 517–40.

Abrams, P.A. (2010). Implications of flexible foraging for interspecific interactions: lessons from simple models. *Functional Ecology*, **24**, 7–17.

Abrams, P.A. (2015). Why ratio dependence is (still) a bad model of predation. *Biological Reviews*, **90**, 794–814.

Abrams, P.A., Brassil C.E., and Holt, R.D. (2003). Dynamics and responses to mortality rates of competing predators undergoing predator-prey cycles. *Theoretical Population Biology*, **64**, 163–76.

Abrams, P.A. and Matsuda, H. (1997). Prey adaptation as a cause of predator–prey cycles. *Evolution*, **51**, 1742–50.

Abrams, P.A., Menge, B.A., Mittelbach, G.G., Spiller D., and Yodzis, P. (1996). The role of indirect effects in food webs. In: Polis, G., and Winemiller, P.O., eds. *Food Webs: Integration of Patterns and Dynamics*, pp. 371–95. Chapman and Hall, New York.

Abrams, P.A. and Nakajima, M. (2007). Does competition between resources change the competition between their consumers to mutualism? Variation on two themes by Vandermeer. *American Naturalist*, **170**, 744–57.

Abrams, P.A. and Roth, J.D. (1994). The effects of enrichment of three-species food chains with nonlinear functional responses. *Ecology*, **75**, 1118–30.

Abrams, P.A. and Rowe, L. (1996). The effects of predation on the age and size of maturity of prey. *Evolution*, **50**, 1052–61.

Abrams, P.A., Rueffler, C., and Dinnage, R. (2008). Competition-similarity relationships and the nonlinearity of competitive effects in consumer-resource systems. *American Naturalist*, **172**, 463–74.

Abramsky, Z., Stauss, E., Subach, A., Kotler B.P., and Riechman, A. (1996). The effect of barn owls (*Tyto alba*) on the activity and microhabitat selection of *Gerbillus allenbyi* and *G. pyramidum*. *Oecologia*, **105**, 313–19.

Adler, P.B. (2004). Neutral models fail to reproduce observed species–area and species–time relationships in Kansas grasslands. *Ecology*, **85**, 1265–72.

Adler, P.B., Ellner, S.P., and Levine, J.M. (2010). Coexistence of perennial plants: an embarrassment of niches. *Ecology Letters*, **13**, 1019–29.

Adler, P.B., J. HilleRisLambers, P.C. Kyriakidis, Q.F. Guan and J.M. Levine. (2006a). Climate variability has a stabilizing effect on the coexistence of prairie grasses. *Proceedings of the National Academy of Sciences*, **103**, 12793–8.

Adler, P.B., HilleRisLambers, J., and Levine, J.M. (2006b). Weak effect of climate variability on coexistence in a sagebrush steppe community. *Ecology*, **90**, 3303–12.

Adler, P.B., HilleRisLambers, J., and Levine, J.M. (2007). A niche for neutrality. *Ecology Letters*, **10**, 95–104.

Adler, P.B., *et al.* (2011). Productivity is a poor predictor of plant species richness. *Science*, **333**, 1750–3.

Adler, P.B., *et al.* (2018). Competition and coexistence in plant communities: intraspecific competition is stronger than interspecific competition. *Ecology Letters*, **21**, 1319–29.

Akçakaya, R., Arditi R. and Ginzburg. L.R. (1995). Ratio-dependent predation: an abstraction that works. *Ecology*, **76**, 995–1004.

Akcay, E. and Roughgarden, J. (2007). Negotiation of mutualism: rhizobia and legumes. *Proceedings of the Royal Society of London B*, **274**, 25–32.

Allee, W.C., Emerson, O., Park T., and Schmidt. K. (1949). *Principles of Animal Ecology*. W. B. Saunders, Philadelphia, PA.

Allen, A.P., Brown, J.H., and Gillooly, J.F. (2002). Global biodiversity, biochemical kinetics, and the energetic-equivalence rule. *Science*, **297**, 1545–8.

Allen, A.P. and Gillooly, J.F. (2006). Assessing latitudinal gradient in speciation rates and biodiversity at the global scale. *Ecology Letters*, **9**, 947–54.

Allen, A., Gillooly, J., Savage V., and Brown. J. (2006). Kinetic effects of temperature on rates of genetic divergence and speciation. *Proceedings of the National Academy of Sciences*, **103**, 9130–5.

Allen, A.P. and White. E.P. (2003). Effects of range size on species-area relationships. *Evolutionary Ecology Research*, **5**, 493–9.

Allesina, S. (2011). Predicting trophic relations in ecological networks: a test of the Allometric Diet Breadth Model. *Journal of Theoretical Biology*, **279**, 161–8.

Almeida, A.A. and Mikich, A.B. (2017). Combining plant-frugivore networks for describing the structure of neotropical communities. *Oikos*, **127**, 184–96.

Altshuler, D.L. (2005). Flight performance and competitive displacement of hummingbirds across elevational gradients. *The American Naturalist*, **167**, 216–29.

Amarasekare, P. (2003). Competitive coexistence in spatially structured environments: a synthesis. *Ecology Letters*, **6**, 1109–22.

Ambrose, S.H. (1998). Late Pleistocene human population bottlenecks, volcanic winter, and differentiation of modern humans. *Journal of Human Evolution*, **34**, 623–51.

Anderson, M.J., *et al.* (2011). Navigating the multiple meanings of β diversity: a roadmap for the practicing ecologist. *Ecology Letters*, **14**, 19–28.

Anderson, R.M. and May, R.M. (1978). Regulation and stability of host-parasite population interactions. I. Regulatory processes. *Journal of Animal Ecology*, **47**, 219–47.

Anderson, R.M. and May, R.M. (1991). *Infectious Diseases of Humans: Dynamics and Control*. Oxford University Press, New York.

Andrewartha, H.G. and Birch, L.C. (1954). *The Distribution and Abundance of Animals*. University of Chicago Press, Chicago, IL.

Angert, A.L., Huxman, T.E. Chesson P.L. and Venable. D.L. (2009). Functional tradeoffs determine species coexistence via the storage effect. *Proceedings of the National Academy of Sciences*, **106**, 11641–5.

Antonelli, A. and Sanmartin. I. (2011). Why are there so many plant species in the Neotropics? *Taxon*, **60**, 403–14.

Arditi, R. and Ginzburg. L.R. (1989). Coupling in predator-prey dynamics: ratio dependence. *Journal of Theoretical Biology*, **139**, 311–26.

Arditi, R. and Ginzburg. L.R. (2012). *How Species Interact*. Oxford University Press. New York, NY.

Armstrong, R.A. (1976). Fugitive species: experiments with fungi and some theoretical considerations. *Ecology*, **57**, 953–63.

Armstrong, R.A. and McGehee, R. (1976). Coexistence of two competitors on one resource. *Journal of Theoretical Biology*, **56**, 499–502.

Armstrong, R.A. and McGehee, R. (1980). Competitive exclusion. *American Naturalist*, **115**, 151–70.

Arnold, S.J. (1983). Morphology, performance and fitness. *American Zoologist*, **23**, 347–61.

Arrhenius, O. (1921). Species and area. *Journal of Ecology*, **9**, 95–9.

Astorga, A., Oksanen, J. Luoto, M. Soininen, J. Virtanen R. and Muotka. T. (2012). Distance decay of similarity in freshwater communities: Do macro- and microorganisms follow the same rules? *Global Ecology and Biogeography*, **21**, 365–75.

Atkinson, W. and Shorrocks. B. (1981). Competition on a divided and ephemeral resource: a simulation model. *Journal of Animal Ecology*, **50**, 461–71.

Attayde, J.L. and Ripa, J. (2008). The coupling between grazing and detritus food chains and the strength of trophic cascades across a gradient of nutrient enrichment. *Ecosystems*, **11**, 980–90.

Aunapuu, M., *et al.* (2008). Spatial patterns and dynamic responses of arctic food webs corroborate the exploitation ecosystems hypothesis (EEH). American Naturalist, **171**: 249–62.

Axelrod, R. and Hamilton, W.D. (1981). The evolution of cooperation. *Science*, **211**, 1390–6.

Bagchi, R., Swinfield, T., Gallery, R.E., *et al.* (2010). Testing the Janzen–Connell mechanism: pathogens cause over-compensating density dependence in a tropical tree. *Ecology Letters*, **13**, 1262–9.

Baguette, M. (2004). The classical metapopulation theory and the real, natural world: a critical appraisal. *Basic and Applied Ecology*, **5**, 213–24.

Baird, D. and Milne, H. (1981). Energy flow and the Ythan Estuary, Aberdeenshire, Scotland. *Estuarine and Coastal Shelf Science*, **13**, 455–72.

Baird, D. and Ulanowicz, R.E. (1989). The seasonal dynamics of the Chesapeake Bay ecosystem. *Ecological Monographs*, **59**, 329–64.

Baldwin, B.G. and Sanderson, M.J. (1998). Age and rate of diversification of the Hawaiian silversword alliance (Compositae). *Proceedings of the National Academy of Sciences*, **95**, 9402–6.

Bannar-Martin, K.H., *et al.* (2017). Integrating community assembly and biodiversity to better understand ecosystem function: the Community Assembly and the Functioning of Ecosystems (CAFÉ) approach. *Ecology Letters*, **21**, 167–80.

Barabás, G., Andrea, R.D., and Ostling, A.M. (2013). Species packing in nonsmooth competition models. *Theoretical Ecology*, **6**, 1–19.

Barabás, G., Andrea, R.D., and Stump, S.M. (2018). Chesson's coexistence theory. *Ecological Monographs*, **99**, 1633–43.

Barabás, G., Michalska-Smith, M.J., and Allesina, S. (2016). The effect of intra- and interspecific competition on coexistence in multispecies communities. *American Naturalist*, **188**, E1–E12.

Barabás, G., Pigolotti, S., Gyllenberg, M., Dieckmann, U., and Meszéna. G. (2012). Continuous coexistence or discrete species? A new review of an old question. *Evolutionary Ecology Research*, **14**, 523–54.

Barabasi, A-L. (2009). Scale-free networks: a decade and beyond. Science, **325**: 412–13.

Barnes, A.D., Jochum, M., Lefcheck, J.S., *et al.* (2018). Energy flux: the link between multitrophic biodiversity and ecosystem functioning. *Trends in Ecology and Evolution*, **33**, 186–97.

Bartha, S., Czaran, T., and Scheuring, I. (1997). Spatiotemporal scales of non-equilibrium community dynamics: a methodological challenge. *New Zealand Journal of Ecology*, **21**, 199–206.

Barton, A.D., Pershing, A.J., Litchman, E., *et al.* (2013). The biogeography of marine plankton traits. *Ecology Letters*, **16**, 522–34.

Bascompte, J. (2009). Disentangling the web of life. *Science*, **325**, 416–19.

Bascompte, J. and Jordano, P. (2007). Plant–animal mutualistic networks: The architecture of biodiversity. *Annual Review of Ecology, Evolution, and Systematics*, **38**, 567–93.

Bascompte, J. and Jordano, P. (2013). *Mutualistic Networks*. Princeton University Press, Princeton, NJ.

Bascompte, J., Jordano, P., Melián, C.J., and Olesen, J.M. (2003). The nested assembly of plant–animal mutualistic networks. *Proceedings of the National Academy of Sciences*, **100**, 9383–7.

Bascompte, J., C. Melián, J., and Sala, E. (2005). Interaction strength combinations and the overfishing of a marine food web. *Proceedings of the National Academy of Sciences*, **102**, 5443–7.

Bascompte, J. and Rodríguez-Trelles, F. (1998). Eradication thresholds in epidemiology, conservation biology and genetics. *Journal of Theoretical Biology*, **192**, 415–18.

Baskett, C.A. and Schemske, D.W. (2018). Latitudinal patterns of herbivore pressure in a temperate herb support the biotic interactions hypothesis. *Ecology Letters*, **21**, 578–87.

Bassar, R.D., *et al.* (2010). Local adaptation in Trinidadian guppies alters ecosystem processes. *Proceedings of the National Academy of Sciences*, **107**, 3616–21.

Baum, D. (2008). Reading a phylogenetic tree: the meaning of monophyletic groups. *Nature Education*, **1**(1), 190.

Bawa, K.S. (1990). Plant-pollinator interactions in tropical rain forests. *Annual Review of Ecology and Systematics*, **21**, 399–422.

Beckerman, A.P., Petchey, O.L., and Morin, P.J. (2010). Adaptive foragers and community ecology: Linking individuals to communities and ecosystems. *Functional Ecology*, **24**, 1–6.

Beckerman, A.P., Petchey, O.L., and Warren, P.H. (2006). Foraging biology predicts food web complexity.

Proceedings of the National Academy of Sciences, **103**, 13745–9.

Becks, L., Ellner, S.P., Jones, L.E., and Hairston, N.G., Jr (2010). Reduction of adaptive genetic diversity radically alters eco-evolutionary community dynamics. *Ecology Letters*, **13**, 989–97.

Bednekoff, P.A. (2007). Foraging in the face of danger. In: Stephens, D.W., Brown, J.S., and Ydenberg, R.C., eds. *Foraging: Behavior and Ecology*, pp. 305–29. University of Chicago Press, Chicago, IL.

Begon, M., Townsend, C.R., and Harper, J.L. (2006). *Ecology: From Individuals to Ecosystems*. Blackwell Publishing, Oxford.

Beisner, B.E., Haydon, D.T., and Cuddington, K. (2003). Alternative stable states in ecology. *Frontiers in Ecology and the Environment*, **1**, 376–82.

Beisner, B.E., Peres-Neto, P.R., Lindstrom, E.S., Barnett, A., and Lorena Longhi, M. (2006). The role of environmental and spatial processes in structuring lake communities from bacteria to fish. *Ecology*, **87**, 2985–91.

Bell, G. (2001). Neutral macroecology. *Science*, **293**, 2413–218.

Bell, G. (2017). Evolutionary rescue. *Annual Review of Ecology, Evolution, and Systematics*, **48**, 605–27.

Bell, G. and Gonzalez, A. (2009). Evolutionary rescue can prevent extinction following environmental change. *Ecology Letters*, **12**, 942–8.

Bell, G. and Gonzalez, A. (2011). Adaptation and evolutionary rescue in metapopulations experiencing environmental deterioration. *Science*, **332**, 1327–30.

Belmaker, J., Sekercioglu, C.H., and Jetz, W. (2012). Global patterns of specialization and coexistence in bird assemblages. *Journal of Biogeography*, **39**, 193–203.

Benard, M.F. (2004). Predator-induced phenotypic plasticity in organisms with complex life histories. *Annual Review of Ecology, Evolution, and Systematics*, **35**, 651–73.

Bender, E.A., Case, T.J., and Gilpin, M.E. (1984). Perturbation experiments in community ecology: theory and practice. *Ecology*, **65**, 1–13.

Benke, A.C., Wallace, J.B., Harrison, J.W., and Koebel, J.W. (2001). Food web quantification using secondary production analysis: predaceous invertebrates of the snag habitat of a subtropical river. *Freshwater Biology*, **46**, 329–46.

Benton, M.J. (2009). The Red Queen and the Court Jester: species diversity and the role of biotic and abiotic factors through time. *Science*, **323**, 728–32.

Berlow, E.L. (1999). Strong effects of weak interactions in ecological communities. *Nature*, **398**, 330–4.

Berlow, E.L., Navarrete S.A., Briggs C.J., Power, M.E., and Menge, B.A. (1999). Quantifying variation in the strengths of species interactions. *Ecology*, **80**, 2206–24.

Berlow, E.L., *et al.* (2004). Interaction strengths in food webs: issues and opportunities. *Journal of Animal Ecology*, **73**, 585–98.

Berryman, A.A. (1987). Equilibrium or nonequilibrium: Is that the question? *Bulletin of the Ecological Society of America*, **68**, 500–2.

Bertness, M.D. and Callaway, R.M. (1994). Positive interactions in communities. *Trends in Ecology and Evolution*, **9**, 191–3.

Bertness, M.D. and Hacker, S.D. (1994). Physical stress and positive associations among marsh plants. *American Naturalist*, **144**: 363–72.

Best, R., Anaya-Rojas, J.M., Leal, M.C., Schmid, D.W., Seehausen, O., and Matthews, B. (2017). Transgenerational selection driven by divergent ecological impacts of hybridizing lineages. *Nature Ecology and Evolution*, **1**, 1757–65.

Bever, J.D. (1999). Dynamics within mutualism and the maintenance of diversity: inference from a model of interguild frequency dependence. *Ecology Letters*, **2**, 52–61.

Bever, J.D. (2002). Negative feedback within a mutualism: host-specific growth of mycorrhizal fungi reduces plant benefit. *Proceedings of the Royal Society of London B*, **269**, 2595–601.

Bever, J.D., *et al.* (2010). Rooting theories of plant community ecology in microbial interactions. *Trends in Ecology and Evolution*, **25**, 468–78.

Beverton, R.J.H. and Holt, S.J. (1957). On the dynamics of exploited fish populations. *MAFF Fisheries Investigation of London Series 2*, 19, 1–533.

Bezemer, T.M., and van der Putten, W.H. (2007). Diversity and stability in plant communities. *Nature*, **446**, E6–E7.

Bezerra, E.L.S., Machado, I.C., and Mello, M.A.R. (2009). Pollination networks of oil-flowers: a tiny world within the smallest of all worlds. *Journal of Animal Ecology*, **78**, 1096–101.

Bohannan, B.J.M. and Lenski, R.E. (1997). Effect of resource enrichment on a chemostat community of bacteria and bacteriophage. *Ecology*, **78**, 2303–15.

Bohannan, B.J.M. and Lenski, R.E. (1999). Effect of prey heterogeneity on the response of a model food chain to resource enrichment. *American Naturalist*, **153**, 73–82.

Bohannan, B.J.M. and Lenski, R.E. (2000). The relative importance of competition and predation varies with productivity in a model community. *American Naturalist*, **156**, 329–40.

Bokma, F. and Mönkkönen, M. (2000). The mid-domain effect and the longitudinal dimension of continents. *Trends in Ecology and Evolution*, **15**, 288–9.

Bolker, B., Holyoak, M., Křivan, V., Rowe, L., and Schmitz, O. (2003). Connecting theoretical and empirical studies of trait-mediated interactions. *Ecology*, **84**, 1101–14.

Bolker, B. and Pacala, S.W. (1997). Using moment equations to understand stochastically driven spatial pattern formation in ecological systems. *Theoretical Population Biology*, **52**, 179–97.

Bolker, B.M. and Pacala, S.W. (1999). Spatial moment equations for plant competition: understanding spatial strategies and the advantages of short dispersal. *American Naturalist*, **153**, 575–602.

Bolnick, D.I. and Preisser, E.I. (2005). Resource competition modifies the strength of trait-mediated predator-prey interactions: a meta-analysis. *Ecology*, **86**, 2771–9.

Bond, W.J. (2010). Consumer control by megafauna and fire. In: Terborgh, J. and Estes, J.A., eds. *Trophic Cascades: Predators, Prey, and the Changing Dynamics of Nature*, pp. 275–85. Island Press, Washington, DC.

Bonenfant, C., *et al.* (2009). Empirical evidence of density-dependence in populations of large herbivores. *Advances in Ecological Research*, **41**, 313–57.

Bongers, F., Poorter, L., Hawthorne, W.D., and Sheil, D. (2009). The intermediate disturbance hypothesis applies to tropical forests, but disturbance contributes little to tree diversity. *Ecology Letters*, **12**, 798–805.

Booth, M.G. and Hoeksema, J.D. (2010). Mycorrhizal networks counteract competitive effects of canopy trees on seedling survival. *Ecology*, **91**, 2294–312.

Borenstein, M., Hedges, L.V., Higgins, J.P.T., and Rothstein, H.R. (2009). *Introduction to Meta-Analysis*. John Wiley and Sons, London.

Boucher-Lalonde, V., De Camargo, R.X., Fortin, J-M., *et al.* (2015). The weakness of evidence supporting tropical niche conservatism as a main driver of current richness-temperature gradients. *Global Ecology and Biogeography*, **24**, 795–803.

Box, G.E. and Draper, N.R. (1987). *Empirical model-building and response surfaces*, Wiley, New York, NY.

Bracewell, S.A., Johnston, E.L., and Clark, G.F. (2017). Latitudinal variation in the competition-colonization trade-off reveals rate-mediated mechanisms of coexistence. *Ecology Letters*, **20**: 947–57.

Brassil, C.E. and Abrams, P.A. (2004). The prevalence of asymmetrical indirect effects in two-host-one parasitoid systems. *Theoretical Population Biology*, **66**, 71–82.

Brehm, G., Colwell, R.K., and Kluge, J. (2007). The role of environmental and mid-domain effect on moth species richness along a tropical elevational gradient. *Global Ecology and Biogeography*, **16**, 205–19.

Brokaw, N.V.L. (1985). Gap-phase regeneration in a tropical forest. *Ecology*, **66**, 682–7.

Brönmark, C. and Hansson, L-A. (2005). *The Biology of Lakes and Ponds*. Oxford University Press, Oxford.

Brönmark, C. and Miner, J.G. (1992). Predator-induced phenotypical change in body morphology in crucian carp. *Science*, **258**, 1348–50.

Brönmark, C. and Miner, J.G. (1992). Predator-induced phenotypical change in body morphology in crucian carp. *Science*, **258**, 1348–50.

Bronstein, J.L. (1994). Our current understanding of mutualism. *Quarterly Review of Biology*, **69**, 31–51.

Bronstein, J.L. (2009). The evolution of facilitation and mutualism. *Journal of Ecology*, **97**, 1160–70.

Bronstein, J.L. (2015). *Mutualism*. Oxford University Press, Oxford.

Brook, B.W. and Bradshaw, C.J.A. (2006). Strength of evidence for density dependence in abundance time series of 1198 species. *Ecology*, **87**, 1445–51.

Brooker, R.W. and Callaghan, T. (1998). The balance between positive and negative plant interactions and its relationship to environmental gradients: a model. *Oikos*, **81**, 196–207.

Brooks, D.R. (1985). Historical ecology: a new approach to studying evolution of ecological associations. *Annals of the Missouri Botanical Garden*, **72**, 660–80.

Brose, U., Ehnes, R.B., Rall, B.C., Vucic-Pestic, O., Berlow, E.L., and Scheu, S. (2008). Foraging theory predicts predator–prey energy fluxes. *Journal of Animal Ecology*, **77**, 1072–8.

Brose, U., *et al.* (2016). Predicting the consequences of species loss using size-structured biodiversity approaches. *Biological Reviews*, **92**, 684–97.

Brousseau, P-M., Gravel, D., and Handa, I.T. (2018). Trait matching and phylogeny as predictors of predator-prey interactions involving ground beetles. *Functional Ecology*, **32**, 192–202.

Brown, J.H. (1971). Mountaintop mammals: nonequilibrium insular biogeography. *American Naturalist*, **105**, 467–78.

Brown, J.H. (1975). Geographical ecology of desert rodents. In: Cody, M.L. and Diamond, J.M., eds. *Ecology and Evolution of Communities*, pp. 315–41. Belknap Press of Harvard University Press, Cambridge, MA.

Brown, J.H. (1981). Two decades of homage to Santa Rosalia: toward a general theory of diversity. *American Zoologist*, **21**, 877–88.

Brown, J.H. (2014). Why are there so many species in the tropics? *Journal of Biogeography*, **41**, 8–22.

Brown, J.H., Gillooly, J.F., Allen, A.P., Savage, V.M., and West, G.B. (2004). Toward a metabolic theory of ecology. *Ecology*, **85**, 1771–89.

Brown, J.H. and Kodric-Brown, A. (1977). Turnover rates in insular biogeography: effect of immigration on extinction. *Ecology*, **58**, 445–9.

Brown, J.S. (1988). Patch use as an indicator of habitat preference, predation risk, and competition. *Behavioral Ecology and Sociobiology*, **22**, 37–47.

Brown, J.S. (1992). Patch use under predation risk. I. Models and predictions. *Annales Zoologici Fennici*, **29**, 301–9.

Brown, J.S. (1998). Game theory and habitat selection. In: Dugatkin, L.A. and Reeve, A.K., eds. *Game theory and animal behavior*, pp. 188–220. Oxford University Press, New York, NY.

Brown, J.S. and Kotler, B.P. (2004). Hazardous duty pay and the forging cost of predation. *Ecology Letters*, **7**: 999–1014.

Brown, W.L., Jr. and Wilson, E.O. (1956). Character displacement. *Systematic Zoology*, **5**, 49–64.

Bruno, J.F., Stachowicz, J.J., and Bertness, M.D. (2003). Inclusion of facilitation into ecology theory. *Trends in Ecology and Evolution*, **18**, 119–25.

Burns, J.H. and Strauss, S.Y. (2011). More closely related species are more ecologically similar in an experimental test. *Proceedings of the National Academy of Sciences*, **108**, 5302–7.

Burson, A., Stomp, M., Greenwell, E., Grosse, J., and Huisman, J. (2018). Competition for nutrients and light: testing advances in resource competition with a natural phytoplankton community. *Ecology*, **99**, 1108–18.

Butler, G.J. and Wolkowicz, G.S.K. (1987). Exploitative competition in a chemostat for two complementary, and possibly inhibitory, resources. *Mathematical Biosciences*, **83**, 1–48.

Butler, J.L., Gotelli, N.J., and Ellison, A.M. (2008). Linking the brown and green: nutrient transformation and fate in the Sarracenia microecosystem. *Ecology*, **89**, 898–904.

Butterfield, B.J. (2009). Effects of facilitation on community stability and dynamics: synthesis and future directions. *Journal of Ecology*, **97**, 1192–201.

Butterfield, B.J., Betancourt, J.L., Turner, R.M., and Briggs, J.M. (2010). Facilitation drives 65 years of vegetation change in the Sonoran Desert. *Ecology*, **91**, 1132–9.

Buzas, M.A., Collins, L.S., and Culver, S.J. (2002). Latitudinal difference in biodiversity caused by higher tropical rate of increase. *Proceedings of the National Academy of Sciences*, **99**, 7841–3.

Byrnes, J., Stachowicz, J.J., Hultgren, K.M., Hughes, A.R., Olyarnik, S.V., and Thornbert, C.S. (2006). Predator diversity strengthens trophic cascades in kelp forests by modifying herbivore behaviour. *Ecology Letters*, **9**, 61–71.

Cabana, G. and Rasmussen, J.B. (1994). Modeling food chain structure and contaminant bioaccumulation using stable nitrogen isotopes. *Nature*, **372**, 255–7.

Cáceres, C.E. (1997). Temporal variation, dormancy, and coexistence: a field test of the storage effect. *Proceedings of the National Academy of Sciences*, **94**, 9171–5.

Cadotte, M.W., Fortner, A.M., and Fukami, T. (2006a). The effects of resource enrichment, dispersal, and predation on local and metacommunity structure. *Oecologia*, **149**, 150–7.

Cadotte, M.W., Mai, D.V., Jantz, S., Collins, M.D., Keele, M., and Drake, J.A. (2006b). On testing the competition-colonization trade-off in a multispecies assemblage. *American Naturalist*, **168**, 704–9.

Cadotte, M.W. and Tucker, C.M. (2017). Should environmental filtering be abandoned? *Trends in Ecology & Evolution*, **32**, 429–37.

Cahill, J.F., Kembel, S.W., Lamb, E.G., and Keddy, P.A. (2008). Does phylogenetic relatedness influence the strength of competition among vascular plants? *Perspectives in Plant Ecology, Evolution and Systematics*, **10**, 41–50.

Calcagno, V., Grognard, F., Hamelin, F.M., Wajnberg, E., and Mailleret, L. (2014). The functional response predicts the effect of resource distribution on the optimal movement rate of consumers. *Ecology Letters*, **17**, 1570–9.

Callaway, R.M. (1998). Are positive interactions species-specific? *Oikos*, **82**, 202–7.

Callaway, R.M. (2007). *Positive Interactions and Interdependence in Plant Communities*. Springer, Dordrecht.

Callaway, R.M., Brooker, R.W., Choler, P., Kikvidze, Z., Lortie, C.J., and Michalet, R. (2002). Positive interactions among alpine plants increase with stress. *Nature*, **417**, 844–88.

Callaway, R.M. and Walker, L.R. (1997). Competition and facilitation: A synthetic approach to interactions in plant communities. *Ecology*, **78**, 1958–65.

Camerano, L. (1880). On the equilibrium of living beings by means of reciprocal destruction. In: Levin, S.A., ed. (1994_. *Frontiers in Mathematical Biology* (transl. Cohen, J.E.), pp. 360–80. Springer-Verlag, New York, NY.

Campbell, B.D. and Grime, J.P. (1992). An experimental test of plant strategy theory. *Ecology*, **73**, 15–29.

Campbell, D.R. and Motten, A.F. (1985). The mechanism of competition for pollination between two forest herbs. *Ecology*, **66**, 554–63.

Capon, S.J., Lynch, A.J.J., Bond, N., *et al.* (2015). Regime shifts, thresholds and multiple stable states in freshwater ecosystems: a critical appraisal of the evidence. *Science of the Total Environment*, **534**, 122–30.

Cardinale, B.J. (2011). Biodiversity improves water quality through niche partitioning. *Nature*, **472**, 86–9.

Cardinale, B.J., *et al.* (2007). Impacts of plant diversity on biomass production increase through time because of species complementarity. *Proceedings of the National Academy of Sciences*, **104**, 18123–8.

Cardinale, B.J., *et al.* (2011). The functional role of producer diversity in ecosystems. *American Journal of Botany*, **98**: 572–92.

Carlquist, S., Baldwin, B.G., and Carr, G. (2003). *Tarweeds and Silverswords: Evolution of the Madinae*. Missouri Botanical Garden Press, St. Louis, MO.

Carlson, S.M., Cunningham, C.J., and Westley, P.A.H. (2014). Evolutionary rescue in a changing world. *Trends in Ecology and Evolution*, **29**, 521–30.

Carlson, S.M., Quinn, T.P., and Hendry, A.P. (2011). Eco-evolutionary dynamics in Pacific salmon. *Heredity*, **106**, 438–47.

Carpenter, S.R. and Kitchell, J.F. (1993). *The Trophic Cascade in Lakes*. Cambridge University Press, Cambridge.

Carpenter, S.R., Kitchell, J.F., and Hodgson, J.R. (1985). Cascading trophic interactions and lake productivity. *BioScience*, **35**, 634–9.

Carpenter, S.R., *et al.* (2011). Early warning of regime shifts: a whole-ecosystem experiment. *Science*, **332**, 1079–82.

Carroll, S.P., Hendry, A.P., Reznick, D.N., and Fox, C.W. (2007). Evolution on ecological time-scales. *Functional Ecology*, **21**, 387–93.

Carstensen, D.W., Lessard, J.P., Holt, B.G., Krabbe Borregaard, M., and Rahbek, C. (2013). Introducing the biogeographic species pool. *Ecography*, **36**, 1310–18.

Caruso, C.M. (2002). Influence of plant abundance on pollination and selection on floral traits of *Ipomopsis aggregata*. *Ecology*, **83**, 241–54.

Carvalheiro, L.G., Buckley, Y.M., Ventim, R., Fowler, S.V., and Memmott, J. (2008). Apparent competition can compromise the safety of highly specific biocontrol agents. *Ecology Letters*, **11**, 690–700.

Case, T.J. (1990). Invasion resistance arises in strongly interacting species-rich model competition communities. *Proceedings of the National Academy of Sciences*, **87**, 9610–14.

Case, T.J. (2000. *An Illustrated Guide to Theoretical Ecology*. Oxford University Press, New York.

Caswell, H. (1976). Community structure: a neutral model analysis. *Ecological Monographs*, **46**, 327–54.

Cattin, M-F., Bersier, L-F., Banašek-Richter, C., Baltensperger, R., and Gabriel, J-P. (2004). Phylogenetic constraints and adaptation explain food-web structure. *Nature*, **427**, 835–9.

Cavender-Bares, J., Ackerly, D., Baum, D., and Bazzaz, F. (2004). Phylogenetic overdispersion in Floridian oak communities. *American Naturalist*, **163**, 823–43.

Cavender-Bares, J., Keen, A., and Miles, B. (2006). Phylogenetic structure of Floridian plant communities depends on taxonomic and spatial scale. *Ecology*, **87**, S109–S122.

Cavender-Bares, J., Kozak, K.H., Fine, P.V.A., and Kembel, S.W. (2009. The merging of community ecology and phylogenetic biology. *Ecology Letters*, **12**, 693–715.

Cayford, J.T. and Goss-Custard, J.D. (1990). Seasonal changes in the size-selection of mussels, *Mytilus edulis*, by oystercatchers, *Haematopus ostralegus*: an optimality approach. *Animal Behaviour*, **40**, 609–24.

Ceballos, G., Ehrlich, P.R., and Dirzo, R. (2017). Biological annihilation via the ongoing sixth mass extinction signaled by vertebrate population losses and declines. *Proceedings of the National Academy of Sciences*, **114**(30), E6089–96.

Chaneton, E.J. and Bonsall, M.B. (2000). Enemy-mediated apparent competition: empirical patterns and the evidence. *Oikos*, **88**, 380–94.

Chamberlain, S.A. and Holland, J.N. (2009). Quantitative synthesis of context dependency in ant-plant protection mutualisms. *Ecology*, **90**, 2384–92.

Chao, A. and Jost, L. (2015). Estimating diversity and entropy profiles via discovery rates of new species. *Methods and Ecology and Evolution*, **6**, 873–82.

Chapin, F.S., Walker, L.R., Fastie, C.L. and Sharman, L.C. (1994. Mechanisms of primary succession following deglaciation at Glacier Bay, Alaska. *Ecological Monographs*, **64**, 149–75.

Charnov, E.L. (1976a). Optimal foraging: attack strategy of a mantid. *American Naturalist*, **110**, 141–51.

Charnov, E.L. (1976b). Optimal foraging: the marginal value theorem. *Theoretical Population Biology*, **9**, 129–36.

Chase, J.M. (2003). Experimental evidence for alternative stable equilibria in a benthic pond food web. *Ecology Letters*, **6**, 733–41.

Chase, J.M. (2007). Drought mediates the importance of stochastic community assembly. *Proceedings of the National Academy of Sciences*, **104**, 17430–4.

Chase, J.M. and Leibold, M.A. (2003a). *Ecological Niches: Linking Classical and Contemporary Approaches*. University of Chicago Press, Chicago, IL.

Chase, J.M. and Leibold, M.A. (2003b). Spatial scale dictates the productivity–biodiversity relationship. *Nature*, **416**, 427–30.

Chase, J.M., Leibold, M.A., Downing, A.L., and Shurin, J.B. (2000). The effects of productivity, herbivory, and plant species turnover in grassland food webs. *Ecology*, **81**, 2485–97.

Chase, J.M., *et al.* (2002). The interaction between predation and competition: a review and synthesis. *Ecology Letters*, **5**, 302–15.

Chase, J.M., *et al.* (2005). Competing theories for competitive metacommunities. In: Holyoak, M., Leibold, M.A., and Holt, R.D., eds. *Metacommunities: Spatial Dynamics and Ecological Communities*, pp. 336–54. University of Chicago Press, Chicago, IL.

Chave, J. (2004). Neutral theory and community ecology. *Ecology Letters*, **7**, 241–53.

Chave, J., Muller-Landau, H.C., and Levin, S.A. (2002). Comparing classical community models: Theoretical consequences for patterns of diversity. *American Naturalist*, **159**, 1–23.

Chen, H.W., Shao, K.T., Liu, C.W.J., Lin, W.H., and Liu, W.C. (2011). The reduction of food web robustness by parasitism: fact and artifact. *International Journal of Parasitology*, **41**, 627–34.

Chen, I-C., Hill, J.K., Ohlemüller, R., Roy, D.B., and Thomas, C.D. (2011). Rapid range shifts of species associated with high levels of climate warming. *Science*, **333**, 1024–6.

Chesson, J. (1978). Measuring preference in selective predation. *Ecology*, **59**, 211–15.

Chesson, J. (1983). The estimation and analysis of preference and its relationship to foraging models. *Ecology*, **64**, 1297–304.

Chesson, P.L. (1985). Coexistence of competitors in spatially and temporally varying environments: a look at the combined effects of different sorts of variability. *Theoretical Population Biology*, **28**, 263–87.

Chesson, P.L. (1991). A need for niches? *Trends in Ecology and Evolution*, **6**, 26–8.

Chesson, P.L. (1994). Multispecies competition in variable environments. *Theoretical Population Biology*, **45**, 227–76.

Chesson, P.L. (2000a). Mechanisms of maintenance of species diversity. *Annual Review of Ecology and Systematics*, **31**, 343–66.

Chesson, P.L. (2000b). General theory of competitive coexistence in spatially-varying environments. *Theoretical Population Biology*, **58**, 211–37.

Chesson, P.L. (2011). Ecological niches and diversity maintenance. In: Pavlinov, I.Y., ed. *Research in Biodiversity—Models and Applications*, pp. 43–60. Intech, Rijeka.

Chesson, P. (2018). Updates on mechanisms of maintenance of species diversity. *Journal of Ecology*, **106**, 1773–94.

Chesson, P.L. and Huntly, N. (1997). The role of harsh and fluctuating conditions in the dynamics of ecological systems. *American Naturalist*, **150**, 519–53.

Chesson, P.L. and Kuang, J.J. (2008). The interaction between predation and competition. *Nature*, **456**, 235–8.

Chesson, P.L. and Kuang, J.J. (2010). The storage effect due to frequency-dependent predation in multispecies plant communities. *Theoretical Population Biology*, **78**, 148–64.

Chesson, P.L. and Rees, M. (2007). Commentary on Clark *et al.* (2007): Resolving the biodiversity paradox. *Ecology Letters*, **10**, 659–60.

Chesson, P.L. and Warner, R.R. (1981). Environmental variability promotes coexistence in lottery competitive systems. *American Naturalist*, **117**, 923–43.

Chiba, S. (2004). Ecological and morphological patterns in communities of land snails of the genus *Mandarina* from the Bonin Islands. *Journal of Evolutionary Biology*, **17**, 131–43.

Chisholm, R.A. and Fung, T. (2018). Comment on "Plant diversity increases with the strength of negative density dependence at the global scale". *Science*, **360**(6391), eaar4685.

Chisholm, R.A. and Muller-Landau, H.C. (2011). A theoretical model linking interspecific variation in density dependence to species abundance. *Theoretical Ecology*, **4**, 241–53.

Chisholm, R.A. and Pacala, S.W. (2010). Niche and neutral models predict asymptotically equivalent species abundance distributions in high-diversity ecological communities. *Proceedings of the National Academy of Sciences*, **107**, 15821–5.

Chown, S.L. and Gaston, K.J. (2000). Areas, cradles and museums: the latitudinal gradient in species richness. *Trends in Ecology and Evolution*, **15**, 311–15.

Chu, C. and Adler, P. (2015). Large niche differences emerge at the recruitment stage to stabilize grassland coexistence. *Ecological Monographs*, 85, 373–92.

Clark, D.A., York, P.H., Rasheed, M.A., and Northfield, T.D. (2017). Does biodiversity-ecosystem function literature neglect tropical ecosystems? *Trends in Ecology and Evolution*, **32**, 320–32.

Clark, F., Brook, B.W., Delean, S., Reşit Akçakaya, H., and Bradshaw, C.J. (2010). The theta-logistic is unreliable for modelling most census data. *Methods in Ecology and Evolution*, **1**, 253–62.

Clark, J.S. (2010). Individuals and the variation needed for high species diversity in forest trees. *Science*, **327**, 1129–32.

Clark, J.S. and Agarwal, P.K. (2007). Rejoinder to Clark *et al.* (2007): Response to Chesson and Rees. *Ecology Letters*, **10**: 661–2.

Clark, J.S., Dietze, M., Chakraborty, S., *et al.* (2007). Resolving the biodiversity paradox. *Ecology Letters*, **10**, 647–62.

Clements, F.E. (1916). *Plant Succession: Analysis of the Development of Vegetation*. Carnegie Institute of Washington, Publication no. 242, Washington, DC.

Clements, F.E. (1936). Nature and structure of the climax. *Journal of Ecology*, **24**, 252–84.

Cohen, J.E., Briand, F., and Newman, C.M. (1990). *Community Food Webs: Data and Theory*. Springer-Verlag, New York, NY.

Cohen, J.E., Jonsson, T., and Carpenter, S.R. (2003). Ecological community description using the food web, species abundance, and body size. *Proceedings of the National Academy of Sciences*, **100**, 1781–6.

Cole, K.W. (1971). A consideration of macro-climatic and macro-biotic change in the Ozark Highlands during post-glacial times. *Proceedings of the Arkansas Academy of Science*, **25**, 15–20.

Coley, P.D. (1983). Herbivory and defensive characteristics of tree species in a lowland tropical forest. *Ecological Monographs*, **53**, 209–33.

Collins, S.L. and Glenn, S.M. (1997). Intermediate disturbance and its relationship to within- and between-patch dynamics. *New Zealand Journal of Ecology*, **21**, 103–10.

Colwell, R.K. (1985). The evolution of ecology. *American Zoologist*, **25**, 771–7.

Colwell, R.K. and Fuentes, E.R. (1975). Experimental studies of the niche. *Annual Review of Ecology and Systematics*, **6**, 281–310.

Colwell, R.K. and Futuyma. D.J. (1971). On the measurement of niche breadth and overlap. *Ecology*, **52**, 567–76.

Colwell, R.K. and Hurtt, G.C. (1994). Nonbiological gradients in species richness and a spurious Rapoport effect. *American Naturalist*, **144**, 570–95.

Colwell, R.K., Rahbek, C., and Gotelli, N.J. (2004). The mid-domain effect and species richness patterns: what have we learned so far? *American Naturalist*, **163**, E1–E23.

Colwell, R.K., Rahbek, C., and Gotelli, N.J. (2005). The mid-domain effect: there's a baby in the bathwater. *American Naturalist*, **166**, E149–E154.

Comita, L.S., Condit, R., and Hubbell, S.P. (2007). Developmental changes in habitat associations of tropical trees. *Journal of Ecology*, **95**, 482–92.

Comita, L.S., Muller-Landau, H.C., Aguilar, S., and Hubbell, S.P. (2010). Asymmetric density dependence shapes species abundances in a tropical tree community. *Science*, **329**, 330–2.

Comita, LS., Queenborough, S.A., Murphy, S.J., *et al.* (2014). Testing predictions of the Janzen–Connell hypothesis: a meta-analysis of experimental evidence for distance- and density-dependent seed and seedling survival. *Journal of Ecology*, **102**, 845–56.

Condamine, F.L., Sperling, F.A.H., Wahlberg, N., Rasplus, J-Y., and Kergoat, G.J. (2012). What causes latitudinal gradients in species diversity? Evolutionary processes and ecological constraints on swallowtail biodiversity. *Ecology Letters*, **15**, 267–77.

Condit, R. (1998). *Tropical Forest Census Plots*. Springer-Verlag, Berlin.

Condit, R., *et al.* (2000). Spatial patterns in the distribution of tropical tree species. *Science*, **288**, 1414–18.

Condit, R., *et al.* (2002). Beta-diversity in tropical forest trees. *Science*, **295**, 666–9.

Condon, M.A., Scheffer, S.J., Lewis, M.L., Wharton, R., Adams, D.C., and Forbes, A.A. (2014). Lethal interactions between parasites and prey increase niche diversity in a tropical community. *Science*, **343**, 1240–4.

Connell, J.H. (1961a). The influence of interspecific competition and other factors on the distribution of the barnacle *Chthamalus stellatus*. *Ecology*, **42**, 710–23.

Connell, J.H. (1961b). Effects of competition, predation by *Thais lapillus*, and other factors on natural populations of the barnacle *Balanus balanoides*. *Ecological Monographs*, **31**, 61–104.

Connell, J.H. (1971). On the role of natural enemies in preventing competitive exclusion in some marine animals and in rain forest trees. In: den Boer, P.J., and Gradwell, G.R., eds. *Dynamics of Populations*, pp. 298–312. Center for Agricultural Publications and Documentation, Wageningen.

Connell, J.H. (1978). Diversity in tropical rain forests and coral reefs. *Science*, **199**, 1302–10.

Connell, J.H. and Slatyer, R.O. (1977). Mechanisms of succession in natural communities and their role in community stability and organization. *American Naturalist*, **111**, 1119–44.

Connell, J.H. and Sousa, W.P. (1983). On the evidence needed to judge ecological stability or persistence. *American Naturalist*, **121**, 789–824.

Connolly, S.R., Hughes, T.P., Bellwood, D.R., and Karlson, R.H. (2005). Community structure of corals and reef fishes at multiple scales. *Science*, **309**, 1363–5.

Connor, E.F. and McCoy, E.D. (1979). The statistics and biology of the species–area relationship. *American Naturalist*, **113**, 791–833.

Cooney, C.R., Tobias, J.A., Weir, J.T., Botero, C.A., and Seddon, N. (2017). Sexual selection, speciation and constraints on geographical range overlap in birds. *Ecology Letters*, **20**, 863–71.

Cooper, W.S. (1923). The recent ecological history of Glacier Bay, Alaska. *Ecology*, **4**, 93–128, 223–46, 355–65.

Cornell, H.V. and Harrison, S.P. (2013). Regional effects as important determinants of local diversity in both marine and terrestrial systems. *Oikos*, **122**, 288–97.

Cornell, H.V. and Harrison, S.P. (2014). What are species pools and when are they important? *Annual Review of Ecology, Evolution, and Systematics*, **45**, 45–67.

Cornell, H.V. and Lawton, J.H. (1992). Species interactions, local and regional processes, and limits to the richness of ecological communities: a theoretical perspective. *Journal of Animal Ecology*, **61**, 1–12.

Cornell, H.V., Karlson, R.H., and Hughes, T.P. (2008). Local-regional species richness relationships are linear at very small to large scales in west-central Pacific corals. *Coral Reefs*, **27**, 145–51.

Cornwell, W.K. and Ackerly, D.D. (2009). Community assembly and shifts in plant trait distributions across an environmental gradient in coastal California. *Ecological Monographs*, **79**, 109–26.

Cornwell, W.K., Schwilk, D.W., and Ackerly, D.D. (2006). A trait-based test for habitat filtering: Convex hull volume. *Ecology*, **87**, 1465–71.

Costa-Pereira, R., Araújo, M.S., da Silva Olivier, R., Souza, F.L., and Rudolf, V.H.W. (2018). Prey limitation drives variation in allometric scaling of predator-prey interactions. *American Naturalist*, **192**(4), E139–49.

Cottenie, K. (2005). Integrating environmental and spatial processes in ecological community dynamics. *Ecology Letters*, **8**, 1175–82.

Cottenie, K. and De Meester, L. (2003). Connectivity and cladoceran species richness in a metacommunity of shallow lakes. *Freshwater Biology*, **48**, 823–32.

Cottenie, K. and De Meester, L. (2004). Metacommunity structure: synergy of biotic interactions as selective agents and dispersal as fuel. Ecology 85: 114–19.

Cottenie, K., Michels, N., Nuytten, N., and De Meester, L. (2003). Zooplankton metacommunity structure: regional vs. local processes in highly interconnected ponds. *Ecology*, **84**, 991–1000.

Cowles, H.C. (1899). The ecological relations of the vegetation of sand dunes of Lake Michigan. *Botanical Gazette*, **27**, 95–117, 167–202, 281–308, 361–91.

Coyne, J.A. and Orr, H.A. (2004). *Speciation*. Sinauer Associates, Sunderland, MA.

Crame, J.A. (2001). Taxonomic diversity gradients through geological time. *Diversity and Distributions*, **7**, 175–89.

Crame, J.A. (2009). Time's stamp on modern biogeography. *Science*, **323**, 720–1.

Crame, J.A., McGowan, A.J., and Bell, M.A. (2018). Differentiation of high-latitude and polar marine faunas in a greenhouse world. *Global Ecology and Biogeography*, **27**, 518–37.

Crawley, M.J. (2007). Plant population dynamics. In: May, R. and McLean, A., eds. *Theoretical Ecology*, 3rd edn, pp. 62–83. Oxford University Press, Oxford.

Creel, S. and Christianson, D. (2008). Relationships between direct predation and risk effects. *Trends in Ecology and Evolution*, **23**, 194–201.

Crist, T.O., Veech, J.A., Gering, J.C., and Summerville, K.S. (2003). Partitioning species diversity across landscapes and regions: a hierarchical analysis of alpha, beta, and gamma diversity. *American Naturalist*, **162**, 734–43.

Currie, D.J. (1991). Energy and large-scale patterns of animal-species and plant-species richness. *American Naturalist*, **137**, 27–49.

Currie, D.J. and Paquin, V. (1987). Large-scale biogeographical patterns of species richness of trees. *Nature*, **329**, 326–7.

Currie, D.J., *et al.* (2004). Predictions and tests of climate-based hypotheses of broad-scale variation in taxonomic richness. *Ecology Letters*, **7**, 1121–34.

Curry, B.B., *et al.* (1999). *Quaternary geology, geomorphology, and climatic history of Kane County, Illinois*, Guidebook no. 028. Illinois State Geological Survey, Champaign, IL.

Dakos, V., Kéfi, S., Rietkerk, M., von Nes, E.H., and Scheffer, M. (2011). Slowing down in spatially patterned ecosystems at the brink of collapse. *American Naturalist*, **177**, E153–66.

Damschen, E.I., Haddad, N.M., Orrock, J.L., Tewksbury, J.J., and Levey, D.J. (2006). Corridors increase plant species richness at large scales. *Science*, **313**, 1284–6.

Darcy-Hall, T.L. (2006). Patterns of nutrient and predator limitation of benthic algae in lakes along a productivity gradient. *Oecologia*, **148**, 660–71.

Darcy-Hall, T.L. and Hall, S.R. (2008). Linking limitation to species composition: importance of inter- and intra-specific variation in grazing resistance. *Oecologia*, **155**, 797–808.

Darlington, P.J. (1957). *Zoogeography: The Geographical Distribution of Animals*. John Wiley and Sons, New York, NY.

Darwin, C.R. (1859). *On The Origin of Species by Means of Natural Selection or the Preservation of Favoured Races in the Struggle for Life*. John Murray, London.

Daskalov, G.M., Grishin, A.N., Rodionov, S., and Mihneva, V. (2007). Trophic cascades triggered by overfishing reveal possible mechanisms of ecosystem regime shifts. *Proceedings of the National Academy of Sciences*, **104**, 10518–23.

Davies, K.F., Holyoak, M., Preston, K.A., Offeman, V.A., and Lum, Q. (2009). Factors controlling community structure in heterogeneous metacommunities. *Journal of Animal Ecology*, **78**, 937–44.

Davies, T.J., Savolainen, V., Chase, M.W., Moat, J., and Barraclough, T.G. (2004). Environmental energy and evolutionary rates in flowering plants. *Proceedings of the Royal Society of London B*, **271,** 2195–200.

Davis, J.M., Rosemond, A.D., Eggert, S.L., Cross, W.F., and Wallace, J.B. (2010). Long-term nutrient enrichment decouples predator and prey production. *Proceedings of the National Academy of Sciences*, **107**, 121–6.

Dayton, P.K. (1971). Competition, disturbance, and community organization: the provision and subsequent utilization of space in a rocky intertidal community. *Ecological Monographs*, **41**, 351–89.

Dayton, P.K. (1975). Experimental evaluation of ecological dominance in a rocky intertidal algal community. *Ecological Monographs*, **45**, 137–59.

deKroon, H., *et al.* (2012). Root responses to nutrients and soil biota: drivers of species coexistence and ecosystem productivity. *Journal of Ecology*, **100**, 6–15.

DeLong, J.P., *et al.* (2015). The body size dependence of trophic cascades. *American Naturalist*, **185**, 354–66.

DeMaster, D.P., Trites, A.W., Clapham, P., *et al.* (2006). The sequential megafaunal collapse: Testing with existing data. *Progress in Oceanography*, **68**, 329–42.

de Mazancourt, C., Johnson, E., and Barraclough, T.G. (2008). Biodiversity inhibits species' evolutionary responses to changing environments. *Ecology Letters*, **11**, 380–8.

de Mazancourt, C., Loreau, M., and Dieckmann, U. (2005). Understanding mutualism when there is adaptation to the partner. *Journal of Ecology*, **93**, 305–14.

de Mazancourt, C. and Schwartz, M.W. (2010). A resource ratio theory of cooperation. *Ecology Letters*, **13**, 349–59.

de Mazancourt, C., *et al.* (2013). Predicting ecosystem stability from community composition and biodiversity. *Ecology Letters*, **16**, 617–25.

De Roos, A.M., Persson, L., and McCauley, E. (2003). The influence of size-dependent life-history traits on the structure and dynamics of populations and communities. *Ecology Letters*, **6**, 473–87.

Dehling, D.M., T. Töpfer, H.M. Schaefer, P. Jordano, K. Böhning-Gaese and M. Schleuning. (2014). Functional relationships beyond species richness patterns: trait matching in plant-bird mutualisms across scales. *Global Ecology and Biogeography*, **23**, 1085–93.

Dempster, J.P. (1983). The natural control of populations of butterflies and moths. *Biological Review*, **58**, 461–81.

Denison, R.F. (2000). Legume sanctions and the evolution of symbiotic cooperation by rhizobia. *American Naturalist*, **156**, 567–76.

Dennis, B. and Taper, B. (1994). Density dependence in time series observations of natural populations: estimation and testing. *Ecological Monographs*, **64**, 205–24.

Denslow, J. (1980). Gap partitioning among tropical rainforest trees. *Biotropica* (Supplement), **12**, 47–55.

Derrickson, E.M. and Ricklefs, R.E. (1988). Taxon-dependent diversification of life-history traits and the perception of phylogenetic constraints. *Functional Ecology*, **2**, 417–23.

Des Roches, S., Shurin, J.B., Schluter, D., and Harmon, L.J. (2013). Ecological and evolutionary effects of stickleback on community structure. *PLOS One*, **8**(4), e59644.

Descamps-Julien, B. and Gonzalez, A. (2005). Stable coexistence in a fluctuating environment: an experimental demonstration. *Ecology*, **86**, 2815–24.

Desjardins-Proulx, P. and Gravel, D. (2012). How likely is speciation in neutral ecology? *American Naturalist*, **179**, 137–44.

Dethlefsen, L. and Relman, D.A. (2011). Incomplete recovery and individualized responses of the human distal gut microbiota to repeated antibiotic perturbation. *Proceedings of the National Academy of Sciences*, **108**, 4554–61.

Diamond, J. (1973). Distributional ecology of New Guinea birds. *Science*, **179**, 759–69.

Diamond, J., ed. (1975). *Assembly of Species Communities*. Harvard University Press, Cambridge, MA.

Diaz, S., Symstad, A.J., Chapin, F.S., Wardle, D.A., and Huenneke, L.F. (2003). Functional diversity revealed by removal experiments. *Trends in Ecology and Evolution*, **18**, 140–6.

Dickie, I.A., Koide, R.T., and Steiner, K.C. (2002). Influences of established trees on mycorrhizas, nutrition, and growth of *Quercus rubra* seedlings. *Ecological Monographs*, **72**, 505–21.

Dickie, I.A., Schnitzer, S.A., Reich, P.B., and Hobbie, S.E. (2005). Spatially disjunct effects of co-occurring competition and facilitation. *Ecology Letters*, **8**, 1191–200.

Diehl, S. and Eklöv, P. (1995). Effects of piscivore-mediated habitat use on resources, diet, and growth of perch. *Ecology*, **76**, 1712–26.

Dijkstra, F.A., West, J.B., Hobbie, S.E., Reich, P.B., and Trost, J. (2007). Plant diversity, CO_2, and N influence inorganic and organic N leaching in grasslands. *Ecology*, **88**, 490–500.

Doak, D.F., Bigger, D., Harding, E.K., Marvier, M.A., O'Malley, R.E., and Thomson, D. (1998). The statistical inevitability of stability–diversity relationships in community ecology. *American Naturalist*, **151**, 264–76.

Dobzhansky, Th. (1950). Evolution in the tropics. *American Scientist*, **38**, 209–21.

Dobzhansky, Th. (1964). Biology, molecular and organismic. *American Zoologist*, **4**, 443–52.

Doebeli, M. and Knowlton, N. (1998). The evolution of interspecific mutualisms. *Proceedings of the National Academy of Sciences*, **95**, 8676–80.

Douglas, A.E. 2008. Conflict, cheats and the persistence of symbioses. *New Phytologist*, **177**: 849–58.

Dornelas, M., Gotelli, N.J., McGill, B., *et al.* (2014). Assemblage time series reveal biodiversity change but not systematic loss. *Science*, **344**, 296–9.

Downing, A.L. (2005). Relative effects of species composition and richness on ecosystem properties in ponds. *Ecology*, **86**, 701–15.

Drakare, S., Lennon, J.J., and Hillebrand, H. (2006). The imprint of the geographical, evolutionary and ecological context on species–area relationships. *Ecology Letters*, **9**, 215–27.

Driscoll, D.A. (2007). How to find a metapopulation. *Canadian Journal of Zoology*, **85**, 1031–48.

Dudgeon, S.R., Aronson, R.B., Bruno, J.F., and Precht, W.F. (2010). Phase shifts and stable states on coral reefs. *Marine Ecology Progress Series*, **413**, 201–16.

Duffy, J.E., Godwin, C.M., and Cardinale, B.J. (2017). Biodiversity effects in the wild are common and as strong as key drivers of productivity. *Nature*, **549**, 261–4.

Duffy, J. E., Richardson, J.P., and France, K.E. (2005). Ecosystem consequences of diversity depend on food chain length in estuarine vegetation. *Ecology Letters*, **8**, 301–9.

Dugatkin, L.A. (2009). *Principles of Animal Behavior*, 2nd edn. W. W. Norton & Company, New York, NY.

Dunn, R.R., Colwell, R.K., and Nilsson, C. (2006). The river domain: why are there more species halfway up the river? *Ecography*, **29**, 251–9.

Dunne, J.A. (2006). The network structure of food webs. In: Pascual, M. and Dunne, J.A., eds. *Ecological Networks: Linking Structure to Dynamics in Food Webs*, pp. 27–86. Oxford University Press, Oxford.

Dunne, J.A., *et al.* (2013). Parasites affect food web structure primarily through increased diversity and complexity. *PLoS Biology*, **11**, e1001579.

Durrett, R. and Levin, S. (1994). The importance of being discrete (and spatial). *Theoretical Population Biology*, **46**, 363–94.

Dybzinski, R. and Tilman, D. (2007). Resource use patterns predict long-term outcomes of plant competition for nutrients and light. *American Naturalist*, **170**, 305–18.

Ehrlich, P.R. and Birch, L.C. (1967). The "Balance of Nature" and "Population Control." *American Naturalist*, **101**, 97–107.

Eiserhardt, W.L., Bjorholm, S., Svenning, J-C., Rangel, T.F., and Balslev, H. (2011). Testing the water-energy theory on American palms (Arecaceae) using geographically weighted regression. *PLoS One*, **6**(11), e27027.

Eklof, A., *et al.* (2013). The dimensionality of ecological networks. *Ecology Letters*, **16**, 577–83.

Elahi, R., O'Connor, M.I., Byrnes, J.E., *et al.* (2015). Recent trends in local-scale marine biodiversity reflect community structure and human impacts. *Current Biology*, **25**, 1938–43.

Ellner, S.P. (2013). Rapid evolution: from genes to communities, and back again? *Functional Ecology*, **27**, 1087–99.

Ellner, S.P., Geber, M.A., and Hairston, N.G., Jr (2011). Does rapid evolution matter? Measuring the rate of contemporary evolution and its impacts on ecological dynamics. *Ecology Letters*, **14**, 603–14.

Ellner, S.P., Snyder, R.E., and Adler, P.B. (2016). How to quantify the temporal storage effect using simulations instead of math. *Ecology Letters*, **19**, 1333–42.

Ellwood, M.D.F., Manica, A., and Foster, W.A. (2009). Stochastic and deterministic processes jointly structure tropical arthropod communities. *Ecology Letters*, **12**, 277–84.

Elner, R.W. and Hughes, R.N. (1978). Energy maximization in the diet of the shore crab, *Carcinus maenas. Journal of Animal Ecology*, **47**, 103–16.

Elser, J.J., *et al.* (2000). Nutritional constraints in terrestrial and freshwater food webs. *Nature*, **408**, 578–80.

Elser, J.J., *et al.* (2007). Global analysis of nitrogen and phosphorus limitation of primary producers in freshwater, marine, and terrestrial ecosystems. *Ecology Letters*, **10**, 1135–42.

Elton, C.S. (1927). *Animal Ecology*. Reprint (2001), University of Chicago Press, Chicago, IL.

Elton, C. (1946). Competition and the structure of ecological communities. *Journal of Animal Ecology*, **15**: 54–68.

Elton, C. (1949). Population interspersion: an essay on animal community patterns. *Journal of Ecology*, **37**, 1–23.

Elton, C.S. (1950). *The Ecology of Animals*, 3rd edition. Methuen, London.

Elton, C.S. (1958). *The Ecology of Invasions by Animals and Plants*. Methuen, London.

Emerson, B.C. and Gillespie, R.G. (2008). Phylogenetic analysis of community assembly and structure over space and time. *Trends in Ecology and Evolution*, **23**, 619–30.

Emlen, J.M. (1966). The role of time and energy in food preference. *American Naturalist*, **100**, 611–17.

Emlen, J.M. (1973). *Ecology: An Evolutionary Approach*. Addison-Wesley, New York, NY.

Emmerson, M.C. and Raffaelli, D. (2004). Predator–prey body size, interaction strength and the stability of a real food web. *Journal of Animal Ecology*, **73**, 399–409.

Emmerson, M. and Yearsley, J.M. (2004). Weak interactions, omnivory and emergent food-web properties. *Proceedings of the Royal Society of London B*, **271**, 397–405.

Endler, J.A. (1980). Natural section in color patterns in *Poecilia recticulata. Evolution*, **34**, 76–91.

Engen, S., Lande, R., Walla, T., and DeVries, P.J. (2002). Analyzing spatial structure of communities using the two-dimensional Poisson lognormal species abundance model. *American Naturalist*, **160**, 60–73.

Enquist, B.J. and Niklas, K.J. (2001). Invariant scaling relations across tree-dominated communities. *Nature*, **410**, 655–60.

Eriksson, O. (1997). Clonal life histories and the evolution of seed recruitment. In: de Kroon, H. and Groenendael, V., eds. *The Ecology and Evolution of Clonal Plants*, pp. 211–26. Backhuys, Leiden.

Ernest, S.K.M., Brown, J.H., Thibault, K.M., White, E.P., and Goheen, J.R. (2008). Zero sum, the niche, and metacommunities: long-term dynamics of community assembly. *American Naturalist*, **172**, E257–69.

Erwin, D.H. (2001). Lessons from the past: biotic recoveries from mass extinctions. *Proceedings of the National Academy of Sciences*, **98**, 5399–403.

Erwin, D.H. (2008). Macroevolution of ecosystem engineering, niche construction and diversity. *Trends in Ecology and Evolution*, **23**, 304–10.

Estes, J.A. and Duggins, D. (1995). Sea otters and kelp forest in Alaska: generality and variation in a community ecological paradigm. *Ecological Monographs*, **65**, 75–100.

Estes, J.A., Duggins, D.O., and Rathbun, G.B. (1989). The ecology of extinctions in kelp forest communities. *Conservation Biology*, **3**, 252–64.

Estes, J.A. and Palmisano, J.F. (1974). Sea otters: their role in structuring nearshore communities. *Science*, **185**, 1058–60.

Estes, J.A., Tinker, M.T., Williams, T.M., and Doak, D.F. (1998). Killer whale predation on sea otters linking oceanic and nearshore ecosystems. *Science*, **282**, 473–6.

Estes, J.A., *et al.* (2011). Trophic downgrading of planet Earth. *Science*, **333**, 301–6.

Evans, K.L., Greenwood, J.J.D., and Gaston, K.J. (2005). Dissecting the species-energy relationship. *Proceedings of the Royal Society of London B*, **272**, 2155–63.

Evans, K.L., Newson, S.E., Storch, D., Greenwood, J.J.D., and Gaston, K.J. (2008). Spatial scale, abundance and the species-energy relationship in British birds. *Journal of Animal Ecology*, **77**, 395–405.

Ezard, T.H.G., Côté, S.D., and Pelletier, F. (2009). Eco-evolutionary dynamics: Disentangling phenotypic, environmental and population fluctuations. *Philosophical Transactions of the Royal Society of London B*, **364**, 1491–8.

Fagan, W.F. and Hurd, L.E. (1994). Hatch density variation of a generalist arthropod predator: Population consequences and community impact. *Ecology*, **75**, 2022–32.

Fanin, N., Gundale, M.J., Farrell, M., *et al.* (2017). Consistent effects of biodiversity loss on multifunctionality across contrasting ecosystems. *Nature Ecology and Evolution*, **2**, 269–78.

Fargione, J. and Tilman, D. (2005). Diversity decreases invasion via both sampling and complementary effects. *Ecology Letters*, **8**, 604–11.

Fargione, J., *et al.* (2007). From selection to complimentarity: shifts in the causes of biodiversity–productivity relationships in a long-term biodiversity experiment. *Proceedings of the Royal Society of London B*, **274**, 871–76.

Farnworth, E.G. and Golley, F.B. (1974). *Fragile Ecosystems: Evaluation of Research and Applications in the Neotropics*. Springer-Verlag, New York, NY.

Farrell, B.D., Mitter, C., and Futuyma, D.J. (1992). Diversification at the insect-plant interface. *BioScience*, **42**, 34–42.

Fauset, S., *et al.* (2015). Hyperdominance in Amazonian forest carbon cycling. *Nature Communications*, **6**, 6857.

Fauth, J.E., Bernardo, J., Camara, M., Resitarits, W.J.J., Van Buskirk, J., and McCollum, S.A. (1996). Simplifying the jargon of community ecology: a conceptual approach. *American Naturalist*, **147**, 282–6.

Fedorov, A.A. (1966). The structure of the tropical rain forest and speciation in the humid tropics. *Journal of Ecology*, **54**, 1–11.

Feeny, P. (1976). Plant apparency and chemical defense. *Recent Advances in Phytochemistry*, **10**, 1–40.

Feldman, T.S., Morris, W.F., and Wilson, W.G. (2004). When can two plant species facilitate each other's pollination? *Oikos*, **105**, 197–207.

Felsenstein, J. (1985). Phylogenies and the comparative method. *American Naturalist*, **125**, 1–15.

Feng, Z. and DeAngelis, D.L. (2017). *Mathematical Models of Plant-Herbivore Interactions*. Chapman and Hall, Abingdon.

Feng, Z., Liu, R., and DeAngelis, D.L. (2008). Plant-herbivore interactions mediated by plant toxicity. *Theoretical Population Biology*, **73**, 449–59.

Ficetola, G.F., Mazel, F., and Thuiller, W. (2017). Global determinants of zoogeographic boundaries. *Nature Ecology and Evolution*, **1**(4), 89

Field, R., *et al.* (2009). Spatial species-richness gradients across scales: a meta-analysis. *Journal of Biogeography*, **36**, 32–147.

Fine, P.E.M. (1975). Vectors and vertical transmission—epidemiologic perspective. *Annals of the New York Academy of Sciences*, **266**, 173–94.

Fine, P.V.A. (2015). Ecological and evolutionary drivers of geographic variation in species diversity. *Annual Review of Ecology, Evolution and Systematics*, **46**, 369–92.

Fine, P.V.A. and Ree, R.H. (2006). Evidence for a time-integrated species–area effect on the latitudinal gradient in tree diversity. *American Naturalist*, **168**, 796–804.

Firn, J., House, A.P.N., and Buckley, Y.M. (2010). Alternative states models provide an effective framework for invasive species control and restoration of native communities. *Journal of Applied Ecology*, **47**, 96–105.

Fischer, A.G. (1960). Latitudinal variations in organic diversity. *Evolution*, **14**, 64–81.

Fisher, R.A. (1930). *The Genetical Theory of Natural Selection*. Clarendon Press, Oxford.

Fisher, R.A., Corbet, A.S., and Williams, C.B. (1943). The relation between the number of species and the number of individuals in a random sample of an animal population. *Journal of Animal Ecology*, **12**, 42–58.

Fontaine, C., *et al.* (2011). The ecological and evolutionary implications of merging different types of networks. *Ecology Letters*, **14**, 1170–81.

Foote, M. (1997). The evolution of morphological diversity. *Annual Review of Ecology and Systematics*, **28**, 129–52.

Forbes, S.A. (1887). The lake as a microcosm. *Bulletin of the Scientific Association (Peoria, IL)*, **1887**, 77–87.

Forister, M.A. and Feldman, C.R. (2011). Phylogenetic cascades and the origins of tropical diversity. *Biotropica*, **43**, 270–8.

Fortunel, C., Paine, C., Fine, P.V., Kraft, N.J., and Baraloto, C. (2014). Environmental factors predict community functional composition in Amazonian forests. *Journal of Ecology*, **102**, 145–55.

Fortunel, C., Valencia, R., Wright, S.J., Garwood, N.C., and Kraft, N.J. (2016). Functional trait differences influence neighbourhood interactions in a hyperdiverse Amazonian forest. *Ecology Letters*, **19**, 1062–70.

Fox, B.J. (1987). Species assembly and the evolution of community structure. *Evolutionary Ecology*, **1**, 201–13.

Fox, B.J. and Brown, J.H. (1993). Assembly rules for functional groups in North American desert rodent communities. *Oikos*, **67**, 358–70.

Fox, J.W. (2007). Testing the mechanisms by which source-sink dynamics alter competitive outcomes in a model system. *American Naturalist*, **170**, 396–408.

France, K.E. and Duffy, J.E. (2006). Consumer diversity mediates invasion dynamics at multiple trophic levels. *Oikos*, **113**, 515–29.

Francis, A.P. and Currie, D.J. (2003). A globally consistent richness-climate relationship for angiosperms. *American Naturalist*, **161**, 523–36.

Franco, A.C. and Nobel, P.S. (1989). Effect of nurse plants on the microhabitat and growth of cacti. *Journal of Ecology*, **77**, 870–86.

Frank, K.T., Petrie, B., Choi, J.S., and Leggett, W.C. (2006). Reconciling differences in trophic controls in mid-latitude marine ecosystems. *Ecology Letters*, **9**, 1096–105.

Fraser, L.H., *et al.* (2015). Worldwide evidence of a unimodal relationship between productivity and plant species richness. *Science*, **349**, 302–5.

Frederickson, A.G. and Stephanopoulos, G. (1981). Microbial competition. *Science*, **213**, 972–9.

Frederickson, M.E. (2013). Rethinking mutualism stability: cheaters and the evolution of sanctions. *Quarterly Review of Biology*, **88**, 269–95.

Freestone, A.L. and Osman, R.W. (2011). Latitudinal variation in local interactions and regional enrichment shape patterns of marine community diversity. *Ecology*, **92**, 208–17.

Freestone, A.L., Osman, R.W., Ruiz, G.M., and Torchin, M.E. (2011). Stronger predation in the tropics shapes species richness patterns in marine communities. *Ecology*, **92**, 983–93.

Fretwell, S.D. and Lucas, H.L. (1970). On territorial behavior and other factors influencing habitat distribution in birds. I. Theoretical development. *Acta Biotheoretica*, **19**, 16–36.

Fridley, J.D., *et al.* (2007). The invasion paradox: reconciling pattern and process in species invasions. Ecology, **88**, 3–17.

Froese, R. and Pauly, D. (2000). *FishBase 2000: Concepts Designs and Data Sources*. WorldFish, Timor-Leste.

Fryer, G. and Iles, T.D. (1972). *The Cichlid Fishes of the Great Lakes of Africa: Their Biology and Evolution*. Oliver and Boyd, Edinburgh.

Fryxell, J.M. and Lundberg, P. (1994). Diet choice and predator–prey dynamics. *Evolutionary Ecology*, **8**, 407–21.

Fung, T., Seymour, R.M., and Johnson, C.R. (2011). Alternative stable states and phase shifts in coral reefs under anthropogenic stress. *Ecology*, **92**, 967–82.

Fussmann, G.F., Ellner, S.P., Hairston, N.G., Jr, Jones, L.E., Shertzer, K.W., and Yoshida, T. (2005). Ecological and evolutionary dynamics of experimental plankton communities. *Advances in Ecological Research*, **37**, 221–43.

Fussmann, G.F., Loreau, M., and Abrams, P.A. (2007). Eco-evolutionary dynamics of communities and ecosystems. *Functional Ecology*, **21**, 465–77.

Futuyma, D.J. (2010). Evolutionary constraint and ecological consequences. *Evolution*, **64**, 1865–84.

Gamfeldt, L., Hillebrand, H., and Jonsson, P.R. (2008). Multiple functions increase the importance of biodiversity for overall ecosystem functioning. *Ecology*, **89**, 1223–31.

Gamfeldt, L. and Roger, F. (2017). Revisiting the biodiversity-ecosystem multifunctionality relationship. *Nature Ecology and Evolution*, **1**, 0168

Garay, J., Cressman, R., Xu, F., Varga, Z., and Cabello, T. (2015). Optimal forager against ideal free distributed prey. *American Naturalist*, **186**, 111–22.

García-Callejas, D., Molowny-Horas, R., and Araújo, M.B. (2018). Multiple interactions networks: toward more realistic descriptions of the web of life. *Oikos*, **127**, 5–22.

Gascoigne, J., Berec, L., Gregory, S., and Courchamp, F. (2009). Dangerously few liaisons: a review of mate-finding Allee effects. *Population Ecology*, **51**, 355–72.

Gaston, K.J. (2000). Global patterns in biodiversity. *Nature*, **405**, 220–7.

Gause, G.F. (1934). *The Struggle for Existence*. Williams and Wilkins, Baltimore, MD.

Gause, G.F. (1936). The principles of biocoenology. *Quarterly Review of Biology*, **11**, 320–36.

Gause, G. (1937). Experimental populations of microscopic organisms. *Ecology*, **18**, 173–9.

Gause, G.F. and Witt, A.A. (1935). Behavior of mixed populations and the problem of natural selection. *American Naturalist*, **69**, 596–609.

Gavrilets, S. and Losos, J.B. (2009). Adaptive radiation: contrasting theory with data. *Science*, **323**, 732–7.

Gerhold, P., Cahill, J.F., Winter, M., Bartish, I.V., and Prinzing, A. (2015). Phylogenetic patterns are not proxies of community assembly mechanisms (they are far better). *Functional Ecology*, **29**, 600–14.

Gerhold, P., *et al.* (2011). Phylogenetically poor plant communities receive more alien species, which more easily coexist with natives. *American Naturalist*, **177**, 668–80.

Germain, R.M., Weir, J.T., and Gilbert, B. (2016). Species coexistence: macroevolutionary relationships and the contingency of historical interactions. *Proceedings of the Royal Society of London B*, **283**, 20160047.

Ghazoul, J. (2006). Floral diversity and the facilitation of pollination. *Journal of Ecology*, **94**, 295–304.

Ghedini, G., White, C.R., and Marshall, D.J. (2017). Does energy flux predict density-dependence? An empirical field test. *Ecology*, **98**, 3166–26.

Gilbert, B. and Lechowicz, M.J. (2004). Neutrality, niches, and dispersal in a temperate forest understory. *Proceedings of the National Academy of Sciences*, **101**, 7651–56.

Gilbert-Norton, L., Wilson, R., Stevens, J.R., and Beard, K.H. (2010). A meta-analytic review of corridor effectiveness. *Conservation Biology*, **24**, 660–8.

Gillespie, R.G. (2015). Island time and the interplay between ecology and evolution in species diversification. *Evolutionary Applications*, **9**(1), 53–73.

Gilliam, J.F. (1982). Foraging under mortality risk in size-structured populations. PhD dissertation, Michigan State University, East Lansing, MI.

Gilliam, J.F. and Fraser, D.F. (1987). Habitat selection when foraging under predation hazard: a model and a test with stream-dwelling minnows. *Ecology*, **68**, 1856–62.

Gillies, C.S. and St. Clair, C.C. (2008). Riparian corridors enhance movement of a forest specialist bird in fragmented tropical forest. *Proceedings of the National Academy of Sciences*, **105**, 19774–9.

Gillman, L.N., Ross, H.A., Keeling, J.D., and Wright, S.D. (2009). Latitude, elevation and the tempo of molecular evolution in mammals. *Proceedings of the Royal Society of London B*, **276**, 3353–9.

Gillman, L.N., McBride, P., Keeling, D.J., Ross, H.A., and Wright, S.D. (2010). Are rates of molecular evolution in mammals substantially accelerated in warmer environments? Reply. *Proceedings of the Royal Society of London B*, **278**, 1294–7.

Gillman, L.N. and Wright, S.D. (2006). The influence of productivity on the species richness of plants: a critical assessment. *Ecology*, **87**, 1234–43.

Gilpin, M.E. and Ayala, F.J. (1973). Global models of growth and competition. *Proceedings of the National Academy of Sciences*, **70**, 3590–3.

Ginzburg, L.R. and Akçakaya, H.R. (1992). Consequences of ratio-dependent predation for steady-state properties of ecosystems. *Ecology*, **73**, 1536–43.

Gittings, T. and Giller, P.S. (1998). Resource quality and the colonisation and succession of coprophagous dung beetles. *Ecography*, **21**, 581–92.

Gleason, H.A. (1917). The structure and development of plant association. *Bulletin of the Torrey Botany Club*, **44**, 463–81.

Gleason, H.A. (1922). On the relation between species and area. *Ecology*, **3**, 158–62.

Gleason, H.A. (1926). The individualistic concept of the plant association. *Torrey Botanical Club Bulletin*, **53**, 7–26.

Gleason, H.A. (1927). Further views on the succession concept. *Ecology*, **8**, 299–326.

Gleeson, S.K. and Wilson, D.S. (1986). Equilibrium diet: optimal foraging and prey coexistence. *Oikos*, **46**, 139–44.

Godfray, H.C.J. and Hassell, M.P. (1992). Long time series reveal density dependence. *Nature*, **359**, 673–4.

Godoy, O., Kraft, N.J.B., and Levine, J.M. (2014). Phylogenetic relatedness and the determinants of competitive outcomes. *Ecology Letters*, **17**, 836–44.

Gomulkiewicz, R. and Holt, R.D. (1995). When does evolution by natural selection prevent extinction? *Evolution*, **49**, 201–7.

Gonzalez, A., Cardinale, B.J., Allington, G.R.H., et al. (2016). Estimating local biodiversity change: a critique of papers claiming no net loss of local diversity. *Ecology*, **97**, 1949–60.

Gonzalez, A. and Loreau, M. (2009). The causes and consequences of compensatory dynamics in ecological communities. *Annual Review of Ecology, Evolution, and Systematics*, **40**, 393–414.

Gotelli, N., Shimadzu, H., Dornelas, M., McGill, B., Moyes, F., and Magurran, A.E. (2017). Community-level regula-

tion of temporal trends in biodiversity. *Science Advances*, **3**, e1700315.

Gotelli, N.J. (2008). *A Primer of Ecology*, 4th edn. Sinauer Associates, Sunderland, MA.

Gotelli, N.J. and Colwell, R.K. (2001). Quantifying biodiversity: Procedures and pitfalls in the measurement and comparison of species richness. *Ecology Letters*, **4**, 379–91.

Gotelli, N.J. and Graves, G.R. (1996). *Null Models in Ecology*. Smithsonian Institution Press, Washington, DC.

Gotelli, N.J. and McCabe, D.J. (2002). Species co-occurrence: a meta-analysis of JM Diamond's assembly rules model. *Ecology*, **83**, 2091–6.

Gotelli, N.J. and McGill, B.J. (2006). Null versus neutral models: What's the difference? *Ecography*, **29**, 793–800.

Gotelli, N.J. and Ulrich, W. (2012). Statistical challenges in null model analysis. *Oikos*, **121**, 171–80.

Gotelli, N.J., et al. (2009). Patterns and causes of species richness: A general simulation model for macroecology. *Ecology Letters*, **12**, 873–86.

Gouhier, T.C., Menge, B.A., and Hacker, S.D. (2011). Recruitment facilitation can promote coexistence and buffer population growth in metacommunities. *Ecology Letters*, **14**, 1201–10.

Govaert, L., Pantel, J.H., and De Meester, L. (2016). Eco-evolutionary partitioning metrics: assessing the importance of ecological and evolutionary contributions to population and community change. *Ecology Letters*, **19**, 839–53.

Grace, J.B., et al. (2007). Does species diversity limit productivity in natural grassland communities? *Ecology Letters*, **10**, 680–9.

Grace, J.B., et al. (2016). Integrative modeling reveals mechanisms linking productivity and plant species richness. *Nature*, **529**, 390–3.

Graham, A. (2011). The age and diversification of terrestrial New World ecosystems through Cretaceous and Cenozoic time. *American Journal of Botany*, **98**, 336–51.

Graham, C.H., Parra, J.L., Rahbek C., and McGuire. J.A. (2009). Phylogenetic structure in tropical hummingbird communities. *Proceedings of the National Academy of Sciences*, **106**, 19673–8.

Grant, P.R. (1972). Convergent and divergent character displacement. *Biological Journal of the Linnean Society*, **4**, 39–68.

Grant, P.R. (1986). *Ecology and Evolution of Darwin's Finches*. Princeton University Press, Princeton, NJ.

Grant, P.R. (1986). Interspecific competition in fluctuating environments. In: Diamond, J. and Case, J.M., eds. *Community Ecology*, pp. 173–91. Harper and Row, New York, NY.

Grant, P.R. and Grant, B.R. (2008). *How and Why Species Multiply: The Radiation of Darwin's Finches*. Princeton University Press, Princeton, NJ.

Gravel, D., Canham, C.D., Beaudet, M., and Messier, C. (2006). Reconciling niche and neutrality: the continuum hypothesis. *Ecology Letters*, **9**, 399–409.

Gravel, D., Poisot, T., Albouy, C., Velez, L., and Mouillot, D. (2013). Inferring food web structured from predator–prey body size relationships. *Methods in Ecology and Evolution*, **4**, 1083–90.

Gray, J.S. (1979). Pollution-induced changes in populations. *Philosophical Transactions of the Royal Society of London B*, **286**, 545–61.

Gray, R. (1986). Faith and foraging. In: Kamil, A.C., Krebs, J.R., and Pulliam, H.R., eds, *Foraging Behavior*, pp. 69–140. Plenum Press, New York.

Green, J.L. and Plotkin, L.B. (2007). A statistical theory for sampling species abundances. *Ecology Letters*, **10**, 1037–45.

Green, J.L., *et al.* (2004). Spatial scaling of microbial eukaryote diversity. *Nature*, **432**, 747–50.

Greene, D.F. and Johnson, E.A. (1993). Seed mass and dispersal capacity in wind-dispersed diaspores. *Oikos*, **67**, 69–74.

Greene, D.F. and Johnson, E.A. (1994). Estimating the mean annual seed production of trees. *Ecology*, **75**, 642–7.

Gregory, S.D., Bradshaw, C.J.A., Brook, B.W., and F. Courchamp. (2010). Limited evidence for the demographic Allee effect from numerous species across taxa. *Ecology*, **91**, 2151–61.

Grime, J.P. (1973a). Competitive exclusion in herbaceous vegetation. *Nature*, **242**, 344–7.

Grime, J.P. (1973b). Control of species diversity in herbaceous vegetation. *Journal of Environmental Management*, **1**, 151–67.

Grime, J.P. (1977). Evidence for the existence of three primary strategies in plants and its relevance to ecological and evolutionary theory. *American Naturalist*, **111**, 1169–94.

Grime, J.P. (1997). Biodiversity and ecosystem function: The debate deepens. *Science*, **277**, 1260–1.

Grime, J.P., Hunt, R., and Krzanowski, W.J. (1987). Evolutionary physiological ecology of plants. In: P. Calow, ed. *Evolutionary Physiological Ecology*, pp. 105–25. Cambridge University Press, Cambridge.

Grinnell, J. (1917). The niche-relationship of the California thrasher. *Auk*, **34**, 427–33.

Grman, E., Robinson, T.M.P., and Klausmeier, C.A. (2012). Ecological specialization and trade affect the outcome of negotiations in mutualism. *American Naturalist*, **179**, 567–81.

Gross, K. (2008). Positive interactions among competitors can produce species-rich communities. *Ecology Letters*, **11**, 929–36.

Gross, K. and Cardinale, B.J. (2005). The functional consequences of random vs. ordered species extinctions. *Ecology Letters*, **8**, 409–18.

Gross, K. and Cardinale, B.J. (2007). Does species richness drive community production or vice versa? Reconciling historical and contemporary paradigms in competitive communities. *American Naturalist*, **170**, 207–20.

Gross, K., B.J. Cardinale, J.W. Fox, *et al.* (2014). Species richness and the temporal stability of biomass production: a new analysis of recent biodiversity experiments. *American Naturalist*, **183**, 1–12.

Gross, K. and Snyder-Beattie, A. (2016). A general, synthetic model predicting biodiversity gradients from environmental geometry. *American Naturalist*, **188**, E85–E97.

Gross, K.L. and Werner P.A. (1982). Colonizing abilities of "biennial" plant species in relation to ground cover: Implications for their distributions in a successional sere. *Ecology*, **63**, 921–31.

Grover, J.P. (1990). Resource competition in a variable environment: phytoplankton growing according to Mondo's model. *American Naturalist*, **136**, 771–89.

Grover, J.P. (1997). *Resource Competition*. Chapman and Hall, London.

Gruner, D.S., *et al.* (2008). A cross-system synthesis of consumer and nutrient resource control on producer biomass. *Ecology Letters*, **11**, 740–55.

Guerrero-Ramírez, N.R., *et al.* (2017). Diversity-dependent temporal divergence of ecosystem functioning in experimental ecosystems. *Nature Ecology and Evolution*, **1**, 1639–42.

Guimarães, P.R., Jr, Rico-Gray, V., dos Reis, S.F., and Thompson, J.N. (2006). Asymmetries in specialization in ant–plant mutualistic networks. *Proceedings of the Royal Society of London B*, **273**, 2041–7.

Guimarães, P.R., Jr, Sazima, C., dos Reis, S.F., and Sazima, I. (2007). The nested structure of marine cleaning symbiosis: is it like flowers and bees? *Biology Letters*, **3**, 51–4.

Guimerà, R., Stouffer, D.B., Sales-Pardo, M., Leicht, E.A., Newman, M.E.J., and Amaral, L.A.N. (2010). Origin of compartmentalization in food webs. *Ecology*, **91**, 2941–51.

Gunderson, L.H. (2000). Ecological resilience—in theory and application. *Annual Review of Ecology and Systematics*, **31**, 425–39.

Gurevitch, J., Morrow, L.L., Wallace, A., and Walsh, J.S. (1992). A meta-analysis of competition in field experiments. *American Naturalist*, **140**, 539–72.

Hacker, S.D. and Bertness, M.D. (1999). Experimental evidence for factors maintaining plant species diversity in a New England salt marsh. *Ecology*, **80**, 2064–73.

Haddad, N.M., *et al.* (2014). Potential negative ecological effects of corridors. *Conservation Biology*, **28**, 1178–87.

Hairston, N.G., Jr, Ellner, S.P., Geber, M.A., Yoshida, T., and Fox. J.A. (2005). Rapid evolution and the convergence of ecological and evolutionary time. *Ecology Letters*, **8**, 1114–27.

Hairston, N.G., Jr, Smith, F.E., and Slobodkin, L.B. (1960). Community structure, population control, and competition. *American Naturalist*, **94**, 421–5.

Hall, S.R., Shurin, J.B., Diehl, S., and Nisbet, R.M. (2007). Food quality, nutrient limitation of secondary production, and the strength of trophic cascades. *Oikos*, **116**, 1128–43.

Hammill, E., Petchey, O.L., and Anholt, B.R. (2010). Predator functional response changed by inducible defenses in prey. *American Naturalist*, **176**, 723–31.

Handa, I.T., Harmsen, R., and Jefferies, R.L. (2002). Patterns of vegetation change and the recovery potential of degraded areas in a coastal marsh system of the Hudson Bay lowlands. *Journal of Ecology*, **90**, 86–99.

Hanly, P. and Mittelbach, G.G. (2017). The influence of dispersal on the realized trajectory of a pond metacommunity. *Oikos*, **126**, 1269–80.

Hanly, P.J., Mittelbach, G.G., and Schemske, D.W. (2017). Speciation and the latitudinal diversity gradient: insights from the global distribution of endemic fish. *American Naturalist*, **189**, 604–15.

Hanski, I. (1990). Density dependence, regulation and variability in animal populations. *Philosophical Transactions of the Royal Society of London B*, **330**, 141–50.

Hanski, I. (1994). A practical model of metapopulation dynamics. *Journal of Animal Ecology*, **63**, 151–62.

Hanski, I. (1997). Metapopulation dynamics: from concepts and observations to predictive models. In: Hanski, I. and Gilpin, M., eds. *Metapopulation Biology: Ecology, Genetics, and Evolution*, pp. 69–91. Academic Press, San Diego, CA.

Hanski, I. and Gilpin, M. (1991). Metapopulation dynamics: brief history and conceptual domain. *Biological Journal of the Linnean Society*, **42**, 3–16.

Hanski, I. and Kuussaari, M. (1995). Butterfly metapopulation dynamics. In: Cappuccino, N. and Price, P.W., eds. *Population Dynamics: New Approaches and Synthesis*, pp. 149–72. Academic Press, San Diego, CA.

Hanski, I., Kuussaari, M., and Niemenen, M. (1994). Metapopulation structure and migration in the butterfly *Melitaea cinxia*. *Ecology*, **75**, 747–62.

Hanski, I. and Ranta, E. (1983). Coexistence in a patchy environment: three species of daphnia in rock pools. *Journal of Animal Ecology*, **52**, 263–79.

Hanski, I. and Simberloff, D. (1997). The metapopulation approach, its history, conceptual domain and application to conservation. In: Hanski, I. and Gilpin, M.E., eds. *Metapopulation Biology: Ecology, Genetics, and Evolution*, pp. 5–26. Academic Press, San Diego, CA.

Hargreaves, A.L., Samis, K.E., and Eckert, C.G. (2014). Are species' range limits simply niche limits writ large? A review of transplant experiments beyond the range. *American Naturalist*, **183**, 157–73.

Harmon, L.J. and Harrison, S. (2015). Species diversity is dynamic and unbounded at local and continental scales. *American Naturalist*, **185**, 584–93.

Harmon, L.J., Matthews, B., Des Roches, S., Chase, J.M., Shurin, J.B., and Schluter, D. (2009). Evolutionary diversification in stickleback affects ecosystem functioning. *Nature*, **458**, 1167–70.

Harper, J.L., Lovell, P.H., and Moore, K.G. (1970). The shapes and sizes of seeds. Annual Review of Ecology and Systematics, **1**, 327–56.

Harpole, W.S., *et al.* (2011). Nutrient co-limitation of primary producer communities. *Ecology Letters*, **14**, 852–62.

Harrison, S. (1991). Local extinction in a metapopulation context: an empirical evaluation. *Biological Journal of the Linnean Society*, **42**, 73–88.

Harrison, S. (2008). Commentary on Stohlgren *et al.* (2008): The myth of plant species saturation. *Ecology Letters*, **11**, 322–4.

Harrison, S. and Cappuccino, N. (1995). Using density-manipulation experiments to study population regulation. In: Cappuccino, N. and Price, P.W., eds. *Population Dynamics: New Approaches and Synthesis*, pp. 131–47. Academic Press, San Diego, CA.

Harrison, S. and Cornell, H.V. (2008). Towards a better understanding of the regional causes of local community richness. *Ecology Letters*, **11**, 969–79.

Harrison, S. and Grace, J.B. (2007). Biogeographical affinity helps explain productivity–richness relationships at regional and local scales. *American Naturalist*, **170**, S5–S15.

Hartman, A.L., Lough, D.M., Barupal, D.K., *et al.* (2009). Human gut microbiome adopts an alternative state following small bowel transplantation. *Proceedings of the National Academy of Sciences*, **106**, 17187–92.

Harvell, C.D. (1990). The ecology and evolution of inducible defenses. *Quarterly Review of Biology*, **65**, 323–40.

Harvey, P.H., Read, A.F., and Nee, S. (1995). Why ecologists need to be phylogenetically challenged. *Journal of Ecology*, **83**, 535–6.

Hassell, M.P. (1978). *The Dynamics of Arthropod Predator–Prey Systems*. Princeton University Press, Princeton, NJ.

Hastings, A. (1980). Disturbance, coexistence, history, and competition for space. *Theoretical Population Biology*, **18**, 363–73.

Hastings, A. (2003). Metapopulation persistence with age-dependent disturbance or succession. *Science*, **301**, 1525–6.

Hatton, I.A., McCann, K.S., Fryxell, J.M., *et al.* (2015). The predator-prey power law: Biomass scaling across terrestrial and aquatic biomes. *Science*, **249**, aac6284.

Hautier, Y., Tilman, D., Isbell F., Seabloom, E.W., Borer, E.T., and Reich, P.B. (2015). Anthropogenic environmental changes affect ecosystem stability via biodiversity. *Science*, **348**, 336–40.

Hawkins, B.A. (1992). Parasitoid–host food webs and donor control. *Oikos*, **65**, 159–62.

Hawkins, B.A. (2010). Multiregional comparisons of the ecological and phylogenetic structure of butterfly species richness gradients. *Journal of Biogeography*, **37**, 647–56.

Hawkins, B.A. and Diniz-Filho, J.A.F. (2002). The mid-domain effect cannot explain the diversity gradient of Nearctic birds. *Global Ecology and Biogeography*, **11**, 419–26.

Hawkins, B.A., Diniz-Filho, J.A.F., Jaramillo, C.A., and Soeller, S.A. (2007). Climate, niche conservatism, and the global bird diversity gradient. *American Naturalist*, **170**, S16–S27.

Hawkins, B.A., *et al.* (2003). Energy, water, and broad-scale geographic patterns of species richness. *Ecology*, **84**, 3105–17.

Hawkins, C.P., Norris, R.H., Hogue, J.N., and Feminella, J.W. (2000). Development and evaluation of predictive models for measuring the biological integrity of streams. *Ecological Applications*, **10**, 1456–77.

Hay, M.E. (1986). Associational plant defenses and the maintenance of species-diversity—turning competitors into accomplices. *American Naturalist*, **128**, 617–41.

He, F., Gaston, K.J., Conner, E.F., and Srivastava, D.S. (2005). The local-regional relationship: Immigration, extinction, and scale. *Ecology*, **86**, 360–5.

He, X. and Kitchell, J.F. (1990). Direct and indirect effects of predation on a fish community: a whole-lake experiment. *Transactions of the American Fisheries Society*, **119**, 825–35.

Hector, A. and Bagchi, R. (2007). Biodiversity and ecosystem multifunctionality. *Nature*, **448**, 188–90.

Hector, A., Bazeley-White, E., Loreau, M., Otway, S., and Schmid, B. (2002). Overyielding in grassland communities: Testing the sampling effect hypothesis with replicated biodiversity experiments. *Ecology Letters*, **5**, 502–11.

Hector, A., *et al.* (1999). Plant diversity and productivity experiments in European grasslands. *Science*, **286**, 1123–7.

Hector, A., *et al.* (2011). BUGS in the analysis of biodiversity experiments: Species richness and composition are of similar importance for grassland productivity. *PLoS One*, **6(3)**, e17434.

Heemsbergen, D.A., Berg, M.P., Loreau, M., van Hal, J.R., Faber, J.H., and Verhoef, H.A. (2004). Biodiversity effects on soil processes explained by interspecific functional dissimilarity. *Science*, **306**, 1019–20.

Hegland, S.J., Grytnes, J-A., and Totland, Ø. (2009). The relative importance of positive and negative interactions for pollinator attraction in a plant community. *Ecological Research*, **24**, 929–36.

Heil, M. and McKey, D. (2003). Protective ant-plant interactions as model systems in ecological and evolutionary research. *Annual Review of Ecology, Evolution, and Systematics*, **34**, 425–53.

Helmus, M.R., Savage, K., Diebel, M.W., Maxted, J.T., and Ives, A.R. (2007). Separating the determinants of phylogenetic community structure. *Ecology Letters*, **10**, 917–25.

Henderson, P.A. and Magurran, A.E. (2014). Direct evidence that density-dependent regulation underpins the temporal stability of abundant species in a diverse animal community. *Proceedings of the Royal Society of London B*, **281**, 20141336.

Hendriks, M., Mommer, L., de Caluwe, H., Smit-Tiekstra, A.E., van der Putten, W.H., and de Kroon, H. (2013). Independent variations of plant and soil mixtures reveal soil feedback effects on plant community overyielding. *Journal of Ecology*, **101**, 287–97.

Hendry, A.P. (2017). *Eco-evolutionary Dynamics*. Princeton University Press, Princeton, NJ.

Hendry, A.P. and Green, D.M. (2017). Eco-evolutionary dynamics in cold blood. *Copeia*, **105**, 441–50.

Herben, T., Mandák, B., Bímova, K., and Münzbergova, Z. (2004). Invasibility and species richness of a community: a neutral model and a survey of published data. *Ecology*, **85**, 3223–33.

Herre, E.A., Knowlton, N., Mueller, U.G., and Rehner, S.A. (1999). The evolution of mutualisms: Exploring the paths between conflict and cooperation. *Trends in Ecology and Evolution*, **14**, 49–53.

Hill, M.O. (1973). Diversity and evenness: a unifying notation and its consequences. *Ecology*, **54**, 427–32.

Hillebrand, H. (2004). On the generality of the latitudinal diversity gradient. *American Naturalist*, **163**, 192–211.

Hillebrand, H. (2005). Regressions of local on regional diversity do not reflect the importance of local interactions or saturation of local diversity. *Oikos*, **110**, 195–8.

Hillebrand, H. and Blenckner, T. (2002). Regional and local impact of species diversity—from pattern to process. *Oecologia*, **132**, 479–91.

Hillebrand, H. and Shurin, J.B. (2005). Biodiversity and aquatic food webs. In: Belgrano, A., Scharler, U.M., Dunne, J.A., and Ulanowicz, R.E., eds. *Aquatic Food Webs: An Ecosystem Approach*, pp. 184–97. Oxford University Press, Oxford.

Hillebrand, H., Bennett, D.M., and Cadotte, M.W. (2008). Consequences of dominance: a review of evenness effects on local and regional ecosystem processes. *Ecology*, **89**, 1510–20.

Hillebrand, H., *et al.* (2017). Biodiversity change is uncoupled from species richness trends; consequences for conservation and monitoring. *Journal of Applied Ecology*, **55**, 169–84.

HilleRisLambers, J., Adler, P., Harpole, W., Levine, J., and Mayfield, M. (2012). Rethinking community assembly through the lens of coexistence theory. *Annual Review of Ecology, Evolution, and Systematics*, **43**, 227–48.

Hiltunen, T., Hairston, N.G., Jr, Hooker, G., Jones, L.E., and Ellner, S.P. (2014). A newly discovered role of evolution in previously published consumer-resource dynamics. *Ecology Letters*, **17**, 915–23.

Hoegh-Guldberg, O., Hughes, L., McIntyre, S., *et al.* (2008). Assisted colonization and rapid climate change. *Science*, **321**, 345–6.

Hoeksema, J.D. and Bruna, E.M. (2015). Context-dependent outcomes of mutualistic interactions. In: Bronstein, J.L., ed. *Mutualism*, pp. 181–202. Oxford University Press, Oxford.

Hoeksema, J.D. and Schwartz, M.W. (2003). Expanding comparative-advantage biological market models: Contingency of mutualism on partner's resource requirements and acquisition trade-offs. *Proceedings of the Royal Society of London B*, **270**, 913–19.

Holland, J.N. (2015). Population ecology of mutualism. In Bronstein, J.L., ed. *Mutualism*, pp. 133–58. Oxford University Press, Oxford, UK.

Holland, J.N., Okuyama, T., and DeAngelis, D.L. (2006). Comment on "Asymmetric coevolutionary networks facilitate biodiversity maintenance." *Science*, **313**, 5795.

Holling, C.S. (1959). The components of predation as revealed by a study of small mammal predation on the European pine sawfly. *Canadian Entomologist*, **91**, 293–320.

Holling, C.S. (1973). Resilience and stability of ecological systems. *Annual Review of Ecology and Systematics*, **4**, 1–23.

Holt, B.G., *et al.* (2013). An update of Wallace's zoogeographic regions of the world. *Science*, **339**, 74–8.

Holt, R.D. (1977). Predation, apparent competition, and structure of prey communities. *Theoretical Population Biology*, **12**, 197–229.

Holt, R.D. (1984). Spatial heterogeneity, indirect interactions, and the coexistence of prey species. *American Naturalist*, **124**, 377–406.

Holt, R.D. (1985). Density-independent mortality, non-linear competitive interactions, and species coexistence. *Journal of Theoretical Biology*, **116**, 479–93.

Holt, R.D. (1996). Community modules. In Begon, M., Gange, A., and Brown, V., eds. *Multitrophic Interactions*, pp. 333–48. Chapman and Hall, London.

Holt, R.D. (2002). Food webs in space: on the interplay of dynamic instability and spatial process. *Ecological Research*, **17**, 261–73.

Holt, R.D. (2010). Towards a trophic island biogeography: reflections on the interface of island biogeography and food web ecology. In: Losos, J.B. and R.E. Ricklefs (eds) *The Theory of Island Biography Revisited*, pp. 143–85. Princeton University Press, Princeton, NJ.

Holt, R.D. and Bonsall, M.B. (2017). Apparent competition. *Annual Review of Ecology, Evolution and Systematics*, **48**, 447–71.

Holt, R.D., Grover, J., and Tilman, D. (1994). Simple rules of interspecific dominance in systems with exploitative and apparent competition. *American Naturalist*, **144**, 741–71.

Holt, R.D. and Lawton, J.H. (1994). The ecological consequences of shared natural enemies. *Annual Review of Ecology and Systematics*, **25**, 495–520.

Holyoak, M. (1993). The frequency of detection of density dependence in insect orders. *Ecological Entomology*, **18**, 339–47.

Holyoak, M., Leibold, M.A., Mouquet, N., Holt, R.D., and Hoopes, M.F. (2005). Metacommunities: A framework for large-scale community ecology. In: Holyoak, M., Leibold, M.A., and Holt, R.D., eds. *Metacommunities: Spatial Dynamics and Ecological Communities*, pp. 30–1. University of Chicago Press, Chicago, IL.

Hone, J. (1999). On rate of increase (*r*): patterns of variation in Australian mammals and implications for wildlife management. *Journal of Applied Ecology*, **36**, 709–18.

Hooper, D.U. and Vitousek, P.M. (1997). The effects of plant composition and diversity on ecosystem processes. *Science*, **277**, 1302–5.

Hooper, D.U. and Vitousek, P.M. (1998). Effects of plant composition and diversity on nutrient cycling. *Ecological Monographs*, **68**, 121–49.

Hooper, D.U., *et al.* (2005). Effects of biodiversity on ecosystem functioning: a consensus of current knowledge. *Ecological Monographs*, **75**, 3–35.

Hoopes, M.F., Holt, R.D. and Holyoak, M. (2005). The effects of spatial processes on two species interactions. In: Holyoak, M., Leibold, M.A., and Holt, R.D., eds. *Metacommunities: Spatial Dynamics and Ecological Communities*, pp. 35–67. University of Chicago Press, Chicago, IL.

Horn, H.S. and MacArthur, R.H. (1972). Competition among fugitive species in a harlequin environment. *Ecology*, **53**, 749–52.

Hortal, J., Triantis, K.A., Meiri, S., Thebault, E., and Sfenthourakis, S. (2009). Island species richness increases with habitat diversity. *American Naturalist*, **174**, E205–17.

Horton, T.R., Bruns, T.D., and Parker, T. (1999). Ectomycorrhizal fungi associated with *Arctostaphylos* contribute to *Pseudotsuga menziesii* establishment. *Canadian Journal of Botany*, **77**, 93–102.

Houlahan, J.E., *et al.* (2007). Compensatory dynamics are rare in natural ecological communities. *Proceedings of the National Academy of Sciences*, **104**, 3273–7.

Houston, A.I., MacNamara, J.M., and Hutchinson, J.M.C. (1993). General results concerning the trade-off between gaining energy and avoiding predators. *Philosophical Transactions of the Royal Society of London B*, **341**, 375–97.

Howe, H.F. and Westley, L.C. (1986). Ecology of pollination and seed dispersal. In: Crawley, M.J., ed. *Plant Ecology*, pp. 185–215. Blackwell, London.

Howeth, J.G. and Leibold, M.A. (2010). Species dispersal rates alter diversity and ecosystem stability in pond metacommunities. *Ecology*, **91**, 2727–41.

Hsu, S-B., Cheng, K-S., and Hubbell, S.P. (1981). Exploitative competition of microorganisms for two complementary nutrients in continuous cultures. *SIAM Journal of Applied Mathematics*, **41**, 422–44.

Hsu, S-B., Hubbell, S.P., and Waltman, P. (1977). A mathematical theory for single-nutrient competition in continuous cultures of microorganisms. *SIAM Journal of Applied Mathematics*, **32**, 366–83.

Hsu, S-B., Hubbell, S.P., and Waltman, P. (1978). A contribution to the theory of competing predators. *Ecological Monographs*, **48**, 337–49.

Hubbell, S.P. (2001). *The Unified Neutral Theory of Biodiversity and Biogeography*. Princeton University Press, Princeton, NJ.

Hubbell, S.P. (2006). Neutral theory and the evolution of ecological equivalence. *Ecology*, **87**, 1387–98.

Hubbell, S.P. and Foster, R.B. (1983). Diversity of canopy trees in a Neotropical forest and implications for conservation. In: Sutton, S., Whitmore, T.C., and Chadwick, A., eds. *Tropical Rain Forest: Ecology and Management*, pp. 25–41. Blackwell, Oxford.

Hubbell, S.P. and Foster, R.B. (1986). Biology, chance and history and the structure of tropical rain forest tree communities. In: Diamond, J. and Case, J.M., eds. *Community Ecology*, pp. 314–29. Harper and Row, New York, NY.

Hughes, T.P., Graham, N.A.J., Jackson, J.B.C., Mumby, P.J., and Steneck, R.S. (2010). Rising to the challenge of sustaining coral reef resilience. *Trends in Ecology and Evolution*, **25**, 633–42.

Hülsmann, L. and Hartig, F. (2018). Comment on "Plant diversity increases with the strength of negative density dependence at the global sclae". *Science*, **360**, eaar 2435.

Humboldt, A., von. (1808). *Ansichten der Natur mit wissenschftlichen Erlauterungen*. J. G. Cotta, Tubingen.

Hunter, M.L., Jr (2007). Climate change and moving species: Furthering the debate on assisted colonization. *Conservation Biology*, **21**, 1356–8.

Hurlbert, S.H. (1971). The nonconcept of species diversity: a critique and alternative parameters. *Ecology*, **52**, 577–86.

Hurlbert, S.H. (1978). The measurement of niche overlap and some relatives. *Ecology*, **59**, 67–77.

Hurtt, G.C. and Pacala, S.W. (1995). The consequences of recruitment limitation: Reconciling chance, history, and competitive differences between plants. *Journal of Theoretical Biology*, **176**, 1–12.

Huss, M., Howeth, J.G., Osterman, J.I., and Post, D.M. (2014). Intraspecific phenotypic variation among alewife populations drives parallel phenotypic shifts in bluegill. *Proceedings of the Royal Society of London B*, **281**, 20140275.

Huston, M.A. (1979). A general hypothesis of species diversity. *American Naturalist*, **113**, 81–101.

Huston, M.A. (1997). Hidden treatments in ecological experiments: re-evaluating the ecosystem function of biodiversity. *Oecologia*, **108**, 449–60.

Huston, M.A. and DeAngelis, D.L. (1994). Competition and coexistence: the effects of resource transport and supply rates. *American Naturalist*, **144**, 954–77.

Hutchings, J.A. and Reynolds, J.D. (2004). Marine fish population collapses: consequences for recovery and extinction risk. *BioScience*, **54**, 297–309.

Hutchison, D.W. and Templeton, A.R. (1999). Correlation of pairwise genetic and geographic distance measures: inferring the relative influences of gene flow and drift on the distribution of genetic variability. *Evolution*, **53**, 1898–914.

Hutchinson, G.E. (1951). Copepodology for the ornithologist. *Ecology*, **32**, 571–7.

Hutchinson, G.E. (1957). Concluding remarks. *Cold Spring Harbor Symposium on Quantitative Biology*, **22**, 415–27.

Hutchinson, G.E. (1959). Homage to Santa Rosalia, or why are there so many kinds of animals? *American Naturalist*, **93**, 145–59.

Hutchinson, G.E. (1961). The paradox of the plankton. *American Naturalist*, **95**, 137–45.

Hutchinson, G.E. (1965). *The Ecological Theater and the Evolutionary Play*. Yale University Press, New Haven, CT.

Hutchinson, G.E. (1978). *An Introduction to Population Ecology*. Yale University Press, New Haven.

Hutto, R.L., McAuliffe, J.R., and Hogan, L. (1986). Distributional associates of the saguaro (*Carnegiea gigantea*). *Southwest Naturalist*, **31**, 469–76.

Iknayan, K.J. and Beissinger, S.R. (2018). Collapse of a desert bird community over the past century driven by climate change. *Proceedings of the National Academy of Sciences*, **115**, 8597–602.

Ings, T.C., *et al.* (2009). Ecological networks—beyond food webs. *Journal of Animal Ecology*, **78**, 253–69.

Isbell, F., *et al.* (2011). High plant diversity is needed to maintain ecosystem services. *Nature*, **477**, (199–202.

Ives, A.R., Cardinale, B.J., and Snyder, W.E. (2005). A synthesis of subdisciplines: predator–prey interactions, and biodiversity and ecosystem functioning. *Ecology Letters*, **8**, 102–16.

Ives, A.R. and Carpenter, S.R. (2007). Stability and diversity of ecosystems. *Science*, **317**, 58–62.

Ives, A.R. and Dobson, A.P. (1987). Antipredator behavior and the population dynamics of simple predator–prey systems. *American Naturalist*, **130**, 1431–42.

Ives, A.R. and May, R.M. (1985). Competition within and between species in a patchy environment: relations between microscopic and macroscopic models. *Journal of Theoretical Biology*, **115**, 65–92.

Jablonski, D. (1993). The tropics as a source of evolutionary novelty through geological time. *Nature*, **364**, 142–4.

Jablonski, D. (2008). Biotic interactions and macroevolution: Extensions and mismatches across scales and levels. *Evolution*, **62**, 715–39.

Jablonski, D., Huang, S., Roy, K., and Valentine, J.W. (2017). Shaping the latitudinal diversity gradient: new perspectives from a synthesis of paleobiology and biogeography. *American Naturalist*, **189**, 1–12.

Jablonski, D., Roy, K., and Valentine, J.W. (2006). Out of the tropics: evolutionary dynamics of the latitudinal diversity gradient. *Science*, **314**, 102–6.

Jabot, F. and Chave, J. (2009). Inferring the parameters of the neutral theory of biodiversity using phylogenetic information and implications for tropical forests. *Ecology Letters*, **12**, 239–48.

Jacobson, B. and Peres-Neto, P.R. (2010). Quantifying and disentangling dispersal in metacommunities: how close have we come? How far is there to go? *Landscape Ecology*, **25**, 495–507.

Jakobsson, A. and Eriksson, O. (2003). Trade-offs between dispersal and competitive ability: A comparative study of wind-dispersed Asteraceae forbs. *Evolutionary Ecology*, **17**, 233–46.

Jansson, R., Rodriguez-Castañeda, G., and Harding, L.E. (2013). What can multiple phylogenies say about the latitudinal diversity gradient? A new look at the tropical conservatism, out of the tropics, and diversification rate hypotheses. *Evolution*, **67**, 1741–55.

Janzen, D.H. (1966). Coevolution of mutualism between ants and acacias in Central America. *Evolution*, **20**, 249–75.

Janzen, D.H. (1967). Why mountain passes are higher in the tropics. *American Naturalist*, **101**, 233–49.

Janzen, D.H. (1970). Herbivores and the number of tree species in tropical forests. *American Naturalist*, **104**, 501–28.

Jaramillo, C., Rueda, J.J., and Mora, G. (2006). Cenozoic plant diversity in the Neotropics. *Science*, **311**, 1893–6.

Jenkins, D.G. and Buikema, A.L.J. (1998). Do similar communities develop in similar sites? A test with zooplankton structure and function. *Ecological Monographs*, **68**, 421–43.

Jeschke, J.M., Kopp, M., and Tollrian, R. (2004). Consumer-food systems: Why type I functional responses are exclusive to filter feeders. *Biological Reviews*, **79**, 337–49.

Jetz, W. and Fine, P.V.A. (2012). Global gradients in vertebrate diversity predicted by historical area-productivity dynamics and contemporary environment. *PLoS Biology*, **10**, e1001292.

Jetz, W. and Rahbek, C. (2001). Geometric constraints explain much of the species richness pattern in African birds. *Proceedings of the National Academy of Sciences*, **98**, 5661–6.

Jetz, W., G.H. Thomas, J.B. Joy, K. Hartman and A. Mooers. (2012). The global diversity of birds in space and time. *Nature*, **492**, 444–8.

Jia, S., Wang, X., Yuan, Z., *et al.* (2018). Global signal of top-down control of terrestrial plant communities by herbivores. *Proceedings of the National Academy of Sciences*, **115**, 6237–42.

Johnson, D.M. and Crowley, P.H. (1980). Habitat and seasonal segregation among coexisting odonate larvae. *Odonatologica*, **9**, 297–308.

Johnson, N.C. (2010). Resource stoichiometry elucidates the structure and function of arbuscular mycorrhizas across scales. *New Phytologist*, **185**, 631–47.

Johnson, N.C., Graham, J.H., and Smith, F.A. (1997). Functioning of mycorrhizal associations along the mutualism-parasitism continuum. *New Phytologist*, **135**, 575–85.

Johnson, N.C., Wilson, G.W.T., Wilson, J.A., Miller, R.M., and Bowker, M.A. (2015). Mycorrhizal phenotypes and the Law of the Minimum. *New Phytologist*, **205**, 1473–84.

Jones, C.G., Lawton, J.H., and Shachak, M. (1994). Organisms as ecosystem engineers. *Oikos*, **69**, 373–86.

Jones, C.G., Lawton, J.H., and Shachak, M. (1997). Ecosystem engineering by organisms: why semantics matters. *Trends in Ecology and Evolution*, **12**, 275.

Jones, L.E., Becks L., Ellner S.P., Hairston, N.G., Jr, Yoshida, T., and Fussmann, G.F. (2009). Rapid contemporary evolution and clonal food web dynamics. *Philosophical Transactions of the Royal Society of London B*, **364**, 1579–91.

Jones, L.E. and Ellner, S.P. (2007). Effects of rapid prey evolution on predator–prey cycles. *Journal of Mathematical Biology*, **55**, 541–73.

Jonsson, T., Kaartinen, R., Jonsson, M., and Bommarco, R. (2018). Predictive power of food web models based on body size decreases with trophic complexity. *Ecology Letters*, **21**: 702–12.

Jordano, P. (2000). Fruits and frugivory. In: M. Fenner, ed. *Seeds: The Ecology of Regeneration in Natural Plant Communities*, pp. 125–66. Commonwealth Agricultural Bureau International, Wallingford.

Jost, L. (2006). Entropy and diversity. *Oikos*, **113**, 363–75.

Jump, A.S. and Peñuelas, J. (2005). Running to stand still: adaptation and the response of plants to rapid climate change. *Ecology Letters*, **81**, 1010–20.

Kalmar, A. and Currie, D.J. (2006). A global model of island biogeography. *Global Ecology and Biogeography*, **15**, 72–81.

Kalmar, A. and Currie, D.J. (2007). A unified model of avian species richness on islands and continents. *Ecology*, **88**, 1309–21.

Kardol, P., Fanin, N., and Wardle, D.A. (2018). Long-term effects of species loss on community properties across contrasting ecosystems. *Nature*, **557**, 710-713.

Kaunzinger, C.M.K. and Morin, P.J. (1998). Productivity controls food-chain properties in microbial communities. *Nature*, **395**, 495–97.

Kattge, J., Diaz, S., Lavorel, S., *et al.* (2011). TRY–a global database of plant traits. *Global Change Biology*, **17**, 2905–35.

Keddy, P.A. (1989). *Competition*. Chapman and Hall, London.

Keddy, P.A. (1992). Assembly and response rules: two goals for predictive community ecology. *Journal of Vegetation Science*, **3**, 157–64.

Kéfi, S. (2008). Reading the signs: spatial vegetation patterns, arid ecosystems, and desertification. Ph.D. dissertation, Utrecht University.

Kéfi, S., Berlow, E.L., Wieters, E.A., *et al.* (2015). Network structure beyond food webs: mapping non-trophic and trophic interactions on Chilean rocky shores. *Ecology*, **96**, 291–303.

Kéfi, S., Miele, V., Wieters, E.A., Navarrete, S.A., and Berlow, E.L. (2016). How structured is the entangled bank? The surprisingly simple organization of multiplex ecological networks leads to increased persistence and resilience. *PLoS Biology*, **14**, e1002527.

Kéfi, S., Rietkerk, M., Alados, C.L., *et al.* (2007a). Spatial vegetation patterns and imminent desertification in Mediterranean arid ecosystems. *Nature*, **449**, 213–17.

Kéfi, S., Rietkerk, M., van Baalen, and M. Loreau, M. (2007b). Local facilitation, bistability and transitions in arid ecosystems. *Theoretical Population Biology*, **71**, 367–79.

Kempton, R.A. (1979). The structure of species abundance and measurement of diversity. *Biometrics*, **35**, 307–21.

Kerkhoff, A.J., Moriarty, P.E., and Weiser, M.D. (2014). The latitudinal species richness gradient in new World woody angiosperms is consistent with the tropical conservatism hypothesis. *Proceedings of the National Academy of Sciences*, **111**, 8125–30.

Kermack, W.O. and McKendrick, A.G. (1927). A contribution to the mathematical theory of epidemics. *Proceedings of the Royal Society of London A*, **115**, 700–21.

Keymer, J.E., Marquet, P.A., Velasco-Hernandez, J.X., and Levin, S.A. (2000). Extinction thresholds and metapopulation persistence in dynamic landscapes. *American Naturalist*, **156**, 478–94.

Kiessling, W., Simpson, C., and Foote, M. (2010). Reefs as cradles of evolution and sources of biodiversity in the Phanerozoic. *Science*, **327**, 196–8.

Kilham, S.S. (1986). Dynamics of Lake Michigan natural phytoplankton communities in continuous cultures along a Si:P loading gradient. *Canadian Journal of Fisheries and Aquatic Sciences*, **43**, 351–60.

Kimura, M. (1968). Evolutionary rate at the molecular level. *Nature*, **217**, 624–6.

Kimura, M. and Ohta, T. (1969). The average number of generations until fixation of a mutant gene in a finite population. *Genetics*, **61**, 763–71.

Kingsland, S. (1985). *Modeling Nature*. University of Chicago Press, Chicago, IL.

Kingsolver, J.G. and Diamond, S.E. (2011). Phenotypic selection in natural populations: what limits directional selection? *American Naturalist*, **177**, 346–57.

Kingsolver, J.G., *et al.* (2001). The strength of phenotypic selection in natural populations. *American Naturalist*, **157**, 245–61.

Kinzig, A.P., Ryan, R., Etienne, M., Allison, H., Elmqvist, T., and Walker, B.H. (2006). Resilience and regime shifts: assessing cascading effects. *Ecology and Society*, **11**. 20 (Online).

Kisel, Y. and Barraclough, T.G. (2010). Speciation has a spatial scale that depends on levels of gene flow. *American Naturalist*, **175**, 316–34.

Kislaliogu, M. and Gibson, R.N. (1976). Prey handling time and its importance in food selection by the 15-spined stickleback *Spinachia spinachia*. *Journal of Experimental Marine Biology and Ecology*, **25**, 151–8.

Kleidon, A. and Mooney, H.A. (2000). A global distribution of biodiversity inferred from climatic constraints: results for a process-based modeling study. *Global Change Biology*, **6**, 507–23.

Kleijn, D., *et al.* (2015). Delivery of crop pollination services is an insufficient argument for wild pollinator conservation. *Nature Communications*, **6**, 7414.

Klironomas, J.N. (2003). Variation in plant response to native and exotic arbuscular mycorrhizal fungi. *Ecology*, **84**, 2292–301.

Klironomos, J.N. (2002). Feedback with soil biota contributes to plant rarity and invasiveness in communities. *Nature*, **417**, 67–70.

Knape, J. and de Valpine, P. (2012). Are patterns of density dependence in the Global Population Dynamics Database driven by uncertainty about population abundance? *Ecology Letters*, **15**, 17–23.

Kneitel, J.M. and Chase, J.M. (2004). Trade-offs in community ecology: Linking spatial scales and species coexistence. *Ecology Letters*, **7**, 69–80.

Kneitel, J.M. and Miller, T.E. (2003). Dispersal rates affect species composition in metacommunities of *Sarracenia purpurea* inquilines. *American Naturalist*, **162**, 165–71.

Koch, A.L. (1974). Competitive coexistence of two predators utilizing the same prey under constant environmental conditions. *Journal of Theoretical Biology*, **44**, 387–95.

Koleff, P., Gaston, K.A., and Lennon, J.J. (2003). Measuring beta diversity for presence-absence data. *Journal of Animal Ecology*, **72**, 367–82.

Kondoh, M., Kato, S., and Sakato, Y. (2010). Food webs are built up with nested subwebs. *Ecology*, **91**, 3123–30.

Kondoh, M. and Ninomiya, K. (2009). Food-chain length and adaptive foraging. *Proceedings of the Royal Society of London B*, **276**, 3113–21.

Kotler, B.P and Brown, J.S. (2007). Community ecology. In: Stephens, D.W., Brown, J.S., and Ydenberg, R.C., eds.

Foraging: Behavior and Ecology, pp. 397–434. University of Chicago Press, Chicago, IL.

Kraft, N.J.B. and Ackerly, D.D. (2010). Functional trait and phylogenetic tests of community assembly across spatial scales in an Amazonian forest. *Ecological Monographs*, **80**, 401–22.

Kraft, N.J. and Ackerly, D.D. (2014). The assembly of plant communities. In: R. Monson, ed. *The Plant Sciences—Ecology and the Environment*, pp. 67–88. Springer-Verlag, Berlin.

Kraft, N.J., Godoy, O., and Levine, J.M. (2015). Plant functional traits and the multidimensional nature of species coexistence. *Proceedings of the National Academy of Sciences*, **112**, 797–802.

Kraft, N.J., Valencia, R., and Ackerly, D.D. (2008). Functional traits and niche-based tree community assembly in an Amazonian forest. *Science*, **322**, 580–2.

Kraft, N.J.B., Cornwell, W.K., Webb, C.O., and Ackerly, D.D. (2007). Trait evolution, community assembly, and the phylogenetic structure of ecological communities. *American Naturalist*, **170**, 271–83.

Kraft, N.J.B., et al. (2011). Disentangling the drivers of β diversity along latitudinal and elevational gradients. *Science*, **333**, 1755–8.

Kramer, A.M., Dennis, B., Liebhold, A.M., and Drake, J.M. (2009). The evidence for Allee effects. *Population Ecology*, **51**, 341–54.

Kramer, A.M., Sarnelle, O., and Knapp, R.A. (2008). Allee effect limits colonization success of sexually reproducing zooplankton. *Ecology*, **89**, 2760–9.

Krause, A.E., Frank, K.A., Mason, D.M., Ulanowicz, R.E., and Taylor, W.W. (2003). Compartments revealed in food-web structure. *Nature*, **426**, 282–5.

Krebs, C.J. (1995). Two paradigms of population regulation. *Wildlife Research*, **22**, 1–10.

Krebs, J.R. (1978). Optimal foraging: decision rules for predators. In: Krebs, J.R. and Davies, N.B., eds. *Behavioural Ecology*, pp. 22–63. Blackwell Scientific, Oxford.

Krebs, J.R., Erichsen, J.T., Webber, M.I., and Charnov, E.L. (1977). Optimal prey-selection by the great tit (*Parus major*). *Animal Behaviour*, **25**, 30–8.

Krebs, J.R., Ryan, J.C., and Charnov, E.L. (1974). Hunting by expectation of optimal foraging? A study of patch use by chickadees. *Animal Behaviour*, **22**, 953–64.

Kreft, H. and Jetz, W. (2007). Global patterns and determinants of vascular plant diversity. *Proceedings of the National Academy of Sciences*, **104**: 5925–30.

Kreft, H. and Jetz, W. (2010). A framework for delineating biogeographical regions based on species distributions. *Journal of Biogeography*, **37**, 2029–53.

Krenek, L. and Rudolf, V.H.W. (2014). Allometric scaling of indirect effects: body size ratios predict non-consumptive effects in multi-predator systems. *Journal of Animal Ecology*, **83**, 1461–8.

Křivan, V. and Eisner, J. (2006). The effect of the Holling type II functional response on apparent competition. *Theoretical Population Biology*, **70**, 421–30.

Krug, A.Z., Jablonski, D., and Valentine, J.W. (2009). Signature of the End-Cretaceous mass extinction in the modern biota. *Science*, **323**, 767–71.

Kuang, J.J. and Chesson, P.L. (2008). Predation-competition interactions for seasonally recruiting species. *American Naturalist*, **171**, E119–33.

Kuang, J.J. and Chesson, P.L. (2010). Interacting coexistence mechanisms in annual plant communities: Frequency-dependent predation and the storage effect. *Theoretical Population Biology*, **77**, 56–70.

Kuris, A.M., et al. (2008). Ecosystem energetic implications of parasite and free-living biomass in three estuaries. *Nature*, **454**, 515–18.

Lack, D. (1947). *Darwin's Finches*. Cambridge University Press, Cambridge.

Lafferty, K.D., Dobson, A.P., and Kuris, A.M. (2006). Parasites dominate food web links. *Proceedings of the National Academy of Sciences*, **103**, 11211–16.

Lafferty, K.D. and Kuris, A.M. (2009). Parasites reduce food web robustness because they are sensitive to secondary extinction as illustrated by an invasive estuarine snail. *Philosophical Transactions of the Royal Society of London series B*, **364**, 1659–63.

Lafferty, K.D. and Suchanek, T.H. (2016). Revisiting Paine's 1966 sea star removal experiment, the most-cited empirical article in the *American Naturalist*. *American Naturalist*, **188**, 365–78.

Lafferty, K.D., et al. (2008). Parasites in food webs: the ultimate missing links. *Ecology Letters*, **11**, 533–46.

Laigle, I., Aubin, I., Digel, C., Brose, U., Boulangeat, I., and Gravel, D. (2018). Species traits as drivers of food web structure. *Oikos*, **127**, 316–26

Laiolo, P.J. Seoane, Obeso, J.R., and Iilera, J.C. (2017). Ecological divergence among young lineages favours sympatry, but convergence among old ones allows coexistence in syntopy. *Global Ecology and Biogeography*, **26**, 601–8.

Lamanna, C., et al. (2014). Functional trait space and the latitudinal diversity gradient. *Proceedings of the National Academy of Sciences*, **111**, 13745–50.

LaManna, J.A., Walton, M.L., Turner, B.L., and Myers, J.A. (2016). Negative density dependence is stronger in resource-rich environments and diversifies communities when stronger for common but not rare species. *Ecology Letters*, **19**, 657–67.

LaManna, J.A., et al. (2017). Plant diversity increases with the strength of negative density dependence at the global scale. *Science*, **356**, 1389–92.

Lau, J.A. (2012). Evolutionary indirect effects of biological invasions. *Oecologia*, **170**, 171–81.

Laughlin, D.C., Joshi, C., Bodegom, P.M., Bastow, Z.A., and Fulé, P.Z. (2012). A predictive model of community assembly that incorporates intraspecific trait variation. *Ecology Letters*, **15**, 1291–9.

Laughlin, D.C. and Messier, J. (2015). Fitness of multidimensional phenotypes in dynamic adaptive landscapes. *Trends in Ecology & Evolution*, **30**, 487–96.

Lampert, W. (1987). Vertical migration of freshwater zooplankton: Indirect effects of vertebrate predators on algal communities. In: Kerfoot, W.C. and Sih, A., eds. *Predation: Direct and Indirect Impacts on Aquatic Communities*, pp. 291–9. University Press of New England, Hanover, NH.

Lande, R. (1987). Extinction thresholds in demographic models of territorial populations. *American Naturalist*, **130**, 624–35.

Lande, R. (1988a). Demographic models of the northern spotted owl. *Oecologia*, **75**, 601–7.

Lande, R. (1988b). Genetics and demography in biological conservation. *Science*, **241**, 1455–60.

Lande, R., Engen, S., and Saether, B.E. (2003). *Stochastic Population Dynamics in Ecology and Conservation*. Oxford University Press, Oxford.

Laska, M.S. and Wootton, J.T. (1998). Theoretical concepts and empirical approaches to measuring interaction strength. *Ecology*, **79**, 461–76.

Latham, R.E. and Ricklefs, R.E. (1993). Continental comparisons of temperate-zone tree species diversity. In: Ricklefs, R.E. and Schluter, D., eds. *Species Diversity in Ecological Communities: Historical and Geographical Perspectives*, pp. 294–314. University of Chicago Press, Chicago, IL.

Laughlin, D.C., Joshi, C., Bodegom, P.M., Bastow, Z.A., and Fulé, P.Z. (2012). A predictive model of community assembly that incorporates intraspecific trait variation. *Ecology Letters*, **15**(11), 1291–9.

Laundré, J.W., Hernández, L., Medina, P.L., *et al.* (2017). The landscape of fear: the missing link to understand top-down and bottom-up controls of prey abundance. *Ecology*, **95**, 1141–52.

Lavergne, S., Mouquet, N., Thuiller, W., and Ronce, O. (2010). Biodiversity and climate change: Integrating evolutionary and ecological responses of species and communities. *Annual Review of Ecology, Evolution, and Systematics*, **41**, 321–50.

Laverty, T.M. (1992). Plant interactions for pollinator visits: A test of the magnet species effect. *Oecologia*, **89**, 502–8.

Law, R. and Morton, R.D. (1996). Permanence and the assembly of ecological communities. *Ecology*, **77**, 762–75.

Lawrence, P.A. (2011). "The heart of research is sick." Interview by J. Garwood. *Lab Times*, **2–2011**, 24–31.

Lawton, J.H. (1999). Are there general laws in ecology? *Oikos*, **84**, 177–92.

Lawton, J.H., Nee, S., Letcher, A.J., and Harvey, P.H. (1994). Animal distributions; patterns and processes. In: Edwards, P.J., May, R.M., and Webb, N.R., eds. *Large-Scale Ecology and Conservation Biology*, pp. 41–58. Blackwell Scientific, Oxford.

Lazarus E.D. and McGill, B.J. (2014). Pushing the pace of tree species migration. *PLoS One*, **9**(8), e105380.

Lebrija-Trejos, E., Pérez-García, E.A., Meave J.A., Bongers, F., and Poorter, L. (2010). Functional traits and environmental filtering drive community assembly in a species-rich tropical system. *Ecology*, **91**, 386–98.

Lechowicz, M.J. and Bell, G. (1991). The ecology and genetics of fitness in forest plants. II. Microspatial heterogeneity of the edaphic environment. *Journal of Ecology*, **79**, 687–96.

Lefcheck, J.S., *et al.* (2015). Biodiversity enhances ecosystem multifunctionality across trophic levels and habitats. *Nature Communications*, **6**, 6936.

Lehman, C. and Tilman, D. (2000). Biodiversity, stability, and productivity in competitive communities. *American Naturalist*, **156**, 534–52.

Lehtonen, J. and Jaatinen, K. (2016). Safety in numbers: the dilution effect and other drivers of group life in the face of danger. *Behavioral Ecology and Sociobiology*, **70**, 449–58.

Leibold, M.A. (1989). Resource edibility and the effects of predators and productivity on the outcome of trophic interactions. *American Naturalist*, **134**, 922–49.

Leibold, M.A. (1996). A graphical model of keystone predators in food webs: Trophic regulation of abundance, incidence, and diversity patterns in communities. *American Naturalist*, **147**, 784–812.

Leibold, M.A. (1997). Do nutrient-competition models predict nutrient availabilities in limnetic ecosystems? *Oecologia*, **110**, 132–42.

Leibold, M.A. and Chase, J.M. (2018). *Metacommunity Ecology*. Princeton University Press, Princeton, NJ.

Leibold, M.A., Chase, J.M., Shurin, J.B., and Downing, A.L. (1997). Species turnover and the regulation of trophic structure. *Annual Review of Ecology and Systematics*, **28**, 467–94.

Leibold, M.A., Economo, E.P., and Peres-Neto, P. (2010). Metacommunity phylogenetics: Separating the roles of environmental filters and historical biogeography. *Ecology Letters*, **13**, 1290–9.

Leibold, M.A., Hall, S.R., Smith, V.H., and Lytle, D.A. (2017). Herbivory enhances the diversity of primary producers in pond ecosystems. *Ecology*, **98**, 48–56.

Leibold, M.A., Holt, R.D., and Holyoak, M. (2005). Adaptive and coadaptive dynamics in metacommunities. In: Holyoak, M., Leibold, M.A., and Holt, R.D., eds. *Metacommunities: Spatial Dynamics and Ecological Communities*, pp. 439–64. University of Chicago Press, Chicago, IL.

Leibold, M.A. and McPeek, M.A. (2006). Coexistence of the niche and neutral perspectives in community ecology. *Ecology*, **87**, 1399–410.

Leibold, M.A. and Norberg, J. (2004). Biodiversity in meta-communities: plankton as complex adaptive systems? *Limnology and Oceanography*, **49**, 1278–89.

Leibold, M.A. and Wootton, J.T. (2001). Introduction to *Animal Ecology*, by Charles Elton. University of Chicago Press, Chicago, IL.

Leibold, M.A., *et al.* (2004). The metacommunity concept: a framework for multi-scale community ecology. *Ecology Letters*, **7**, 601–13.

Leigh, E.G. (1999). *Tropical Forest Ecology*. Oxford University Press, Oxford.

Leigh, E.G. (2007). Neutral theory: a historical perspective. *Journal of Evolutionary Biology*, **20**, 2075–91.

Leigh, E.G. and Rowell, T.E. (1995). The evolution of mutualism and other forms of harmony at various levels of biological organization. *Ecologie*, **26**, 131–58.

Leigh, E.G., Davidar, P., Dick, C.W., *et al.* (2004). Why do some tropical forests have so many species of trees? *Biotropica*, **36**, 447–73.

León, J.A. and Tumpson, D.B. (1975). Competition between two species for two complementary or substitutable resources. *Journal of Theoretical Biology*, **50**, 185–201.

Lepš, J. (2004). What do the biodiversity experiments tell us about consequences of plant species loss in the real world? *Basic and Applied Ecology*, **5**, 529–34.

Lessard, J-P., Belmaker, J., Myers, J.A., Chase, J.M., and Rahbek, C. (2012). Inferring local ecological processes amid species pool influences. *Trends in Ecology and Evolution*, **27**, 600–7.

Lessels, C.M. (1995). Putting resource dynamics into continuous input ideal free distribution models. *Animal Behaviour*, **49**, 487–94.

Letten, A.D., Dhami, M.K., Ke, P-J., and Fukami, T. (2018). Species coexistence through simultaneous fluctuation-dependent mechanisms. *Proceedings of the National Academy of Sciences*, **115**, 6745–50.

Letten, A.D., Ke, P.J., and Fukami, T. (2017). Linking modern coexistence theory and contemporary niche theory. *Ecological Monographs*, **87**, 161–77.

Levin, S.A. (1974). Disturbance, patch formation, and community structure. *Proceedings of the National Academy of Sciences*, **71**, 2744–7.

Levin, S.A. (1998). Ecosystems and the biosphere as complex adaptive systems. *Ecosystems*, **1**, 431–6.

Levine, J.M. (2000). Species diversity and biological invasions: Relating local process to community pattern. *Science*, **288**, 852–4.

Levine, J.M., Adler, P.B., and Yelenik, S.G. (2004). A meta-analysis of biotic resistance to exotic plant invasions. *Ecology Letters*, **7**, 975–89.

Levine, J.M. and HilleRisLambers, J. (2009). The importance of niches for the maintenance of species diversity. *Nature*, **461**, 254–7.

Levine, J.M. and Rees, M. (2002). Coexistence and relative abundance in annual plant assemblages: The roles of competition and colonization. *American Naturalist*, **160**, 452–67.

Levins, R. (1968). *Evolution in Changing Environments*. Princeton University Press, Princeton, NJ.

Levins, R. (1969). Some demographic and genetic consequences of environmental heterogeneity for biological control. *Bulletin of the Entomological Society of America*, **15**, 237–40.

Levins, R. (1970). Extinction. In: Gerstenhaber, M., ed. *Some Mathematical Problems in Biology*, pp. 75–107. American Mathematical Society, Providence, RI.

Levins, R. (1979). Coexistence in a variable environment. *American Naturalist*, **114**, 765–83.

Levins, R. and Culver, D. (1971). Regional coexistence of species and competition between rare species. *Proceedings of the National Academy of Sciences*, **68**, 1246–8.

Levitan, C. (1987). Formal stability analysis of a planktonic freshwater community. In: Kerfoot, W.C. and Sih, A., eds. *Predation: Direct and Indirect Impacts on Aquatic Communities*, pp. 71–100. University Press of New England, Hanover, NH.

Lewin, R. (1989). Biologists disagree over bold signature of nature. *Science*, **244**, 527–8.

Lewinsohn, T.M., Prado, P.I., Jordano, P., Bascompte, J., and Olesen, J.M. (2006). Structure in plant-animal interaction assemblages. *Oikos*, **113**, 174–84.

Lewontin, R.C. (1969). The meaning of stability. *Brookhaven Symposium on Biology*, **22**, 13–23.

Lewontin, R.C. (2000). *The Triple Helix, Gene, Organisms, and Environment*. Harvard University Press, Cambridge, MA.

Liang, J., *et al.* (2016). Positive biodiversity-productivity relationship predominant in global forests. *Science*, **354**, doi: 10.1126/science.aaf8957.

Lichter, J. (2000). Colonization constraints during primary succession on coastal Lake Michigan sand dunes. *Journal of Ecology*, **88**, 825–39.

Lima, S.L. (1995). Back to the basics of anti-predatory vigilance: the group-size effect. *Animal Behaviour*, **49**, 11–20.

Lima, S. (1998). Stress and decision making under the risk of predation: recent developments from behavioral, reproductive, and ecological perspectives. *Advances in the Study of Behavior*, **27**, 215–90.

Lindeman, R.L. (1942). The trophic-dynamic aspect of ecology. *Ecology*, **23**, 399–417.

Litchman, E. and Klausmeier, C.A. (2008). Trait-based community ecology of phytoplankton. *Annual Review of Ecology, Evolution, and Systematics*, **39**, 615–39.

Liu, Y., Yu, S., Xie, Z-P., and Staehelin, C. (2012). Analysis of negative plant-soil feedback in a subtropical monsoon forest. *Journal of Ecology*, **100**, 1019–28.

Livingston, G., Matias, M., Calcagno, V., *et al.* (2012). Competition-colonization dynamics in experimental bacterial metacommunities. *Nature Communications*, **3**, 1234.

Loehle, C. and Eschenbach, W. (2012). Historical bird and terrestrial mammal extinction rates and causes. *Diversity and Distributions*, **18**, 84–91.

Loeuille, N. and Loreau, M. (2005). Evolutionary emergence of size-structured food webs. *Proceedings of the National Academy of Sciences*, **102**, 5761–6.

Lohbeck, M., Bongers, F., Martinez-Ramos, M., and Poorter, L. (2016). The importance of biodiversity and dominance for multiple ecosystem functions in a human-modified tropical landscape. *Ecology*, **97**, 2772–9.

Long, Z.T., Bruno, J.F., and Duffy, J.E. (2007). Biodiversity mediates productivity through different mechanisms at adjacent trophic levels. *Ecology*, **88**, 2821–9.

Longo, G.O., Hay, M.E., Ferreira, C.E.L., and Floeter, S.R. (2018). Trophic interactions across 61 degrees of latitude in the Western Atlantic. *Global Ecology and Biogeography* (in press).

Lonsdale, W.M. (1999). Global patterns of plant invasions and the concept of invasibility. *Ecology*, **80**, 1522–36.

Loreau, M. (2000). Biodiversity and ecosystem functioning: recent theoretical advances. *Oikos*, **91**, 3–17.

Loreau, M. (2010a). *From Populations to Ecosystems*. Princeton University Press, Princeton, NJ.

Loreau, M. (2010b). Linking biodiversity and ecosystems: towards a unifying ecological theory. *Philosophical Transactions of the Royal Society of London B*, **365**, 49–60.

Loreau, M. and de Mazancourt, C. (2008). Species synchrony and its drivers: neutral and nonneutral community dynamics in fluctuating environments. *American Naturalist*, **172**, E48–66.

Loreau, M. and Hector, A. (2001). Partitioning selection and complementarity in biodiversity experiments. *Nature*, **412**, 72–6.

Losos, J.B. (2008a). Phylogenetic niche conservatism, phylogenetic signal and the relationship between phylogenetic relatedness and ecological similarity between species. *Ecology Letters*, **11**, 995–1007.

Losos, J.B. (2008b). Rejoinder to Wiens (2008): phylogenetic niche conservatism, its occurrence and importance. *Ecology Letters*, **11**, 1005–7.

Losos, J.B. (2009). *Lizards in an Evolutionary Tree: Ecology and Adaptive Radiation of Anoles*. University of California Press, Berkeley, CA.

Losos, J.B. (2010). Adaptive radiation, ecological opportunity, and evolutionary determinism. *American Naturalist*, **175**, 623–39.

Losos, J.B. (2011). Seeing the forest for the trees: the limitations of phylogenetics in comparative biology. *American Naturalist*, **177**, 709–27.

Losos, J.B. and Ricklefs, R.E. (2009). Adaptation and diversification on islands. *Nature*, **457**, 830–6.

Loss, S.R., Terwilliger, L.A., and Peterson, A.C. (2010). Assisted colonization: Integrating conservation strategies in the face of climate change. *Biological Conservation*, **144**, 92–100.

Lotka, A.J. (1920). Analytical note on certain rhythmic relations in organic systems. *Proceedings of the National Academy of Sciences*, **6**, 410–15.

Lotka, A.J. (1925). *Elements of Physical Biology*. Williams and Wilkins, Baltimore, MD.

Louette, G. and De Meester, L. (2007). Predation and priority effects in experimental zooplankton communities. *Oikos*, **116**, 419–26.

Louette, G., De Meester, L., and Declerck, S. (2008). Assembly of zooplankton communities in newly created ponds. *Freshwater Biology*, **53**, 2309–20.

Lovejoy, T.E. (1986). Species leave the ark one by one. In: Norton, B.G., ed. *The Preservations of Species: The Value of Biological Diversity*, pp. 13–27. Princeton University Press, Princeton, NJ.

Lubchenco, J. (1978). Plant species diversity in a marine intertidal community: importance of herbivore food preferences and algal competitive abilities. *American Naturalist*, **112**, 23–39.

Ludwig, D. and Rowe, L. (1990). Life history strategies for energy gain and predator avoidance under time constraints. *American Naturalist*, **135**, 686–707.

Lui, X., Liang, M., Etienne, R.S., Wang, Y., Staehelinm C., and Yu, S. (2012). Experimental evidence for a phylogenetic Janzen–Connell effect in a subtropical forest. *Ecology Letters*, **15**, 111–18.

Lundberg, P. (1988). Functional response of a small mammalian herbivore: the disc equation revisited. *Journal of Animal Ecology*, **57**, 999–1006.

Ma, B.O., Abrams, P.A., and Brassil, C.E. (2003). Dynamic versus instantaneous models of diet choice. *American Naturalist*, **162**, 668–84.

MacArthur, R.H. (1955). Fluctuations of animal populations and a measure of community stability. *Ecology*, **36**, 533–6.

MacArthur, R.H. (1965). Patterns of species diversity. *Biological Reviews*, **40**, 510–33.

MacArthur, R.H. (1968). The theory of the niche. In: Lewontin, R.C., ed. *Population Biology and Evolution*, pp. 159–76. Syracuse University Press, Syracuse, NY.

MacArthur, R.H. (1969). Species packing, and what interspecies competition minimizes. *Proceedings of the National Academy of Sciences*, **64**, 1369–71.

MacArthur, R.H. (1970). Species packing and competitive equilibrium for many species. *Theoretical Population Biology*, **1**, 1–11.

MacArthur, R.H. (1972). *Geographical Ecology*. Harper and Row, New York, NY.

MacArthur, R.H., Diamond, J., and Karr, J.M. (1972). Density compensation in island faunas. *Ecology*, **53**, 330–42.

MacArthur, R.H. and Levins, R. (1967). The limiting similarity, convergence, and divergence of coexisting species. *American Naturalist*, **101**, 377–85.

MacArthur, R.H. and Pianka, E.R. (1966). On optimal use of a patchy environment. *American Naturalist*, **100**, 603–9.

MacArthur, R.H. and Wilson, E.O. (1967). *The Theory of Island Biogeography*. Princeton University Press, Princeton, NJ.

Machac, A. and Graham, C.H. (2017). Regional diversity and diversification in mammals. *American Naturalist*, **189**, E1–E13.

Mack, M.C. and D'Antonio, C.M. (1998). Impacts of biological invasions on disturbance regimes. *Trends in Ecology and Evolution*, **13**, 195–8.

Mack, M.C., D'Antonio, C.M., and Ley, R.E. (2001). Alteration of ecosystem nitrogen dynamics by exotic plants: a case study of C_4 grasses in Hawaii. *Ecological Applications*, **11**, 1323–35.

Mackay, R.L. and Currie, D.J. (2001). The diversity–disturbance relationship: is it generally strong and peaked? *Ecology*, **82**, 3479–92.

MacLeod, K.J., Krebs, C.J., Boonstra, R., and Sheriff, M.J. (2018). Fear and lethality in snowshoe hares: the deadly effects of non-consumptive predation risk. *Oikos*, **127**, 375–80.

Madenjian, C.P., Schloesser, D.W., and Krieger, K.A. (1998). Population models of burrowing mayfly recolonization in Western Lake Erie. *Ecological Applications*, **8**, 1206–12.

Maestre, F.T., Callaway, R.M., Valladares, F., and Lortie, C.J. (2009). Refining the stress-gradient hypothesis for competition and facilitation in plant communities. *Journal of Ecology*, **97**, 199–205.

Maestre, F.T. and Cortina, J. (2004). Do positive interactions increase with abiotic stress? A test from a semi-arid steppe. *Proceedings of the Royal Society of London B Supplement*, **271**, S331–3.

Maestre, F.T., Valladares, F., and Reynolds, J.F. (2005). Is the change of plant-plant interactions with abiotic stress predictable? A meta-analysis of field results in arid environments. *Journal of Ecology*, **93**, 748–57.

Maglianesi, M.A., Blüthgen, N., Böhning-Gaese, K., and Schleuning, M. (2014). Morphological traits determine specialization and resource use in plant-hummingbird networks in the neotropics. *Ecology*, **95**, 3325–34.

Magurran, A.E. (2004). *Measuring Biological Diversity*. Blackwell Publishing, Oxford.

Magurran, A.E. (2005). *Evolutionary Ecology: The Trinidadian Guppy*. Oxford University Press, New York.

Magurran, A.E. and Henderson, P.A. (2003). Explaining the excess of rare species in natural species abundance distributions. *Nature*, **422**, 714–16.

Magurran, A.E., Deacon, A.E., Moyes, F., *et al.* (2018). Divergent biodiversity change in ecosystems. *Proceedings of the National Academy of Sciences*, **115**, 1843–7.

Magurran, A.E. and McGill, B.J. (2011). *Biological Diversity: Frontiers in Measurement and Assessment*. Oxford University Press, Oxford.

Mahler, D.L., Revell, L.J., Glor, R.E., and Losos, J.B. (2010). Ecological opportunity and the rate of morphological evolution in the diversification of Greater Antillean anoles. *Evolution*, **64**, 2731–45.

Malthus, T.R. (1789). *An Essay on the Principle of Population*. J. Johnson, London.

Mangan, S.A., Schnitzer, S.A., Herre, E.A., *et al.* (2010). Negative plant–soil feedback predicts tree species relative abundance in a tropical forest. *Nature*, **466**, 752–5.

Manly, B. (1974). A model for certain types of selection experiments. *Biometrics*, **30**, 281–94.

Manly, B. (1985). *The Statistics of Natural Selection on Animal Populations*. Chapman and Hall, London.

Mäntylä, E., Klemola, T., and Laaksonen, T. (2011). Birds help plants: A meta-analysis of top-down trophic cascades caused by avian predators. *Oecologia*, **165**, 143–51.

Marcogliese, D.J. and Cone, D.K. (1997). Food webs: a plea for parasites. *Trends in Ecology and Evolution*, **12**, 320–5.

Marks, C.O. and Lechowicz, M.J. (2006). Alternative designs and the evolution of functional diversity. *American Naturalist*, **167**, 55–66.

Marin, J. and Hedges, S.B. (2016). Time best explains global variation in species richness of amphibians, birds and mammals. *Journal of Biogeography*, **43**, 1069–79.

Maron, J. and Marler, M. (2007). Native plant diversity resists invasion at both low and high resource levels. *Ecology*, **88**, 2651–61.

Maron, J., Marler, M., Klironomos, J.N., and Cleveland, C.C. (2011). Soil fungal pathogens and the relationship between plant diversity and productivity. *Ecology Letters*, **14**, 36–41.

Marquard, E., *et al.* (2009). Plant species richness and functional composition drive overyielding in a six-year grassland experiment. *Ecology*, **90**, 3290–302.

Marshall, S.D., Walker, S.E., and Rypstra, A.L. (2000). A test for a differential colonization and competitive ability in two generalist predators. *Ecology*, **81**, 3341–9.

Martin, K.L. and Kirkman, L.K. (2009). Management of ecological thresholds to re-establish disturbance-maintained herbaceous wetlands of the south-eastern USA. *Journal of Applied Ecology*, **46**, 906–14.

Martinez, A.E., Parra, E., Collado, L.F., and Vredenburg, V.T. (2017). Deconstructing the landscape of fear in stable multi-species societies. *Ecology*, **98**, 2447–55.

Martinez, N.D. (1992). Constant connectance in community food webs. *American Naturalist*, **139**, 1208–18.

Martins, E.L.P. (2000). Adaptation and the comparative method. *Trends in Ecology & Evolution*, **15**, 296–9.

Matthiessen, B. and Hillebrand, H. (2006). Dispersal frequency affects local biomass production by controlling local diversity. *Ecology Letters*, **9**, 652–62.

May, R.M. (1971). Stability in multispecies community models. *Mathematical Biosciences*, **12**, 59–79.

May, R.M. (1972). Will a large complex system be stable? *Nature*, **238**, 413–14.

May, R.M. (1973a). *Stability and Complexity in Model Ecosystems*. Princeton University Press, Princeton, NJ.

May, R.M. (1973b). Stability in randomly fluctuating versus deterministic environments. *American Naturalist*, **107**, 621–50.

May, R.M. (1975). Patterns of species abundance and diversity. In: Cody, M.L. and Diamond, J., eds. *Ecology and Evolution of Communities*, pp. 81–120. Belknap Press of Harvard University, Cambridge, MA.

May, R.M. (1977a). Mathematical models and ecology: past and future. In: Goulden, C.E., ed. *Changing Scenes in the Natural Sciences*, Special Publication no. 12, pp. 189–202. Academy of Natural Sciences, Philadelphia, PA.

May, R.M. (1977b). Thresholds and breakpoints in ecosystems with a multiplicity of stable states. *Nature*, **269**, 471–7.

May, R.M. (1981). Models for two interacting populations. In: May, R.M., ed. *Theoretical Ecology: Principles and Applications*, pp. 78–104. Sinauer Associates, Sunderland, MA.

May, R.M. and Anderson, R.M. (1978). Regulation and stability of host-parasite population interactions. II. Destabilizing processes. *Journal of Animal Ecology*, **47**, 249–67.

May, R.M. and MacArthur, R.H. (1972). Niche overlap as a function of environmental variability. *Proceedings of the National Academy of Sciences*, **69**, 1109–13.

May, R.M. and Seger, J. (1986). Ideas in ecology. *American Scientist*, **74**, 256–67.

May, R.M. and Watts, C.H. (2009). The dynamics of predator-prey and resource-harvester systems. In: E.J. Crawley, ed. *The Population Biology of Predators, Parasites and Diseases*, pp. 431–57. Blackwell Publishing Ltd., London.

Mayfield, M.M. and Levine, J.M. (2010). Opposing effects of competitive exclusion on the phylogenetic structure of communities. *Ecology Letters*, **13**, 1085–93.

Maynard Smith, J. (1974). *Models in Ecology*. Cambridge University Press, Cambridge.

McCain, C.M. (2007). Area and mammalian elevational diversity. *Ecology*, **88**, 76–86.

McCann, K.S. (2000). The diversity–stability debate. *Nature*, **405**, 228–33.

McCann, K.S., Hastings, A., and Huxel, G.R. (1998). Weak trophic interactions and the balance of nature. *Nature*, **395**, 794–8.

McCauley, E. and Briand, F. (1979). Zooplankton grazing and phytoplankton species richness: Field tests of the predation hypothesis. *Limnology and Oceanography*, **24**, 243–52.

McCauley, E., Murdoch, W.W., and Watson, S. (1988). Simple models and variation in plankton densities among lakes. *American Naturalist*, **132**, 383–403.

McCauley, S.J., Rowe, L., and Fortin, M.J. (2011). The deadly effects of "nonlethal" predators. *Ecology*, **92**, 2043–8.

McCook, L.J. (1999). Macroalgae, nutrients and phase shifts on coral reefs: Scientific issues and management consequences for the Great Barrier Reef. *Coral Reefs*, **18**, 357–67.

McGill, B.J. (2003). A test of the unified neutral theory of biodiversity. *Nature*, **422**, 881–5.

McGill, B.J. (2005). A mechanistic model of a mutualism and its ecological and evolutionary dynamics. *Ecological Modelling*, **187**, 413–25.

McGill, B.J. (2011a). Linking biodiversity patterns to autocorrelated random sampling. American *Journal of Botany*, **98**, 481–502.

McGill, B.J. (2011b). Species abundance distributions. In: Magurran, A.E. and McGill, B.J., eds. *Biological Diversity*, pp. 105–22. Oxford University Press, Oxford.

McGill, B.J. (2013). A macroecological approach to the equilibrial vs. nonequilibrial debate using bird populations and communities. In: Rohde, K., ed. *The Balance of Nature and Human Impact*, pp. 103–18. Cambridge University Press, Cambridge.

McGill, B.J., Enquist, B.J., Weiher, E., and Westoby, M. (2006). Rebuilding community ecology from functional traits. *Trends in Ecology and Evolution*, **21**, 178–85.

McGill, B.J., Hadly, E.A., and Maurer, B.A. (2005). Community inertia of Quaternary small mammal assemblages in North America. *Proceedings of the National Academy of Sciences*, **102**, 16701–6.

McGill, B.J., Maurer, B.A., and Weiser, M.D. (2006). Empirical evaluation of neutral theory. *Ecology*, **87**, 1411–23.

McGill, B.J. and Mittelbach, G.G. (2006). An allometric vision and motion model to predict prey encounter rates. *Evolutionary Ecology Research*, **8**, 1–11.

McGill, B.J. and Nekola, J.C. (2010). Mechanisms in macroecology: AWOL or purloined letter? Towards a pragmatic view of mechanism. *Oikos*, **119**, 591–603.

McGill, B.J., et al. (2007). Species abundance distributions: moving beyond single prediction theories to integration within an ecological framework. *Ecology Letters*, **10**, 995–1015.

McGrady-Steed, J., Harris, P.M., and Morin, P.J. (1997). Biodiversity regulates ecosystem predictability. *Nature*, **390**, 162–5.

McHugh, P., McIntosh, A.R., and Jellyman, P. (2010). Dual influences of ecosystem size and disturbance on food chain length in streams. *Ecology Letters*, **13**, 881–90.

McIntosh, R.P. (1980). The background and some current problems of theoretical ecology. *Synthese*, **43**, 195–255.

McIntosh, R.P. (1985). *The Background of Ecology: Concept and Theory*. Cambridge University Press, Cambridge.

McIntosh, R.P. (1987). Pluralism in ecology. *Annual Review of Ecology and Systematics*, **18**, 321–41.

McKinney, M.L. (1997). Extinction vulnerability and selectivity: combining ecological and paleontological views. *Annual Review of Ecology and Systematics*, **28**, 495–516.

McLachlan, J.S., Hellman, J.J., and Schwartz, M.W. (2007). A framework for debate of assisted migration in an era of climate change. *Conservation Biology*, **21**, 297–302.

McNamara, J.M. and Houston, A.I. (1994). The effect of a change in foraging options on intake rate and predation rate. *American Naturalist*, **144**, 978–1000.

McNaughton, S.J. (1977). Diversity and stability of ecological communities: a comment on the role of empiricism in ecology. *American Naturalist*, **111**, 515–25.

McNaughton, S.J. (1993). Biodiversity and stability of grazing ecosystems. In: Shulze, E-D. and Mooney, H.A., eds. *Biodiversity and Ecosystem Function*, pp. 361–83. Springer-Verlag, Berlin.

McPeek, M.A. (1998). The consequences of changing the top predator in a food web: a comparative experimental approach. *Ecological Monographs*, **68**, 1–23.

McPeek, M.A. (2004). The growth/predation risk trade-off: so what is the mechanism? *American Naturalist*, **163**, E88–E111.

McPeek, M.A. (2007). The macroevolutionary consequences of ecological differences among species. *Palaeontology*, **50**, 111–29.

McPeek, M.A. (2008). The ecological dynamics of clade diversification and community assembly. *American Naturalist*, **172**, E270–84.

McPeek, M.A. (2017). *Evolutionary Community Ecology*. Princeton University Press, Princeton, NJ.

McPeek, M.A. and Peckarsky, B.L. (1998). Life histories and the strengths of species interactions: combining mortality, growth, and fecundity effects. *Ecology*, **79**, 867–79.

McPeek, M.A., Shen, L., and Farid, H. (2008). The correlated evolution of 3-dimensional reproductive structures between male and female damselflies. *Evolution*, **63**, 73–83.

Meiners, S.J. (2007). Apparent competition: an impact of exotic shrub invasion on tree regeneration. *Biological Invasions*, **9**, 849–55.

Meire, P.M. and Ervynck, A. (1986). Are oystercatchers (*Haemotopus ostralegus*) selecting the most profitable mussels (*Mytilus edulis*)? *Animal Behaviour*, **34**, 1427–35.

Melián, C.J. and Bascompte, J. (2002). Complex networks: two ways to be robust. *Ecology Letters*, **5**, 705–8.

Melián, C.J. and Bascompte, J. (2004). Food web cohesion. *Ecology*, **85**, 352–8.

Melián, C.J., Bascompte, J., Jordano, P., and Křivan, V. (2009). Diversity in a complex ecological network with two interaction types. *Oikos*, **118**, 122–30.

Menge, B.A. (1995). Indirect effects in marine rocky intertidal interaction webs: patterns and importance. *Ecological Monographs*, **65**, 21–74.

Menge, B.A., Berlow, E.L., Blanchette, C.A., Navarrete, S.A., and Yamada, S.B. (1994). The keystone species concept—variation in interaction strength in a rocky intertidal habitat. *Ecological Monographs*, **64**, 249–86.

Menge, B.A. and Sutherland, W.J. (1987). Community regulation: variation in disturbance, competition, and predation in relation to environmental stress and recruitment. *American Naturalist*, **130**, 730–57.

Menge, B.A., et al. (2011). Potential impact of climate-related changes is buffered by differential responses to recruitment and interactions. *Ecological Monographs*, **81**, 493–509.

Menges, E.S. (1990). Population viability analysis for an endangered plant. *Conservation Biology*, **4**, 52–62.

Merriam, G. (1991). Corridors in restoration of fragmented landscapes. In: Saunders, D.A., Hobbs, R.J., and Ehrlich, P.R., eds. *Nature Conservation 3: Reconstruction of Fragmented Ecosystems*, pp. 71–87. Surrey Beatty and Sons, Chipping Norton, NSW.

Messier, J., McGill, B.J., and Lechowicz, M.J. (2010). How do traits vary across ecological scales? A case for trait-based ecology. *Ecology Letters*, **13**, 838–48.

Meszéna, G., Gyllenberg, M., Pásztor, L., and Metz, J.A.J. (2006). Competitive exclusion and limiting similarity: a unified theory. *Theoretical Population Biology*, **69**, 68–87.

Meyer, A.L.S. and Wiens, J.J. (2017). Estimating diversification rates for higher taxa: BAMM can give problematic estimates of rates and rate shifts. *Evolution*, **72**, 39–53.

Meyer, S.T., et al. (2016). Effects of biodiversity strengthen over time as ecosystem functioning declines at low and increases at high biodiversity. *Ecosphere*, **7**(12), e01619.

Michels, E., Cottenie, K., Neys, L., and De Meester, L. (2001). Zooplankton on the move: first results on the quantification of dispersal in a set of interconnected ponds. *Hydrobiologia*, **442**, 117–26.

Mikkelson, G.M., McGill, B.J., Beaulieu, S., and Beukema, P.L. (2011). Multiple links between species diversity and temporal stability in bird communities across North America. *Evolutionary Ecology Research*, **13**, 361–72.

Milchunas, D.G. and Lauenroth, W.K. (1993). Quantitative effects of grazing on vegetation and soils over a global range of environments. *Ecological Monographs*, **63**, 327–66.

Miller, K.M. and McGill, B.J. (2018). Land use and life history limit migration capacity of eastern tree species. *Global Ecology and Biogeography*, **27**, 57–67.

Miller, M.P. and Vincent, E.R. (2008). Rapid natural selection for resistance to an introduced parasite of rainbow trout. *Evolutionary Applications*, **1**, 336–41.

Miller, T.E., *et al.* (2005). A critical review of twenty years' use of the resource-ratio theory. *American Naturalist*, **165**, 339–448.

Miriti, M.N. (2006). Ontogenetic shift from facilitation to competition in a desert shrub. *Journal of Ecology*, **94**, 973–9.

Mitchell, W.A. and Valone, T.J. (1990). The optimization research program: studying adaptations by their function. *Quarterly Review of Biology*, **65**, 43–52.

Mittelbach, G.G. (1981). Foraging efficiency and body size: a study of optimal diet and habitat use by bluegills. *Ecology*, **62**, 1370–86.

Mittelbach, G.G. (2002). Fish foraging and habitat choice: a theoretical perspective. In: Hart, P.J.B. and Reynolds, J.D., eds. *Handbook of Fish Biology and Fisheries*, Volume 1 *Fish Biology*, pp. 251–66. Blackwell, Oxford.

Mittelbach, G.G. (2010). Understanding species richness-productivity relationships: The importance of meta-analysis. *Ecology*, **91**, 2540–4.

Mittelbach, G.G. (2012). *Community Ecology*, 1st edn. Sinauer Associates, Sunderland, MA.

Mittelbach, G.G. and Chesson, P.L. (1987). Predation risk: indirect effects on fish populations. In: Kerfoot, W.C. and Sih, A., eds. *Predation: Direct and Indirect Impacts on Aquatic Communities*, pp. 315–32. University Press of New England, Hanover, NH.

Mittelbach, G.G., Garcia, E.A., and Taniguchi, Y. (2006). Fish reintroductions reveal smooth transitions between lake community states. *Ecology*, **87**, 312–18.

Mittelbach, G.G. and Osenberg, C.W. (1993). Stage-structured interactions in bluegill: consequences of adult resource variation. *Ecology*, **74**, 2381–94.

Mittelbach, G.G., Osenberg, C.W., and Leibold, M.A. (1988). Trophic relations and ontogenetic niche shifts in aquatic organisms. In: Ebenman, B. and Persson, L., eds. *Size-Structured Populations: Ecology and Evolution*, pp. 217–35. Springer-Verlag, Berlin.

Mittelbach, G.G. and Schemske, D.W. (2015). Ecological and evolutionary perspectives on community assembly. *Trends in Ecology and Evolution*, **30**, 241–7.

Mittelbach, G.G., Turner, A.M., Hall, D.J., Rettig, J.E., and Osenberg, C.W. (1995). Perturbation and resilience in an aquatic community: a long-term study of the extinction and reintroduction of a top predator. *Ecology*, **76**, 2347–60.

Mittelbach, G.G., *et al.* (2001). What is the observed relationship between species richness and productivity? *Ecology*, **82**, 2381–96.

Mittelbach, G.G., *et al.* (2007). Evolution and the latitudinal diversity gradient: speciation, extinction and biogeography. *Ecology Letters*, **10**, 315–31.

Moles, A.T., Falster, D.S., Leishman, M.R., and Westoby, M. (2004). Small-seeded species produce more seeds per square metre of canopy per year, but not per individual per lifetime. *Journal of Ecology*, **92**, 384–96.

Monod, J. (1950). La technique de culture continue, théorie et applications. *Annales d'Institut Pasteur*, **79**, 390–410.

Montoya, J.M., Woodward, G., Emmerson, M., and Solé, R.V. (2009). Press perturbations and indirect effects in real food webs. *Ecology*, **90**, 2426–33.

Mooney, H. and Ehleringer, J. (2003). *Photosynthesis in Plant Ecology*. Crawley, M.J. ed., pp. 1–27. Blackwell, Oxford.

Moore, J. (2002). *Parasites and the Behavior of Animals*. Oxford University Press, Oxford.

Moore, J.C. and de Ruiter, P.C. (2012). *Energetic Food Webs: An Analysis of Real and Model Ecosystems*. Oxford University Press, Oxford, UK.

Moore, J.C., *et al.* (2004). Detritus, trophic dynamics and biodiversity. *Ecology Letters*, **7**, 584–600.

Morales, C.L. and Traveset, A. (2009). A meta-analysis of impacts of alien vs. native plants on pollinator visitation and reproductive success of co-flowering native plants. *Ecology Letters*, **12**, 716–28.

Morlon, H. (2014). Phylogenetic approaches for studying diversification. *Ecology Letters*, **17**, 508–25.

Motomura, I. (1932). On the statistical treatment of communities (in Japanese). *Zoological Magazine Tokyo*, **44**, 379–83.

Mouquet, N. and Loreau, M. (2003). Community patterns in source-sink metacommunities. *American Naturalist*, **162**, 544–57.

Müller, C.B. and Godfray, H.C.J. (1997). Apparent competition between two aphid species. *Journal of Animal Ecology*, **66**, 57–64.

Muller-Landau, H.C. (2008). Colonization-related trade-offs in tropical forests and their role in the maintenance of plant species diversity. In: Carson, W.P. and Schnitzer, S.A., eds. *Tropical Forest Community Ecology*, pp. 182–95. Wiley-Blackwell, Chichester.

Mumby, P.J., Steneck, R.S., and Hastings, A. (2013). Evidence for and against the existence of alternate attractors on coral reefs. *Oikos*, **122**, 481–91.

Münkemüller, T., Boucher, F.C., Thuiller, W., and Lavergne, S. (2015). Phylogenetic niche conservatism – common pitfalls and ways forward. *Functional Ecology*, **29**, 627–39.

Murdoch, W.W. (1966). "Community structure, population control, and competition"—a critique. *American Naturalist*, **100**, 219–26.

Murdoch, W.W. (1969). Switching in general predators: experiments on predator specificity and stability of prey populations. *Ecological Monographs*, **39**, 335–54.

Murdoch, W.W. (1970). Population regulation and population inertia. *Ecology*, **51**, 497–502.

Murdoch, W.W. (1991). The shift from an equilibrium to a non-equilibrium paradigm in ecology. *Bulletin of the Ecological Society of America*, **72**, 49–51.

Murdoch, W.W. (1994). Population regulation in theory and practice. *Ecology*, **75**, 271–87.

Murdoch, W.W., Briggs, C.J., and Nisbet, R.M. (2003). *Consumer-Resource Dynamics*. Princeton University Press, Princeton, NJ.

Murdoch, W.W. and Walde, S.J. (1989). Analysis of insect population dynamics. In: Grubb, P.J. and Whittaker, J.B., eds. *Towards a More Exact Ecology*, pp. 113–40. Blackwell, Oxford.

Murphy, G.E.P. and Romanuk, T.N. (2014). A meta-analysis of declines in local species richness from human disturbances. *Ecology and Evolution*, **4**, 91–103.

Naeem, S., Thompson, L.J., Lawler, S.P., Lawton, J.H., and Woodfin, R.M. (1994). Declining biodiversity can alter the performance of ecosystems. *Nature*, **368**, 734–7.

Narwani, A., Alexandrou, M.A., Oakley, T.H., Carroll, I.T., and Cardinale, B.J. (2013). Experimental evidence that evolutionary relatedness does not affect the ecological mechanisms of coexistence in freshwater green algae. *Ecology Letters*, **16**,1373–81.

Navarrete, S.A. and Menge, B.A. (1996). Keystone predation and interaction strength: interactive effects of predators on their main prey. *Ecological Monographs*, **66**, 409–29.

Nee, S. (1994). How populations persist. *Nature*, **367**, 123–4.

Nee, S. (2005). The neutral theory of biodiversity: do the numbers add up? *Functional Ecology*, **19**, 173–6.

Nee, S. and May, R.M. (1992). Dynamics of metapopulations: habitat destruction and competitive coexistence. *Journal of Animal Ecology*, **61**, 37–40.

Nee, S., May, R.M., and Hassell, M.P. (1997). Two-species metapopulation models. In: Hanski, I. and Gilpin, M., eds. *Metapopulation Biology*, pp. 123–47. Academic Press, San Diego, CA.

Neff, J.C., Hobbie, S.E., and Vitousek, P.M. (2003). Nutrient and mineralogical control on dissolved organic C, N and P fluxes and stoichiometry in Hawaiian soils. *Biogeochemistry*, **51**, 283–302.

Neill, W.E. (1974). The community matrix and interdependence of the competition coefficients. *American Naturalist*, **108**, 399–408.

Nekola, J. C. and P. S. White. (1999. The distance decay of similarity in biogeography and ecology. Journal of Biogeography **26**: 867–78.

Neuwald, J.L. and Templeton, A.R. (2013). Genetic restoration in the eastern collared lizard under prescribed woodland burning. *Molecular Ecology*, **22**, 366–3679.

Newbold, T., *et al.* (2015). Global effects of land use on local terrestrial biodiversity. *Nature*, **520**, 45–50.

Nicholson, A.J. (1957). The self-adjustment of populations to change. *Cold Spring Harbor Symposia on Quantitative Biology*, **22**, 153–72.

Niinemets, Ü. (2001). Global-scale climatic controls of leaf dry mass per area, density, and thickness in trees and shrubs. *Ecology*, **82**, 453–69.

Niklaus, P.A., Kandeler, E., Leadley, P.W., Schmid, B., Tscherko, D., and Körner, C. (2001). A link between plant diversity, elevated CO_2 and soil nitrate. *Oecologia*, **127**, 540–8.

Nisbet, R.M. and Gurney, W.S.C. (1982). *Modeling Fluctuating Populations*. John Wiley & Sons, New York, NY.

Noë, R. and Hammerstein, P. (1994). Biological markets: supply and demand determine the effect of partner choice in cooperation, mutualism and mating. *Behavioral Ecology and Sociobiology*, **35**, 1–11.

Noonburg, E.G. and Byers, J.E. (2005). More harm than good: when invader vulnerability to predators enhances impact on native species. *Ecology*, **86**, 2555–60.

Norberg, J. (2000). Resource-niche complementarity and autotrophic compensation determines ecosystem-level responses to increased cladoceran species richness. *Oecologia*, **112**, 264–72.

Norbury, G. (2001). Conserving dryland lizards by reducing predator-mediated apparent competition and direct competition with introduced rabbits. *Journal of Applied Ecology*, **38**, 1350–61.

Northfield, T.D., Snyder, G.B., Ives, A.R., and Snyder, W.E. (2010). Niche saturation reveals resource partitioning among consumers. *Ecology Letters*, **13**, 338–48.

Noy-Meir, I. (1975). Stability of grazing systems: an application of predator-prey graphs. *Journal of Ecology*, **63**, 459–81.

Nuñez, C.I., Raffaele, E., Nuñez, M.A., and Cuassolo, F. (2009). When do nurse plants stop nursing? Temporal changes in water stress levels in *Austrocedrus chilensis* growing within and outside shrubs. *Journal of Vegetation Science*, **20**, 1064–71.

Nyström, M., Graham, N.A.J., Lokrantz, J., and Norström, A. (2008). Capturing the cornerstones of coral reef resilience: linking theory to practice. *Coral Reefs*, **27**, 795–809.

O'Brien, E.M. (1998). Water-energy dynamics, climate, and prediction of woody plant species richness: an interim general model. *Journal of Biogeography*, **20**, 181–98.

O'Connor, M.I., *et al.* (2016). A general biodiversity-function relationship is mediated by trophic level. *Oikos*, **126**, 18–31.

O'Gorman, E.J. and Emmerson, M.C. (2009). Perturbations to trophic interactions and the stability of complex food webs. *Proceedings of the National Academy of Sciences*, **106**, 13393–8.

O'Meara, B.C. (2012). Evolutionary inferences from phylogenetics: a review of methods. Annual *Review of Ecology, Evolution, and Systematics*, **43**, 267–85.

Odum, E.P. (1953). *Fundamentals of Ecology*. Saunders, Philadelphia, PA.

Odum, E.P. (1959). *Fundamentals of Ecology*. Saunders, Philadelphia, PA.

Oelmann, Y., *et al.* (2007). Soil and plant nitrogen pools as related to plant diversity in an experimental grassland. *Soil Science Society of America Journal*, **71**, 720–9.

Oksanen, L., Fretwell, S.D., Arruda, J., and Niemelä, P. (1981). Exploitation ecosystems in gradients of primary productivity. *American Naturalist*, **118**, 240–61.

Oksanen, T., Power, M.E., and Oksanen, L. (1995). Ideal free habitat selection and consumer-resource dynamics. *American Naturalist*, **146**, 565–85.

Okuyama, T. and Bolker, B.M. (2007). On quantitative measures of indirect interactions. *Ecology Letters*, **10**, 264–71.

Oliver, M., Luque-Larena, J.J., and Lambin, X. (2009). Do rabbits eat voles? Apparent competition, habitat heterogeneity and large-scale coexistence under mink predation. *Ecology Letters*, **12**, 1201–9.

Olson, J.S. (1958). Rates of succession and soil changes on southern Lake Michigan sand dunes. *Botanical Gazette*, **119**, 125–70.

Olszewski, T.D. (2004). A unified mathematical framework for the measurement of richness and evenness within and among multiple communities. *Oikos*, **104**, 377–87.

Orrock, J.L., Witter, M.S., and Reichman, O.J. (2008). Apparent competition with an exotic plant reduces native plant establishment. *Ecology*, **89**, 1168–74.

Osenberg, C.W. and Mittelbach, G.G. (1996). The relative importance of resource limitation and predator limitation in food chains. In: Polis, G.A. and Winemiller, K.O., eds. *Food Webs: Integration of Patterns and Dynamics*, pp. 134–48. Chapman and Hall, New York, NY.

Osenberg, C.W., Sarnelle, O., and Cooper, S.D. (1997). Effect size in ecological experiments: the application of biological models in meta-analysis. *American Naturalist*, **150**, 798–812.

Östman, Ö. and Ives, A.R. (2003). Scale-dependent indirect interactions between two prey species through a shared predator. *Oikos*, **102**, 505–14.

Pacala, S.W. and Tilman, D. (1994). Limiting similarity in mechanistic and spatial models of plant competition in heterogeneous environments. *American Naturalist*, **143**, 222–57.

Paine, R.T. (1966). Food web complexity and species diversity. *American Naturalist*, **100**, 65–75.

Paine, R.T. (1969). A note on trophic complexity and community stability. *American Naturalist*, **103**, 91–3.

Paine, R.T. (1980). Food webs, linkage interaction strength, and community infrastructure. *Journal of Animal Ecology*, **49**, 667–85.

Paine, R.T. (1992). Food-web analysis through field measurement of per capita interaction strength. *Nature*, **355**, 73–5.

Pajunen, V.I. and Pajunen, I. (2003). Long-term dynamics in rock pool *Daphnia* metapopulations. *Ecography*, **26**: 731–8.

Pake, C.E. and Venable, D.L. (1995). Is coexistence of Sonoran Desert annuals mediated by temporal variability in reproductive success? *Ecology*, **76**, 246–61.

Palkovacs, E.P. and Post, D.M. (2009). Experimental evidence that phenotypic divergence in predators drives community divergence in prey. *Ecology*, **90**, 300–5.

Palmer, M.W. (1994). Variation in species richness: towards a unification of hypotheses. *Folia Geobotanica*, **29**, 511–30.

Palmer, T.M., Stanton, M.L., Young, T.P., Goheen, J.R., Pringle, R.M., and Karban, R. (2008). Breakdown of an ant-plant mutualism follows the loss of large herbivores from an African savanna. *Science*, **319**, 192–5.

Palmer, T.M., Pringle, E.G., Stier, A., and Holt, R.D. (2015). Mutualism In a community context. In Bronstein, J.L., ed. *Mutualism*, pp. 159–80. Oxford University Press, Oxford.

Pangle, K.L., Peacor, S.D., and Johannsson, O.E. (2007). Large nonlethal effects of an invasive invertebrate predator on zooplankton population growth rate. *Ecology*, **88**, 402–12.

Parker, S. and Huryn, A. (2006). Food web structure and function in two arctic streams with contrasting disturbance. *Freshwater Biology*, **51**, 1249–63.

Parmesan, C. (2006). Ecological and evolutionary responses to recent climate change. *Annual Review of Ecology, Evolution, and Systematics*, **37**, 637–69.

Parmesan, C. and Yohe, G. (2003). A globally coherent fingerprint of climate change impacts across natural systems. *Nature*, **421**, 37–42.

Parr, C.L., Dunn R.R., Sanders N.J., *et al.* (2017). GlobalAnts: a new database on the geography of ant traits (Hymenoptera: Formicidae). *Insect Conservation and Diversity*, **10**, 5–20.

Partel, M., Laanisto, L., and Zobel, M. (2007). Contrasting plant productivity–diversity relationships across latitude: the role of evolutionary history. *Ecology*, **88**, 1091–7.

Partel, M. and Zobel, M. (2007). Dispersal limitation may result in the unimodal productivity–diversity relationship: a new explanation for a general pattern. *Journal of Ecology*, **95**, 90–4.

Pascual, M. and Dunne, J.A., eds. (2006). *Ecological Networks: Linking Structure to Dynamics in Food Webs*. Oxford University Press, Oxford.

Pastore, A.I., Prather, C.M., Gornish, E.S, Ryan, W.H., Ellis, R.D., and Miller, T.E. (2014). Testing the competition-colonization trade-off with a 32-year study of a saxicolous lichen community. *Ecology*, **95**, 306–15.

Patil, G.P. and Taillie, C. (1982). Diversity as a concept and its measurement. *Journal of the American Statistical Association*, **77**, 548–61.

Patrick, R. (1967). The effect of invasion rate, species pool, and size of area on the structure of the diatom community. *Proceedings of the National Academy of Sciences*, **58**, 1335–42.

Pausas, J.G. and Verdú, M. (2010). The jungle of methods for evaluating phenotypic and phylogenetic structure of communities. *BioScience*, **60**, 614–25.

Pawar, S., Dell, A.I and Savage, V.M. (2012). Dimensionality of consumer search space drives trophic interaction strengths. *Nature*, **486**, 485–9.

Payne, J.L., Bush, A.M., Heim, N.A., Knope, M.L., and McCauley, D.J. (2016). Ecological selectivity of the emerging mass extinction in the oceans. *Science*, **353**, 1284–6.

Peacor, S.D. (2002). Positive effect of predators on prey growth rate through induced modifications of prey behaviour. *Ecology Letters*, **5**, 77–85.

Peacor, S.D. and Werner, E.E. (1997). Trait mediated indirect interactions in a simple aquatic food web. *Ecology*, **78**, 1146–56.

Pearman, P.B., Guisan, A., Broennimann, O., and Randin, C.F. (2008). Niche dynamics in space and time. *Trends in Ecology and Evolution*, **23**, 149–58.

Peckarsky, B.L., Cowan C.A., Penton, M.A., and Anderson, C. (1993). Sublethal consequences of predator-avoidance by stream-dwelling mayfly larvae to adult fitness. *Ecology*, **74**, 1836–46.

Peckarsky, B.L., McIntosh, A.R., Taylor, B.W., and Dahl, J. (2002). Predator chemicals induce changes in mayfly life history traits: a whole-stream manipulation. *Ecology*, **83**, 612–18.

Peckarsky, B.L., Taylor, B.W., McIntosh, A.R., McPeek, M.A., and Lytle, D.A. (2001). Variation in mayfly size at metamorphosis as a developmental response to risk of predation. *Ecology*, **82**, 740–57.

Pedruski, M.T. and Arnott, S.E. (2011). The effects of habitat connectivity and regional heterogeneity on artificial pond metacommunities. *Oecologia*, **166**, 221–8.

Pennell, M.W. and Harmon, L.J. (2013). An integrative view of phylogenetic comparative methods: connections to population genetics, community ecology, and paleobiology. *Annals of the New York Academy of Sciences*, **1289**, 90–105.

Pennings, S.C., Seling, E.R., Houser, L.T., and Bertness, M.D. (2003). Geographic variation in positive and negative interactions among salt marsh plants. *Ecology*, **84**, 1527–38.

Perälä, T. and Kuparinen, A. (2017). Detection of Allee effects in marine fishes: analytical biases generated by data availability and model selection. *Proceedings of the Royal Society of London, B*, **284**, 20171284.

Pereira, H.M., *et al.* (2010). Scenarios for global biodiversity in the 21st century. *Science*, **330**, 1496–501.

Perlman, D.L. and Adelson, G. (1997). *Biodiversity: Exploring Values and Priorities in Conservation*. Blackwell Scientific, Cambridge, MA.

Persson, L. (1993). Predator-mediated competition in prey refuges: the importance of habitat dependent prey resources. *Oikos*, **68**, 12–22.

Persson, L., Amundsen, P-A., De Roos, A.M., Klemetsen, A., Knudsen, R., and Primicerio, R. (2007). Culling prey promotes predator recovery—alternative states in a whole-lake experiment. *Science*, **316**, 1743–6.

Pervez, A. and Omkar. (2005). Functional responses of coccinellid predators: An illustration of a logistic approach. *Journal of Insect Science*, **5**. Available at: insectscience.org/5.5 (accessed 7 December 2018).

Petchey, O.L., Beckerman, A.P., Riede, J.O., and Warren, P.H. (2008). Size, foraging, and food web structure. *Proceedings of the National Academy of Sciences*, **105**, 4191–6.

Petchey, O.L., Beckerman, A.P., Riede, J.O., and Warren, P.H. (2011). Fit, efficiency, and biology: Some thoughts on judging food web models. Journal of Theoretical Biology **279**: 169–71.

Petchey, O.L., Downing, A.L., Mittelbach, G.G., *et al.* (2004). Species loss and the structure and functioning of multitrophic aquatic systems. *Oikos* **104**, 467–78.

Petchey, O.L., Evans, K.L., Fishburn, I.S., and Gaston, K.J. (2007). Low functional diversity and no redundancy in British avian assemblages. *Journal of Animal Ecology*, **76**, 977–85.

Petermann, J.S., Fergus, A.J.F., Roscher, C., Turnbull, L.A., Weigelt, A., and Schmid, B. (2010). Biology, chance, or history? The predictable reassembly of temperate grassland communities. *Ecology*, **91**, 408–21.

Peters, R.H. (1983). *The Ecological Implications of Body Size*. Cambridge University Press, Cambridge.

Petraitis, P.S. (2013). *Multiple Stable States in Natural Ecosystems*. Oxford University Press, Oxford, UK.

Pettersson, L. and Brönmark, C. (1997). Trading off safety against food: State dependent habitat choice and foraging in crucian carp. *Oecologia*, **95**, 353–7.

Pfisterer, A.B. and Schmid, B. (2002). Diversity-dependent production can decrease the stability of ecosystem functioning. *Nature*, **416**, 84–6.

Phillimore, A.B. and Price, T.D. (2008). Density-dependent cladogenesis in birds. *PLoS Biology*, **6(3)**, e71.

Phillips, O.M. (1973). The equilibrium and stability of simple marine biological systems. I. Primary nutrient consumers. *American Naturalist*, **107**, 73–93.

Phillips, O.M. (1974). The equilibrium and stability of simple marine systems. II. Herbivores. *Archiv fur Hydrobiologie*, **73**, 310–33.

Pianka, E.R. (1966). Latitudinal gradients in species diversity: a review of concepts. *American Naturalist*, **100**, 33–46.

Pielou, E.C. (1975). *Ecological Diversity*. John Wiley & Sons, New York, NY.

Pierce, G.J. and Ollason, J.G. (1987). Eight reasons why optimal foraging theory is a complete waste of time. *Oikos*, **49**, 111–18.

Pigot, A.L., Tobias, J.A., and Jetz, W. (2016). Energetic constraints on species coexistence in birds. PLoS Biology, **14**(3), e1002407.

Pilosof, S., Porter, M.A., Pascual, M., and Kéfi, S. (2017). The multilayer nature of ecological networks. *Nature Ecology and Evolution*, **1**, 0101.

Pimm, S.L. (1984). The complexity and stability of ecosystems. *Nature*, **307**, 321–6.

Pimm, S.L. (1992). *Food Webs*. Chapman and Hall, London.

Pimm, S.L., Jones, H.L., and Diamond, J. (1988). On the risk of extinction. *American Naturalist*, **132**, 757–85.

Pimm, S.L. and Lawton, J.H. (1977). On the number of trophic levels. *Nature*, **268**, 329–31.

Pimm, S.L., Lawton, J.H., and Cohen, J.E. (1991). Food web patterns and their consequences. *Nature*, **350**, 669–74.

Pitman, N.C.A., *et al.* (2002). A comparison of tree species diversity in two upper Amazonian forests. *Ecology*, **83**, 3210–24.

Platt, W.J. and Weiss, I.M. (1977). Resource partitioning and competition within a guild of fugitive prairie plants. *American Naturalist*, **111**, 479–513.

Polis, G. (1991). Complex trophic interactions in deserts: an empirical critique of food web theory. *American Naturalist*, **138**, 123–55.

Polis, G.A. and Strong, D.R. (1996). Food web complexity and community dynamics. *American Naturalist*, **147**, 813–46.

Pollard, E., Lakhani, K.H., and Rothery, P. (1987). The detection of density-dependence from a series of annual censuses. *Ecology*, **68**, 2046–55.

Post, D.M. (2002). The long and short of food-chain length. *Trends in Ecology and Evolution*, **17**, 269–77.

Post, D.M., Pace, M., and Hairston, N.G., Jr (2000). Ecosystem size determines food-chain length in lakes. *Nature*, **405**: 1047–9.

Post, D.M. and Palkovacs, E.P. (2009). Eco-evolutionary feedbacks in community and ecosystem ecology: Interactions between the ecological theatre and the evolutionary play. *Philosophical Transactions of the Royal Society of London B*, **364**, 1629–40.

Post, D.M., Palkovacs, E.P., Schielke, E.G., and Dodson, S.I. (2008). Intraspecific phenotypic variation in a predator affects community structure and cascading trophic interactions. *Ecology*, **89**: (2019–32.

Powell, K.I., Chase, J.M., and Knight, T.M. (2013). Invasive species have scale-dependent effects on diversity by altering species-area relationships. *Science*, **339**, 316–18.

Powell, M.G. (2007). Latitudinal diversity gradients for brachiopod genera during late Palaeozoic time: Links

between climate, biogeography and evolutionary rates. *Global Ecology and Biogeography*, **16**, 519–28.

Powell, M.G. (2009). The latitudinal diversity gradient of brachiopods over the past 530 million years. *Journal of Geology*, **117**, 585–94.

Power, M.E. (1984). Habitat quality and the distribution of algae-grazing catfish in a Panamanian stream. *Journal of Animal Ecology*, **53**, 357–74.

Power, M.E. (1990). Effects of fish in river food webs. *Science*, **250**, 811–14.

Power, M.E. (1997). Ecosystem engineering by organisms: why semantics matters—reply. *Trends in Ecology and Evolution*, **12**, 275–6.

Power, M.E., Mathews, W.J., and Stewart, A.J. (1985). Grazing minnows, piscivorous bass, and stream algae: dynamics of a strong interaction. *Ecology*, **66**, 1448–56.

Power, M.E., *et al.* (1996). Challenges in the quest for keystones. *BioScience*, **46**, 609–20.

Pratt, H.D. (2005). *The Hawaiian Honeycreepers; Drepanididae*. Oxford University Press, Oxford.

Preisser, E.I., Bolnick, D.I., and Benard, M.F. (2005). Scared to death? The effects of intimidation and consumption in predator–prey interactions. *Ecology*, **86**, 501–9.

Preston, F.W. (1948). The commonness, and rarity, of species. *Ecology*, **29**, 254–83.

Preston, F.W. (1960). Time and space and the variation of species. *Ecology*, **41**, 611–27.

Preston, F.W. (1962). The canonical distribution of commonness and rarity: Part I. *Ecology*, **43**, 185–215.

Price, P.W. (1980). *Evolutionary Biology of Parasites*. Princeton University Press, Princeton, NJ.

Price, T.D. (2008). *Speciation in Birds*. Roberts and Company, Greenwood Village, CO.

Price, T.D., Hooper, D.M., Buchanan, C.D., *et al.* (2014). Niche filling slows the diversification of Himalayan songbirds. *Nature*, **509**, 222–5.

Price, T.D., Phillimore, A.B., Awodey, M., and Hudson, R. (2010). Ecological and geographical influences on the allopatric phase of island speciation. In: Grant, P.R. and Grant, B.R., eds. *In Search of the Causes of Evolution: From Field Observations to Mechanisms*, pp. 251–81. Princeton University Press, Princeton, NJ.

Primack, R., *et al.* (2018). Biodiversity gains? The debate on changes in local- vs global-scale species richness. *Biological Conservation*, **219**, A1–A3.

Prior, K.M. and Palmer, T.M. (2018). Economy of scale: third partner strengthens a keystone ant-plant mutualism. *Ecology*, **99**, 334–46.

Ptacnik, R., Andersen, T., Brettum, P., Lepistö, L., and Willén, E. (2010). Regional species pools control community saturation in lake phytoplankton. *Proceedings of the Royal Society of London B*, **277**, 3755–64.

Ptacnik, R., Solimini, A.G., Anderson, T., *et al.* (2008). Diversity predicts stability and resource use efficiency in natural phytoplankton communities. *Proceeding of the National Academy of Sciences*, **105**, 5134–8.

Pu, Z., Cortez, M.H., and Jiang, L. (2017). Predator–prey coevolution drives productivity-richness relationships in planktonic systems. *American Naturalist*, **189**, 28–42.

Pulido-Santacruz, P. and Weir, J.T. (2016). Extinction as a driver of avian latitudinal diversity gradients. *Evolution*, **70**, 860–72.

Pulliam, H.R. (1974). On the theory of optimal diets. *American Naturalist*, **108**, 59–75.

Pulliam, H.R. (1985). Foraging efficiency, resource partitioning, and the coexistence of sparrow species. *Ecology*, **66**, 1829–36.

Pulliam, H.R. (1988). Sources, sinks and population regulation. *American Naturalist*, **132**, 652–61.

Pulliam, H.R. (2000). On the relationship between niche and distribution. *Ecology Letters*, **3**, 349–61.

Purvis, A., Gittleman, J.L., Cowlishaw, G., and Mace, G.M. (2000). Predicting extinction risk in declining species. *Proceedings of the Royal Society of London B*, **267**, 1947–52.

Pyke, G.H. (1984). Optimal foraging theory: a critical review. *Annual Review of Ecology and Systematics*, **15**, 523–75.

Qian, H., Shengbin, C., Mao, L., and Zhiyun, O. (2013). Drivers of β-diversity along latitudinal gradients revisited. *Global Ecology and Biogeography*, **22**, 659–70.

Rabosky, D.L. and Goldberg, E.E. (2015). Model inadequacy and mistaken inferences of trait-dependent speciation. *Systematic Biology*, **64**, 340–55.

Rabosky, D.L. and Huang, H. (2016). A robust semi-parametric test for detecting trait-dependent diversification. *Systematic Biology*, **65**, 181–93.

Rabosky, D.L. and Hurlbert, A.H. (2015). Species richness at continental scales is dominated by ecological limits. *American Naturalist*, **185**, 572–83.

Rabosky, D.L. and Lovette, I.J. (2008). Density-dependent diversification in North American wood warblers. *Proceedings of the National Academy of Sciences*, **275**, 2363–71.

Rabosky, D.L. and Matute, D.R. (2013). Macroevolutionary speciation rates are decoupled from the evolution of intrinsic reproductive isolation in Drosophila and birds. *Proceedings of the National Academy of Sciences*, **110**, 15345–59.

Rabosky, D.L., Title, P.O., and Huang, H. (2015). Minimal effects of latitude on present-day speciation rates in New World birds. *Proceedings of the Royal Society B*, **282**, 20142889.

Raffaelli, D. and Hall, S.R. (1996). Assessing the relative importance of trophic links in food webs. In: Polis, G.A. and Winemiller, K.O., eds. *Food Webs: Integration of Patterns and Dynamics*, pp. 185–91. Chapman and Hall, New York, NY.

Rand, T.A. and Louda, S.M. (2004). Exotic weed invasion increases the susceptibility of native plants attack by a biocontrol herbivore. *Ecology*, **85**, 1548–54.

Ranta, E., Kaitala, V., Fowler, M.S., Laakso, J., Ruokolainen, L., and O'Hara, R. (2008). Detecting compensatory dynamics in competitive communities under environmental forcing. *Oikos*, **117**, 1907–11.

Rasmussen, J.B., Rowan, D.J., Lean, R.S. and Carey, J.H. (1990). Food chain structure in Ontario lakes determines PCB levels in lake trout (*Salvelinus namaycush*) and other pelagic fish. *Canadian Journal of Fisheries and Aquatic Sciences*, **47**, 2030–8.

Rathcke, B. (1988). Competition and facilitation among plants for pollination. In: Real, L., ed. *Pollination Biology*, pp. 305–29. Academic Press, New York, NY.

Read, D.J., Koucheki, H.K. and Hodgson, J.R. (1976). Vesicular-arbuscular mycorrhiza in natural vegetation systems. I. The occurrence of infection. *New Phytologist*, **77**, 641–54.

Reed, D.H., O'Grady, J.J., Brook, B.W., Ballou, J.D., and Frankham, R. (2003). Estimates of minimum viable population sizes for vertebrates and factors influencing those estimates. *Biological Conservation*, **113**, 23–34.

Rees, M. (1995). Community structure in sand dune annuals: is seed weight a key quantity? *Journal of Ecology*, **83**, 857–64.

Rees, M., Grubb, P.J., and Kelly, D. (1996). Quantifying the impact of competition and spatial heterogeneity on the structure and dynamics of a four-species guild of winter annuals. *American Naturalist*, **147**, 1–32.

Reich, P.B., Tilman, D., Isbell, F., *et al.* (2012). Impacts of biodiversity loss escalate through time as redundancy fades. *Science*, **336**, 589–92.

Reichman, O.J. and Seabloom, E.W. (2002). Ecosystem engineering: a trivialized concept? *Trends in Ecology and Evolution*, **17**, 308.

Rejmanek, M. (2003). The rich get richer—responses. *Frontiers in Ecology and the Environment*, **1**, 122–3.

Resasco, J., Haddad, N.M., Orrock, J.L., *et al.* (2014). Landscape corridors can increase invasion by an exotic species and reduce diversity of native species. *Ecology*, **95**, 2033–9.

Rex, M.A., Crame, J.A., Stuart, C.T., and Clark, A. (2005). Large-scale biogeographic patterns in marine mollusks: A confluence of history and productivity? *Ecology*, **86**, 2288–97.

Reynolds, P.L. *et al.* (2018). Latitude, temperature, and habitat complexity predict predation pressure in eelgrass beds across the Northern Hemisphere. *Ecology*, **99**, 29–35.

Rezende, E.L., Albert, E.M., Fontuna, M.A., and Bascompte, J. (2009). Compartments in a marine food web associated with phylogeny, body mass, and habitat structure. *Ecology Letters*, **12**, 779–88.

Rezende, E.L., Lavabre, J.E., Guimarães, P.R., Jr, Jordano, P., and Bascompte, J. (2007). Non-random coextinctions in phylogenetically structured mutualistic networks. *Nature*, **448**, 925–8.

Reznick, D.N. and Ricklefs, R.E. (2009). Darwin's bridge between microevolution and macroevolution. *Nature*, **457**, 837–42.

Ricciardi, A. and Simberloff, D. (2009). Assisted colonization is not a viable conservation strategy. *Trends in Ecology and Evolution*, **24**, 248–53.

Rich, J.L. and Currie, D.J. (2018). Are North American bird species' geographic ranges mainly determined by climate? *Global Ecology and Biogeography*, **27**, 461–73.

Richardson, H. and Verbeek, N.A.M. (1986). Diet selection and optimization by Northwestern crows feeding on Japanese little-neck clams. *Ecology*, **67**, 1219–26.

Ricklefs, R.E. (1987). Community diversity: relative roles of regional and local processes. *Science*, **235**, 167–71.

Ricklefs, R.E. (2004). A comprehensive framework for global patterns in biodiversity. *Ecology Letters*, **7**, 1–15.

Ricklefs, R.E. (2006). The unified neutral theory of biodiversity: do the numbers add up? *Ecology*, **87**, 1424–31.

Ricklefs, R.E. (2007). Estimating diversification rates from phylogenetic information. *Trends in Ecology and Evolution*, **22**, 601–10.

Ricklefs, R.E. (2011). A biogeographic perspective on ecological systems: some personal reflections. *Journal of Biogeography*, **38**, 2045–56.

Ricklefs, R.E. and Jenkins. D.G. (2011). Biogeography and ecology: Towards the integration of two disciplines. *Philosophical Transactions of the Royal Society of London B*, **366**, 2438–48.

Ricklefs, R.E. and Schluter, D., editors. (1993). *Species Diversity in Ecological Communities: Historical and Geographical Perspectives*. University of Chicago Press, Chicago, IL.

Ricklefs, R.E. and Starck, J.M. (1996). Applications of phylogenetically independent contrasts: a mixed progress report. *Oikos*, **77**, 167–72.

Ricklefs, RE. and Travis, J. (1980). A morphological approach to the study of avian community organization. *Auk*, **97**, 321–38.

Ricker, W.E. (1954). Stock and recruitment. *Journal of the Fisheries Research Board of Canada*, **11**, 559–623.

Rietkerk, M., Dekker, S.C., deRuiter, P.C., and van de Koppel, J. (2004). Self-organized patchiness and catastrophic shifts in ecosystems. *Science*, **305**, 1926–9.

Ripple, W.J. and Beschta, R. (2007). Hardwood tree decline following large carnivore loss on the Great Plains. *Frontiers in Ecology and the Environment*, **5**, 241–6.

Robinson, G.R., Quinn, J.F., and Stanton, M.L. (1995). Invasibility of experimental habitat islands in a California winter annual grassland. *Ecology*, **76**, 786–94.

Roelants, K., Hass, A., and Bossuyt, F. (2011). Anuran radiations and the evolution of tadpole morphospace. *Proceedings of the National Academy of Sciences*, **108**, 8731–6.

Rohde, K. (1992). Latitudinal gradients in species diversity: the search for the primary cause. *Oikos*, **65**: 514–27.

Rohr, R.P., Saavedra, S., and Bascompte, J. (2014). On the structural stability of mutualistic systems. *Science*, **345**, 1253497.

Rolland, J., Condamine, F.L., Jiguet, F., and Morlon, H. (2014). Faster speciation and reduced extinction in the tropics contribute to the mammalian latitudinal diversity gradient. *PLoS Biology*, **12**, e1001775.

Rominger, A.J., *et al.* (2016). Community assembly on isolated islands: macroecology meets evolution. *Global Ecology and Biogeography*, **25**, 769–80.

Rooney, N. and McCann, K.S. (2012). Integrating food web diversity, structure and stability. *Trends in Ecology and Evolution*, **27**, 40–6.

Rooney, N., McCann, K.S., Gellner, G., and Moore, J.C. (2006). Structural asymmetry and stability of diverse food webs. *Nature*, **442**, 265–9.

Rosà R., Rizzoli, A., Ferrari, N., and Pugliese, A. (2006). Models for host–macroparasite interactions in micromammals. In: Morand, S., Krasnov, B.R., and Poulin, R., eds. *Micromammals and Macroparasites: From Evolutionary Ecology to Management*, pp. 319–48. Springer-Verlag, Berlin.

Rosemond, A.D., Mulholland, P.J., and Elwood, J.W. (1993). Top-down and bottom-up control of stream periphyton—effects of nutrients and herbivores. *Ecology*, **74**, 1264–80.

Rosenzweig, M.L. (1971). Paradox of enrichment: Destabilization of exploitation ecosystems in ecological time. *Science*, **171**, 385–7.

Rosenzweig, M.L. (1973). Exploitation in three trophic levels. *American Naturalist*, **107**, 275–94.

Rosenzweig, M.L. (1977). Aspects of biological exploitation. *Quarterly Review of Biology*, **52**, 371–80.

Rosenzweig, M.L. (1991). Habitat selection and population interactions: the search for mechanism. *American Naturalist*, 137(Supplement), S5–S28.

Rosenzweig, M.L. (1995). *Species Diversity in Space and Time*. Cambridge University Press, Cambridge.

Rosenzweig, M.L. and Abramsky, Z. (1986). Centrifugal community organization. *Oikos*, **46**, 339–48.

Rosenzweig, M.L. and MacArthur, R.H. (1963). Graphical representation and stability conditions of predator–prey interactions. *American Naturalist*, **97**, 209–23.

Rosindell, J., Cornell, S.J., Hubbell, S.P., and Etienne, R.S. (2010). Protracted speciation revitalizes the neutral theory of biodiversity. *Ecology Letters*, **13**, 716–27.

Roslin, T., *et al.* (2017). Higher predation risk for insect prey at low latitudes and elevations. *Science*, **356**, 742–4.

Rossberg, A.G., Matsuda, H., Amemiya, T., and Itoh, K. (2006). Some properties of the speciation model for food-web structure—mechanisms for degree distribution and intervality. *Journal of Theoretical Biology*, **238**, 401–15.

Rothhaupt, K.O. (1988). Mechanistic resource competition theory applied to laboratory experiments with zooplankton. *Nature*, **333**, 660–2.

Rothhaupt, K.O. (1996). Laboratory experiments with a mixotrophic chrysophyte and obligately phagotrophic and phototrophic competitors. *Ecology*, **77**, 716–24.

Roughgarden, J. (1974). Species packing and the competition function with illustrations from coral reef fish. *Theoretical Population Biology*, **5**, 163–86.

Roughgarden, J. (1975). Evolution of a marine symbiosis—a simple cost–benefit model. *Ecology*, **56**, 1201–8.

Rowe, L. and Ludwig, D. (1991). Size and timing of metamorphosis in complex life cycles: time constraints and variation. *Ecology*, **72**, 413–27.

Roxburgh, S.H., Shea, K., and Wilson, J.B. (2004). The intermediate disturbance hypothesis: Patch dynamics and mechanisms of species coexistence. *Ecology*, **85**, 359–71.

Roy, K., Jablonski, D. and Valentine, J.W. (1996). Higher taxa in biodiversity studies: patterns from Eastern Pacific marine molluscs. *Philosophical Transactions of the Royal Society of London B*, **351**, 1605–13.

Rudolf, V.H.W. and Lafferty, K.D. (2011). Stage structure alters how complexity affects stability of ecological networks. *Ecology Letters*, **14**, 75–9.

Rudman, S.M. and Schluter, D. (2016). Ecological impacts of reverse speciation in threespine stickleback. *Current Biology*, **26**, 490–5.

Runkle, J.R. (1989). Synchrony of regeneration, gaps, and latitudinal differences in tree species diversity. *Ecology*, **79**, 546–7.

Saavedra, S., Rohr, R.P., Olesen, J.M., and Bascompte, J. (2016). Nested species interactions promote feasibility over stability during the assembly of a pollinator community. *Ecology and Evolution*, **6**, 997–1007.

Sachs, J.L., Mueller, U.G., Wilcox, T.P., and Bull, J.J. (2004). The evolution of cooperation. *Quarterly Review of Biology*, **79**, 136–60.

Sala, E. and Graham, M.H. (2002). Community-wide distribution of predator–prey interaction strength in kelp forests. *Proceedings of the National Academy of Sciences*, **99**, 3678–83.

Sale, P.F. (1977). Maintenance of high diversity in coral reef fish communities. *American Naturalist*, **111**, 337–59.

Samani, P. and Bell, G. (2010). Adaptation of experimental yeast populations to stressful conditions in relation to population size. *Journal of Evolutionary Biology*, **23**, 791–6.

Sandel, B., Arge, L., Dalsgaard, B., *et al.* (2011). The influence of late quaternary climate-change velocity on species endemism. *Science*, **334**, 660–4.

Sanders, H.L. (1969). Benthic marine diversity and the stability-time hypothesis. In: Woodwell, G.M. and Smith, H.H., eds. *Diversity and Stability in Ecological Systems*, Brookhaven Symposium no. 22, pp. 71–81. US Department of Commerce, Springfield, VA.

Sarnelle, O. and Knapp, R.A. (2004). Zooplankton recovery after fish removal: limitations of the egg bank. *Limnology and Oceanography*, **49**, 1382–92.

Sarnelle, O. and Wilson, A.E. (2008). Type III functional response in *Daphnia*. *Ecology*, **89**, 1723–32.

Sauve, A.M., Thébault, E., Pocock, M.J.O., and Fontaine, C. (2016). How plants connect pollination and herbivory networks and their contribution to community stability. *Ecology*, **97**, 908–17.

Schaffer, W.M. and Kot, M. (1985). Do strange attractors govern ecological systems? *BioScience*, **35**, 342–50.

Scheffer, M. (1998). *Ecology of Shallow Lakes*. Chapman and Hall, London.

Scheffer, M. (2009). *Critical Transitions in Nature and Society*. Princeton University Press, Princeton, NJ.

Scheffer, M. and Carpenter, S.R. (2003). Catastrophic regime shifts in ecosystems: linking theory to observation. *Trends in Ecology and Evolution*, **18**, 648–56.

Scheffer, M., Carpenter, S.R., Foley J.A., Folke, C., and Walker, B. (2001). Catastrophic shifts in ecosystems. *Nature*, **413**, 591–6.

Scheffer, M. and Van Nes, E.H. (2007). Shallow lakes theory revisited: various alternative regimes driven by climate, nutrients, depth and lake size. *Hydrobiologia*, **584**, 455–66.

Scheffer, M., *et al.* (2009). Early-warning signals for critical transitions. *Nature*, **461**, 53–9.

Schemske, D.W. (1981). Floral convergence and pollinator sharing in two bee-pollinated tropical herbs. *Ecology*, **62**, 946–54.

Schemske, D.W. (2002). Tropical diversity: patterns and processes. In: Chazdon, R., and Whitmore, T., eds. *Ecological and Evolutionary Perspectives on the Origins of Tropical Diversity: Key Papers and Commentaries*, pp. 163–73. University of Chicago Press, Chicago, IL.

Schemske, D.W., Mittelbach, G.G., Cornell, H.V., Sobel, J.M., and Roy, K. (2009). Is there a latitudinal gradient in the importance of biotic interactions? *Annual Review of Ecology, Evolution, and Systematics*, **40**, 245–69.

Scherer-Lorenzen, M., Palmborg, C., Prinz, A., and Schulze, E.D. (2003). The role of plant diversity and composition for nitrate leaching in grasslands. *Ecology*, **84**, 1539–52.

Schindler, D.E., Armstrong, J.B., and Reed, T.E. (2015). The portfolio concept in ecology and evolution. *Frontiers in Ecology and the Environment*, **13**, 257–63.

Schindler, D.E., Hilborn, R., Chasco, B., *et al.* (2010). Population diversity and the portfolio effect in an exploited species. *Nature*, **465**, 609–12.

Schlesinger, W.H., Reynolds, J.F., Cunningham G.L., *et al.* (1990). Biological feedbacks in global desertification. *Science*, **247**, 1043–8.

Schleuning, M., *et al.* (2014). Ecological, historical and evolutionary determinants of modularity in weighted seed-dispersal networks. *Ecology Letters*, **17**, 454–63.

Schluter, D. (1990). Species-for-species matching. *American Naturalist*, **136**, 560–8.

Schluter D. (1994). Experimental evidence that competition promotes divergence in adaptive radiation. *Science*, **266**, 798–801.

Schluter, D. (2000). *The Ecology of Adaptive Radiation*. Oxford University Press, New York, NY.

Schluter, D. (2016). Speciation, ecological opportunity, and latitude. *American Naturalist*, **187**, 1–18

Schluter, D. and Grant, P.R. (1984). Determinants of morphological patterns in communities of Darwin's finches. *American Naturalist*, **123**, 175–96.

Schluter, D. and Pennell, M.W. (2017). Speciation gradients and the distribution of biodiversity. *Nature*, **546**, 48–55.

Schluter, D., Price, T.D., and Grant, P.R. (1985). Ecological character displacement in Darwin finches. *Science*, **227**, 1056–9.

Schmid, B. (2002). The species richness–productivity controversy. *Trends in Ecology and Evolution*, **17**, 113–14.

Schmitt, R.J. (1987). Indirect interactions between prey: apparent competition, predator aggregation, and habitat segregation. *Ecology*, **68**, 1887–97.

Schmitt, R.J. and Holbrook, S.J. (2003). Mutualism can mediate competition and promote coexistence. *Ecology Letters*, **6**, 898–902.

Schmitz, O. (1998). Direct and indirect effects of predation and predation risk in old-field interactions webs. *American Naturalist*, **151**, 327–42.

Schmitz, O. (2003). Top predator control of plant biodiversity and productivity in an old-field ecosystem. *Ecology Letters*, **6**, 156–63.

Schmitz, O. (2004). Perturbation and abrupt shift in trophic control of biodiversity and productivity. *Ecology Letters*, **7**, 403–9.

Schmitz, O. (2006). Predators have large effects on ecosystem properties by changing plant diversity not plant biomass. *Ecology*, **87**, 1432–7.

Schmitz, O. (2008). Effects of predator hunting mode on grassland ecosystem function. *Science*, **319**, 952–4.

Schmitz, O. (2010). *Resolving Ecosystem Complexity*. Princeton University Press, Princeton, NJ.

Schmitz, O., Hambäck, P.A., and Beckerman, A.P. (2000). Trophic cascades in terrestrial systems: a review of the effects of carnivore removals on plants. *American Naturalist*, **155**, 141–53.

Schmitz, O., Křivan, V., and Ovadia, O. (2004). Trophic cascades: The primacy of trait-mediated indirect interactions. *Ecology Letters*, **7**, 153–63.

Schnitzer, S.A., *et al.* (2011). Soil microbes drive the classic plant diversity–productivity pattern. *Ecology*, **92**, 296–303.

Schoener, T.W. (1971). Theory of feeding strategies. *Annual Review of Ecology and Systematics*, **2**, 369–404.

Schoener, T.W. (1974). Competition and the form of habitat shift. *Theoretical Population Biology*, **6**, 265–307.

Schoener, T.W. (1976). Alternative to Lotka-Volterra competition: models of intermediate complexity. *Theoretical Population Biology*, **10**, 309–33.

Schoener, T.W. (1986). Kinds of ecological communities: ecology becomes pluralistic. In: Diamond, J. and Case, T.J., eds. *Community Ecology*, pp. 467–79. New York: Harper & Row.

Schoener, T.W. (1989). Food webs from the small to the large. *Ecology*, **70**, 1559–89.

Schoener, T.W. (1993). On the relative importance of direct versus indirect effects in ecological communities. In: Kawanabe, H., Cohen, J.E., and Iwasaki, K., eds. *Mutualism and Community Organization: Behavioral, Theoretical and Food Web Approaches*, pp. 365–411. Oxford University Press, Oxford.

Schoener, T.W. (2011). The newest synthesis: understanding the interplay of evolutionary and ecological dynamics. *Science*, **331**, 426–9.

Schoener, T.W. and Spiller, D. (1999). Indirect effects in an experimentally staged invasion by a major predator. American Naturalist **153**: 347–58.

Schröder, A., Persson, L., and De Roos, A.M. (2005). Direct experimental evidence for alternative stable states: a review. *Oikos*, **110**: 3–19.

Schtickzelle, N. and Quinn, T.P. (2007). A metapopulation perspective for salmon and other anadromous fish. *Fish and Fisheries*, **8**, 297–14.

Schupp, E.W. (1995). Seed–seedling conflicts, habitat choice, and patterns of plant recruitment. *American Journal of Botany*, **82**, 399–409.

Schwartz, M.W. and Hoeksema, J.D. (1998). Specialization and resource trade: biological markets as a model of mutualisms. *Ecology*, **79**, 1029–38.

Schwarzer, G., Carpenter, J.R., and Rücker, G. (2015). *Meta-Analysis with R*. Springer, Berlin.

Scurlock, J.M.O. and Olson, R.J. (2002). Terrestrial net primary productivity—a brief history and a new worldwide database. *Environmental Reviews*, **10**, 91–109.

Seabloom, E.W., Harpole, W.S., Reichman, O.J., and Tilman, D. (2003). Invasion, competitive dominance, and resource use by exotic and native California grassland species. *Proceedings of the National Academy of Sciences*, **100**, 13384–9.

Seabloom, E.W., Kinkel, L., Borer, E.T., Hautier, Y., Montgomery, R.A., and Tilman, D. (2017). Food webs obscure the strength of plant diversity effects on primary production. *Ecology Letters*, **20**, 505–12.

Sears, A.L.W. and Chesson, P.L. (2007). New methods for quantifying the spatial storage effect: an illustration with desert annuals. *Ecology*, **88**, 2240–7.

Seehausen, O. (2006). African cichlid fish: a model system in adaptive radiation research. *Proceedings of the Royal Society of London B*, **273**, 1987–98.

Seifert, R.P. and Seifert, F.H. (1976). A community matrix analysis of *Heliconia* insect communities. *American Naturalist*, **110**, 461–83.

Sexton, J.P., McIntyre, P.J., Angert, A.L., and Rice, K.J. (2009). Evolution and ecology of species range limits. *Annual Review of Ecology, Evolution, and Systematics*, **40**, 415–36.

Shea, K., Roxburgh, S.H., and Rauschert, E.S.J. (2004). Moving from pattern to process: Coexistence mechanisms under intermediate disturbance regimes. *Ecology Letters*, **7**, 491–508.

Shelford, V.E. (1913). *Animal Communities in Temperate America*. University of Chicago Press, Chicago, IL.

Shiono, T., Kusumoto, B., Yasuhara, M., and Kubota, Y. (2018). Roles of climate niche conservatism and range dynamics in woody plant diversity pattern through the Cenozoic. *Global Ecology and Biogeography*, **27**, 865–74.

Shipley, B. and Dion, J. (1992). The allometry of seed production in herbaceous angiosperms. *American Naturalist*, **139**, 467–83.

Shipley, B., Vile, D., and Garnier, E. (2006). From plant traits to plant communities: A statistical mechanistic approach to biodiversity. *Science*, **314**, 812–14.

Shmida, A. and Ellner, S.P. (1984). Coexistence of plant species with similar niches. *Vegetatio*, **58**, 29–55.

Shmida, A. and Wilson, M.V. (1985). Biological determinants of species diversity. *Journal of Biogeography*, **12**, 1–20.

Shorrocks, B. and Sevenster, J.G. (1995). Explaining local species diversity. *Proceedings of the Royal Society of London, B*, **260**, 305–9

Shreve, F. (1951). *Vegetation of the Sonoran Desert*, Publication no. 591. Carnegie Institution, Washington, DC.

Shurin, J.B. (2000). Dispersal limitation, invasion resistance, and the structure of pond zooplankton communities. *Ecology*, **81**, 3074–86.

Shurin, J.B., Cottenie, K., and Hillebrand, H. (2009). Spatial autocorrelation and dispersal limitation in freshwater organisms. *Oecologia*, **159**, 151–9.

Shurin, J.B., Havel, J.E., Leibold, M.A., and Pinel-Alloul, B. (2000). Local and regional zooplankton species richness: a scale-independent test for saturation. *Ecology*, **81**, 3062–73.

Shurin, J.B., Markel, R.W., and Matthews, B. (2010). Comparing trophic cascades across ecosystems. In: Terborgh, J., and Estes, J.A., eds. *Trophic Cascades: Predators, Prey, and the Changing Dynamics of Nature*, pp. 319–35. Island Press, Washington, DC.

Shurin, J.B. and Srivastava, D.S. (2005). New perspectives on local and regional diversity: beyond saturation. in Holyoak, M., Leibold, M.A., and Holt, R.D., eds. *Metacommunities: Spatial Dynamics and Ecological Communities*, pp. 399–417. University of Chicago Press, Chicago, IL.

Shurin, J.B., *et al.* (2002). A cross-ecosystem comparison of the strength of trophic cascades. *Ecology Letters*, **5**, 785–91.

Sibly, R.M., Barker, D., Denham, M.C., Hone, J., and Pagel, M. (2005). On the regulation of populations of mammals, birds, fish, and insects. *Science*, **309**, 607–10.

Sibly, R.M. and Hone, J. (2002). Population growth rate and its determinants: an overview. *Philosophical Transactions of the Royal Society of London B*, **357**, 1153–70.

Siepielski, A.M., Hung, K-L., Bein, E.E.B., and McPeek, M.A. (2010). Experimental evidence for neutral community dynamics governing an insect assemblage. *Ecology*, **91**, 847–57.

Sih, A. and Christensen, B. (2001). Optimal diet theory: When does it work and why does it fail? *Animal Behaviour*, **61**, 379–90.

Silvertown, J., Holtier, S., Johnson, J., and Dale, P. (1992). Cellular automaton models of interspecific competition for space—the effect of pattern on process. *Journal of Ecology*, **80**, 527–33.

Silvertown, J., McConway, K., Gowing, D., *et al.* (2006). Absence of phylogenetic signal in the niche structure of meadow plant communities. *Proceedings of the Royal Society of London B*, **273**, 39–44.

Simberloff, D. (1976). Experimental zoogeography of islands: effects of island size. Ecology, **57**, 629–48.

Simberloff, D. and Conner, E.F. (1981). Missing species combinations. *American Naturalist*, **118**, 215–39.

Simberloff, D. and Cox, J. (1987). Consequences and costs of conservation corridors. *Conservation Biology*, **1**, 63–71.

Simberloff, D., Farr, J.A., Cox, J., and Mehlman, D.W. (1992). Movement corridors: conservation bargains or poor investments? *Conservation Biology*, **6**, 493–504.

Simms, E.L. and Taylor, D.L. (2002). Partner choice in nitrogen-fixation mutualisms of legumes and rhizobia. *Integrative and Comparative Biology*, **42**, 369–80.

Simms, E.L., Taylor D.L., Povich J., *et al.* (2006). An empirical test of partner choice mechanisms in a wild legume–rhizobium interaction. *Proceedings of the Royal Society of London B*, **273**, 77–81.

Simon, T.N., *et al.* (2017). Local adaptation in Trinidadian guppies alters stream ecosystem structure at landscape scales despite high environmental variability. *Copeia*, **105**, 504–13.

Šímová, I., Storch, D., Keil, P., Boyle, B., Phillips, O.L., and Enquist, B.J. (2011). Global species-energy relationship in forest plots: Role of abundance, temperature and species climatic tolerances. *Global Ecology and Biogeography*, **20**, 842–56.

Šímová, I., Violle, C., Kraft, N.J., *et al.* (2015). Shifts in trait means and variances in North American tree assemblages: species richness patterns are loosely related to the functional space. *Ecography*, **38**, 649–58.

Simpson, G.G. (1953). *The Major Features of Evolution*. Columbia University Press, New York, NY.

Sinclair, T.R.E., Mduma, S., and Brashares, J.S. (2003). Patterns of predation in a diverse predator–prey system. *Nature*, **425**, 288–90.

Sites, J.W. Jr. (2013). Extinction, reintroduction, and restoration of a lizard meta-population equilibrium in the Missouri ozarks. *Molecular Ecology*, **22**, 3653–5.

Sjögren, P. (1988). Metapopulation biology of *Rana lessonae* Camerano on the northern periphery of its range. *Acta Universitatis Upsaliensis*, **157**, 1–35.

Sjögren Gulve, P. (1991). Extinctions and isolation gradients in metapopulations: the case of the pool frog (*Rana lessonae*). *Biological Journal of the Linnean Society*, **42**, 135–47.

Skellam, J.G. (1951). Random dispersal in theoretical populations. *Biometrika*, **38**, 196–218.

Slatkin, M. (1993). Isolation by distance in equilibrium and non-equilibrium populations. *Evolution*, **47**, 264–79.

Slik, J.W.F., *et al.* (2015). An estimate of the number of tropical tree species. *Proceedings of the National Academy of Sciences*, **112**, 7472–7.

Slobodkin, L.B. (1961). *Growth and Regulation of Animal Populations*. Holt, Rinehart and Winston, New York, NY.

Sluijs, A., *et al.* (2006). Subtropical Arctic Ocean temperatures during the Palaeocene/Eocene thermal maximum. *Nature*, **441**, 610–13.

Smith, F.A., Grace, E.J., and Smith, S.E. (2009). More than a carbon economy: nutrient trade and ecological sustainability in facultative arbuscular mycorrhizal symbiosis. *New Phytologist*, **182**, 347–58.

Smith, M.D. and Knapp, A.K. (2003). Dominant species maintain ecosystem function with non-random species loss. *Ecology Letters*, **6**, 509–17.

Soininen, J. (2014). A quantitative analysis of species sorting across organisms and ecosystems. *Ecology*, **95**, 3284–92.

Soininen, J., McDonald, R., and Hillebrand, H. (2007). The distance decay of similarity in ecological communities. *Ecography*, **30**, 3–12.

Solan, M., Cardinale, B.J., Downing, A.L., Engelhardt, K.A.M., Ruesink, J.L., and Srivastava, D.S. (2004). Extinction and ecosystem function in the marine benthos. *Science*, **306**, 1177–80.

Sommer, U. (1986). Phytoplankton competition along a gradient of dilution rates. *Oecologia*, **68**, 503–6.

Sommer, U. (1989). The role of resource competition in phytoplankton succession. In Sommer, U., ed. *Plankton Ecology: Succession in Plankton Communities*, pp. 57–170. Springer, New York, NY.

Song, C., Rohr, R.P., and Saavedra, S. (2017). Why are some plant-pollinator networks more nested than others? *Journal of Animal Ecology*, 86, 1417–24.

Sousa, W.P. (1979a). Disturbance in marine intertidal boulder fields: the nonequilibrium maintenance of species diversity. *Ecology*, **60**, 1225–39.

Sousa, W.P. (1979b). Experimental investigations of disturbance and ecological succession in a rocky intertidal community. *Ecological Monographs*, **49**, 227–54.

Spehn, E.M., *et al.* (2002). The role of legumes as a component of biodiversity in a cross-European study of grassland biomass nitrogen. *Oikos*, **98**, 205–18.

Spiller, D. and Schoener, T.W. (1990). A terrestrial field experiment showing the impact of eliminating top predators on foliage damage. *Nature*, **347**, 469–72.

Springer, A., *et al.* (2003). Sequential megafaunal collapse in the North Pacific Ocean: an ongoing legacy of industrial whaling? *Proceedings of the National Academy of Sciences*, **100**, 12233–28.

Srinivasan, U.T., Dunne, J.A., Harte, J., and Martinez, N.D. (2007). Response of complex food webs to realistic extinction sequences. *Ecology*, **88**, 671–82.

Stachowicz, J.J., Bruno, J.F., and Duffy, J.E. (2007). Understanding the effects of marine biodiversity on communities and ecosystems. *Annual Review of Ecology, Evolution, and Systematics*, **38**, 739–66.

Stachowicz, J.J., Fried, H., Osman, R.W., and Whitlatch, R.B. (2002). Biodiversity, invasion resistance, and marine ecosystem function: reconciling pattern and process. *Ecology*, **83**, 2575–90.

Stachowicz, J.J., Whitlatch, R.B., and Osman, R.W. (1999). Species diversity and invasion resistance in a marine ecosystem. *Science*, **286**, 1577–8.

Stebbins, G.L. (1974). *Flowering Plants: Evolution above the Species Level*. Belknap Press of Harvard University Press, Cambridge, MA.

Steiner, C.F. (2001). The effects of prey heterogeneity and consumer identity on the limitation of trophic-level biomass. *Ecology*, **82**, 2495–506.

Steiner, C.F., Long, Z.T., Krumins, J.A., and Morin, P.J. (2006). Population and community resilience in multitrophic communities. *Ecology*, **87**, 996–1007.

Steneck, R.S., Grahman, M.H., Bourque B.J., *et al.* (2002). Kelp forest ecosystems: biodiversity, stability, resilience and future. *Environmental Conservation*, **29**, 436–59.

Stephens, D.W., Brown, J.S., and Ydenberg, R.C. (2007). *Foraging: Behavior and Ecology*. University of Chicago Press, Chicago, IL.

Stephens, D.W. and Krebs, J.R. (1986). *Foraging Theory*. Princeton University Press, Princeton, NJ.

Stephens, P.A., Sutherland, W.J., and Freckleton, R.P. (1999). What is the Allee effect? *Oikos*, **87**, 185–90.

Sterner, R.W. and Elser, J.J. (2002). *Ecological Stoichiometry*. Princeton University Press, Princeton, NJ.

Stewart, B.S., Yochem P.K., LeBoeuf B.J., *et al.* (1994). History and present status of the northern elephant seal population. In Le Boeuf, B.J. and Laws, R.M., eds. *Elephant Seals: Population Ecology, Behavior, and Physiology*, pp. 29–48. University of California Press, Berkeley, CA.

Stich, H.B. and Lampert, W. (1981). Predator evasion as an explanation of diurnal vertical migration by zooplankton. *Nature*, **293**, 396–8.

Stohlgren, T.J., Flather, C., Jarnevich, C.S., Barnett, D.T., and Kartesz, J. (2008). Rejoinder to Harrison (2008): the myth of plant species saturation. *Ecology Letters*, **11**, 324–6.

Stohlgren, T.J., *et al.* (1999). Exotic plant species invade hot spots of native plant diversity. *Ecological Monographs*, **69**, 25–46.

Stoll, P. and Prati, D. (2001). Intraspecific aggregation alters competitive interactions in experimental plant communities. *Ecology*, **82**, 319–27.

Stomp, M., Huisman, J., De Jongh, F., *et al.* (2004). Adaptive divergence in pigment composition promotes phytoplankton biodiversity. *Nature*, **432**, 104–7.

Stomp, M., Huisman, J., Mittelbach, G.G., Litchman, E., and Klausmeier, C.A. (2011). Large-scale biodiversity patterns in freshwater phytoplankton. *Ecology*, **92**, 2096–107.

Storch, D., Bohdalková, E., and Okie, J. (2018). The more-individuals hypothesis revisited: the role of community abundance in species richness regulation and the productivity-diversity relationship. *Ecology Letters*, **21**(6), 920–37.

Storch, D., *et al.* (2006). Energy, range dynamics and global species richness patterns: reconciling mid-domain effects and environmental determinants of avian diversity. *Ecology Letters*, **9**, 1308–20.

Stouffer, D.B. (2010). Scaling from individuals to networks in food webs. *Functional Ecology*, **24**, 44–51.

Stouffer, D.B. and Bascompte, J. (2011). Compartmentalization increases food-web persistence. *Proceedings of the National Academy of Sciences*, **108**, 3648–52.

Strauss, S.Y. (2014). Ecological and evolutionary responses in complex communities: implications for invasions and eco-evolutionary feedbacks. *Oikos*, **123**, 257–66.

Strauss, S.Y., Lau, J.A., and Carroll, S.P. (2006a). Evolutionary responses of natives to introduced species: what do introductions tell us about natural communities? *Ecology Letters*, **9**, 357–74.

Strauss, S.Y., Lau, J.A., Schoener, T.W., and Tiffin, P. (2008). Evolution in ecological field experiments: implications for effect size. *Ecology Letters*, **11**, 199–207.

Strauss, S.Y., Webb, C.O., and Salamin, N. (2006b). Exotic taxa less related to native species are more invasive. *Proceedings of the National Academy of Sciences*, **103**, 5841–5.

Strong, D.R. (1986). Density vagueness: abiding the variance in the demography of real populations. In: Diamond, J. and Case, T.J., eds. *Community Ecology*, pp. 257–68. Harper and Row, New York, NY.

Strong, D.R. (1992). Are trophic cascades all wet? Differentiation and donor-control in speciose ecosystems. *Ecography*, **73**, 747–54.

Strong, D.R., Szyska, L.A., and Simberloff, D. (1979). Test of community-wide character displacement against null hypotheses. *Evolution*, **33**, 897–913.

Suding, K.N., Gross, K.L., and Houseman, G.R. (2004). Alternative states and positive feedbacks in restoration ecology. *Trends in Ecology and Evolution*, **19**, 46–53.

Suding, K.N. and Hobbs, R.J. (2009). Threshold models in restoration and conservation: a developing framework. *Trends in Ecology and Evolution*, **24**, 271–9.

Suding, K.N., Miller, A.E., Bechtold, H., and Bowman, W.D. (2006). The consequence of species loss to ecosystem nitrogen cycling depends on community compensation. *Oecologia*, **149**, 141–9.

Sugihara, G. (1980). Minimal community structure: an explanation of species abundance patterns. *American Naturalist*, **116**, 770–87.

Supp, S.R., Koons, D.N., and Ernest, S. (2015). Using life history trade-offs to understand core-transient structuring of a small mammal community. *Ecosphere*, **6**, 1–15.

Sutherland, J.P. (1974). Multiple stable points in natural communities. *American Naturalist*, **108**, 859–73.

Swenson, N.G. (2011). The role of evolutionary processes in producing biodiversity patterns, and the interrelationships between taxonomic, functional and phylogenetic biodiversity. *American Journal of Botany*, **98**, 472–80.

Swenson, N.G. and Enquist, B.J. (2007). Ecological and evolutionary determinants of a key plant functional trait: wood density and its community-wide variation across latitude and elevation. *American Journal of Botany*, **94**, 451–95.

Swenson, N.G., Enquist, B.J., Pither, J., *et al.* (2012). The biogeography and filtering of woody plant functional diversity in North and South America. *Global Ecology and Biogeography*, **21**, 798–808.

Swenson, N.G., Enquist, B.J., Pither, J., Thompson, J., and Zimmerman, J.K. (2006). The problem and promise of scale dependency in community phylogenetics. *Ecology*, **87**, 2418–24.

Swenson, N.G., Enquist, B.J., Thompson, J., and Zimmerman, J.K. (2007). The influence of spatial and size scales on phylogenetic relatedness in tropical forest communities. *Ecology*, **88**, 1770–80.

Swenson, N.G. and Weiser, M.D. (2014). On the packing and filling of functional space in eastern American tree assemblages. *Ecography*, **37**, 1056-62.

Swenson, N.G., Weiser, M.D., Mao, L., *et al.* (2016). Constancy in functional space across a species richness anomaly. *American Naturalist*, **187**, E83–92.

Symstad, A.J. and Tilman, D. (2001). Diversity loss, recruitment limitation, and ecosystem functioning: lessons learned from a removal experiment. *Oikos*, **82**, 424–35.

Symstad, A.J., *et al.* (2003). Long-term and large-scale perspectives on the relationship between biodiversity and ecosystem functioning. *BioScience*, **53**, 89–98.

Szava-Kovats, R.C., Ronk, A., and Pärtel, M. (2013). Pattern without bias: local–regional richness relationship revisited. *Ecology*, **94**, 1986–92.

Takimoto, G. and Post, D.M. (2013). Environmental determinants of food-chain length: a meta-analysis. *Ecological Research*, **28**, 675–81.

Takimoto, G., Spiller, D., and Post, D.M. (2008). Ecosystem size, but not disturbance, determines food-chain length on islands of the Bahamas. *Ecology*, **89**, 3001–7.

Takimoto, G., Post, D.M., Spiller, D.A., and Holt, R.D. (2012). Effects of productivity, disturbance, and ecosystem size on food-chain length: insights from a metacommunity model of intraguild predation. *Ecological Research*, **27**, 481–93.

Tanner, J., Hughes, T.P., and Connell, J.H. (2009). Community-level density dependence: an example from a shallow coral assemblage. Ecology, **90**: 506–16.

Tansley, A.G. (1923). *Practical Plant Ecology: A Guide for Beginners in Field Study of Plant Communities*. Allen and Unwin, London.

Tansley, A.G. (1939). *The British Islands and Their Vegetation*. Cambridge University Press, Cambridge.

Taylor, B.W. and Irwin, R.E. (2004). Linking economic activities to the distribution of exotic plants. *Proceedings of the National Academy of Sciences*, **101**, 17725–30.

Tegner, M.J. and Dayton, P.K. (1991). Sea urchins, El Niños, and the long term stability of southern California kelp forest communities. *Marine Ecology Progress Series*, **77**, 49–63.

Templeton, A.R., Brazeal, H., and Neuwald, J.L. (2011). The transition from isolated patches to a metapopulation in the eastern collared lizard in response to prescribed fires. *Ecology*, **92**, 1736–47.

Temperton, V.M., Mwangi, P.N., Scherer-Lorenzen, M., Schmid, B., and Buchmann, N. (2007). Positive interactions between nitrogen-fixing legumes and four different neighboring species in a biodiversity experiment. *Oecologia*, **151**, 190–205.

Templeton, A.R., Shaw, K., Routman, E., and Davis, S.K. (1990). The genetic consequences of habitat fragmentation. *Annuals of the Missouri Botanical Garden*, **77**, 13–27.

Terborgh, J. (1973). On the notion of favorableness in plant ecology. *American Naturalist*, **107**, 481–501.

Terborgh, J. and Estes, J.A., editors. (2010). *Trophic Cascades: Predators, Prey, and the Changing Dynamics of Nature*. Island Press, Washington, DC.

Terborgh, J., Foster, R.B., and Nuñez, V. (1996). Tropical tree communities: a test of the nonequilibrium hypothesis. *Ecology*, **77**, 561–7.

Terborgh, J., Holt, R.D., and Estes, J.A. (2010). Trophic cascades: what they are, how they work, and why they matter. In: Terborgh, J. and Estes, J.A., eds. *Trophic Cascades: Predators, Prey, and the Changing Dynamics of Nature*, pp. 1–18. Island Press, Washington DC.

Terborgh, J., *et al.* (2001). Ecological meltdown in predator-free forest fragments. *Science*, **294**, 1923–6.

terHorst, C.P., Miller, T.E., and Levitan, D.R. (2010). Evolution of prey in ecological time reduces the effect size of predators in experimental microcosms. *Ecology*, **91**, 629–36.

Ter Steege, H., Pitman, N.C., Phillips, O.L., *et al.* (2006). Continental-scale patterns of canopy tree composition and function across Amazonia. *Nature*, **443**, 444–7.

Terry, J.C.D., Morris, R.J., and Bonsall, M.B. (2017). Trophic interaction modifications: an empirical and theoretical framework. *Ecology Letters*, **20**, 1219–30.

Tewksbury, J.J. and Lloyd, J.D. (2001). Positive interactions under nurse-plants: spatial scale, stress gradients and benefactor size. *Oecologia*, **127**, 425–34.

Thébault, E. and Fontaine, C. (2010). Stability of ecological communities and the architecture of mutualistic and trophic networks. *Science*, **329**, 853–6.

Thebault, E., Huber, V., and Loreau, M. (2007). Cascading extinctions and ecosystem functioning: contrasting effects of diversity depending on food web structure. *Oikos*, **116**, 163–73.

Thierry, A., Petchey, O.L., Beckerman, A.P., Warren, P.H., and Williams, R.J. (2011). The consequences of size dependent foraging for food web topology. *Oikos*, **120**, 493–502.

Thomas, C.D., Bodsworth E.J., Wilson R.J., *et al.* (2001). Ecological and evolutionary processes at expanding range margins. *Nature*, **411**, 577–81.

Thompson, J.N. (1999). Specific hypotheses on the geographic mosaic of coevolution. *American Naturalist*, **153**, S1–S14.

Thompson, K., Rickard, L.C., Hodkinson, D.J., and Rees, M. (2002). Seed dispersal, the search for trade-offs. In: Bullock, J.M., Kenward, R.E., and Hails, R.S., eds. *Dispersal Ecology*, pp. 152–72. Blackwell, Oxford.

Thomson, J.D. (1978). Effects of stand composition on insect visitation in two-species mixtures of *Hieracium*. *American Midland Naturalist*, **100**, 431–40.

Thrall, P.H., Hochberg, M.E., Burdon, J.J., and Bever, J.D. (2007). Coevolution of symbiotic mutualists and parasites in a community context. *Trends in Ecology and Evolution*, **22**, 120–6.

Thuiller, W., Lavorel, S., Araújo, M.B., Sykes, M.T., and Prentice, I.C. (2005). Climate change threats to plant diversity in Europe. *Proceedings of the National Academy of Sciences*, **102**, 8245–50.

Tielbörger, K. and Kadmon, R. (2000). Temporal environmental variation tips the balance between facilitation and interference in desert plants. *Ecology*, **81**, 1544–53.

Tilman, D. [published as D. Titman]. (1976). Ecological competition between algae: experimental confirmation of resource-based competition theory. *Science*, **192**, 463–6.

Tilman, D. (1977). Resource competition between planktonic algae: experimental and theoretical approach. *Ecology*, **58**, 338–48.

Tilman, D. (1980). A graphical-mechanistic approach to competition and predation. *American Naturalist*, **116**, 362–93.

Tilman, D. (1981). Tests of resource competition theory using four species of Lake Michigan algae. *Ecology*, **62**, 802–25.

Tilman, D. (1982). *Resource Competition and Community Structure*. Princeton University Press, Princeton, NJ.

Tilman, D. (1994). Competition and biodiversity in spatially structured habitats. *Ecology*, **75**, 2–16.

Tilman, D. (2001). An evolutionary approach to ecosystem functioning. *Proceedings of the National Academy of Sciences*, **98**, 10979–80.

Tilman, D. (2007a). Resource competition and plant traits: a response to. *Journal of Ecology*, **95**, 231–4.

Tilman, D. (2007b). Interspecific competition and multispecies coexistence. In: May, R.M. and McLean, A., eds. *Theoretical Ecology: Principles and Applications*, 3rd edn, pp. 84–97. Oxford University Press, Oxford.

Tilman, D. and Downing, A.L. (1994). Biodiversity and stability in grasslands. *Nature*, **367**, 363–5.

Tilman, D., Isbell, F., and Cowles, J.M. (2014). Biodiversity and ecosystem functioning. *Annual Review of Ecology, Evolution and Systematics*, **45**, 471–93.

Tilman, D., Knops, J., Wedin, D., Reich, P.B., Ritchie, M., and Siemann, E. (1997). The influence of functional diversity and composition on ecosystem processes. *Science*, **277**, 1300–2.

Tilman, D., Lehman, C.L., and Bristow, C.E. (1998). Diversity–stability relationships: statistical inevitability or ecological consequence? *American Naturalist*, **151**, 277–82.

Tilman, D., May, R.M., Lehman, C.L., and Nowak, M.A. (1994). Habitat destruction and the extinction debt. *Nature*, **371**, 65–6.

Tilman, D., Reich, P.R., and Isbell, F. (2012). Biodiversity impacts ecosystem productivity as much as resources, disturbance, or herbivory. *Proceedings of the National Academy of Sciences*, **109**, 10394–7.

Tilman, D., Reich, P.B., and Knops, J.M.H. (2006). Biodiversity and ecosystem stability in a decade-long grassland experiment. *Nature*, **441**, 629–32.

Tilman, D., Reich, P.B., Knops, J., Wedin, D., Mielke, T., and Lehman, C. (2001). Diversity and productivity in a long-term grassland experiment. *Science*, **294**, 843–5.

Tilman, D. and Pacala, S. (1993). The maintenance of species richness in plant communities. In: Ricklefs, R.E. and Schluter, D., eds. *Species Diversity in Ecological Communities: Historical and Geographical Perspectives*, pp. 13–25. University of Chicago Press, Chicago, IL.

Tilman, D. and Wedin, D. (1991a). Plant traits and resource reduction for five grasses growing on a nitrogen gradient. *Ecology*, **72**, 685–700.

Tilman, D. and Wedin, D. (1991b). Dynamics of nitrogen competition between successional grasses. *Ecology*, **72**, 1038–49.

Tilman, D., Wedin, D., and Knops, J.M.H. (1996). Productivity and sustainability influenced by biodiversity in grassland ecosystems. *Nature*, **379**, 718–20.

Tinghitella, R.M. (2008). Rapid evolutionary change in a sexual signal: genetic control of the mutation "flatwing" that renders male field crickets (Teleogryllus oceanicus) mute. *Heredity*, **100**, 261–7.

Tokeshi, M. (1999). *Species Coexistence*. Blackwell Scientific, Oxford.

Tollrian, R. and Harvell, C.D., eds. (1999). *The Ecology and Evolution of Inducible Defenses*. Princeton University Press, Princeton, NJ.

Townsend, C.R., Thompson, R.M., McIntosh, A.R., Kilroy, C., Edwards, E., and Scarsbrook, M. (1998). Disturbance, resource supply, and food-web architecture in streams. *Ecology Letters*, **1**, (200–9.

Tracy, C.R. and George, T.L. (1992). On the determinants of extinction. *American Naturalist*, **139**, 102–22.

Tredennick, A.T., *et al.* (2016). Comment on "Worldwide evidence of a unimodel relationship between productivity and species richness." *Science*, **351**, 457.

Trussell, G.C., Matassa, C.M., and Ewanchuk, P.J. (2017). Moving beyond linear food chains: trait-mediated indirect interactions in a rocky intertidal food web. *Proceedings of the Royal Society of London B*, **284**, 20162590.

Tuomisto, H., Ruokolainen, K., and Yli-Halla, M. (2003). Dispersal, environment, and floristic variation of western Amazonian forests. *Science*, **299**, 241–4.

Turchin, P. (1995). Population regulation: old arguments and a new synthesis. In: Cappuccino, N. and Price, P.W., eds. *Population Dynamics: New Approaches and Synthesis*, pp. 19–40. Academic Press, San Diego, CA.

Turchin, P. (1999). Population regulation: a synthetic view. *Oikos*, **84**, 153–9.

Turchin, P. (2003). *Complex Population Dynamics: A Theoretical/ Empirical Synthesis*. Princeton University Press, Princeton, NJ.

Turchin, P. and Batzli, G.O. (2001). Availability of food and the population dynamics of arvicoline rodents. *Ecology*, **82**, 1521–34.

Turchin, P. and Taylor, A.D. (1992). Complex dynamics in ecological time series. *Ecology*, **73**, 289–305.

Turgeon, J., Stokes, R., Thum, R.A., Brown, J.M., and McPeek, M.A. (2005). Simultaneous Quaternary radiations of three damselfly clades across the Holarctic. *American Naturalist*, **165**, E78–E107.

Turnbull, L.A., Rees, M., and Crawley, M.J. (1999). Seed mass and the competition/colonization trade-off: a sowing experiment. *Journal of Ecology*, **87**, 899–912.

Turner, A.M. and Mittelbach, G.G. (1990). Predator avoidance and community structure: interactions among piscivores, planktivores, and plankton. *Ecology*, **71**, 2241–54.

Ulrich, W.A., Baselga, A., Kusumoto, B., Shiono, T., Tuomisto, H., and Kubota, Y. (2017). The tangled link between β- and γ-diversity: a Narcissus effect weakens statistical inferences in null model analyses of diversity patterns. *Global Ecology and Biogeography*, **26**, 1–5.

Ulrich, W., Ollik, M., and Ugland, K.I. (2010). A meta-analysis of species-abundance distributions. *Oikos*, **119**, 1149–55.

Umaña, M.N., Zhang, C., Cao, M., Lin, L. and Swenson, N.G. (2017). A core-transient framework for trait-based community ecology: an example from a tropical tree seedling community. *Ecology Letters*, **20**, 619–28.

Uriarte, M., Swenson, N.G., Chazdon R.L., *et al.* (2010). Trait similarity, shared ancestry and the structure of neighbourhood interactions in a subtropical wet forest: implications for community assembly. *Ecology Letters*, **13**, 1503–14.

Usinowicz, J., Chang-Yang C-H., Chen Y-Y., *et al.* (2017). Temporal coexistence mechanisms contribute to the latitudinal gradient in forest diversity. *Nature*, **550**, 105–8.

Usinowicz, J., Wright, S.J., and Ives, A.R. (2012). Coexistence in tropical forests through asynchronous variation in annual seed production. *Ecology*, **93**, 2073–84.

Valdovinos, F.S., Ramos-Jiliberto, R., Garay-Narváez, L., Urbani, P., and Dunne, J.A. (2010). Consequences of adaptive behaviour for the structure and dynamics of food webs. *Ecology Letters*, **13**, 1546–59.

Valentine, J.W. and Jablonski, D. (2010). Origins of marine patterns of biodiversity: Some correlates and applications. *Palaeontology*, **53**, 1203–10.

Valone, T.J. and Schutzenhofer, M.R. (2007). Reduced rodent biodiversity destabilizes plant populations. *Ecology*, **88**, 26–31.

Vamosi, S.M., Heard S.B., Vamosi, J.C., and Webb, C.O. (2009). Emerging patterns in the comparative analysis of phylogenetic community structure. *Molecular Ecology*, **18**, 572–92.

van Bael, S. and Pruett-Jones, S. (1996). Exponential population growth of Monk Parakeets in the United States. *Wilson Bulletin*, **108**, 584–8.

van Moorsel, S.J., Schmid, M.W., Hahl, T., Zuppinger-Dingley, D., and Schmid, B. (2018). Selection in response to community diversity alters plant performance and functional traits. *Perspectives in Plant Ecology, Evolution and Systematics*, **33**, 51–61.

van der Heijden, M.G.A., Bardgett, R.D., and van Straalen, N.M. (2008). The unseen majority: soil microbes as drivers of plant diversity and productivity in terrestrial ecosystems. *Ecology Letters*, **11**, 296–310.

van der Heijden, M.G.A. and Horton, T.R. (2009). Socialism in soil? The importance of mycorrhizal fungal networks for facilitation in natural ecosystems. *Journal of Ecology*, **97**, 1139–50.

Vandermeer, J.H. (1970). The community matrix and the number of species in a community. *American Naturalist*, **104**, 73–83.

Vandermeer, J.H. (1972). On the covariance of the community matrix. *Ecology*, **53**, 187–9.

Vandermeer, J.H. (1989). *The Ecology of Intercropping*. Cambridge University Press, Cambridge.

Vandermeer, J.H. and Goldberg, D.E. (2013). *Population Ecology: First Principles*, 2nd edn. Princeton University Press, Princeton, NJ.

van der Merwe, M. and Brown, J.S. (2008). Mapping the landscape of fear of the cape ground squirrel (*Xerus inauris*). *Journal of Mammalogy*, **89**, 1162–9.

Van Der Windt, H.J. and Swart, J.A.A. (2008). Ecological corridors, connecting science and politics: the case of the Green River in the Netherlands. *Journal of Applied Ecology*, **45**, 124–32.

Vander Zanden, M.J. and Rasmussen, J.B. (1996). A trophic position model of pelagic food webs: impact on contaminant bioaccumulation in lake trout. *Ecological Monographs*, **66**, 451–77.

Van Nes, E.H. and Scheffer, M. (2004). Large species shifts triggered by small forces. *American Naturalist*, **164**, 255–66.

Vanormelingen, P., Cottenie, K., Michels, E., Muylaert, K., Vyverman, W., and De Meester, L. (2008). The relative importance of dispersal and local processes in structuring phytoplankton communities in a set of highly interconnected ponds. *Freshwater Biology*, **53**, 2170–83.

Vanschoenwinkel, B., DeVries, C., Seaman, M., and Brendonck, L. (2007). The role of metacommunity processes in shaping invertebrate rock pool communities along a dispersal gradient. *Oikos*, **116**, 1255–66.

van Veen, F.J., Morris, R.J., and Godfray, H.C.J. (2006). Apparent competition, quantitative food webs, and the structure of phytophagous insect communities. *Annual Review of Entomology*, **51**, 187–208.

van Veen, F.J.F., Muller, C.B., Pell, J.K., and Godfray, H.C.J. (2008). Food web structure of three guilds of natural enemies: predators, parasitoids and pathogens of aphids. *Journal of Animal Ecology*, **77**, 191–200.

Vázquez, D.P., Morris, W.F., and Jordano, P. (2005). Interaction frequency as a surrogate for the total effect of animal mutualists on plants. *Ecology Letters*, **8**, 1088–94.

Vázquez, D.P., Ramos-Jiliberto, R., Urbani, P., and Valdovinos, F.S. (2015). A conceptual framework for studying the strength of plant-animal mutualistic interactions. *Ecology Letters*, **18**, 385–400.

Veech, J.A. (2006). A probability-based analysis of temporal and spatial co-occurrence in grassland birds. *Journal of Biogeography*, **33**, 2145–53.

Vellend, M. (2010). Conceptual synthesis in community ecology. *Quarterly Review of Biology*, **85**, 183–206.

Vellend, M. (2016). *The Theory of Ecological Communities*. Princeton University Press, NJ.

Vellend, M., Baeten, L., Becker-Scarpitta A., *et al.* (2017a). Plant biodiversity change across scales during the Anthropocene. *Annual Review of Plant Biology*, **68**, 563–86.

Vellend, M., L. Baeten, I.H. Myers-Smith, S.C., *et al.* (2013). Global meta-analysis reveals no net change in local-scale plant biodiversity over time. *Proceedings of the National Academy of Sciences*, **110**, 19456–9.

Velland, M., *et al.* (2017b). Estimates of local biodiversity change over time stand up to scrutiny. *Ecology*, **98**, 583–90.

Verboom, J., Schotman, A., Opdam, P., and Metz, J.A.J. (1991). European nuthatch metapopulations in a fragmented agricultural landscape. *Oikos*, **61**, 149–56.

Verhulst, P.F. (1838). Notice sur la loi que la population suit dans son accroissement. *Correspondances Mathématiques et Physiques*, **10**, 113–21.

Vidal, M.C. and Murphy, S.M. (2017). Bottom-up and top-down effects on terrestrial insect herbivores: a meta-analysis. *Ecology Letters*, **21**, 138–50.

Violle, C., Nemergut, D.R., Pu, Z., and Jiang, L. (2011). Phylogenetic limiting similarity and competitive exclusion. *Ecology Letters*, **14**, 782–7.

Vogt, R.J., Romanuk, T.N., and Kolasa, J. (2006). Species richness–variability relationships in multi-trophic aquatic microcosms. *Oikos*, **113**, 55–66.

Volkov, I., Banavar, J.R., Hubbell, S.P., and Maritan, A. (2003). Neutral theory and relative species abundance in ecology. *Nature*, **424**, 1035–7.

Volterra, V. (1926). Variazioni e fluttuazioni del numero d'individui in specie animali conviventi. *Memoria della*

Regia Accademia Nazionale dei Lincei, **ser 6**(2) (1926): 31–113. Reprinted (1931) in Chapman, R.N., *Animal Ecology*, McGraw-Hill, New York.

Volterra, V. (1931). Variations and fluctuations of the number of individuals in animal species living together. In Chapman, R.N., ed. *Animal Ecology*, pp. 409–48. McGraw-Hill Publishers, New York, NY.

Vos, M., Vershoor, A., Kooi, B., Wäckers, F., DeAngelis, D., and Mooij, W. (2004). Inducible defenses and trophic structure. *Ecology*, **85**, 2783–94.

Wade, P.R., ver Hoef, J.M., and DeMaster, D.P. (2009). Mammal-eating killer whales and their prey—trend data for pinnipeds and sea otters in the North Pacific Ocean do not support the sequential megafaunal collapse hypothesis. *Marine Mammal Science*, **25**, 737–47.

Wake, D. and Vredenburg, V. (2008). Are we in the midst of the 6th mass extinction? A view from the world of amphibians. *Proceedings of the National Academy of Sciences*, **105**, 11466–73.

Walker, L.R. and del Moral, R. (2003). *Primary Succession and Landscape Restoration*. Cambridge University Press, Cambridge.

Walker, M.K. and Thompson, R.M. (2010). Consequences of realistic patterns of biodiversity loss: an experimental test from the intertidal zone. *Marine and Freshwater Research*, **61**, 1015–22.

Wallace, A.R. (1876). *The Geographical Distribution of Animals*. Harper and Brothers, New York, NY.

Wallace, A.R. (1878). *Tropical Nature and Other Essays*. Macmillan, New York, NY.

Walter, H. and Breckle, S-W. (2002). *Walter's Vegetation of the Earth: the Ecological Systems of the Geo-biosphere*. Springer, Berlin.

Walters, A. and Post, D.M. (2008). An experimental disturbance alters fish size structure but not food chain length in streams. *Ecology*, **89**, 3261–67.

Ward, C.L. and McCann, K.S. (2017). A mechanistic theory for aquatic food chain length. *Nature Communications*, **8**, 2028.

Ward, C., McCann, K.S., and Rooney, N. (2015). HSS revisited: multi-channel processes mediate trophic control across a productivity gradient. *Ecology Letters*, **18**, 1190–7.

Wardle, D.A. (1999). Is "sampling effect" a problem for experiments investigating biodiversity ecosystem function relationships? *Oikos*, **87**, 403–7.

Wardle, D.A. (2016). Do experiments exploring plant diversity-ecosystem functioning relationships inform how biodiversity loss impacts natural ecosystems? *Journal of Vegetation Science*, **27**, 646–53.

Wardle, D.A., Bardgett, R.D., Callaway, R.M., and Van der Putten, W.H. (2011). Terrestrial ecosystem responses to species gains and losses. *Science*, **332**, 1273–7.

Wardle, D.A. and Zackrisson, O. (2005). Effects of species and functional group loss on island ecosystem properties. *Nature*, **435**, 806–10.

Warren, B.H., *et al.* (2015). Islands as model systems in ecology and evolution: prospects fifty years after MacArthur-Wilson. *Ecology Letters*, **18**, 200–17.

Warren, D.L., Cardillo, M., Rosauer, D.F., and Bolnick, D.I. (2014). Mistaking geography for biology: inferring processes from species distributions. *Trends in Ecology and Evolution*, **29**, 572–80.

Warren, D.L., Glor, R.E., and Turelli, M. (2008). Environmental niche equivalency versus conservatism: quantitative approaches to niche evolution. *Evolution*, **62**, 2868–83.

Warren, P.H. (1996). Dispersal and destruction in a multiple habitat system: an experimental approach using protist communities. *Oikos*, **77**, 317–25.

Watkins, J.E., Cardelus, C., Colwell, R.K., and Moran, R.C. (2006). Species richness and distribution of ferns along an elevational gradient in Costa Rica. *American Journal of Botany*, **93**, 73–83.

Watson, H.C. (1859). *Cybele Britannica, or British Plants and their Geographical Relations*. Longman and Company, London.

Watson, S., McCauley, E., and Downing, J.A. (1992). Sigmoid relationships between production and biomass among lakes. *Canadian Journal of Fisheries and Aquatic Sciences*, **45**, 915–20.

Watts, D.J. and Strogatz, S.H. (1998). Collective dynamics of "small-world" networks. *Nature*, **393**, 440–2.

Webb, C.O., Ackerly, D.D., McPeek, M.A., and Donoghue, M.J. (2002). Phylogenies and community ecology. *Annual Review of Ecology and Systematics*, **33**, 475–505.

Wedin, D. and Tilman, D. (1993). Competition among grasses along a nitrogen gradient: initial conditions and mechanisms of competition. *Ecological Monographs*, **63**, 199–229.

Weese, D.J., Heath, K.D., Dentinger, B.T.M., and Lau, J.A. (2015). Long-term nitrogen addition causes the evolution of less-cooperative mutualists. *Evolution*, **69**, 631–42.

Weiher, E. (2011). A primer of trait and functional diversity. *Biological Diversity: Frontiers in Measurement and Assessment*. A. E. Magurran and B. J. McGill. Oxford, Oxford University Press: 175–93.

Weiher, E. and Keddy, P.A. (1995). Assembly rules, null models, and trait dispersion: New questions from old patterns. Oikos, **74**: 159–64.

Weir, J.T. and Price, T.D. (2011). Limits to speciation inferred from times to secondary sympatry and ages of hybridizing species along a latitudinal gradient. *American Naturalist*, **177**, 462–9.

Weir, J. T. and Schluter, D. (2007). The latitudinal gradient in recent speciation and extinction rates of birds and mammals. *Science*, **315**, 1574–6.

Weisser, W.W., *et al.* (2017). Biodiversity effects on ecosystem functioning in a 15-year grassland experiment: patterns, mechanisms, and open questions. *Basic and Applied Ecology*, **23**, 1–73.

Went, F.W. (1942). The dependence of certain annual plants on shrubs in Southern California deserts. *Bulletin of the Torrey Botanical Club*, **69**, 100–14.

Werner, E.E. (1986). Amphibian metamorphosis: growth rate, predation risk, and the optimal size at transformation. *American Naturalist*, **128**, 319–41.

Werner, E.E. and Anholt, B.R. (1993). Ecological consequences of the trade-off between growth and mortality rates mediated by foraging activity. *American Naturalist*, **142**, 242–72.

Werner, E.E. and Gilliam, J.F. (1984). The ontogenetic niche and species interactions in size-structured populations. *Annual Review of Ecology and Systematics*, **15**, 393–425.

Werner, E.E., Gilliam J.F., Hall, D.J., and Mittelbach, G.G. (1983). An experimental test of the effects of predation risk on habitat use in fish. *Ecology*, **64**, 1540–8.

Werner, E.E. and Hall, D.J. (1974). Optimal foraging and the size selection of prey by the bluegill sunfish (*Lepomis macrochirus*). *Ecology*, **55**, 1042–52.

Werner, E.E. and McPeek, M.A. (1994). The roles of direct and indirect effects on the distributions of two frog species along an environmental gradient. *Ecology*, **75**, 1368–82.

Werner, E.E. and Peacor, S.D. (2003). A review of trait-mediated indirect interactions in ecological communities. *Ecology*, **84**, 1083–100.

West, S.A., Kiers, E.T., Simms, E.L. and Denison, R.F. (2002). Nitrogen fixation and the stability of the legume–rhizobium mutualism. *Proceedings of the Royal Society of London B*, **269**, 685–94.

Westoby, M., Leishman, M.R., and Lord, J.M. (1995). On misinterpreting the "phylogenetic correction." *Journal of Ecology*, **83**, 531–4.

Westoby, M., Leishman, M.R., and Lord, J.M. (1996). Comparative ecology of seed size and dispersal. *Philosophical Transactions of the Royal Society of London B*, **351**, 1309–18.

White, E.P., *et al.* (2006). A comparison of the species/time relationship across ecosystems and taxonomic groups. *Oikos*, **112**, 185–95.

Whitmore, T.C. (1989). Canopy gaps and the two major groups of forest trees. *Ecology*, **70**, 536–8.

Whittaker, R.H. (1956). Vegetation of the Great Smoky Mountains. *Ecological Monographs*, **26**, 1–80.

Whittaker, R.H. (1960). Vegetation of the Siskiyou Mountains, Oregon and California. *Ecological Monographs*, **30**, 279–338.

Whittaker, R.H. (1972). Evolution and measurement of species diversity. *Taxon*, **21**, 213–51.

Whittaker, R.J. (2010). Meta-analysis and mega-mistakes: calling time on meta-analysis of the species richness–productivity relationship. *Ecology*, **91**, 2522–33.

Whittaker, R.J., Nogues-Bravo, D., and Araujo, M.B. (2007). Geographical gradients of species richness: a test of the water-energy conjecture of Hawkins *et al.* (2003) using European data for five taxa. *Global Ecology and Biogeography*, **16**, 76–89.

Wiens, J.J. (2008). Commentary on Losos (2008): niche conservatism deja vu. *Ecology Letters*, **11**, 1004–5.

Wiens, J.J. and Donoghue, M.J. (2004). Historical biogeography, ecology and species richness. *Trends in Ecology and Evolution*, **19**, 639–44.

Wiens, J.J. and Graham, C.H. (2005). Niche conservatism: Integrating evolution, ecology, and conservation biology. *Annual Review of Ecology, Evolution, and Systematics*, **36**, 519–39.

Wiens, J.J., Sukumaran, J., Pyron, R.A., and Brown, R.M. (2009). Evolutionary and biogeographic origins of high tropical diversity in old world frogs (Ranidae). *Evolution*, **63**, 1217–31.

Wiens, J.J. *et al.* (2010). Niche conservatism as an emerging principle in ecology of evolutionary biology. *Ecology Letters*, **13**, 1310–24.

Williams, C. (1947). The generic relations of species in small ecological communities. *Journal of Animal Ecology*, **16**, 11–18.

Williams, C.B. (1964). *Patterns in the Balance of Nature*. Academic Press, London.

Williams, R.J. and Martinez, N.D. (2000). Simple rules yield complex food webs. *Nature*, **404**, 180–3.

Williams, T.M., Estes, J.A., Doak, D.F., and Springer, A. (2004). Killer appetites: assessing the role of predators in ecological communities. *Ecology*, **85**, 3373–84.

Williamson, M. (1972). *The Analysis of Biological Populations*. Edward Arnold, London.

Willis, A.J. and Memmott, J. (2005). The potential for indirect effects between a weed, one of its biocontrol agents and native herbivores: a food web approach. *Biological Control*, **35**, 299–306.

Wilsey, B.J. and Potvin, C. (2000). Biodiversity and ecosystem functioning: importance of species evenness in an old field. *Ecology*, **81**, 887–92.

Wilson, J.B. and Agnew, A.D.Q. (1992). Positive feedback switches in plant communities. *Advances in Ecological Research*, **23**, 263–336.

Wilson, J.B., Spijkerman, E., and Huisman, J. (2007). Is there really insufficient support for Tilman's R^* concept? A comment on Miller *et al. American Naturalist*, **169**, 700–6.

Winfree, R., Fox, J.W., Williams, N.M., Reilly, J.R., and Cariveau, D.P. (2015). Abundance of common species, not species richness, drives delivery of a real-world ecosystem service. *Ecology Letters*, **18**, 626–35.

Winfree, R., Reilly, J.R., Bartomeus, I., Cariveau, D.P., Williams, N.M., and Gibbs, J. (2018). Species turnover promotes the importance of bee diversity for crop pollination at regional scales. *Science*, **359**, 791–3.

Winfree, R. (2018). How does biodiversity relate to ecosystem functioning in natural ecosystems? In: Holt, B., Tilman, D., and Dobson, A.P., eds, *Unsolved Problems in Ecology*. Princeton University Press, Princeton, NJ (in press).

Wiser, S.K., Allen, R.B., Clinton, P.W., and Platt, K.H. (1998). Community structure and forest invasion by an exotic herb over 23 years. *Ecology*, **79**, 2071–81.

Wisheu, I.C. (1998). How organisms partition habitats: different types of community organization can produce identical patterns. *Oikos*, **83**, 246–58.

Woiwod, I.P. and Hanski, I. (1992). Patterns of density dependence in moths and aphids. *Journal of Animal Ecology*, **61**, 619–29.

Wojdak, J.M. and Mittelbach, G.G. (2007). Consequences of niche overlap for ecosystem functioning: an experimental test with pond grazers. *Ecology*, **88**, 2072–83.

Wolda, H. (1989). The equilibrium concept and density dependence tests: what does it all mean? *Oecologia*, **81**, 430–2.

Wolda, H. and Dennis, B. (1993). Density dependent tests, are they? *Oecologia*, **95**, 581–91.

Wollkind, D.J. (1976). Exploitation in three trophic levels: an extension allowing intraspecies carnivore interaction. *American Naturalist*, **110**, 431–47.

Wollrab, S., Diehl, S., and DeRoos, A.M. (2012). Simple rules describe bottom-up and top-down control in food webs with alternative energy pathways. *Ecology Letters*, **15**, 935–46.

Woodward, G., Ebenman, B., Emmerson, M., *et al.* (2005). Body size in ecological networks. *Trends in Ecology and Evolution*, **20**, 402–9.

Woodward, G., *et al.* (2010). Individual-based food webs: species identity, body size and sampling effects. *Advances in Ecological Research*, **43**, 209–65.

Wootton, J.T. (1994). Putting the pieces together: testing the independence of interactions among organisms. *Ecology*, **75**, 1544–51.

Wootton, J.T. (1997). Estimates and tests of per capita interaction strength: diet, abundance, and impact of intertidally foraging birds. *Ecological Monographs*, **67**, 45–64.

Wootton, J.T. (2005). Field parameterization and experimental test of the neutral theory of biodiversity. *Nature*, **433**, 309–12.

Wootton, J.T. and Downing, A.L. (2003). Understanding the effects of reduced biodiversity: a comparison of two approaches. In: Kareiva, P. and Levin, S.A., eds. *The Importance of Species*, pp. 85–104. Princeton University Press, Princeton, NJ.

Wootton, J.T. and Emmerson, M. (2005). Measurement of interaction strength in nature. *Annual Review of Ecology and Systematics*, **36**, 419–44.

Wootton, J.T. and Power, M.E. (1993). Productivity, consumers, and the structure of a river food chain. *Proceedings of the National Academy of Sciences*, **90**, 1384–7.

Wright, D.H. (1983). Species-energy theory: an extension of species-area theory. *Oikos*, **41**, 496–506.

Wright, D.H., Keeling, J., and Gilman, L. (2006). The road from Santa Rosalia: a faster tempo of evolution in tropical climates. *Proceedings of the National Academy of Sciences*, **103**, 7718–22.

Wright, I.J., Reich, P.B., Westoby, M., *et al.* (2004). The world-wide leaf economics spectrum. *Nature*, **428**, 821–7.

Wright, J.P. and Jones, C.G. (2006). The concept of organisms as ecosystem engineers ten years on: progress, limitations, and challenges. *BioScience*, **56**, 203–9.

Wright, S., Gray, R.D., and Gardner, R.C. (2003). Energy and the rate of evolution: inferences from plant rDNA substitution rates in the western Pacific. *Evolution*, **57**, 2893–8.

Yang, L.H. and Rudolf, V.H.W. (2010). Phenology, ontogeny and the effects of climate change on the timing of species interactions. *Ecology Letters*, **13**, 1–10.

Ydenerg, R.C., Brown, J.S., and Stephens, D.W. (2007). Foraging: an overview. In: Stephens, D.W., Brown, J.S., and Ydenberg, R.C., eds. *Foraging: Behavior and Ecology*, pp. 1–28. University of Chicago Press, Chicago, IL.

Yenni, G., Adler, P.B., and Ernest, S. (2012). Strong self-limitation promotes the persistence of rare species. *Ecology*, **93**, 456–61.

Yenni, G., Adler, P.B., and Ernest, S. (2017). Do persistent rare species experience stronger negative frequency dependence than common species? *Global Ecology and Biogeography*, **26**, 513–23.

Yodzis, P. (1988). The indeterminacy of ecological interactions as perceived through perturbation experiments. *Ecology*, **69**, 508–15.

Yodzis, P. (1989). *Introduction to Theoretical Ecology*. Harper and Row, New York, NY.

Yodzis, P. (1998). Local trophodynamics and the interaction of marine mammals and fisheries in the Benguela ecosystem. *Journal of Animal Ecology*, **67**, 635–58.

Yodzis, P. (2001). Must top predators be culled for the sake of fisheries? *Trends in Ecology and Evolution*, **16**, 78–84.

Yoon, I., Williams, R.J., Levine, E., Yoon, S., Dunne, J.A., and Martinez, N.D. (2004). Webs on the web (WoW): 3D visualization of ecological networks on the WWW for collaborative research and education. *Proceedings of the IS&T/SPIE Symposium on Electronic Imaging, Visualization and Data Analysis*, **5295**, 124–32.

Young, H.S., McCauley, D.J., Dunbar, R.B., Hutson, M.S., Ter-Kuile, A.M., and Dirzo, R. (2013). The roles of productivity and ecosystem size in determining food chain length in tropical terrestrial ecosystems. *Ecology*, **94**, 692–701.

Yoshida, T., Jones, L.E., Ellner, S.P., Fussmann, G.F., and Hairston, N.G., Jr (2003). Rapid evolution drives ecological dynamics in a predator–prey system. *Nature*, **424**, 303–6.

Yu, D.W., Wilson, H.B., and Pierce, N.E. (2001). An empirical model of species coexistence in a spatially structured environment. *Ecology*, **82**, 1761–71.

Yu, D.W., Wilson, H.B., Frederickson M.E., *et al.* (2004). Experimental demonstration of species coexistence enabled by dispersal limitation. *Journal of Animal Ecology*, **73**, 1102–14.

Zavaleta, E.S., Pasari, J.R., Hulvey, K.B., and Tilman, G.D. (2010). Sustaining multiple ecosystem functions in grassland communities requires higher biodiversity. *Proceedings of the National Academy of Sciences*, **107**, 1443–6.

Zhang, C., Yang, J., Sha, L., *et al.* (2017). Lack of phylogenetic signals within environmental niches of tropical tree species across life stages. *Scientific Reports*, **7**, 42007.

Zillio, T. and Condit, R. (2007). The impact of neutrality, niche differentiation and species input on diversity and abundance distributions. *Oikos*, **116**, 931–40.

Zimmermann, F., Ricard, D., and Heino, M. (2018). Density regulation in Northeast Atlantic fish populations: Density dependence is stronger in recruitment than in somatic growth. *Journal of Animal Ecology*, **87**, 672–81.

Zobel, M. (1997). The relative of species pools in determining plant species richness: an alternative explanation of species coexistence? *Trends in Ecology & Evolution*, **12**, 266–9.

Zuk, M., Rotenberry, R.T., and Tinghitella, R.M. (2006). Silent night: adaptive disappearance of a sexual signal in a parasitized population of field crickets. *Biology Letters*, **2**, 521–4.

Zuckerkandl, E. and Pauling, L. (1965). Evolutionary divergence and convergence in proteins. In: Bryson, V. and Vogel, H.J., eds. *Evolving Genes and Proteins*, pp. 97–166. Academic Press, New York, NY.

Zuppinger-Dingley, D., Schmid, B., *et al.* (2014). Selection for niche differentiation in plant communities increases biodiversity effects. *Nature*, **515**, 108–11.

Züst, T. and Agrawal, A.A. (2017). Trade-offs between plant growth and defense against insect herbivory: an emerging mechanistic synthesis. *Annual Review of Ecology, Evolution and Systematics*, **68**, 513–34.

Author index

Subject index

Page numbers in italics refer to figures and tables